THE ENVIRONMENTAL CASE

Fifth Edition

For Judith Layzer
An inspiration for the study and practice of environmental policy

THE ENVIRONMENTAL CASE

Translating Values Into Policy

Fifth Edition

Judith A. Layzer
Massachusetts Institute of Technology

Sara R. Rinfret
University of Montana

FOR INFORMATION:

CQ Press
An Imprint of SAGE Publications, Inc.
2455 Teller Road
Thousand Oaks, California 91320
E-mail: order@sagepub.com

SAGE Publications Ltd.
1 Oliver's Yard
55 City Road
London EC1Y 1SP
United Kingdom

SAGE Publications India Pvt. Ltd.
B 1/I 1 Mohan Cooperative Industrial Area
Mathura Road, New Delhi 110 044
India

SAGE Publications Asia-Pacific Pte. Ltd.
18 Cross Street #10-10/11/12
China Square Central
Singapore 048423

Printed in the United States of America

Library of Congress Cataloging-in-Publication Data

Names: Layzer, Judith A., author.

Title: The environmental case : translating values into policy / Judith A. Layzer, Massachusetts Institute of Technology, Sara R. Rinfret, University of Montana.

Description: Fifth Edition. | Los Angeles : SAGE | CQ PRESS, [2019] | Previous edition: 2016. | Includes bibliographical references and index.

Identifiers: LCCN 2019006737 | ISBN 9781506396965 (Paperback : alk. paper)

Subjects: LCSH: Environmental policy—United States—Case studies.

Classification: LCC GE180 .L39 2019 | DDC 363.7/05610973—dc23
LC record available at https://lccn.loc.gov/2019006737

This book is printed on acid-free paper.

SFI label applies to text stock

Acquisitions Editor: Scott Greenan; Anna Villarruel
Editorial Assistant: Lauren Younker
Production Editor: Jyothi Sriram
Copy Editor: Lana Todorovic-Arndt
Typesetter: C&M Digitals (P) Ltd.
Proofreader: Barbara Coster
Indexer: Judy Hunt
Cover Designer: Candice Harman
Marketing Manager: Jennifer Jones

19 20 21 22 23 10 9 8 7 6 5 4 3 2 1

BRIEF CONTENTS

DETAILED CONTENTS

PREFACE

The idea for this book was born when Professor Layzer was a graduate student in political science at the Massachusetts Institute of Technology (MIT) in the mid-1990s. She worked with her adviser, Stephen Meyer, to develop a new undergraduate course in environmental politics and policy. After Professor Layzer searched extensively for material, they realized that, although many informative and readable texts on the history, substance, and efficacy of various environmental policies were available, few described environmental politics in action. Moreover, hardly any seemed to recognize the fundamental disagreements that make environmental controversies so intractable or to describe the patterns that emerge as one looks at the policymaking process across a broad range of issues. Professor Layzer set out to craft a set of case studies that would simultaneously convey the drama of U.S. environmental politics and furnish useful insights into the policymaking process. Professor Layzer's primary goal, then, was not to assess outcomes from a normative perspective, but to make some provocative claims about how environmental policies evolve, from the way we decide which problems are worthy of the government's attention to the mechanisms we choose to address them.

The organization of this book emerged and cases were added that captured new aspects of environmental politics or policy, such as a new matter of concern—wetlands or species extinction, for example—or a novel political dynamic, such as landscape-scale planning. The division between pollution control and natural resource management was obvious; the third category—new issues, new politics—was less intuitive but sought to capture some of the most important contemporary puzzles in the realm of environmental policymaking. For the fifth edition, I continue Professor Layzer's tradition and focus on revising much of her work to date post-Trump. The fifth edition highlights how climate change is the crisis of our time and demands a global response. The cases that have been removed from previous editions are available in electronic form at the CQ Press website (http://custom.cqpress.com).

The cases focus on the importance of values, ideas, and information—particularly as they affect problem definition and solution characterization—in policymaking. Although this edition like others does not follow a clear analytic structure, each narrative is sufficiently self-explanatory so that economists, sociologists, historians, and others interested in the book would not feel confined by political science or policy jargon. Moreover, the goal is to make the cases accessible not only to students and scholars, but also to activists, policymakers, journalists, and others interested in environmental politics. Each case introduces the reader to fascinating characters and events and provides a foundation for more in-depth study of the issues raised. Taken as a whole, the book provides both a portrait of and an analytic framework for U.S. environmental policymaking, from the local level to the international level, that should be valuable to anyone concerned about the environment.

ACKNOWLEDGMENTS

Writing this book would not have been possible without the support of MIT, Middlebury College, and the University of Montana. MIT funded the development of the first set of cases. In particular, MIT Energy Initiative (MITEI) funded the development of several cases. Each institution has generously provided support for student research assistants, many of whom labored to bring the book to fruition. At Middlebury, Katrina O'Brien, Lee Rowland, Sarah Weston, and Jess Widay provided research assistance for the first edition. Kelly Cavazos (MIT), Nicole Fenton (University of Vermont), Corey Ferguson (Bowdoin), Jessie Schaffer (MIT), and Anita Yip (Wellesley) helped with data collection and fact checking. MIT graduate students contributed to this endeavor as well. Xixi Chen and Kate Van Tassel tracked down sources and helped with proofreading. Kim Foltz collected source material. Abbie Emison cheerfully hunted for obscure references and searched out elusive information. And Kate Dineen (MIT) was a careful fact checker and figure maker. Several University of Montana students spent a vast amount of time for an independent research project, assisting with the revisions for the fifth edition: Ian Harris, Alyssa Draper, Taeson Josephson, and Shane St. Onge. A significant thank you to Alyssa Draper, Ian Harris, and Shane St. Onge, graduate students within the Master of Public Administration Program at the University of Montana, who spent countless hours delving into research to strengthen this manuscript.

We would also like to thank our colleagues, who have scrutinized various chapters. Colleagues Christopher Bosso (Northeastern University) and Christopher Klyza (Middlebury College) read and made insightful comments on the introduction to the first edition. In addition, reviewers Gordon Bennett (University of Texas at Austin), Mark Lubell (Florida State University), Stuart Shulman (Drake University), Denise Scheberle (UC-Denver), Michelle Pautz (University of Dayton), and Stacy VanDeveer (University of New Hampshire) provided detailed and thoughtful reviews. We appreciate the feedback of those who commented on the strengths and weaknesses of the first edition in preparation for the second: Richard Andrews (University of North Carolina at Chapel Hill), Tina L. Bertrand (McMurray University), Brian Cook (Clark University), Irasema Coronado (University of Texas at El Paso), David L. Feldman (University of Tennessee), Mary Hague (Bucknell University), Timothy Lehman (Rocky Mountain College), Robert J. Mason (Temple University), Joseph Rish (King's College), and Frederick Van Geest (Dordt College). In preparation for the third edition, the following people responded to our queries about how they use the book: Robert Bartlett (University of Vermont), Russell Brenneman (Trinity College), Kristin Campbell (University of Buffalo,

SUNY), Karen McCurdy (Georgia Southern University), Leigh Raymond (Purdue University), and Rusty Russell (Tufts University). In preparation for the fourth edition, the feedback was provided by Robert Bartlett (University of Vermont), Leigh Raymond (Purdue University), and Erika Weinthal (Duke University). Finally, the following people provided thorough and helpful reviews for the current edition of this book: Clement Loo (University of Minnesota, Morris), Seth Victor (Montclair State University), Erinn G. Ryen (Wells College), and Ted Greenhalgh (University of Nevada, Las Vegas).

We are extremely indebted to SAGE acquisitions editor Scott Greenan, who has handled the book with great professionalism. He has managed the revision process with great diplomacy—his dedication to this project is noteworthy and greatly appreciated. A thank you also goes to Kate Scheiman for keeping us on target and moving forward.

Most importantly, the fifth edition is dedicated to Judith Layzer. We have learned so much from her work and desire for us to understand environmental policy through the lens of storytelling. Thank you to Michael Kraft for encouraging me to carry the torch and continue Judith Layzer's remarkable work. I am honored to continue Professor Layzer's *Environmental Case*. For me, Professor Layzer's work is profound because of its ability to transform our classrooms across the globe. In this vein, I continue her tradition of richly crafted case studies in the fifth edition. Together, we share a commitment to sharing the stories that have transformed U.S. environmental policy to date.

ABOUT THE AUTHORS

Judith A. Layzer was professor of environmental policy in the Department of Urban Studies and Planning at the Massachusetts Institute of Technology (MIT) until her death in 2015. She earned a PhD in political science at MIT. After four years at Middlebury College in Vermont, she returned to MIT, where she taught courses in science and politics in environmental policymaking, ecosystem-based management, food systems and the environment, urban sustainability, energy and environmental politics, and public policy.

Layzer's research focused on several aspects of U.S. environmental politics, including the roles of science, values, and storytelling in environmental politics, as well as on the effectiveness of different approaches to environmental planning and management. A recent project asked: Do urban sustainability initiatives significantly reduce cities' ecological footprints? And which aspects of "green cities" are most effective at reducing cities' environmental impacts? In addition to *The Environmental Case*, Layzer was the author of *Natural Experiments: Ecosystem-Based Management and the Environment* (2008) and *Open for Business: Conservatives' Opposition to Environmental Regulation* (2012).

Layzer was an athlete as well as a scholar. In addition to having finished five Boston marathons, she shared nine national championship titles and one world championship trophy with her teammates on Lady Godiva, formerly Boston's premier women's Ultimate Frisbee team.

Judith A. Layzer
1961–2015

Sara R. Rinfret is an associate professor of public administration and policy, chair of the Department of Public Administration and Policy, and director of the University of Montana's Master of Public Administration Program. She earned a PhD in political science from Northern Arizona University and MPA from the Ohio State University. Her research and teaching center on environmental and regulation, committed to sharing with the field and students how policies come to life by understanding agency policymaking (rulemaking). Her research focuses on how interest groups' influence impacts agency rulemaking and the collaborative responses of federal environmental and natural resource agencies. Rinfret's research and teaching on public policymaking more broadly attempt to expose our students to moving beyond us-versus-them debates. Because of this approach, she coauthored one of the first-ever female-led public policy textbooks with Denise Scheberle and Michelle Pautz—*Public Policy: A Concise Introduction*, CQ Press.

CHAPTER ONE

A POLICYMAKING FRAMEWORK

Defining Problems and Portraying Solutions in U.S. Environmental Politics

Environmental politics concerns "how humanity organizes itself to relate to the nature that sustains it."[1] Because human life depends on well-functioning natural systems, one might think environmental protection would be uncontroversial. Yet bitter disputes have erupted over proposals to preserve undeveloped land, save endangered species, protect or restore ecosystems, clean up toxic dumps and spills, reduce air and water pollution, conserve energy, mitigate human-caused changes in the global climate, and ensure an equitable distribution of environmental hazards. These issues have become prominent in American politics, taking their place alongside more conventional social, economic, and foreign policy concerns, so it is essential to understand their political dynamics. Furthermore, although the policy process has many generic features, environmental policymaking has a host of distinct attributes and therefore warrants its own analytic niche. The goal of this book, then, is to illuminate how the American political system grapples with the environment as a particular object of public action.

This introductory chapter begins by laying out the following two-part argument: (1) environmental policy conflicts almost always concern fundamental differences in values, and (2) the way problems are defined and solutions depicted plays a central role in shaping how those values get translated into policies. The chapter goes on to describe the contributions of the system's major actors—policymakers, advocates, experts, and the media—in defining environmental problems, formulating and characterizing solutions, and ultimately making decisions. It then introduces a number of concepts that help explain the process by which environmental policy is made. The result is a general framework, the elements of which are treated in greater depth in the cases that follow. The chapter concludes by explaining the rationale behind the selection, organization, and presentation of the volume's sixteen case studies. Each of the cases is interesting in its own right; in combination, they offer important lessons for anyone who wants to understand why environmental policy controversies turn out the way they do.

TWO CRITICAL FEATURES OF U.S. ENVIRONMENTAL POLICYMAKING

Nearly all environmental policy disputes are, at heart, contests over values. To the casual observer, these conflicts may appear to revolve around arcane technical issues, but in fact almost all of them involve a fundamental disagreement over how humans ought to interact with the natural world. Even though environmental policy disputes are rooted in conflicting beliefs, the participants in those contests rarely make value-based arguments. Instead, they define problems and characterize their solutions in terms of science, economics, and risk. Because value differences divide participants, environmental policy conflicts are rarely resolved by appeals to reason; no amount of technical information is likely to convert adversaries in such disputes.[2]

The Clash of Values at the Heart of Environmental Policymaking

The participants in environmental debates fall into two broad camps based on entrenched differences in their beliefs about the appropriate relationship between humans and the natural world. Although each side incorporates a wide range of perspectives, for analytic purposes we can categorize them as environmentalists and cornucopians.

Environmentalists. Environmentalism is not a single philosophy but a congeries of beliefs with several roots. Environmental values, in one form or another, have been part of American culture and politics since before the arrival of white settlers on the North American continent. In fact, some contemporary environmentalists trace their values to the spiritual beliefs of Native Americans.[3] But most historians date the origins of American environmentalism to the late eighteenth- and early nineteenth-century Romantics and Transcendentalists, an elite community of artists and writers who celebrated wild nature as a source of spiritual renewal and redemption. They believed that only by preserving untrammeled wilderness could the nation ensure that intact landscapes remained to replenish the weary soul. In the 1830s, George Catlin, a painter who traveled frequently in the West, was the first to plead for the establishment of a national park to preserve land in its "pristine beauty and wildness" for future generations.[4] Twenty years later, Henry David Thoreau deplored the wholesale clearing of land for farming and moved to a cabin at Walden Pond in search of a more "simple" and "natural" life. Thoreau also emphasized the importance of preserving wild nature for building character; in his famous essay "Walking," he wrote, "Hope and the future for me are not in lawns and cultivated fields, not in towns and cities, but in the impervious and quaking swamps."[5] In 1911 John Muir, founder of the Sierra Club and an ardent and prolific advocate of wilderness preservation, described nature as a "window opening into heaven, a mirror reflecting the Creator."[6] A half-century later, the federal government embedded the preservationist philosophy in laws such as the Wilderness Preservation Act (1964), the Wild and Scenic Rivers Act (1968), and the National Trails Act (1968).

A second form of environmental concern, conservationism, accompanied the Progressive movement that emerged at the turn of the twentieth century. Unlike preservationists, who wanted to set aside swaths of undisturbed nature, conservationists advocated the prudent use of natural resources. As historian Samuel Hays points out, conservationists adhered to the "gospel of efficiency."[7] They were intent on managing the nation's coal, oil, timber, grassland, and water according to scientific principles to ensure their availability in the long run. (Most of the applied science disciplines that emerged during the Progressive Era were geared toward increasing natural resource yields, not preserving ecosystem health.[8]) Conservationists like Gifford Pinchot, for example, deplored the wasteful cut-and-run logging practices of private timber companies and feared that American industrialists would appropriate and squander the nation's natural resources unless government stepped in and planned for their orderly exploitation. It was this concern that drove the federal government to set aside forest reserves and, in 1905, create the U.S. Forest Service to manage those lands for the public benefit.

Although strains of preservationist and conservationist thought pervade contemporary environmental debates, a third strand of environmentalism emerged after World War II—one more concerned with fighting pollution and protecting biological diversity than with preserving pristine natural areas or managing natural resources efficiently. Ideas about the interdependence of human beings and nature derived from the scientific discipline of ecology, which focuses on the study of living organisms and their environment, are the primary wellsprings of modern environmentalism.[9] In the late 1940s naturalist and forester Aldo Leopold sowed the seeds of the contemporary environmental movement with his book *A Sand County Almanac*, which developed a "land ethic" based on principles of interrelatedness and stability. According to Leopold, "All ethics . . . rest upon a single premise: that the individual is a member of a community of interdependent parts." Therefore, "a thing is right when it tends to preserve the stability and beauty of the biotic community. It is wrong when it tends otherwise."[10]

Another foundation of postwar environmentalism is the limits-to-growth thesis espoused by Massachusetts Institute of Technology (MIT) biophysicist Donella Meadows and her coauthors in 1972. Based on a new mathematical method called system dynamics, this perspective recognized the importance of relationships and feedback loops in complex systems.[11] According to the limits-to-growth argument, the human population is outrunning the Earth's capacity to support it. For some, recognition of the Earth's limited carrying capacity led to skepticism about economic growth and interest in a steady-state economy.[12] Other adherents of this view are not opposed to economic growth altogether, but rather, they advocate growth that is "sustainable" and therefore does not come at the expense of future generations. They point out, however, that an unregulated market system invariably leads to unsustainable levels of production and consumption.

Although it has common origins, the contemporary environmental movement is far from monolithic. For example, deep ecologists distinguish themselves from mainstream environmentalists, whose environmental beliefs they regard as superficial. Deep ecology is ecocentric: whereas anthropocentric perspectives treat humans as morally superior to other forms of life on the basis of our capacity for language and complex thought, ecocentric perspectives treat the world as "an intrinsically dynamic, interconnected web of relations in which there are no absolutely discrete entities and no absolute dividing lines between the living and the nonliving, the animate and the inanimate, or the human and the nonhuman."[13] As this quote makes clear, deep ecology rests on a premise of "biospherical egalitarianism"—that is, the inherent and equal value of all living things. Moreover, deep ecologists believe that human quality of life depends on maintaining a deep connection to, rather than simply a respectful relationship with, other forms of life. It is worth noting that deep ecology is not logically derived from ecology, nor does it depend for substantiation on the results of scientific investigation. Instead, as philosopher Arne Naess explains, "To the ecological field worker, the equal right to live and blossom is an intuitively clear and obvious value axiom."[14]

Even among more mainstream environmentalists, there are major differences, as the cases that follow make abundantly clear. In particular, pragmatic environmentalists seek to promote the adoption of new technologies that will reduce the overall environmental impact of human society but do little to change the overall structure of the global economy. By contrast, more idealistic environmentalists believe that attaining sustainability requires a complete political and economic transformation. They emphasize changes in behavior, not just technology, to produce a more just and environmentally robust political-economic system.

Cornucopians. Unlike environmentalists, cornucopians (or Prometheans) place a preeminent value on economic growth.[15] In sharp contrast to environmentalism, the term *cornucopian* suggests abundance, even limitlessness. Adherents of this perspective believe that environmental restrictions threaten their economic well-being or the economic health of their community. They also fear that such restrictions entail unacceptable limits by government on individual freedom.

Cornucopians have boundless confidence in humans' ability to devise technological solutions to resource shortages.[16] Best known among the cornucopians are economists Julian Simon and Herman Kahn, who say, "We are confident that the nature of the physical world permits continued improvement in humankind's economic lot in the long run, indefinitely."[17] A theme of the cornucopian literature is that the kinds of doomsday forecasts made by environmentalists never come to pass. Resource shortages may arise in the short run, but the lesson of history, according to Simon and Kahn, is that "the nature of the world's physical conditions and the resilience in a well-functioning economic and social system enable us to overcome such problems, and the solutions usually leave us better off than if the problem had never arisen."[18]

In addition to being technological optimists, cornucopians place enormous value on individual liberty—defined as the freedom to do as one wishes without interference. Some proponents of this philosophy contend that environmentalists are actually Socialists disguising their rejection of markets and preference for government control over the means of production as concern about the environment.[19] Cornucopians criticize environmental regulations not only for limiting individual freedom, but also for taking out of the economy resources that would otherwise be used productively. Reasoning that affluence leads to demands for better health and a cleaner environment, they propose that the best way to protect the environment is to ensure that individuals can pursue material prosperity.[20] According to some who hold this worldview, the role of government is to assign property rights in the Earth's resources and let the market dictate allocations of the goods and services that flow from these resources—a philosophy known as free-market environmentalism.

Cornucopians regard their perspective as logical, rational, and optimistic; by contrast, they see environmentalists as sentimental and irrationally pessimistic. They particularly eschew ecocentric philosophies that elevate plants, animals, and even nonliving entities to the level of humans. Instead, they adopt a view of the world in which "people may not sit above animals and plants in any metaphysical sense, but clearly are superior in their placement in the natural order." Therefore, "decent material conditions must be provided for all of the former before there can be long-term assurance of protection for the latter."[21]

In short, the schism between environmentalists and cornucopians arises out of different worldviews. That said, environmentalists are a diverse lot, ranging from those who believe that all life has value to those who yearn for simpler, less harried times to those with practical concerns about the impact of pollution on human health or quality of life. There is similar variation among cornucopians: some place a higher value on economic growth than they do on the aesthetic or moral importance of the natural world; others are avid outdoor enthusiasts who simply have more faith in individuals' than in government's ability to protect natural amenities. But the heterogeneity of environmentalism and cornucopianism should not obscure the fundamental value differences that underpin environmental controversies. Only by recognizing such profound disagreements can we understand why environmental policymaking is rarely a straightforward technical matter of acknowledging a problem and devising an appropriate solution. Moreover, the extent to which participants' fundamental values diverge is the best clue to how intractable a conflict will be: controversies that involve more consensual values like human health are typically less polarized than disputes over biodiversity conservation, where value differences are often vast and sometimes irreconcilable.

How Activists Define Problems and Characterize Their Solutions to Gain Political Support

Because the values of activists on both sides are entrenched, environmental politics consists largely of trying to gain the support of the unaware or undecided rather than trying to convert the already committed. As political scientist E. E. Schattschneider observed, "The outcome of every conflict is determined by the extent to which the audience becomes involved in it."[22] To attract sympathizers, advocates define problems strategically in ways they think will resonate with a majority of the public.[23] Defining a problem in politics is a way of simplifying a complex reality; it involves framing information to draw attention to some elements of a problem, while obscuring or minimizing others.[24] Problem definition also entails explaining cause and effect, identifying victims and villains, and assigning responsibility for remediation.[25]

By changing which aspect of a problem the public focuses on, advocates can raise (or lower) its visibility and thereby get it onto (or keep it off of) the political agenda.[26] Participants in an environmental policy controversy compete ferociously to provide the authoritative explanation for a problem because "causal stories are essential political instruments for shaping alliances and for settling the distribution of benefits and costs."[27] Participants also compete to predict a problem's consequences because they know that fear of loss or harm is likely to galvanize the public. Finally, by authoritatively defining a problem, advocates can limit the range and type of solutions the public and policymakers are likely to regard as plausible. And, as Schattschneider also pointed out, "The definition of the alternatives is the supreme instrument of power."[28]

When polled, a large majority of Americans profess their support for the goals of the environmental movement.[29] These poll numbers make baldly antienvironmental rhetoric generally unacceptable in political discourse.[30] During the early 1980s and again in the 1990s, conservative Republicans experimented with antienvironmental rhetoric, but doing so backfired, provoking a proenvironmental backlash. Therefore, cornucopians know they must define environmental problems in ways that make their points in a subtle and indirect manner, rather than overtly denying the importance of environmental problems. (A prominent exception to this general rule is climate change.) The competition between environmentalists and cornucopians to define an environmental problem thus revolves around three attributes: the scientific understanding of the problem, the economic costs and benefits of proposed solutions, and the risks associated with action or inaction. Because each of these is speculative, advocates can choose assessments and projections that are most consistent with their values. They then frame that information—using symbolic language and numbers, as well as strategically crafted causal stories—to emphasize either environmental or economic risk, depending on their policy objectives.

Translating Scientific Explanations Into Causal Stories. The primary battleground in any environmental controversy is the scientific depiction of the cause, consequences, and magnitude of a problem. Scientists are often the first to identify environmental problems or to certify their seriousness. Furthermore, scientific claims carry particular weight because science has enormous cultural authority in the United States. Rather than providing a clear and authoritative explanation, however, science leaves considerable latitude for framing because, as a general rule, the scientific understanding of an environmental problem is uncertain.

Most scientific research on natural systems involves practitioners in multiple disciplines, many of which are relatively new, working at the frontiers of scientific knowledge. Moreover, scientists' ability to measure the causes and consequences of environmental phenomena is limited, both technologically and financially, and in the case of human health effects, by ethical considerations. Most important, few environmental problems can be simulated in laboratory experiments: they involve complex interactions among factors for which it is difficult or impossible to control. In the early stages of research, therefore, a wide range of uncertainty surrounds explanations of a problem's causes, consequences, and magnitude. Over time, even as additional research opens up further lines of inquiry and exposes new uncertainties, the boundaries of the original uncertainty tend to narrow.

An example of the process of building scientific knowledge about an environmental problem is the way that understanding of atmospheric ozone depletion advanced, beginning in the 1970s. The stratospheric ozone layer absorbs UV-B radiation, thereby regulating the Earth's temperature and protecting plants, animals, and people from excessive radiation. During the 1970s, scientists developed several theories to explain the observed reduction in stratospheric ozone over the poles. Some scientists were concerned about aircraft emissions of nitrogen oxides; others suggested that nitrogen-based fertilizers or fallout from nuclear weapons tests might be the primary culprits. In 1974, chemists Sherwood Rowland and Mario Molina proposed that chlorine-containing compounds, such as chlorofluorocarbons, destroy the ozone layer through a series of complex, solar-induced chemical reactions. Over time this theory superseded its rivals and gained broad acceptance because it was consistent with evidence gathered using a variety of techniques. Subsequently, as researchers learned more about stratospheric chemistry, estimates of ozone loss became more accurate and the mechanisms by which it occurs more accurately specified.[31]

Unfortunately, the time period within which scientists converge on and refine an explanation is usually considerably longer than the time available to policymakers for choosing a solution. The reason is that the identification of a potential problem by scientists almost invariably prompts the mobilization of interests that are concerned about it and demand an immediate government response. The norms of scientific investigation—particularly those of deliberate and thorough study, rigorous peer review and criticism, and forthright expression of uncertainty—create opportunities for proponents of new policies, as well as for defenders of the status quo, to portray the problem in ways that are compatible with their own values and policy preferences. In most cases, advocates of more protective environmental policies publicize the worst-case scenarios hypothesized by scientists, overstate the certainty of scientific knowledge, and press for an early and stringent—or precautionary—policy response to avert catastrophe. By contrast, opponents of such policies typically emphasize the uncertain state of current knowledge or, if there is a strong scientific consensus that an environmental problem is genuine, highlight dissenting views within the scientific community as to its magnitude, causes, or consequences.[32]

Shifting Attention to the Economic Costs and Benefits. As the scientific consensus around the explanation for an environmental problem grows, opponents of protective policies turn to economic arguments. In particular, cornucopians emphasize (and environmentalists downplay) the economic costs of policies to address the problem. Like scientific explanations of cause and effect, the costs of regulation are highly uncertain, and projections vary widely depending on the assumptions used and the time horizon considered. For example, analysts disagree on the number of jobs likely to be lost as the direct result of an environmental regulation and diverge even more dramatically on the number of collateral jobs—in restaurants, banks, and other service industries—that will disappear. They make different assumptions about future levels of economic growth, the extent and pace of technological adaptation to a regulation, and the likelihood and extent of offsetting effects, such as the establishment or growth of new industries. As is true of scientists, given a choice among equally defensible assumptions, economists select premises that reflect their worldviews, so it is not surprising that industry projections of costs associated with a regulation tend to be much higher than projections made by environmentalists, with government estimates typically in the middle.[33]

In addition to debating projections of the cost of environmental policies, competing parties disagree over the desirability of cost-benefit analysis as a decision-making tool. Cost-benefit analysis entails determining the ratio of monetary benefits to costs of addressing a problem; by implication, government should undertake a program only if its benefits outweigh its costs—that is, if the ratio of benefits to costs is greater than one. Economists have developed a host of sophisticated techniques for assessing the costs and benefits of environmental policies, but critics nevertheless contend that cost-benefit analysis is merely a political device for slowing the growth of regulation, not a genuine analytic tool. They point out that estimates of the prospective costs and benefits of environmental policies are inherently biased against environmental protection because judgments about benefits, such as the value of saving wilderness or reducing the likelihood or severity of asthma attacks among sensitive populations, are difficult—if not impossible—to quantify, whereas immediate and tangible costs are easily figured. Moreover, they argue, using cost-benefit analysis as a decision rule eliminates ethical and moral considerations from the political calculus.[34] Regardless of how it is derived or how accurately it reflects a program's value, the number generated by a cost-benefit analysis constitutes a powerful frame because, as Deborah Stone observes, numbers have an aura of credibility and technical neutrality and therefore carry a great deal of political weight.[35]

Dramatizing the Risks of Action or Inaction. Finally, like scientific knowledge about a problem and the economic costs of addressing it, perceptions of the risk associated with action or inaction are subject to framing. Ordinary people do not assess risk based on objective analysis of statistical evidence; rather, they employ heuristics, or inferential rules—what political scientist Howard Margolis calls "habits of mind."[36] Psychologists have identified some inferential rules they believe shape the average person's perception of risks. Using the "availability" heuristic, for example, people judge an event as likely or frequent if instances of it are easy to recall. Therefore, they overestimate the risk of dramatic and sensational events such as airplane crashes, which tend to get abundant media coverage, while underestimating the risk of unspectacular events like car accidents.[37] Psychologists also point out that the public incorporates factors besides expected damages—the measure used by experts—into their assessment of risk. Among those factors are whether the risk is taken voluntarily, its immediacy, its familiarity, the extent of control one has over the risky situation, the severity of the possible consequences, and the level of dread the risk evokes.[38] The public's sensitivity to these factors explains why environmentalists are more successful at drawing attention to problems whose effects appear immediate and catastrophic than to those whose impacts are more remote and mundane.

Psychologists have found that other aspects of the way risk is framed can have a dramatic impact on risk perceptions as well. First, people value the same gains differently, depending on the reference point. For instance, they value the increment from $10 to $20 more highly than the increase from $110 to $120. Second, people are more concerned about losses than about gains of the same magnitude; that is, they fear losing $10 more than they value gaining $10. Third, people tend to overweight low probabilities and underweight moderate and high probabilities, so they worry more about rare occurrences (major oil spills) than common events (routine leakage of oil from pipelines and tankers).[39] Recognizing the importance of these elements of framing to the way the public perceives risk, both sides in an environmental contest define a problem as a loss from an already low status quo and overstate the likelihood of low probability but potentially disastrous outcomes. The difference is that environmentalists tend to emphasize the environmental or human-health risks of inaction, whereas cornucopians minimize environmental risks and focus on a policy's potential economic costs.

In sum, the hallmark of a successful environmental policy campaign is the ability of its organizers to define a problem and characterize its solution in a compelling way, in terms of the scientific explanation, the costs of regulation, and the risks associated with action or inaction. The side that succeeds in crafting the authoritative problem definition has an enormous advantage because the way people think and talk about a policy problem determines which solutions they are likely to embrace. In other words, those who furnish the prevailing problem definition are well-positioned to translate their values into policy.

MAJOR ACTORS IN ENVIRONMENTAL POLICYMAKING

The question, then, is why does the framing contest play out the way it does? Actors both inside and outside government influence the fate of competing problem definitions and solutions. Government officials must choose whether to address an environmental problem and, if so, how they will do it. Advocates on both sides try to influence that decision, adjusting their tactics to be consistent with the incentives and constraints of the institution making the decision. Their success depends heavily on the support of experts and the media's coverage of the issue.

Government Decision Makers

The decision makers in the national environmental policymaking process are the president and members of Congress who formulate legislation, the executive branch officials who interpret and administer the laws, and the judges who review their implementation by agencies. In various combinations, these actors determine whether environmental policy becomes more protective, permissive, or remains the same.

Legislative Actors. Legislative actors, the president and Congress, decide which problems government will address and establish the basic goals of public policy. In reaching their decisions, members of Congress want to make good public policy and attain the respect of their peers.[40] But they are also deeply concerned with the views of their constituents because they must be reelected if they hope to pursue their policy goals. Reelection concerns prompt legislators to support policies that distribute benefits to their constituents and oppose policies that threaten to impose direct, visible costs.[41]

The president and congressional leaders have powerful incentives to take on public policy issues of national, rather than simply district- or state-level, concern. Presidents, because they are elected by a national constituency and want to establish legacies, are attentive to broad public policy goals. Presidents can initiate action to address a problem by sending bills to Congress, by using the bully pulpit to convince the public. (They can also prevent action by vetoing legislation or simply failing to signal its importance.) Similarly, legislative leaders (and aspirants) seek opportunities to demonstrate their stewardship; in addition, their visibility both among the public and the political elite tends to elicit a sense of responsibility for public affairs.[42]

Even if the president and congressional leaders think an issue is important, they are not likely to expend political resources on it unless they perceive it to be widely salient. Similarly, to calculate their wiggle room on an issue, rank-and-file legislators must ascertain its salience among their constituents. An issue's salience can be difficult to discern, but one straightforward indicator is polling data. (According to sociologist Riley Dunlap, "The mere expression of supportive opinion in a scientific survey or informal poll . . . can be a vital resource" for groups hoping to bring about or block policy change.[43]) Although survey evidence can convey broad public preferences, it can also be misleading because the wording of questions and the order in which they are asked can yield different responses. In addition, polling results often contain internal contradictions; for example, a single poll may show overwhelming support for environmental preserves, such as wilderness areas and wild-and-scenic rivers, while simultaneously revealing a ferocious mistrust of government.[44] Moreover, district- and state-level polling on individual issues is rarely available. Most important, however, surveys have difficulty detecting how much people actually care about a problem (the intensity of their concern), their willingness to trade off one value for another, or the extent to which abstract values translate into support for concrete proposals.[45]

Nonetheless, politicians rely on a host of other indicators of an issue's salience. Because they garner media coverage, rallies and protests have been a mainstay of political activists; such public demonstrations are the simplest and most direct way for people with few political resources to transmit their concerns to elected officials. Other activities such as phoning, writing letters, sending e-mails, and posting Facebook messages and tweets also convey salience. Note, however, that creating the perception that an issue is salient is not a one-way street that runs from the public to politicians; legislators who want to promote a particular policy shape their constituents' views using language that is crafted to generate public support.[46]

Bureaucracy. Although less visible than legislators, agencies play a critical role in environmental policymaking because they implement the laws passed by Congress. Doing so involves choosing the scientific and economic models and projections that underpin administrative regulations, crafting the wording of those rules and the implementation guidelines that accompany them, and monitoring and enforcing compliance. Throughout this process, agency personnel have substantial discretion to modify policy goals.[47] In exercising their discretion, they bring ample political resources to bear, including their longevity, expertise, and established relationships with organized interests and members of Congress.

At the same time, whether it is environmentalist or cornucopian, an agency's ability to pursue its preferred goals is constrained in several ways. One institutional feature that limits administrators' flexibility is their mission and organizational culture.[48] An agency's mission is its original mandate, and its organizational culture consists of the norms and standard operating procedures that have evolved over time. For example, some agencies, such as the Forest Service and the Bureau of Land Management, were founded to conserve natural resources for human consumption. For a long time, the natural resource management agencies were staffed by professionals, such as foresters and range managers, whose expertise lay in maximizing natural resource yields. As a result, these agencies' standard operating procedures emphasized resource extraction—often at the expense of environmental protection. On the other hand, the Environmental Protection Agency (EPA) was created to prevent and clean up pollution, and the orientation of its professionals as well as its standard operating procedures reflect this disposition. The EPA's decisions tend to be relatively protective and often impose substantial costs on industry.

The preferences of a federal agency's organized clientele, the nature and extent of its congressional oversight, and the direction given by the president and the president's political appointees also circumscribe bureaucratic choices.[49] For example, organized interests dissatisfied with an agency's behavior can lobby sympathetic members of Congress, who in turn can exert pressure on agency officials by holding hearings, threatening budget cuts or reductions in statutory authority, or simply contacting agency officials directly to express their disfavor. Presidents likewise can impose their will on agencies by appointing directors whose views are consistent with their own. Presidents may also use the White House Office of Management and Budget or other administrative devices to bring an agency into line. In short, when implementing their statutory mandates, agency officials must navigate cautiously to avoid antagonizing powerful interests and their allies in Congress or the White House. Failure to do so almost always moves the battle to the courts.

The Judiciary. The federal courts have the authority to review agency decisions to determine whether they are consistent with congressional intent and, in this way, can circumscribe an agency's ability to pursue environmentally protective (or permissive) policies. The Administrative Procedures Act allows courts to invalidate decisions that lack "substantial evidence" or are "arbitrary and capricious." Moreover, many environmental statutes allow the courts to strike down an agency decision if they cannot discern a reasonable connection between the chosen course of action and the supporting record. The courts increased the potential for environmental litigation substantially in the early 1970s when they expanded the concept of "standing" to permit almost any group to challenge agency regulations in federal court, regardless of whether its members are directly affected by the agency's activities.[50] Congress also encouraged environmentalists' use of the courts by inserting provisions in environmental laws explicitly granting citizens and citizen organizations the right to sue not only polluters that violate the law, but also agencies that fail to implement or enforce their statutory mandates with sufficient zeal.[51]

Like legislators and agencies, judges may be environmentalists or cornucopians, but they also face institutional constraints when evaluating an agency's decisions: they must base their reasoning on precedent, as well as on the actual wording and legislative history of a statute. Judges have debated how closely to scrutinize agency decisions—in particular, whether to examine the reasoning or simply the procedures followed—but regardless of the standard applied, the courts habitually require agencies to document a comprehensible justification for their decisions.[52] As a result, litigation has become "an especially potent resource for making transparent the values, biases, and social assumptions that are embedded in many expert claims about physical and natural phenomena."[53]

Federalism and State and Local Decision Makers. The discussion so far has focused on national policymaking, but the politics of an issue depends in part on whether it is addressed at the local, state, or national level. Although there are many similarities among them, each of these arenas has distinctive features, which in turn have implications for the balance of power among environmentalists and cornucopians. For example, environmental advocates traditionally have felt disadvantaged at the state and local levels. One reason is that state and local officials tend to be deeply concerned with economic development and the need to attract and retain industry.[54] They are especially susceptible to threats by industry that it will take its capital (and jobs) elsewhere if the state or locality is insufficiently accommodating. Although economists have vigorously debated the extent to which industry actually moves in response to stringent environmental regulations, industry threats are effective nonetheless.[55] Another reason environmentalists tend to prefer to operate at the federal level is that, historically, some states and most local governments have lacked the technical capacity necessary to analyze complex environmental problems and therefore to distinguish among effective and ineffective solutions. The extent to which this is true varies widely among the states, however. Overall, as environmental policy scholar Mary Graham points out, states' technical capacities have improved dramatically since the 1970s, as has their propensity to address environmental problems.[56]

As environmental regulation has become more contentious at the national level, some environmentalists have become more interested in addressing environmental problems at the state and local levels on the grounds that place-based solutions are likely to be more effective and durable than approaches devised by Congress or federal agencies. Environmentalists have also turned to the states when the federal government resists acting; in doing so, they often prompt industry to demand federal standards to avoid a patchwork of state-level regulations. And finally, both environmentalists and cornucopians have tried to capitalize on a distinctive mode of decision making at the state level: placing an issue directly on the ballot. Critics charge that such "direct democracy" is simply another opportunity for wealthy interests to promote their agendas.[57] But supporters say ballot initiatives provide citizens the chance to check overreaching or unresponsive legislatures.[58] In practice, environmental policymaking almost always involves multiple levels of government: because the United States has a federal system of government, most national policies are implemented by the states, and political events at one level often affect decisions made at another.

Actors Outside of Government

Although politicians, administrators, and judges make decisions, actors outside of government create the context in which those decisions are made. In particular, organized interests that advocate for particular solutions play a major role because they make strategic choices about which venue to compete in, selecting the one they expect will be most hospitable to their goals.[59] Having chosen the arena, they select from a variety of tactics to influence government decision making. Critical to the success or failure of their efforts are experts, who provide the arguments and empirical support for advocates' positions, and the media, which may promote, reject, or modify the frames imposed by advocates.

Advocacy Organizations. Like government decision makers, advocates in environmental policy debates generally fall into one of two camps: environmentalists, who support more environmentally protective policies, and cornucopians, who endorse less restrictive environmental policies. Advocacy groups on both sides are diverse in terms of funding sources and membership.[60] Some rely heavily on foundations or even the federal government for funding, while others raise most of their money from individual members. Some of the national environmental organizations, such as the Sierra Club and the Wilderness Society, are well established; some, like the Audubon Society, have adopted a federated structure in which state-level branches have significant autonomy. By contrast, community-based environmental groups may be ephemeral, springing up to address a single problem and then disbanding. Similarly, a host of long-standing trade associations, such as the U.S. Chamber of Commerce and the National Association of Manufacturers, oppose efforts to make environmental policies more protective, and myriad local groups have formed to challenge laws and regulations that they believe infringe on private property rights.

Although individual groups tend to specialize, advocates are most effective when they form broad coalitions. The cohesiveness of a coalition over time is a major determinant of its political effectiveness. Coalitions that cannot maintain a united front tend to fare poorly, particularly in the legislative arena.[61] Even highly cohesive coalitions may be fleeting; however, if they are temporarily united by common policy goals, connections among them may not last beyond a single policy battle.

As important as building coalitions is selecting the appropriate tactics for defining a problem in the venue of choice. Advocates know elected officials' perception that a problem is salient is an important determinant of whether they will attend to it. Conventional tactics, such as direct—or inside—lobbying and contributing money to a political campaign, remain important ways to exercise influence in the legislative arena, but "outside lobbying"—that is, raising public awareness of issues and stimulating grassroots mobilization—has become increasingly prominent as an advocacy tool, particularly among business groups.[62] Public mobilization is challenging because much of the public has only a vague understanding of individual environmental issues and relies on cognitive shortcuts and cues to know what to think.[63] Advocates, therefore, rely heavily on stories and symbols to define problems in ways that raise their salience. By contrast, in the administrative and judicial arenas, public opinion and mobilization play a lesser role, and reasoned argument plays a larger one. To persuade bureaucrats and judges to adopt their preferred solution, advocates need to muster more sophisticated theoretical and empirical evidence in support of their definition of a problem.

Experts. Whatever tactics they adopt, advocates rely on experts and the research they generate to buttress their claims about the causes and consequences of an environmental problem. In environmental politics, experts include scientists, economists, lawyers, and policy analysts with specialized knowledge of environmental problems and policies. These individuals can work in academic departments, think tanks, foundations, interest groups, and government agencies. All of them are at the center of what political scientist John Kingdon calls "policy communities," where solutions to public policy problems are devised and the technical justifications for those solutions developed.[64] Both the public and policymakers tend to give greater weight to the views of experts than to those of outright advocates; as political scientist Benjamin Page and his coauthors observe, experts are considered credible because of their perceived objectivity, particularly when "complex technical questions affect the merits of policy alternatives."[65]

Experts, however, are not neutral purveyors of "facts." Most policy-relevant questions require experts to make value judgments based on uncertain data.[66] For example, policymakers might ask experts to ascertain a "safe" level of benzene in the environment. In conducting such a calculation, a chemical industry scientist is likely to make benign assumptions about benzene's hazards, consistent with her cornucopian values, while an academic scientist is likely to adopt more precautionary premises consistent with her environmental values.[67] Similarly, when asked about the economic impacts of benzene regulations, an industry economist is likely to base her projections on more pessimistic assumptions than is a government or academic economist.

The Media. Finally, the media—television, radio, newspapers, news magazines, and the Internet—are critical to determining the success or failure of competing advocates' efforts to define a problem. Writing more than eighty years ago, Walter Lippmann likened news coverage to "the beam of a searchlight that moves restlessly about, bringing one episode and then another out of darkness into vision."[68] There is substantial evidence that "the media may not only tell [the public] what to think about, they may also tell us how and what to think about it, and even what to do about it."[69] For most people the media are the only source of information about the environment, so what the media choose to focus on and the nature of their coverage are crucial to shaping public opinion.[70] Scholars have also found that the way the media frame issues significantly affects how the public attributes responsibility for problems.[71] (Naturally, people are not mere sponges for views expressed in the press; they make sense of the news in the context of their own knowledge and experience.[72])

Media coverage of environmental issues affects policymakers—not just through its impact on the public, but directly as well. There is little evidence that media-driven public opinion is a strong force for policy change, but news stories can prompt an elite response, even without a strong public reaction.[73] This response can occur when policymakers, particularly legislators, become concerned that media coverage *will* affect public opinion over time and so act preemptively. The relationship between the media and policymakers is not unidirectional: policymakers influence the media as well as react to it; in fact, policymakers are often more aware of their own efforts to manipulate media coverage than of the media's influence on their policy choices.[74]

Because the media have such a profound impact, the way they select and portray environmental news can have serious consequences for problem definition. Above all, the media focus on stories that are "newsworthy"—that is, dramatic and timely. Sudden, violent events with immediate and visceral consequences—such as oil spills, floods, and toxic releases—are far more likely to make the headlines than are ongoing problems such as species loss.[75] Furthermore, because journalists face short deadlines, they rely on readily available information from stable, reliable sources—such as government officials, industry, and organized interest groups—and the complexity of environmental science only reinforces this tendency.[76] Finally, in presenting the information gleaned from their sources, most reporters attempt to balance competing points of view, regardless of the relative support for either side; few journalists provide critical analysis to help readers sort out conflicting claims.[77] In their quest for balance, however, journalists tend to overstate extreme positions.[78]

In the 1990s a new wrinkle on the media and politics developed with the advent of Fox News and political talk radio, as well as blogs and news-aggregating websites. Once dominated by a handful of network TV channels and newspapers, the media are now fragmented and, increasingly, polarized.[79] This polarization has been exacerbated with the use of social media, and the contentiousness of environmental policy disputes have made it more difficult for politicians to discern politically palatable positions.

THE ENVIRONMENTAL POLICYMAKING PROCESS

Many scholars have found it helpful to model the process by which advocates, experts, the media, and decision makers interact to create policy as a series of stages:

- agenda setting—getting a problem on the list of subjects to which policymakers are paying serious attention
- alternative formulation—devising the possible solutions to the problem
- decision making—choosing from among the possible alternatives the approach that government will take to address the problem
- implementation—translating a policy decision into concrete actions
- evaluation—assessing those actions for their consistency with a policy's goals[80]

Distinguishing among these steps in the policymaking process can be fruitful analytically. At the same time, scholars generally acknowledge that in reality, the process is rarely as orderly as this linear model suggests.

That the policymaking process is not linear does not mean it is inexplicable; in fact, John Kingdon has developed a useful framework that captures its main attributes. He portrays policymaking as a process in which three "streams" flow independently. In the first stream, people in and around government concentrate on a set of problems; in the second, policy communities made up of experts, journalists, and bureaucrats initiate and refine proposals; and in the third, political events, such as a change of administration or an interest group campaign, occur.[81] In general, legislative and administrative policymakers engage in routine decision making; wary of major change, with its unpredictable political fallout, they tend to prefer making incremental modifications to existing policies.[82] A substantial departure from the status quo is likely only when the three streams merge, as a compelling problem definition and an available solution come together under hospitable political conditions. Such a convergence rarely just happens, however; usually policy entrepreneurs must actively link their preferred solution to a problem when a window of opportunity opens.

Policy Windows and Major Policy Change

Major policy changes are likely to occur only when a window of opportunity opens for advocates to promote their pet solutions.[83] In environmental policymaking, such a policy window may open as the result of a legal decision that forces legislators or administrators to reexamine a policy. A crisis or focusing event, such as a chemical accident, a smog event, or the release of a major scientific report, can also create a chance for action by providing powerful new evidence of the need for a policy and briefly mobilizing public opinion. A recurring event that can alter the dynamics of an issue is turnover of pivotal personnel: the replacement of a congressional committee chair, Speaker of the House, Senate majority leader, president, or agency director. Even more routine events, such as a legislative reauthorization or an administrative rulemaking deadline, occasionally present an opportunity for policy change.

Once a window has opened, policy may or may not change. Although some objective features define a policy window, advocates must recognize it to take advantage of it. And even if they accurately perceive an opportunity, advocates have a limited time to capitalize. A policy window may close because decision makers enact a policy. Alternatively, action may stall, in which case advocates may be unwilling to invest additional time, energy, or political capital in the endeavor. Finally, newsworthy events in other policy realms may divert public attention, causing a window to shut prematurely. Opponents of policy change recognize that policy windows open infrequently and close quickly and that both participants and the public have limited attention spans, so they try to delay action by studying an issue or by another expedient until the pressure for change subsides. As Kingdon observes, supporters of the status quo take advantage of the fact that "the longer people live with a problem, the less pressing it seems."[84]

The Role of Policy Entrepreneurs in "Softening Up" and "Tipping"

Given the advantage held by supporters of the status quo, it is clear that advocates of policy change must do more than simply recognize an opportunity; they must also recruit one or more policy entrepreneurs to promote their cause. Policy entrepreneurs are individuals willing to invest their political resources—time, energy, reputation, money—in linking a problem to a solution and forging alliances among disparate actors to build a majority coalition.[85] Policy entrepreneurs must be adept at discovering "unfilled needs" and linking them to solutions, willing to bear the risks of investing in activities with uncertain consequences, and skilled at coordinating the activities of individuals and groups.[86] In addition, policy entrepreneurs must be ready to "ride the wave" when a policy window opens.[87] They must have lined up political allies, prepared arguments, and generated favorable public sentiment in preparation for the moment a decision-making opportunity presents itself.

Among the most important functions policy entrepreneurs perform while waiting for a policy window to open is "softening up" policy solutions, both in the expert communities whose endorsement is so important to the credibility of a policy and among the larger public. Softening up involves getting people used to a new idea and building acceptance for it.[88] Policy entrepreneurs have a variety of means to soften up policy solutions: they can give speeches, write scholarly and popular articles, give briefings to policymakers, compose editorials and press releases, and teach the new approach in classrooms. Over time, a consensus builds, particularly within a policy community, around a short list of ideas for solving a problem. Eventually, there is a broader convergence on a single approach, a phenomenon known as "tipping." At the tipping point, support for an idea is sufficiently widespread that it seems to take on a life of its own.[89]

The Importance of Process and History

To affect policymaking, solutions that have diffused through the policy community and the public must catch on among decision makers as well. In a traditional decision-making process, participants in a policy debate typically begin by staking out an extreme position and holding fast to it. At this point, bargaining and persuasion commence, and participants try to build a winning coalition by accommodating as many interests as possible. Once it becomes apparent that one side is going to prevail, even the holdouts recognize that, if they do not join in, they will have no say in the final decision. This creates a bandwagon effect: as Kingdon explains, "Once an issue seems to be moving, everybody with an interest in the subject leaps in, out of fear that they will be left out."[90]

In an adversarial process, however, a problem is never really solved; each decision is simply one more step in a never-ending contest. To avoid protracted appeals, policymakers are turning with increasing frequency to nonadversarial processes—such as regulatory negotiation, mediation, and consensus building—to reach agreement on policy solutions. According to their proponents, such processes enhance the quality of participation by including a broader array of stakeholders and promoting deliberation, a search for common ground, and a spirit of cooperation. Proponents also believe that solutions arrived at collaboratively are likely to be more effective and enduring than those attained under adversarial processes.[91] On the other hand, critics fear that such approaches disadvantage environmentalists and result in watered-down solutions.[92]

Regardless of whether they are resolved through adversary or collaborative processes, policy battles are rarely fought on a clean slate; their dynamics are influenced by existing policies and the configuration of advocates around the status quo. Institutions—defined as both formal organizations and the rules and procedures they embody—are resistant to change, even in the face of compelling new ideas.[93] In addition, advocates adjust their tactics over time in response to judgments about their effectiveness and that of their opponents—a phenomenon known as political learning. Sometimes advocates even adjust their beliefs, as a result of policy learning. As political scientist Paul Sabatier observes, such learning tends to concern the efficacy of particular policy mechanisms, rather than the nature or importance of a problem.[94]

Changing Policy Gradually

Most of the mechanisms described above relate to major policy change. But policy can also change incrementally. As political scientist Jacob Hacker observes, sometimes those who seek to challenge popular institutions "may find it prudent not to attack such institutions directly. Instead, they may seek to shift those institutions' ground-level operations, prevent their adaptation to shifting circumstances, or build new institutions on top of them."[95] Members of Congress have a variety of low-profile means to challenge the status quo. They can communicate directly with high-level agency officials or adjust an agency's budget to reflect their own preferences. One of the most effective low-profile tactics legislators can employ is to attach a rider—a nongermane amendment—to must-pass legislation. Such riders can prohibit agency spending on particular activities, forbid citizen suits or judicial review of any agency decision, or make other, more substantive policy adjustments. The president also has a variety of tools with which to change policy quietly and unilaterally: executive orders, proclamations, presidential memoranda, and presidential signing statements.[96]

Agency personnel have numerous low-profile options for modifying policy as well. They can change the way a law is implemented by instituting new rules or repealing or substantially revising existing ones; they can also expedite or delay a rulemaking process. In formulating a rule, they can choose to consult with (or ignore) particular interests or to infuse the rulemaking process with a new analytic perspective. They can alter the implementation of a rule by adjusting the agency's budget; increasing, cutting, or reorganizing personnel; taking on new functions or privatizing functions previously performed by the bureaucracy; hiring and promoting, or demoting and transferring, critical personnel; creating new advisory bodies or altering the membership of existing panels, or adjusting the rules by which such groups reach closure; adjusting agency operating procedures through internal memos and unpublished directives; and reducing or increasing the aggressiveness with which criminal and civil violations are pursued.[97]

The common feature of low-profile policy challenges is that it is difficult to garner publicity for them and therefore to make them salient. The more arcane they are, the more difficult it is to mobilize resistance. And if successful, low-profile challenges can result in "gradual institutional transformations that add up to major historical discontinuities."[98]

CASE SELECTION

How the policymaking process translates values into policy over time, abruptly or gradually or both, is the subject of the next fifteen chapters. The cases in these chapters not only introduce a variety of environmental issues, but also capture most aspects of the environmental policymaking process. They cover disputes from all regions of the country; offer examples of local, national, and international politics; and focus on problems that are of great concern to those who attend to this policy area. The cases are organized into three parts: regulating polluters, natural resource management, and new issues. Although these divisions reflect topical differences, they are somewhat arbitrary; at least some of the cases fit in multiple categories. In addition, there are many similarities across the cases. All of them illuminate the impact of participants' values on how problems are defined and solutions characterized, while each case also highlights a small number of more particular attributes of environmental policymaking, expanding on various aspects of the framework sketched above. As a collection, the cases provide a comprehensive foundation for understanding the way the American political system handles environmental issues.

Regulating Polluters

When the issue of pollution burst onto the national political scene in the 1960s and early 1970s, the public responded with overwhelming interest and concern about dirty air and water as well as toxic waste. Chapter 2 describes how the impact of public mobilization on the legislative process led to the formation of the EPA and the passage of the Clean Air and Clean Water acts. The case makes clear that when political leaders perceive an issue as widely salient, they often compete to craft a strong legislative response. Agencies trying to implement laws forged under such circumstances, however, are likely to encounter a host of practical obstacles. Chapter 3, which relates the story of toxic waste dumping at Love Canal, reveals the extent to which local and state governments historically have resisted confronting pollution issues because of technical incapacity and concerns about economic development. It also demonstrates the impact of media coverage on politics: alarming news stories can prompt citizen mobilization, and coverage of local groups' claims can nationalize an issue and produce a strong policy response. In addition, this chapter makes clear that scientific experts rarely can resolve environmental policy controversies and may, in fact, exacerbate them. The case of the Chesapeake Bay restoration (Chapter 4) illustrates demands by scientists and environmentalists for ecosystem-scale solutions, as well as the need to coordinate the activities of multiple agencies and political jurisdictions. The final case in this section (Chapter 5) reveals that sometimes legislative leaders can break a political logjam by proposing a policy tool that facilitates new coalitions, as was done with tradable sulfur-dioxide permits in the effort to address acid rain under the Clean Air Act.

History, Changing Values, and Natural Resource Management

Natural resource issues have a much longer history in American politics than do pollution issues. The distinctive feature of natural resource policymaking in the United States is the extent to which it is shaped by the legacy of past policies. Strong traditions drive decision making regarding public lands. The first chapter in this section, Chapter 6, details the ongoing dynamics of the controversy over drilling for oil in Alaska's Arctic National Wildlife Refuge from 1977 to 2018. This particular case illuminates the intractable nature of conflicts between wilderness and natural resource development. It also illustrates the way competing advocates use information selectively and adopt evocative language—particularly symbols and metaphors—to define problems. After a long fought battle by environmentalists, the Trump administration opened for drilling, the impacts left unknown. Chapter 7 explores federal grazing policy, one of the nation's least publicized natural resource issues: how to manage the arid rangeland of the West. This case provides an opportunity to observe the impact of past policies on current decision making. More specifically, it shows how those who benefit from those policies form symbiotic relationships with policymakers. In particular, this chapter explores how past policies surrounding grazing fees led to a heated standoff with local law enforcement in 2014 in Oregon and an eventual presidential pardon. Nonetheless, Chapter 7 illuminates how the failure by advocates to arouse public concern about an issue helps to perpetuate the status quo.

By contrast, Chapter 8, "Jobs Versus the Environment: Saving the Northern Spotted Owl," makes clear that attracting public attention to a natural resource concern using legal leverage and outside lobbying campaigns can overwhelm historically entrenched interests and change the course of policymaking on an issue. Interestingly, in this case, science and scientists also played pivotal roles in transforming an agency's approach to decision making. "Playground or Paradise? Snowmobiles in Yellowstone National Park" (Chapter 9) looks at the bitter conflict over allowing motorized vehicles to explore how issues of recreational access can divide those who claim to value the nation's parklands. As in the preceding cases, an initial decision to allow an activity in the park spawned economic interests and a perceived "right" to access. But this case shows how agency decision makers can bring about major policy adjustments, prompting dissenters to shop for friendly venues in which to appeal.

The final two cases in this section concern the impacts of extracting resources from the ocean. Chapter 10 ("Crisis and Recovery in the New England Fisheries") illustrates how government's efforts to manage the commons (in this case, New England's cod, haddock, and flounder) can exacerbate the free-rider problem, especially when commercial interests dominate the regulatory process. The case also demonstrates the critical role that litigation can play in disrupting established policymaking patterns. The chapter concludes by outlining some of the novel approaches that governments are exploring to manage common-property resources more effectively, as well as the sources of resistance to adopting such solutions. The question becomes what happens to these approaches if an overhaul of federal legislation occurs. "The Deepwater Horizon Disaster: The High Cost of Offshore Oil" (Chapter 11) provides a look at how the federal government responds to disasters caused by our reliance on fossil fuels—in this instance, a massive oil spill a mile offshore in the Gulf of Mexico. The case makes vividly apparent the extent to which federal agencies can become dependent on and entwined with the companies they are supposed to regulate. This phenomenon is exacerbated by the growing unwillingness of federal policymakers to support strict regulation of industry and to provide the funding necessary for truly effective, independent oversight.

New Issues, New Politics

Chapter 12 examines one of the most complex and divisive commons problems facing the world today: climate change. The case reveals how the Paris Accord set the tone for global policy on climate change. The efforts are highlighted by the fifth Intergovernmental Panel on Climate Change (IPCC) special report on global warming and the United States' fourth national climate assessment. These reports both call for immediate action. Despite these calls for action, U.S. domestic climate policy efforts are obstructed by science–policy debates at the federal level. State-level policymaking may serve as tactical adjustments for the lack of federal policy. Furthermore, the 2019 Democratic-controlled House of Representatives shows promise for the "Green New Deal"—an economic stimulus approach to combat climate change.

A related challenge is that of producing clean energy, which will be central to mitigating the harm inflicted by climate change. Chapter 13 investigates the story of Cape Wind, slated to be one of the nation's first offshore wind farms and how it suffered a very slow death. This case lays bare the difficulties of comparing the costs associated with conventional fuels with the expenses of building new alternative energy plants. It also illustrates the siting difficulties facing alternative energy facilities, which occupy more space than their fossil fuel counterparts and are often proposed for places that have well-defended aesthetic or ecological value. And it highlights how advocates on both sides of a siting dispute capitalize on process requirements to delay action when they perceive that doing so works to their advantage.

"Fracking Wars: Local and State Responses to Unconventional Shale Gas Development" (Chapter 14) explores the way states and localities are managing the exploitation of shale gas using a combination of hydraulic fracturing and horizontal drilling. This relatively new technique—known as high-volume hydraulic fracturing or, more colloquially, as fracking—enables gas drillers to work in areas unaccustomed to the disruption and potential pollution caused by natural resource development. It has prompted a variety of state and local responses—from outright bans to a variety of regulations aimed at reducing the impacts of fracking. The case focuses on the tensions that arise between localities trying to preserve their quality of life and state governments aiming to ensure the orderly development of natural resources and promote economic development.

Chapters 15 and 16 address some of the challenges associated with the pursuit of urban sustainability. "Making Trade-Offs: Urban Sprawl and the Evolving System of Growth Management in Portland, Oregon" (Chapter 15) concerns efforts by states and localities to manage the diffuse and complex issue of urban sprawl. It elucidates the origins of Oregon's land-use planning framework and the role of civic engagement in implementing that framework in the City of Portland. Importantly, it also takes up how Portland responded to a powerful ideological challenge by taking the concerns of critics seriously and adding flexibility to the regulatory framework. Chapter 16 documents how hurricanes post-Katrina can be viewed through the lens of environmental justice. Simply put, this chapter considers how government policy has exacerbated the vulnerability of many U.S. communities to disaster, the extent to which disaster disproportionately affects poor and minority communities, and the many obstacles to recovery, restoration, and resilience.

As all of the cases make clear, many of the same political forces are at play regardless of the decision-making approach adopted or the type of policy mechanism proposed. Even as new ideas emerge and gain traction, institutionalized ideas and practices continue to limit the pace of policy change. Moreover, underlying value differences among participants remain, so the ability to define problems and characterize their solutions persuasively continues to be a critical source of influence.

GETTING THE MOST OUT OF THE CASES

This chapter has provided a cursory introduction to the burgeoning field of environmental politics and policymaking. The cases that follow deepen and make concrete the concepts introduced here, highlighting in particular how participants use language to define problems in ways consistent with their values. As you read, you will notice similarities and differences among cases. I encourage you to look for patterns, generate hypotheses about environmental politics, and test those hypotheses against the other instances of environmental policymaking that you study, read about in the newspaper, or become involved in. Questions posed at the end of each case may help you think more deeply about the issues raised in it and generate ideas for further research.

NOTES

1. John S. Dryzek and David Schlosberg, eds., *Debating the Earth: The Environmental Politics Reader* (New York: Oxford University Press, 1998), 1.
2. Paul Sabatier makes this point about policy controversies more generally in Paul A. Sabatier, "An Advocacy Coalition Framework of Policy Change and the Role of Policy-Oriented Learning Therein," *Policy Sciences* 21 (1988): 129–168.
3. See, for example, Black Elk, "Native Americans Define the Natural Community," in *American Environmentalism*, 3d ed., ed. Roderick Frazier Nash (New York: McGraw-Hill, 1990), 13–16. Critics argue that environmentalists have romanticized Native Americans and other early peoples, pointing out that they too were capable of hunting species to extinction; that the tribes of eastern North America deliberately influenced the range and abundance of countless wild plant species through their food-gathering and land-use practices; and that Native Americans made great use of fire and, in doing so, altered the landscape on a sweeping scale. They contend that much of what we now see as "natural" is in fact the result of human alteration during earlier time periods. See, for example, Stephen Budiansky, *Nature's Keepers: The New Science of Nature Management* (New York: Free Press, 1995).
4. George Catlin, "An Artist Proposes a National Park," in *American Environmentalism*, 31–35.
5. Henry David Thoreau, "Walking," in *Excursions, The Writings of Henry David Thoreau*, Vol. IX, Riverside ed., 11 vols. (Boston: Houghton Mifflin, 1893).
6. Quoted in Roderick Nash, *Wilderness and the American Mind*, 3d ed. (New Haven: Yale University Press, 1982), 125.
7. Samuel Hays, *Conservation and the Gospel of Efficiency* (Cambridge, Mass.: Harvard University Press, 1959).
8. Reed F. Noss and Allen Y. Cooperrider, *Saving Nature's Legacy: Protecting and Restoring Biodiversity* (Washington, D.C.: Island Press, 1994).
9. The relationship between ecology and environmentalism was reciprocal. According to historian Peter Bowler, although a distinct science of ecology emerged in the

1890s, ecology did not flower as a discipline until the 1960s with the proliferation of environmental concern. See Peter Bowler, *Norton History of Environmental Sciences* (New York: Norton, 1992).

10. Aldo Leopold, *A Sand County Almanac* (New York: Oxford University Press, 1948), 239, 262.

11. Donella Meadows et al., *The Limits to Growth* (New York: Universe, 1972).

12. Herman Daly, *Steady-State Economics* (Washington, D.C.: Island Press, 1991). Daly's work draws heavily on the insights developed by economists Nicholas Georgescu-Roegen and Kenneth Boulding. See Nicholas Georgescu-Roegen, *The Entropy Law and the Economic Process* (Cambridge, Mass.: Harvard University Press, 1971); and Kenneth Boulding, "The Economics of the Coming Spaceship Earth," in *Environmental Quality in a Growing Economy*, ed. Henry Ed Jarrett (Baltimore: Johns Hopkins University Press/Resources for the Future, 1966), 3–14.

13. Robyn Eckersley, *Environmentalism and Political Theory* (Albany: State University of New York Press, 1992).

14. Arne Naess, "The Shallow and the Deep, A Long-Range Ecology Movement: A Summary," *Inquiry* 16 (1983): 95.

15. Dryzek and Schlosberg coined the term *cornucopians* in *Debating the Earth*. In Greek mythology, Prometheus stole fire from Olympus and gave it to humans. To punish him for this crime, Zeus chained him to a rock and sent an eagle to eat his liver, which in turn regenerated itself each day. The term Prometheans, therefore, suggests a belief in the endless regenerative capacity of the earth.

16. Aaron Wildavsky and Karl Dake provide another label: hierarchists. They say, "Hierarchists . . . approve of technological processes and products, provided their experts have given the appropriate safety certifications and the applicable rules and regulations are followed." See Aaron Wildavsky and Karl Dake, "Theories of Risk Perception: Who Fears What and Why?" *Daedalus* 119 (Fall 1990): 41–60.

17. Julian L. Simon and Herman Kahn, *The Resourceful Earth* (New York: Blackwell, 1984), 3.

18. Ibid.

19. Mary Douglas and Aaron Wildavsky, *Risk and Culture* (Berkeley: University of California Press, 1982).

20. See, for example, Aaron Wildavsky, *But Is It True?* (Cambridge, Mass.: Harvard University Press, 1995); Michael Fumento, *Science Under Siege* (New York: Quill, 1993); and Bjorn Lomborg, *The Skeptical Environmentalist: Measuring the Real State of the World* (New York: Cambridge University Press, 2001).

21. Gregg Easterbrook, *A Moment on the Earth* (New York: Viking, 1995), 649.

22. E. E. Schattschneider, *The Semisovereign People* (New York: Holt, Rinehart, and Winston, 1960), 189.

23. Political scientists use a variety of terms—*problem definition*, *issue definition*, and *issue framing*—to describe essentially the same phenomenon. Although we use the terms interchangeably, we refer readers to Deborah Stone's *Policy Paradox*, whose discussion of problem definition and its political impact is the most precise we have found. See Deborah Stone, *Policy Paradox: The Art of Political Decision Making*, rev. ed. (New York: Norton, 2002). See also David Rochefort and Roger Cobb, eds., *The Politics of Problem Definition* (Lawrence: University Press of Kansas, 1994).

24. Donald A. Schon and Martin Rein, *Frame Reflection: Toward the Resolution of Intractable Policy Controversies* (New York: Basic Books, 1994).

25. Stone, *Policy Paradox*.

26. Frank R. Baumgartner and Bryan D. Jones, *Agendas and Instability in American Politics* (Chicago: University of Chicago Press, 1993).

27. Stone, *Policy Paradox*, 189.

28. Schattschneider, *The Semisovereign People*, 66.

29. Deborah Guber, *The Grassroots of a Green Revolution: Polling America on the Environment* (Cambridge, Mass.: The MIT Press, 2003); Willett Kempton, James S. Boster, and Jennifer A. Hartley, *Environmental Values in American Culture* (Cambridge, Mass.: The MIT Press, 1995); Everett Carll Ladd and Karlyn H. Bowman, *Attitudes Toward the Environment* (Washington, D.C.: AEI Press, 1995).

30. Riley E. Dunlap, "Public Opinion in the 1980s: Clear Consensus, Ambiguous Commitment," *Environment*, October 1991, 10–22.

31. Although a few skeptics continue to challenge theories of ozone depletion, by the early 1990s, most climatologists accepted the conclusions of the World Meteorological Association that "anthropogenic chlorine and bromine compounds, coupled with surface chemistry on natural polar stratospheric particles, are the cause of polar ozone depletion." See Larry Parker and David E. Gushee, "Stratospheric Ozone Depletion: Implementation Issues," CRS Issue Brief for Congress, No. 97003 (Washington, D.C.: Congressional Research Service, January 16, 1998).

32. On occasion, roles are reversed, particularly when the issue at stake is the impact of technology on human health or the environment. For example, scientists have been unable to detect a relationship between electromagnetic fields (EMFs) and cancer, and proponents of EMF regulation are the ones citing minority scientific opinions. This pattern—in which the experts try to reassure a skeptical public that a technology is safe—recurs in the debates over nuclear power, genetically modified food, and vaccines.

33. Eban Goodstein and Hart Hodges, "Behind the Numbers: Polluted Data," *American Prospect*, November 1, 1997, 64–69.

34. Stephen Kelman, "Cost-Benefit Analysis: An Ethical Critique," in *The Moral Dimensions of Public Policy Choice*, ed. John Martin Gillroy and Maurice Wade (Pittsburgh: University of Pittsburgh Press, 1992), 153–164.

35. Stone, *Policy Paradox*.

36. Howard Margolis, *Dealing with Risk: Why the Public and the Experts Disagree on Environmental Risk* (Chicago: University of Chicago Press, 1996).

37. Paul Slovic, Baruch Fischhoff, and Sarah Lichtenstein, "Rating the Risks," in *Readings in Risk*, ed. Theodore Glickman and Michael Gough (Washington, D.C.: Resources for the Future, 1990), 61–75.

38. Ibid. Howard Margolis rejects this explanation for differences between expert and public assessments of risk. He argues instead that the difference turns on "habits of mind"; in particular, in some cases, the public stubbornly perceives only the costs of a technology or activity and is unable to see the benefits and therefore cannot make the appropriate trade-off between the two. See Margolis, *Dealing with Risk*.

39. Amos Tversky and Daniel Kahneman, "The Framing of Decisions and the Psychology of Choice," *Science*, January 30, 1981, 453–458.

40. Richard Fenno, *Congressmen in Committees* (Boston: Little, Brown, 1973).

41. David R. Mayhew, *Congress: The Electoral Connection* (New Haven: Yale University Press, 1974).

42. Timothy J. Conlan, Margaret Wrightson, and David Beam, *Taxing Choices: The Politics of Tax Reform* (Washington, D.C.: CQ Press, 1990); Martha Derthick and Paul J. Quirk, *The Politics of Deregulation* (Washington, D.C.: Brookings Institution, 1985).

43. Riley Dunlap, "Public Opinion in Environmental Policy," in *Environmental Politics and Policy*, ed. James P. Lester (Durham: Duke University Press, 1989), 87.

44. Timothy Egan, "The 1994 Campaign: Western States," *The New York Times*, November 4, 1994.

45. With occasional exceptions, the environment ranks low as Gallup's "most important problem"; in fact, it was mentioned by only 1 percent of those surveyed in January 1995 and 2 percent in 2000. See Everett Carll Ladd and Karlyn H. Bowman, *Attitudes Toward the Environment: Twenty-Five Years After Earth Day* (Washington, D.C.: AEI Press, 1995); Deborah Guber, *The Grassroots of a Green Revolution* (Cambridge, Mass.: The MIT Press, 2003). But because salience measures like the most-important-problem question are headline sensitive, they are untrustworthy guides to how the public will respond to particular policy proposals. See Robert Cameron Mitchell, "Public Opinion and the Green Lobby: Poised for the 1990s?" in *Environmental Policy in the 1990s*, ed. Norman J. Vig and Michael E. Kraft (Washington, D.C.: CQ Press, 1990), 81–99.

46. Lawrence R. Jacobs and Robert Y. Shapiro, *Politicians Don't Pander: Political Manipulation and the Loss of Democratic Responsiveness* (Chicago: University of Chicago Press, 2000).

47. Jeffrey L. Pressman and Aaron Wildavsky, *Implementation* (Berkeley: University of California Press, 1984).

48. James Q. Wilson, *Bureaucracy* (New York: Basic Books, 1989).

49. Herbert Kaufman, *The Administrative Behavior of Federal Bureau Chiefs* (Washington, D.C.: Brookings Institution, 1981).

50. "Standing" is the right to bring a lawsuit. Historically, the courts have granted standing to anyone who can demonstrate that he or she is personally affected by the outcome of a case. *Sierra Club v. Morton* (1972) laid the groundwork for the subsequent broadening of the courts' interpretation of the standing requirement. In the 1980s, however, conservatives began challenging environmentalists' standing to sue in hopes of reducing their ability to use litigation to pursue their policy goals. In the early 1990s, those efforts began to pay off, as the Supreme Court issued rulings that curtailed judges' propensity to grant standing to environmental plaintiffs. See William Glaberson, "Novel Antipollution Tool is Being Upset by Courts," *The New York Times*, June 5, 1999.

51. A second way the courts can influence environmental policymaking is through civil litigation. In tort litigation, judges must decide whether a company is liable for damages caused by a substance or event. To prevail, a plaintiff must prove negligence on the part of a defendant, a burden that can be extraordinarily heavy in the case of chronic exposure to hazardous substances. See, for example, Jonathan Harr, *A Civil Action* (New York: Vintage Books, 1996).

52. Sheila Jasanoff, *Science at the Bar: Law, Science, and Technology in America* (Cambridge, Mass.: Harvard University Press, 1995); David M. O'Brien, *What Process Is Due? Courts and Science-Policy Disputes* (New York: Russell Sage, 1987).

53. Jasanoff, *Science at the Bar*, 20.

54. Paul Peterson, *City Limits* (Chicago: University of Chicago Press, 1981).

55. For a review of the scholarly debate over the "pollution haven hypothesis," see Smita B. Brunnermeier and Arik Levenson, "Examining the Evidence of Environmental Regulations and Industry Location," *Journal of Environment & Development* 13,1 (March 2004): 6–41.

56. Mary Graham, *The Morning After Earth Day: Practical Environmental Politics* (Washington, D.C.: Brookings Institution Press, 1999).

57. See, for example, Peter Shrag, *Paradise Lost: California's Experience, America's Future* (New York: New Press, 1998); David Broder, *Democracy Derailed: Initiative Campaigns and the Power of Money* (New York: Harcourt Brace, 2000); Richard Ellis, *Democratic Delusions* (Lawrence: University Press of Kansas, 2002).

58. David Schmidt, *Citizen Lawmakers: The Ballot Initiative Revolution* (Philadelphia: Temple University Press, 1989); Arthur Lupia and John G. Matsusaka, "Direct Democracy: New Approaches to Old Questions," *Annual Review of Political Science* 7 (2004): 463–482.

59. Baumgartner and Jones, *Agendas and Instability*.

60. For an extensive discussion of the funding, membership, tactics, and goals of major environmental interest groups, see Ronald G. Shaiko, *Voices and Echoes for the Environment* (New York: Columbia University Press, 1999); Christopher J. Bosso, *Environment Inc.: From Grassroots to Beltway* (Lawrence: University Press of Kansas, 2005).

61. Gary Mucciaroni, *Reversals of Fortune: Public Policy and Private Interests* (Washington, D.C.: Brookings Institution, 1995).

62. Ken Kollman, *Outside Lobbying: Public Opinion and Interest Group Strategies* (Princeton, N.J.: Princeton University Press, 1998). In 2015 the Center for Public Integrity released a report that found the amount that the nation's biggest trade associations spent for public relations and advertising firms to persuade the U.S. public dwarfed the amount spent on lobbying federal officials in 2012. See Erin Quinn and Chris Young, "Who Needs Lobbyists? See What Big Business Spends to Win American Minds," Center for Public Integrity (Website), January 15, 2015. Available at http://www.publicintegrity.org/2015/01/15/16596/who-needs-lobbyists-see-what-big-business-spends-win-american-minds.

63. Shanto Iyengar argues that "people are exquisitely sensitive to context when they make decisions, formulate judgments, or express opinions. The manner in which a problem of choice is 'framed' is a contextual cue that may profoundly influence decision outcomes." See Shanto Iyengar, *Is Anyone Responsible? How Television Frames Political Issues* (Chicago: University of Chicago Press, 1991), 11.

64. John Kingdon, *Agendas, Alternatives, and Public Policies*, 2d ed. (New York: HarperCollins, 1995).

65. Benjamin I. Page, Robert Y. Shapiro, and Glenn R. Dempsey, "What Moves Public Opinion," *Media Power in Politics*, 3d ed., ed. Doris A. Graber (Washington, D.C.: CQ Press, 1994), 132.

66. Alvin Weinberg, "Science and Trans-Science," *Minerva* 10 (1970): 209–222.

67. Frances M. Lynn, "The Interplay of Science and Values in Assessing and Regulating Environmental Risks," *Science, Technology, and Human Values* 11 (Spring 1986): 40–50.

68. Walter Lippman, *Public Opinion* (New York: Macmillan, 1922), 229.

69. Maxwell McCombs and George Estrada, "The News Media and the Pictures in Our Heads," in *Do the Media Govern?* ed. Shanto Iyengar and Richard Reeves (Thousand Oaks, Calif.: SAGE, 1997), 247.

70. Maxwell E. McCombs and Donald L. Shaw, "The Agenda-Setting Function of the Press," in *The Emergence of American Political Issues: The Agenda-Setting Function of the Press* (St. Paul, Minn.: West Publishing, 1977), 89–105; Fay Lomax Cook et al., "Media and Agenda Setting Effects on the Public, Interest Group Leaders, Policy Makers, and Policy," *Public Opinion Quarterly* 47 (1983): 16–35.

71. Shanto Iyengar, "Framing Responsibility for Political Issues," in *Do the Media Govern?* 276–282.

72. Doris A. Graber, *Mass Media and American Politics*, 5th ed. (Washington, D.C.: CQ Press, 1997).

73. Ibid. Graber points out that the relative dearth of evidence supporting the claim that media-generated public opinion causes policy change is probably the result of using insufficiently sophisticated methods to detect such effects, not the absence of effects.

74. Kingdon, *Agendas.*

75. Michael R. Greenberg et al., "Risk, Drama, and Geography in Coverage of Environmental Risk by Network T.V.," *Journalism Quarterly* 66 (Summer 1989): 267–276.

76. Herbert Gans, *Deciding What's News* (New York: Pantheon, 1979); Dorothy Nelkin, *Selling Science*, rev. ed. (New York: Freeman, 1995).

77. Nelkin, *Selling Science*.

78. Eleanor Singer, "A Question of Accuracy: How Journalists and Scientists Report Research on Hazards," *Journal of Communication* 40 (Autumn 1990): 102–116.

79. Kathleen Hall Jamieson and Joseph N. Capella, *Echo Chamber: Rush Limbaugh and the Conservative Media Establishment* (New York: Oxford University Press, 2008); Ken Autletta, "Non-Stop News," *New Yorker*, January 25, 2010, 38–47.

80. See, for example, Charles O. Jones, *An Introduction to Public Policy*, 2d ed. (North Scituate, Mass.: Wadsworth, 1984).

81. In *Agendas*, Kingdon suggests that these three activities, or streams, proceed independently of one another. Gary Mucciaroni argues that changes in problem definition, solutions, and political conditions are actually quite closely linked. See Mucciaroni, *Reversals of Fortune*.

82. Incrementalism involves tinkering with policies at the margin rather than engaging in a comprehensive reexamination of each issue. See Charles E. Lindblom, "The Science of Muddling Through," *Public Administration Review* 14 (Spring 1959): 79–88; Aaron Wildavsky, *The Politics of the Budgetary Process*, 3d ed. (Boston: Little, Brown, 1979).

83. Kingdon, *Agendas*.

84. Ibid., 170.

85. Ibid.

86. Michael Mintrom and Sandra Vergari, "Advocacy Coalitions, Policy Entrepreneurs, and Policy Change," *Policy Studies Journal* 24 (1996): 420–434.

87. Kingdon, *Agendas*.

88. Ibid.

89. Malcolm Gladwell, *The Tipping Point: How Little Things Can Make a Big Difference* (Boston: Little, Brown, 2000).

90. Kingdon, *Agendas*, 162.

91. Lawrence Susskind and Jeffrey Cruikshank, *Breaking the Impasse: Consensual Approaches to Resolving Public Disputes* (New York: Basic Books, 1987); Julia Wondolleck and Steven L. Yaffee, *Making Collaboration Work: Lessons from Innovation in Natural Resource Management* (Washington, D.C.: Island Press, 2000).

92. George Cameron Coggins, "Of Californicators, Quislings, and Crazies: Some Perils of Devolved Collaboration," in *Across the Great Divide: Explorations in Collaborative Conservation and the American West*, ed. Philip Brick, Donald Snow, and Sarah Van de Wetering (Washington, D.C.: Island Press, 2001), 163–171; Cary Coglianese, "Is

Consensus an Appropriate Basis for Regulatory Policy?" in *Environmental Contracts*, ed. Eric W. Orts and Kurt Deketelaere (Boston: Kluwer Law International, 2001), 93–113.

93. Paul Pierson, *Politics in Time: History, Institutions, and Social Analysis* (Princeton, N.J.: Princeton University Press, 2004).

94. Sabatier, "An Advocacy Coalition Framework."

95. Jacob S. Hacker, "Privatizing Risk Without Privatizing the Welfare State: The Hidden Politics of Social Policy Retrenchment in the United States," *American Political Science Review* 98,2 (2004): 243–260.

96. Judith A. Layzer, *Open For Business: Conservatives' Opposition to Environmental Regulation* (Cambridge, Mass.: The MIT Press, 2012).

97. Layzer, *Open For Business.*

98. Wolfgang Streeck and Kathleen Thelen, "Introduction: Institutional Change in Advanced Political Economies," in *Beyond Continuity: Institutional Change in Advanced Political Economies*, ed. Wolfgang Streeck and Kathleen Thelen (New York: Oxford University Press, 2005), 8.

PART ONE

REGULATING POLLUTERS

CHAPTER TWO

THE NATION TACKLES AIR AND WATER POLLUTION

The Environmental Protection Agency and the Clean Air and Clean Water Acts

In the twenty-first century, Americans take for granted the importance of federal laws aimed at reducing air and water pollution. But for most of the nation's history, the federal government was practically uninvolved in pollution control. That changed abruptly on July 9, 1970, when President Richard Nixon established the Environmental Protection Agency (EPA). Shortly thereafter, Congress approved two of the nation's most far-reaching federal environmental laws: the Clean Air Act of 1970 and the Federal Water Pollution Control Act of 1972, commonly known as the Clean Water Act. Both laws shifted primary responsibility for environmental protection from the states to the federal government and required federal regulators to take prompt and stringent action to curb pollution.

The surge in environmental policymaking in the early 1970s was not a response to a sudden deterioration in the condition of the nation's air and water. In fact, while some kinds of pollution were getting worse in the late 1960s, other kinds were diminishing as a result of municipal bans on garbage burning and the phasing out of coal as a heating fuel.[1] Instead, what this case reveals is the profound impact that redefining, or reframing, an issue can have on policymaking. As political scientists Frank Baumgartner and Bryan Jones observe, "[If] disadvantaged policy entrepreneurs are successful in convincing others that their view of an issue is more accurate than the views of their opponents, they may achieve rapid success in altering public policy arrangements, even if these arrangements have been in place for decades."[2]

Their observation is valid because if redefining a problem raises its salience—as manifested by widespread public activism, intense and favorable media coverage, and marked shifts in public opinion polls—politicians tend to respond. In particular, a legislator who seeks a leadership role must take positions that appeal to a national constituency and demonstrate a capacity to build winning coalitions. The president—or anyone who aspires to be president—is the one most likely to embrace issues that are widely salient, such as those that promise broad and visible public benefits. So it is not surprising that competition among presidential candidates has been the impetus behind some of the nation's most significant environmental policies. Rank-and-file legislators are also moved by highly salient issues: they jump on the bandwagon in hopes of gaining credit, or at least avoiding blame, for addressing a problem about which the public is intensely concerned.

This case also shows how a focusing event—in this instance, Earth Day—can open a policy window for a leader to promote solutions that policy entrepreneurs have linked to a newly popular framing of an issue. To be successful, policy entrepreneurs must offer solutions that appear likely to address the problem as it has been defined; their solution must be capable of garnering support from a majority legislative coalition. In trying to address air and water pollution, the proposed solutions reflected both longstanding agendas of key congressional players and the concerns of legislators (and their staff) who were newly empowered by the environmental movement. Interestingly, those solutions did not cater to the needs of the business community, mainly because it was not well-organized to lobby effectively on behalf of its interests. The result was two programs with unprecedented regulatory reach.

The implementation of an ambitious new program often encounters serious practical obstacles, however. Whereas legislators can respond to public enthusiasm about an issue, the implementing agencies must cater to "multiple principals"; that is, they must please the president and the congressional committees that oversee and fund them.[3] In addition, these agencies must grapple with the demands of organized interests: agencies depend on the cooperation of those they regulate because they have neither the resources nor the personnel to enforce every rule they issue; moreover, organized interests provide agencies with political support in Congress.[4] The process of implementing environmental legislation is particularly complicated because the agencies administering it operate in a highly fractious context in which the participants have a propensity to take their disagreements to court. As a result of all these forces, and despite provisions aimed at ensuring compliance with their lofty goals, policies that depart dramatically from the status quo rarely achieve the targets set forth in the legislation.

Furthermore, over time, such landmark statutes may become targets for reformers, as has been the case with both the Clean Air and Clean Water acts. The enactment of those laws triggered a mobilization of both business and conservative interests that espoused a cornucopian worldview; they deeply resented the nationalization of pollution control, as well as the stringent rules that accompanied that shift. Critics of both laws amplified their views in the editorial pages of *The Wall Street Journal* and in conservative magazines. Beginning in the 1990s, they extended their reach by taking advantage of new media outlets, like Fox News and conservative talk shows. In the twenty-first century they continue to voice their outrage through social media. Using both direct and low-visibility challenges, critics have sought to dismantle, weaken, or delay implementation of federal pollution-control statutes. Their efforts have been only minimally effective, however, as environmentalists have succeeded in raising the salience of these attacks and galvanizing the public to support fending them off.

BACKGROUND

Until 1970 a patchwork of local, state, and federal laws and institutions aimed to reduce pollution in order to protect public health. Beginning in the mid-1950s the federal government expanded its funding and advisory roles in pollution control, but these policy changes were incremental, and the emphasis on state-level design and enforcement persisted. Because state and local officials were deeply concerned about fostering economic development, and because environmental activists in most states had insufficient clout to challenge economic interests, this arrangement meant that few states undertook serious pollution-control programs.

Early Efforts to Address Air Pollution

The earliest concerns about air pollution in the United States arose in response to the smoke emitted by factories that accompanied industrialization. Chicago and Cincinnati enacted the nation's first clean air laws in 1881. Chicago's ordinance declared that "the emissions of dense smoke from the smokestack of any boat or locomotive or from any chimney anywhere within the city shall be . . . a public nuisance."[5] By 1912 twenty-three of twenty-eight American cities with populations greater than 200,000 had passed similar laws—although these ordinances did little to mitigate air pollution.[6] During World War II, Los Angeles initiated the nation's first modern air pollution program in response to a public outcry about the odors of a wartime industrial plant. The city also placed severe curbs on oil refineries and backyard incinerators.

Industrialization outpaced efforts to control its impacts, however. In 1948 toxic smog in Donora, Pennsylvania, killed 20 people and sickened almost 6,000, afflicting 43 percent of the city's population.[7] Similar incidents occurred in London and Los Angeles in the 1950s. These episodes attracted widespread media coverage, changed both the experts' and the public's perceptions of air pollution from a nuisance to a public health problem, and prompted the federal government to buttress state efforts with financial and research assistance. In 1955 Congress authorized the Public Health Service (PHS), a bureau within the Department of Health, Education and Welfare (HEW), to conduct air pollution research and to help states and educational institutions train personnel and carry out research and control. Upon taking office in 1961, President John F. Kennedy affirmed the importance of the federal government's role, asserting the need for an effective national program to address air pollution.

Then, in November 1962, a four-day inversion produced an air pollution episode in New York believed to have caused eighty deaths.[8] The event rekindled public interest in pollution-control legislation; in response, Congress passed the Clean Air Act of 1963. This legislation expanded HEW's authority to enforce existing state laws, encouraged the development of new state laws, and regulated interstate air pollution. Two years later, the Motor Vehicle Air Pollution Control Act required HEW to establish regulations controlling emissions from all new motor vehicles. And in 1967 Congress passed the Air Quality Act, which required the National

Air Pollution Control Administration, a small division within HEW, to designate regional air quality control areas, issue air quality criteria, and recommend pollution-control techniques. But the new law lacked deadlines and penalties; as a result, by 1970, the federal government had designated less than one-third of the metropolitan air quality regions projected in the statute, and no state had established a complete set of standards for any pollutant.[9]

Early Efforts to Address Water Pollution

The federal government became involved in controlling water pollution as early as the late nineteenth century, but—as with air pollution—legal authority belonged almost entirely to states and localities. In 1899 Congress passed the Rivers and Harbors Act, prohibiting the dumping of refuse that might impede travel in any navigable body of water. In 1912 Congress passed the Public Health Service Act, which authorized studies of waterborne diseases, sanitation, sewage, and the pollution of navigable streams and lakes. Subsequently, the 1924 Federal Oil Pollution Act prohibited ocean-going vessels from dumping oil into the sea (mainly to protect other vessels). These national laws were largely ineffectual, so by the 1940s every state had established its own agency responsible for controlling water pollution. But the powers of these agencies varied widely, and states had no recourse when upstream users polluted rivers that crossed state borders.[10]

In an effort to create a more coherent water pollution policy, Congress passed the Federal Water Pollution Control Act in 1948. This law directed the surgeon general of the PHS to develop a comprehensive program to abate and control water pollution, administer grants-in-aid for building municipal wastewater treatment plants, conduct research, and render technical assistance to states. The law also authorized the surgeon general to enforce antipollution measures in interstate waters, but only with the consent of the affected states.[11] The PHS was unable to manage the federal water pollution program to the satisfaction of either conservation groups or Congress, however, and President Harry S. Truman further hampered the law's implementation by preventing the agency from distributing loans to states and localities for wastewater treatment plants.

To redirect and strengthen HEW's efforts, Congress enacted the Federal Water Pollution Control Act of 1961, which transferred responsibility for water pollution control from the surgeon general to his or her superior, the secretary of HEW. The new law extended federal enforcement to all navigable waters, not just interstate waters, and called for an increase in appropriations for municipal wastewater treatment plants. Four years later, Congress went even further with the Water Quality Act of 1965, which officially created a separate agency, the Federal Water Pollution Control Administration, within HEW. The act gave the states until June 30, 1967, to develop individual water quality standards for drinking water, fish and wildlife, recreation, and agriculture on their interstate navigable waters. In addition, the bill established an explicit national goal: the "prevention, control, and abatement of water pollution." The following year, Sen. Edmund Muskie, D-Maine, proposed, and

Congress passed, a bill that created a $3.55 billion sewage treatment plant construction fund that would distribute money to congressional districts across the country and reflected the pork-barrel politics that dominated congressional decision making.[12] Despite this expansion in federal jurisdiction, three consecutive bureaucratic reorganizations hampered the new water pollution-control agency's ability to exercise its statutory authority, rendering its efforts more apparent than real.

THE CASE

As this history suggests, the pace of federal air and water pollution-control legislation accelerated during the 1960s, but it was the laws passed in the early 1970s that marked the most significant departure from the past. With these laws, the federal government assumed primary responsibility for ensuring that the nation's air and water were cleaned up by instituting strict new pollution-control standards and enforcing compliance by polluters. The impetus for this change was not a sudden or dramatic increase in pollution; rather, it was a redefinition of the problem sparked by widely read environmental writers and the consequent emergence of environmental protection as a popular national cause. Public concern about pollution outran the incremental responses of the 1960s, finally reaching a tipping point and culminating at the end of the decade in a massive Earth Day demonstration. That event, in turn, opened a policy window for advocates of strict pollution-control policies. Politicians, vying for a leadership role and recognizing the popularity of environmentalism, competed for voters' recognition of their environmental qualifications.

Environmentalism Becomes a Popular Cause

In 1962 Rachel Carson published *Silent Spring*, the book that many credit with lighting the fuse that ignited the modern environmental movement. On *The New York Times* bestseller list for thirty-one weeks, Carson's book sparked a firestorm of environmental activism and was soon followed by a series of antipollution tracts, including an influential book published in 1968 by biologist and environmental popularizer Paul Ehrlich titled *The Population Bomb.*

Then, in 1969, a series of highly publicized disasters hit. A Union Oil Company well blew out six miles off the coast of Santa Barbara, California, and for several weeks, oil leaked into the Pacific Ocean at the rate of 20,000 gallons a day, polluting twenty miles of beaches. Cleveland's Cuyahoga River, heavily polluted with oil and industrial chemicals, burst into flames.[13] Mercury scares frightened people away from seafood, and coastal communities closed beaches when raw sewage washed up on shore.

Calls for greater public awareness of the nation's degraded environment in response to these episodes fell on receptive ears. The population was becoming younger and better educated: between 1950 and 1974, the percentage of adults with some college education rose from 13.4 percent to 25.2 percent.[14] Demographic change was coupled with a streak of unprecedented prosperity as the nation's economy burst out of World War II. The emerging generation, finding itself in the midst of this boom, began to worry about the pollution that accompanied rapid growth and urbanization.[15] One indication of the public's growing interest in environmental issues during this time was the explosion of citations under the heading "environment" in *The New York Times* index. In 1955 the word was not even indexed; in 1965 it appeared as a heading but was followed by only two citations; by 1970, however, there were eighty-six paragraphs under the heading.[16]

Celebrating Earth Day 1970

The heightened environmental awareness of the 1960s reached its pinnacle on April 22, 1970, in the national celebration of Earth Day. The demonstration was the brainchild of Sen. Gaylord Nelson, D-Wis., who had a long-standing interest in the environment but felt that few members of Congress shared his concern. After meeting with Paul Ehrlich, Nelson conceived of an environmental teach-in to raise public awareness. He hired Denis Hayes, a twenty-five-year-old Harvard Law School student, to organize the event on a budget of $125,000.[17] Interestingly, the established preservation-oriented groups, such as the Sierra Club, the Audubon Society, and the National Wildlife Federation, played little or no role in Earth Day. In fact, as Shabecoff points out, they were surprised by and unprepared for the national surge in emotion.[18]

Despite the absence of the mainstream environmental groups, Earth Day was a resounding success—an outpouring of social activism comparable to the civil rights and Vietnam War protests. *The New York Times* proclaimed, "Millions Join Earth Day Observances Across the Nation." *Time* magazine estimated that 20 million people nationwide were involved.[19] Organizers claimed that more than 2,000 colleges, 10,000 elementary and high schools, and citizens' groups in 2,000 communities participated in the festivities.[20]

Citizens in every major city and town rallied in support of the message. For two hours New York City barred the internal combustion engine from Fifth Avenue, and thousands thronged the city's fume-free streets; in Union Square, crowds heard speeches and visited booths that distributed information on topics such as air pollution, urban planning, voluntary sterilization, conservation, and wildlife preservation. In Hoboken, New Jersey, a crowd hoisted a coffin containing the names of America's polluted rivers into the Hudson. In Birmingham, Alabama, one of the most polluted cities in the nation, the Greater Birmingham Alliance to Stop Pollution (GASP) held a "right to live" rally. Washington's chapter of GASP distributed forms that pedestrians could use to report buses emitting noxious fumes or smoke to the transit authority.

Students of all ages participated in an eclectic array of events. Fifth graders at the Charles Barrett Elementary School in Alexandria, Virginia, wrote letters to local polluters. Girls from Washington Irving High School in New York collected trash and dragged white sheets along sidewalks to show how dirty they became.

University of New Mexico students collected signatures on a plastic globe and presented it as an "enemy of the Earth" award to twenty-eight state senators accused of weakening an environmental law. At Indiana University female students tossed birth control pills at crowds to protest overpopulation. At the University of Texas in Austin, the campus newspaper came out with a make-believe April 22, 1990, headline that read, "Noxious Smog Hits Houston: 6,000 Dead."

Although it was the target of most Earth Day criticism, even the business community jumped on the Earth Day bandwagon in an effort to improve its image. Rex Chainbelt, Inc., of Milwaukee announced the creation of a new pollution-control division. Reynolds Metal Can Company sent trucks to colleges in fourteen states to pick up aluminum cans collected in "trash-ins" and paid a bounty of one cent for two cans. And Scott Paper announced plans to spend large sums on pollution abatement for its plants in Maine and Washington.

Republican and Democratic politicians alike tried to capitalize on the public fervor as well. Congress stood in recess because scores of its members were participating in Earth Day programs: Senator Muskie addressed a crowd of 25,000 in Philadelphia; Sen. Birch Bayh, D-Ind., spoke at Georgetown University; Sen. George McGovern, D-S.D., talked to students at Purdue University; and Sen. John Tower, R-Texas, addressed members of the oil industry in Houston. Most audiences greeted politicians with suspicion, however. University of Michigan students heckled former interior secretary Stewart Udall until he promised to donate his $1,000 speaker's fee to the school's environmental quality group. Protestors at a rally held by Sen. Charles Goodell, R-N.Y., distributed a leaflet calling his speech "the biggest cause of air pollution." Organizers in the Environmental Action Coalition refused to allow politicians on their platform at all to avoid giving Earth Day a political cast.

The Polls Confirm a Shift in Public Opinion

Public opinion polls confirm that Earth Day marked the emergence of environmentalism as a mass social movement in the United States. Before 1965 pollsters did not even deem pollution important enough to ask about, but by 1970 it had become a major political force. As Table 2.1 shows, over the five-year period leading up to Earth Day, the increase in public awareness of air and water pollution is striking. Survey data gathered between 1965 and 1969 reflected public recognition of pollution, but most people did not identify it as a high priority issue. Then, between the summer of 1969 and the summer of 1970, the public's concern reached a tipping point, and the issue jumped from tenth to fifth place in the Gallup polls. By 1970, the American public perceived pollution as more important than race, crime, and poverty (see Table 2.2). In December 1970, a Harris survey showed that Americans rated pollution as "the most serious problem" facing their communities. According to another Harris poll, conducted in 1971, 83 percent of Americans wanted the federal government to spend more money on air and water pollution-control programs.[21]

Writing in the spring of 1972, poll editor Hazel Erskine summed up the rapid growth in concern about the environment this way: "A miracle of public opinion has

Table 2.1 Public Opinion on Air and Water Pollution, 1965–1970

Q: Compared with other parts of the country, do you think the problem of air/water pollution in your area is very serious or somewhat serious?

Year	Sample Size	Air (%)	Water (%)
1965	2,128	28	35
1966	2,033	48	49
1967	2,000	53	52
1968	2,079	55	58
1969	NA	NA	NA
1970	2,168	69	74

Source: John C. Whitaker, *Striking a Balance: Environment and Natural Resources Policy in the Nixon-Ford Years* (Washington, D.C.: AEI, 1976), 8. Reprinted with the permission of The American Enterprise Institute.

Table 2.2 Most Important Domestic Problems, 1969 and 1971

Q: Aside from the Vietnam War and foreign affairs, what are some of the most important problems facing people here in the United States?

Problem	May 1969 Survey (%)	May 1971 Survey (%)	Significant Changes (%)
Inflation, cost of living, taxes	34	44	10
Pollution, ecology	1	25	24
Unemployment	7	24	17
Drugs, alcohol	3	23	20
Racial problems	39	22	–17
Poverty/welfare	22	20	–2
Crime, lack of law and order	15	19	4
Unrest among young people	6	12	6
Education	5	8	3
Housing	NA	6	NA

Source: John C. Whitaker, *Striking a Balance: Environment and Natural Resources Policy in the Nixon-Ford Years* (Washington, D.C.: AEI, 1976), 8. Reprinted with the permission of The American Enterprise Institute.

been the unprecedented speed and urgency with which ecological issues have burst into the American consciousness. Alarm about the environment sprang from nowhere to major proportions in a few short years."[22] According to historian Samuel Hays, this shift in public opinion was no transient phase, but it reflected a permanent evolution associated with rising standards of living and human expectations. "Environmental politics," he contends, "reflect major changes in American society and values. People want new services from government stemming from new desires associated with the advanced consumer economy that came into being after World War II."[23]

Politicians Respond

WHEN PPL CARE POLITICIANS CARE!

The emergence of broad-based public support for pollution control empowered proponents of more stringent policies, who pressed their demands on Congress and the president, citing the polls and Earth Day as evidence of the salience of environmental problems. To promote more ambitious policies, they capitalized on the competition between President Nixon and aspiring presidential candidate Muskie for control over the issue of environmental protection. The candidates, in turn, raised the stakes by ratcheting up their proposals.

Creating the Environmental Protection Agency. Reflecting their perception of the issue's low salience, neither of the major party's presidential candidates in 1968 made the environment a campaign focus. Instead, both parties concentrated on peace, prosperity, crime, and inflation. Only one of the thirty-four position papers and statements published in the compendium *Nixon Speaks Out* covers natural resources and environmental quality; in another Nixon campaign publication containing speeches, statements, issue papers, and answers to questions from the press, only 5 of 174 pages are devoted to the environment, natural resources, and energy. Nixon staff members did not recall even one question to the candidate about the environment.[24] The campaign of Democrat Hubert Humphrey was equally silent on the subject.

Yet within two years, Nixon's staff had grasped the growing salience of environmental protection and had begun staking out the president's position. In his State of the Union address in January 1970, Nixon made bold pronouncements about the need for federal intervention to protect the environment, saying,

> Restoring nature to its natural state is a cause beyond party and beyond factions. It has become a common cause of all the people of this country. It is the cause of particular concern to young Americans because they more than we will reap the grim consequences of our failure to act on the programs which are needed now if we are to prevent disaster later—clean air, clean water, open spaces. These should once again be the birthright of every American. If we act now they can.[25]

Nixon went on to assert that the nation required "comprehensive new regulation." The price of goods, he said, "should be made to include the costs of producing and disposing of them without damage to the environment."[26] On February 10, Nixon delivered a special message to Congress on environmental quality in which he outlined a thirty-seven-point program encompassing twenty-three separate pieces of legislation and fourteen administrative actions.[27]

On July 9, the president submitted to Congress an executive reorganization plan that proposed the creation of the Environmental Protection Agency and consolidated a variety of federal environmental activities within the new agency. The EPA's principal functions were to establish and enforce pollution-control standards, gather and analyze information about long-standing and newly recognized environmental problems, and recommend policy changes.[28] Ironically, the original impetus for the EPA came not from the environmental community, but from a commission appointed by President Nixon to generate ideas for streamlining the federal bureaucracy. Although the President's Advisory Council on Executive Organization, known as the Ash Council, was composed primarily of business executives, the staff included several environmental policy entrepreneurs. At first, council head Roy Ash favored vesting responsibility for both natural resources and pollution control in a single "super department," a department of natural resources. But council staff worried that such a plan would force environmentalists to compete with better organized and better financed natural resource development interests. They proposed instead an independent agency with jurisdiction over pollution control.[29] Council members also favored establishing an executive agency because creating a regulatory commission would require legislative action and would therefore subject the council's proposals to congressional politics. Furthermore, council members preferred the scientific and technical nature of executive agency decision making and were concerned that a commission would be dominated by legal and adjudicative experts.[30]

President Nixon did not accept all of the Ash Council's recommendations for the EPA, but he retained the central idea: to create an agency devoted to comprehensive environmental protection. The presidential message accompanying Reorganization Plan Number Three clearly reflected the extent to which ecological ideas about the interconnectedness of the natural environment had permeated the political debate about pollution.

The Senate was hospitable to Nixon's proposal and introduced no resolution opposing it.[31] In spite of the objections of some prominent members, the House did not pass a resolution opposing the reorganization either, so on December 2, 1970, the EPA opened its doors.

The Clean Air Act of 1970. One of the first tasks of the new agency was to implement the Clean Air Act Amendments of 1970. This was a particular challenge for the fledgling EPA because the new legislation was much more than an incremental step beyond past policy experience; in fact, it was a radical departure from the approach previously taken by the federal government. Instead of helping the states design air pollution programs, the EPA was to assume primary responsibility for setting air quality standards and for ensuring that the states enforced those standards.

Congress and the president had begun work on the 1970 Clean Air Act months before the Nixon administration established the EPA. Recognizing the rising political cachet of environmentalism and wanting to launch a preemptive strike against Senator Muskie, his likely rival for the presidency, Nixon sent air pollution legislation to Congress in February 1970. Under the bill, HEW would issue stringent motor vehicle emission standards and improve its testing procedures and regulation of fuel composition and additives. To address air pollution from stationary sources (factories and electric utilities), the bill established national air quality standards, accelerated the designation of air quality control regions, and set national emissions standards for hazardous pollutants and particular classes of new facilities.

The administration's proposal fared well in the House of Representatives, where the chamber's bipartisan consensus reflected the rank-and-file members' sensitivity to the prevailing public mood. Under the guidance of Rep. Paul Rogers, D-Fla., the Commerce Committee's Subcommittee on Public Health and Welfare marked up the bill, and the full committee reported out a somewhat stronger version than the original. On June 10, the full House passed the bill 374–1.

The administration bill received a cooler reception in the Senate, where Nixon's presumed presidential rival, Senator Muskie, was the undisputed champion of the environmental cause.[32] On March 4, shortly after the president submitted his bill to the House, Muskie introduced an alternative, the National Air Quality Standards Act of 1970. His objective at the time was to prod agencies to strengthen their implementation of the 1967 act, rather than to initiate a radically different policy. Muskie had spent his Senate career characterizing pollution control as a state responsibility and the domain of experts; as he understood it, the problem lay not in the design of the program but in its implementation.[33] Over the summer, however, Muskie changed his tune. He asked the Public Works Committee's Subcommittee on Air and Water Pollution to draft a new set of amendments containing stringent new provisions including national, rather than regional, standards for major pollutants.

Muskie's change of heart was a clear attempt to reestablish his dominance in the environmental area. Despite his considerable record, not only Nixon but also some prominent environmental advocates had challenged the senator's commitment to environmental protection. A highly critical report by a study group under the direction of Ralph Nader, released in May 1970, characterized Muskie as a weak and ineffectual sponsor of clean air legislation. The report, titled *Vanishing Air*, assailed Muskie as

> the chief architect of the disastrous Air Quality Act of 1967. That fact alone would warrant his being stripped of his title as "Mr. Pollution Control." But the Senator's passivity since 1967 in the face of an ever worsening air pollution crisis compounds his earlier failure. . . . Muskie awakened from his dormancy on the issue of air pollution the day after President Nixon's State of the Union message. . . . In other words, the air pollution issue became vital again when it appeared that the President might steal the Senator's thunder on a good political issue.[34]

Media publicity of the Nader report's charges put Muskie on the defensive, and the Senate's environmental leader felt compelled to "do something extraordinary in order to recapture his [pollution-control] leadership."[35]

In the end, Muskie's subcommittee drafted an air pollution bill more stringent than either the president's original proposal or the House of Representatives' slightly stronger version. It called for nationally uniform air quality standards that ignored economic cost and technological feasibility considerations and were based solely on health and welfare criteria; it required traffic-control plans to eliminate automobile use in parts of some major cities; and it mandated a 90 percent reduction in automotive emissions of carbon monoxide, hydrocarbons, and nitrogen oxides by 1975. In a clear manifestation of the burgeoning popularity of environmental protection, senators got on the bandwagon and endorsed this version of the clean air bill unanimously (73–0) on September 21, 1970.[36]

Because of substantial differences in critical sections of the bill, the House–Senate conference that ensued was protracted, involving at least eight long sessions over a three-month period. The Senate's eight conferees held an advantage over the five from the House because Muskie's prolonged attention to pollution issues had attracted several qualified and committed staffers who had amassed expertise. As a consequence, the final conference report more closely resembled the Senate version of the bill than the House version.

On December 18, both chambers debated and passed the conference report by voice vote, and on December 31, President Nixon signed the Clean Air Act of 1970 into law. Its centerpiece was the requirement that the EPA set both primary and secondary national ambient air quality standards.[37] The states were to submit state implementation plans (SIPs) outlining a strategy for meeting primary standards by 1975 and secondary standards "within a reasonable time." If the EPA determined a SIP to be inadequate, it had to promulgate a plan of its own. The act also targeted some polluters directly: it required automobile producers to reduce the emissions of new cars by 90 percent by 1975, and it required the EPA to set performance standards for all major categories of new stationary sources.

The Clean Water Act of 1972. President Nixon made not only air pollution but also water pollution legislation a pillar of his February 10, 1970, special message to Congress. When Congress failed to address water pollution in the subsequent legislative session, the president moved administratively, using the permit authority granted by the Refuse Act of 1899 to control industrial pollution of waterways. By executive order, Nixon directed the EPA to require industries to disclose the amount and kinds of effluents they were generating before they could obtain a permit to discharge them into navigable waters.[38] When a polluter failed to apply for a permit or violated existing clean water regulations, the EPA referred an enforcement action to the Justice Department.

Neither the permit process nor the enforcement strategy was particularly effective at ameliorating water pollution, however. Although the president endorsed the permit program, Congress was not pleased at being circumvented; state agencies were angry that federal rules superseded their own regulations; and many industries were furious at the sudden demands for discharge information.[39] Compliance was limited: on July 1, 1971, when the first 50,000 applications from water-polluting industries were due, only 30,000 had arrived, and many of them contained incomplete or inaccurate information. The enforcement process, which relied heavily on the overburdened federal court system, was slow and cumbersome.[40] Then, in December 1971, a district court in Ohio dealt the permit program its final blow: it held that the EPA had to draft an environmental impact statement for each permit issued to comply with the recently passed National Environmental Policy Act.[41]

While the EPA muddled through with its interim program, Congress began to debate the future of water pollution policy in earnest. In February 1971, President Nixon endorsed a proposal to strengthen a bill he had submitted to Congress the previous year. The new bill increased the administration's request for annual municipal wastewater treatment financing from $1 billion to $2 billion for three years and established mandatory toxic discharge standards. In addition, it requested authority for legal actions by private citizens to enforce water-quality standards. Refusing to be upstaged by the president, Muskie again seized the opportunity to offer even more stringent legislation. The Senate began hearings in February, and eight months later, Muskie's Public Works Committee reported out the Federal Water Pollution Control Act Amendments. According to Milazzo, the legislation that emerged reflected not just presidential politics or pressure from the public, but also displayed the input of "unlikely environmentalists," including the proponents of economic development, men who designed ballistic missiles, an agency that built dams (the Army Corps of Engineers), and professional ecologists. "In the course of pursuing their own agendas within well-established organizational channels," he says, "these . . . actors . . . took an active interest in water pollution and proceeded to shape how policymakers devised solutions to the problem."[42]

Much to the Nixon administration's dismay, the price tag for the Senate bill was $18 billion, three times the cost of Nixon's proposal. Moreover, the administration found unrealistic the overarching objectives of the Senate bill: that "wherever attainable, an interim goal of water quality which provides for the protection and propagation of fish, shellfish, and wildlife and provides for recreation in and on the water should be achieved by 1981" and that "the discharge of all pollutants into navigable waters would be eliminated by 1985." Finally, the administration considered the Senate bill inequitable, claiming that it imposed a disproportionate burden on industry by singling out those that could not discharge into municipal waste treatment facilities. Despite the president's reservations, on November 2, 1971, the Senate passed Muskie's bill by a vote of 86–0.

Having failed to shift the Senate, the administration focused on the House deliberations, with some qualified success: the House reported out a bill similar to the one proposed by the White House. In contrast to the Senate version, the House bill retained the primacy of the states in administering the water pollution-control program. After meeting forty times between May and September 1972, the House–Senate conferees overcame their differences and produced a bill satisfactory to both chambers. In another extraordinary display of consensus, the Senate passed the conference bill by 74–0, and the House approved it by 366–11.

The compromise was too stringent for the administration, however. It retained both the fishable, swimmable, and zero-discharge goals and the financing provisions that were so objectionable to the president. Furthermore, the bill's timetables and total disregard for economic costs offended the president. So, in a tactical maneuver, Nixon vetoed the Clean Water Act on October 17, the day that Congress was scheduled to adjourn for the year. To Nixon's chagrin, Congress responded with unusual alacrity: less than two hours after the president delivered his veto message, the Senate voted to override the veto by 52–12.[43] The next afternoon, the House followed suit by a vote of 247–23, and the Clean Water Act became law.

The New Environmental Regulations. The Clean Air and Clean Water acts reflected the prevailing definition of pollution, in which industrial polluters (not consumers) were the villains, and citizens (and only secondarily the environment) were the unwitting victims. They also reflected the public's skepticism of corporations' willingness and government bureaucrats' ability to address pollution. Concerns about "regulatory capture," whereby agencies become subservient to the industries they are supposed to monitor, had preoccupied academics for years, but in 1969, political scientist Theodore Lowi popularized the concept in his book *The End of Liberalism*. Lowi criticized Congress for granting agencies broad discretion in order to avoid making hard political trade-offs. He argued that agencies, operating out of the public eye, strike bargains with the interest groups most affected by their policies, rather than implementing policies in ways that serve a broader national interest. Led by Nader, reformers disseminated the concept of regulatory capture. Two reports issued by Nader's Center for the Study of Responsive Law, *Vanishing Air* in 1970 and *Water Wasteland* in 1971, attributed the failures of earlier air and water pollution-control laws to agency capture. More important, they linked that diagnosis to Nader's preferred solution—strict, action-forcing statutes—reasoning that unambiguous laws would limit bureaucrats' ability to pander to interest groups.

Members of Congress got the message: in addition to transferring standard-setting authority from the states to the federal government, the Clean Air and Clean Water acts employed novel regulatory mechanisms—such as strict deadlines, clear goals, and uniform standards—that both minimized the EPA's discretion and restricted polluters' flexibility. For example, the Clean Air Act gave the EPA thirty days to establish health- and welfare-based ambient air quality standards. The states then had nine months to submit their SIPs to the EPA, which had to approve or disapprove them within four months of receipt. The agency was to ensure the achievement of national air quality standards no later than 1977. Similarly, the Clean Water Act specified six deadlines: by 1973, the EPA was supposed to issue effluent guidelines for major industrial categories; within a year, it was to grant permits to all sources of water pollution; by 1977, every source was supposed to have installed the "best practicable" water pollution-control technology; by 1981, the major waterways in the nation were to be suitable for swimming and fishing; by 1983, polluting sources were to install the "best available" technology; and by 1985, all discharges into the nation's waterways were to be eliminated.

Congress also sought to demonstrate its commitment to preventing regulatory capture by incorporating public participation into agency decision making and thereby breaking up regulated interests' monopoly. For example, both the Clean Air and Clean Water acts required the EPA to solicit public opinion during the process of writing regulations. In addition, both laws encouraged public participation by explicitly granting citizens the right to bring a civil suit in federal court against any violator or "against the administrator [of the EPA] where there is alleged a failure of the administrator to perform any act or duty under [the Clean Air Act] which is not discretionary."[44]

Finally, the Clean Air and Clean Water acts of the early 1970s reflected impatience with market forces and a desire to spur the development of new pollution-control technology as well as to encourage businesses to devise innovative new production processes. Both laws included provisions that fostered technology in three ways: by prompting the development of new technology, by encouraging the adoption of available but not-yet-used technology, and by forcing diffusion of currently used technology within an industry. The motor vehicle provisions of the Clean Air Act, for example, forced the development of the catalytic converter. When Congress was debating the 90 percent reduction in tailpipe emissions, the automobile manufacturers strenuously objected that they did not have the technology to meet those standards, but Muskie responded with a flourish that this level of reduction was necessary to protect human health, so companies would have to devise a solution.[45] (As it turned out, carmakers were able to meet the standards relatively easily.) The Clean Water Act, on the other hand, pushed polluters to adopt technology that was already available but not widely used by its initial deadline. In the second phase, however, the act required businesses to meet standards achievable with the best technology available, even if it was not in use at the time.

Implementation: Idealism Tempered

The Clean Air and Clean Water acts were sufficiently grandiose that they would have presented a challenge to any agency, but they were particularly onerous for a brand new one that drew staff from all over the federal government. Not surprisingly, because of the short time allowed for implementing these laws, combined with the haste in which the agency was designed, the EPA did not attain the ideal of interrelatedness outlined by President Nixon; instead, different offices continued to manage pollution in different media. Nor did the EPA fulfill the mandates of the Clean Air and Clean Water acts to virtually eliminate pollution in the nation's air and waterways. Although born in a period of great idealism and bequeathed a clear mission to protect the environment, the EPA had to survive in the highly circumscribed world of practical politics. It had to establish relationships with and reconcile the demands of the president and Congress, and it had to navigate a course in a sea of competing interests, recalcitrant state and local officials, a skeptical media, and an expectant public. In all of these endeavors, the EPA was vulnerable to lawsuits because the statutes compelled it to act quickly and decisively, despite a dearth of scientific and technical information on which to base its decisions and, more important, with which to justify them.

Setting a Course. From its inception, the new EPA was an organizational nightmare, as it comprised

> an uneasy amalgam of staff and programs previously located in 15 separate federal agencies. EPA had a total budget of $1.4 billion. Its 5,743 employees worked in 157 places, ranging geographically from a floating barge off the Florida coast to a water quality laboratory in Alaska. In Washington, D.C., alone there were 2,000 employees scattered across the city in 12 separate office buildings.[46]

The first EPA administrator, William Ruckelshaus, was a thirty-eight-year-old lawyer and former assistant state attorney assigned to the Indiana State Health Department. He was confronted with the awesome tasks of coordinating the disparate offices of the new agency (it lacked a headquarters until 1973), establishing a set of coherent priorities, and carrying out the statutory mission of regulating polluters. From the outset, Ruckelshaus balanced his own approach against the conflicting preferences of the White House and Congress.

Dealing with the White House posed a considerable challenge. Although President Nixon created the EPA and introduced pollution-control legislation, he did so more out of political opportunism than genuine environmental concern.[47] He regarded environmentalism as a fad, but one that promised political rewards. As political writer Mary Graham explains, "Elected with only 43 percent of the popular vote in 1968, Nixon needed to take bold steps to expand his ideological base in order to be reelected in 1972."[48] In truth, Nixon was hostile toward the federal bureaucracy and, as biographer Stephen Ambrose notes, wanted "credit for boldness and innovation without the costs."[49] Nixon instructed White House staff to scrutinize the EPA's activity and block its rulemaking; he also introduced legislation to curtail its authority. Most notably, he established a "quality of life" review under the Office of Management and Budget (OMB) to assess the legal, economic, and budgetary implications of EPA regulations—a mechanism that by 1972 "had become an administration device for obstructing stringent regulations, as the environmental groups had originally feared."[50]

Congress, on the other hand, was a mixed bag of backers and critics. Several members of Congress exhibited a genuine zeal for environmental protection.

Members of the House and Senate subcommittees with jurisdiction over pollution control encouraged Ruckelshaus to enforce the law vigorously. Muskie, in particular, was dogged in his efforts to train national attention on pollution control and thereby hold the EPA's feet to the fire. His subcommittee convened frequent hearings that required Ruckelshaus to explain delays in setting standards. But other members on related committees were more conservative; for example, Rep. Jamie Whitten, D-Miss., chair of the House Appropriations Subcommittee on Agriculture, Environment, and Consumer Protection, controlled the agency's purse strings and was a vocal opponent of strong environmental regulations.[51]

Squeezed between supporters and detractors in Congress and the White House, Ruckelshaus tried to build an independent constituency that would support the fledgling EPA. To establish credibility as an environmentalist and earn public trust, he initiated a series of lawsuits against known municipal and industrial violators of water pollution-control laws. To reinforce his efforts, he promoted the agency in the media, giving frequent press conferences, appearing on talk shows, and making speeches before trade and business associations.

Ruckelshaus had to do more than file lawsuits and woo the media, however; he had to promulgate a series of regulations to meet statutory deadlines, notwithstanding the paucity of scientific and engineering information. Compounding the technical obstacles, the targeted industries resisted agency rulemaking. Although it had been ambushed by the regulatory onslaught of the late 1960s and early 1970s, business quickly adapted to the new political order. Corporations began to emphasize government relations as a fundamental part of their missions: between 1968 and 1978, the number of corporations with public affairs offices in Washington rose from 100 to more than 500.[52] In short, having lost the first round, polluters sought to recapture their dominance over environmentalists at the implementation stage, and with its almost bottomless resources, industry was able to challenge regulations administratively and in the courts.[53]

Implementing the Clean Air Act. Thanks to both their increased political involvement and a shift in public attention, the industries especially hard hit by regulation—automobile, steel, nonferrous smelting, and electric power—all succeeded in winning delays from the EPA. The automobile manufacturers were among those the Clean Air Act singled out most directly. Before the passage of the 1966 National Traffic and Motor Vehicle Safety Act, the automobile was completely unregulated by the federal government. Yet only four years later, the Clean Air Act required carmakers to cut emissions of carbon monoxide, nitrogen oxides, and hydrocarbons by 90 percent within five years. Producers immediately applied for a one-year extension of the deadline, contending that the technology to achieve the standards was not yet available. Ruckelshaus denied their petition on the grounds that the industry had not made "good faith efforts" to achieve the standards. The manufacturers then took their case to the U.S. Court of Appeals for the District of Columbia Circuit, which overturned Ruckelshaus's decision, saying that the agency needed to give economic factors greater weight. Later that year, Ruckelshaus relented and granted a one-year extension.

The power companies, carmakers, and coal and oil producers saw the 1973–1974 energy crisis as opening a policy window to weaken the requirements of the Clean Air Act. Threatening widespread economic dislocation, these energy-related industries pressured Congress and the president into passing the Energy Supply and Environmental Coordination Act of 1974. The act included another one-year extension for hydrocarbon and carbon monoxide emissions from tailpipes and a two-year extension for nitrogen oxide emissions. When a controversy arose over the health effects of emissions of sulfuric acid from catalytic converters, Russell Train, who succeeded Ruckelshaus as EPA administrator in September 1973, granted the carmakers a third extension.[54]

The delays in achieving automotive emission standards left the EPA in an awkward position, however: because of the extensions, states could not rely on cleaner cars to mitigate their pollution problems and so had to reduce dramatically the *use* of automobiles, a politically unappealing prospect. Acknowledging the enormity of their task, Ruckelshaus granted seventeen of the most urbanized states a two-year extension on the transportation control portion of their implementation plans, giving them until 1977 to achieve air quality standards.[55] Although most state officials were pleased, disgruntled environmentalists in California filed suit in federal court to force the EPA to promulgate a transportation control plan (TCP) for Los Angeles. The plaintiffs charged that the Clean Air Act compelled the EPA to draft a plan for any state whose own plan the agency disapproved, not to grant extensions. The court agreed and ordered the agency to prepare a TCP for Los Angeles by January 15, 1973.

The pollution problem in the Los Angeles basin was so severe that, to bring the region into compliance with air quality standards, the EPA had to write a TCP that included gas rationing and mandatory installation of emissions control devices on all cars. Needless to say, such measures were unpopular. Public officials who were supposed to enforce the plan ridiculed it: Mayor Sam Yorty called it "asinine," "silly," and "impossible."[56] State and local officials clearly believed that their constituents supported clean air in the abstract but would not give up their cars to get it.

Contributing to the agency's credibility woes, just two weeks after Ruckelshaus announced the Los Angeles TCP, a federal court found in favor of the Natural Resources Defense Council in its suit to overturn the two-year extensions for states' compliance with the air quality standards. To Ruckelshaus's chagrin, the court ordered him to rescind all seventeen extensions. The states again were faced with a 1975 compliance deadline to be achieved without the benefit of cleaner cars.

As a result, in late 1973 the EPA found itself forced to produce a spate of TCPs for states whose own TCPs the agency had rejected. State officials immediately challenged the plans in court, and in some cases judges were sympathetic, finding that the EPA plans lacked sufficient technical support. But many of the plans went unchallenged, and by spring 1974 the EPA was in another quandary: it had promulgated numerous TCPs the previous year, but the states were not implementing them. Although EPA lawyers believed they had the legal authority to require out-of-compliance areas to institute transportation controls, it was not clear how they would actually force recalcitrant states to do so, and the agency lacked the administrative apparatus to impose the control strategies itself. EPA officials decided to try enforcing a test case in Boston, a logical choice since it already had an extensive mass transit system.

The backlash in Massachusetts was severe, in part because the Boston plan was haphazard and incoherent—a reflection of the agency's lack of information. For example, one regulation required all companies with fifty or more employees to reduce their available parking spaces by 25 percent. The EPA planned to send enforcement orders to 1,500 employers but discovered that only 300 of those on the list actually fit the category, and many of those turned out to be exempt (hospitals, for example). In the end, only seven or eight of the twenty-five eligible employers responded to the EPA's request to cut parking spaces. As time went on, even northeast regional EPA officials became annoyed with the arbitrary assumptions and technical errors embedded in the Boston TCP. For example, EPA analysts had based the carbon monoxide reduction strategy for the entire city on an unusually high reading from an extremely congested intersection, and they based their ozone calculations on a solitary reading from a monitor that had probably malfunctioned.[57]

The City of Boston took the plan to court, and the judge remanded the plan to the agency for better technical justification. Eventually, a chastened EPA rescinded the Boston TCP altogether and issued a replacement that dropped all mandatory traffic and parking restrictions and relied instead on stationary source controls and voluntary vehicle cutbacks. The EPA went on to abandon its attempts to force major cities to restructure their transportation systems, which in turn meant that many remained out of compliance with air quality standards. By 1975, the statutory deadline, not one state implementation plan had received final approval from the EPA.

Implementing the Clean Water Act. Like the Clean Air Act, the Clean Water Act required the EPA to take on powerful industries armed with only scant technical and scientific information. The law's cornerstone, the National Pollutant Discharge Elimination System (NPDES), prohibited the dumping of any wastes or effluents by any industry or government entity without a permit. To implement this provision, the agency had to undertake a massive data collection task: it needed information about the discharges, manufacturing processes, and pollution-control options of 20,000 different industrial polluters operating under different circumstances in a variety of locations.[58] To simplify its task, the EPA divided companies into 30 categories and 250 subcategories on the basis of product, age, size, and manufacturing process. The water program office then created the Effluent Guidelines Division to set industry-by-industry effluent guidelines based on the "best practicable technology" (BPT). The division collected and tabulated information on companies around the country. But it found sufficient variation to make generalizations about a single best technology highly uncertain. While the EPA wrestled with this problem, the Natural Resources Defense Council sued the agency for delay. The court, finding in favor of the plaintiffs, forced the EPA to release guidelines for more than 30 industry categories and 100 subcategories.

Although the permits granted to individual companies were supposed to be based on the BPT guidelines, as a result of delays in issuing those guidelines, the agency dispensed permits to almost all of the "major" polluters before the guidelines had even appeared.[59] Industry seized on this discrepancy to contest the permits in the agency's adjudicatory proceedings. In addition, major companies brought more than 150 lawsuits to challenge the guidelines themselves: the very day the EPA issued guidelines for the chemical industry, DuPont hired a prestigious law firm to sue the agency.[60] Ultimately, the EPA was forced to adopt a more pragmatic and conciliatory relationship with out-of-compliance companies. In response, disappointed environmental groups began to file suits against polluters themselves.

The 1977 Clean Air and Water Act Amendments: Relaxing the Law

With the public's attention elsewhere, in 1977 Congress relaxed the stringent provisions of both laws. The 1977 Clean Air Act Amendments postponed air quality goals until 1982; in areas heavily affected by car emissions, such as California, the act gave the states until 1987 to achieve air quality goals. The amendments also extended the deadline for the 90 percent reduction in automobile emissions—originally set for 1975 and subsequently postponed until 1978—to 1980 for hydrocarbons and 1981 for carbon monoxide. Congress granted the EPA administrator discretionary authority to delay the achievement of auto pollution reduction objectives for carbon monoxide and nitrogen oxides for up to two additional years if the required technology appeared unavailable. In addition, the amendments required that the EPA take into account competing priorities: it had to grant variances for technological innovation and file economic impact and employment impact statements with all new regulations it issued.[61] Moreover, the amendments gave the governor of any state the right to suspend transportation control measures that required gas rationing, reductions in on-street parking, or bridge tolls.[62]

The 1977 Clean Water Act Amendments extended a host of deadlines as well. The amendments gave industries that acted in "good faith" but did not meet the 1977 BPT deadlines until April 1, 1979, instead of July 1, 1977, to meet the standard. In addition, they postponed and modified the best available technology (BAT) requirement that industry was supposed to achieve by 1983. They retained the strict standard for toxic pollutants but modified it for conventional pollutants.[63] This change gave the EPA the flexibility to set standards less stringent than BAT when it determined that the costs of employing BAT exceeded the benefits. Finally, although the amendments retained the objective of zero discharge into navigable waters by 1985, changes in the law eviscerated that goal; the extension of the BPT target and the modification of the BAT target eliminated the connection between zero discharge and a specific abatement program.[64]

Despite these rollbacks, the EPA continued to have formidable regulatory powers. In January 1978, shortly after Congress passed the amendments, President Jimmy Carter submitted his 1979 budget. Although he called for an overall spending increase of less than 1 percent over 1978, he requested an increase of $668 million for EPA programs.[65] That allocation reflected an important shift that had taken place at the EPA: in the months prior to the budget announcement, the agency had made a concerted effort to recast its image from that of protector of flora and fauna to guardian of the public's health. The move was partly to deflect a threatened merger of the EPA with other natural resource agencies, but it also reflected shrewd recognition of congressional support for programs aimed at fighting cancer.[66] The agency's public-relations campaign worked, and by the end of the 1970s, the EPA had become the largest federal regulatory bureaucracy, with more than 13,000 employees and an annual budget of $7 billion.[67]

More Significant Challenges to the EPA and the Clean Air and Clean Water Acts

Although the EPA positioned itself well during the 1970s, in the 1980s and 2016, it encountered more severe challenges: the administrations of Republican presidents Ronald Reagan and Donald J. Trump. President Reagan ran on a platform antagonistic to environmentalists, environmental regulation, and government in general. Upon taking office, he set out to institutionalize his antiregulatory philosophy in the EPA by appointing as EPA administrator Anne Gorsuch, an avowed critic of environmental regulation. Gorsuch proceeded to bring enforcement of the Superfund Act (see "Love Canal," Chapter 3) to a halt. She also reorganized the agency and cut both its budget and staff severely, with the result that "[t]he atmosphere of frenetic activity and organizational ambition that . . . characterized the EPA during the [preceding] years simply dissipated."[68] Her activities eventually provoked a congressional inquiry, and in 1983 she and twenty other appointees resigned in hopes of sparing the president further embarrassment. Her successor, William Ruckelshaus, had more integrity, but the damage to the EPA's credibility was lasting. Moreover, as the 1980s wore on the environment became an increasingly partisan issue, with conservative Republicans taking aim at the nation's environmental statutes and environmentalists and their congressional allies increasingly on the defensive. Subsequent attacks on the EPA's programs and budgets by conservative members of Congress and Republican presidents through the 1990s and 2000s further eroded the agency's ability to implement and enforce the law.

For the most part, the Clean Air Act survived repeated efforts to sabotage it, thanks primarily to strong support from the courts. In 1990, under President George H. W. Bush, Congress approved the last major set of amendments to the Clean Air Act. The new law was filled with additional requirements and deadlines; it also addressed the issue of acid rain, a problem on which the Reagan administration and a divided Congress had delayed action for a decade (see Chapter 5). Subsequent efforts to challenge the law consisted primarily of resistance to the issuance of updated air quality standards. For example, when the Clinton administration proposed more restrictive standards for ground-level ozone and small particulates in 1996, industry groups and their conservative allies launched a full-scale (but ultimately unsuccessful) effort to prevent the new smog and soot standards from taking effect: in 2001, the Supreme Court unanimously rejected the plaintiffs' argument that the EPA should take costs into account when setting air quality standards.

In another battle, the George W. Bush administration sought to weaken New Source Review requirements, which require stationary sources to install state-of-the-art pollution-control equipment when they make substantial renovations to their operations. Again, however, the courts rebuffed the administration's efforts, with the Supreme Court dealing the final blow in 2007, with its decision in *Environmental Defense et al. v. Duke Energy Corp.* In 2018, the Donald J. Trump administration announced it would withdraw a twenty-year-old EPA standard—"once-in-always in"—from the Clean Air Act, simply meaning that major sources (e.g., anything that emits more than ten tons or more per year) of hazardous air pollutants could be reclassified as area sources. The concern is that this would allow more pollutants into the air created by coal-burning smokestacks.[69] The final decision is yet to be decided and awaiting public comment in the *Federal Register*.

The Clean Water Act also held up over time, despite numerous challenges by homebuilders and property rights activists who have taken particular aim at the wetlands permit program established by the EPA and the Army Corps of Engineers under Section 404 of the act. In the mid-1990s, for example, after Republicans gained control of Congress, House Republicans lost no time in trying to revise the Clean Water Act to drastically reduce protection for the nation's wetlands. In spring 1995, Bud Shuster, R-Penn., chair of the House Transportation and Infrastructure Committee, introduced a set of radical revisions to the act, including provisions to restrict federal wetlands protection and compensate landowners whose property values declined more than 20 percent as a result of federal regulations. The proposal infuriated both scientists and environmentalists, who complained that regulated industries had helped draft the legislation and that its standards were inconsistent with the scientific understanding of wetland function. When Maryland Republican Wayne Gilchrest argued that wetlands deserved special protection, Rep. Jimmy Hayes, D-La., responded that the property rights of individuals were more important than ecologically worthless wetlands.[70] In the end, although the House passed Shuster's bill by a vote of 240–185, the Senate refused to adopt a similar measure. Property rights activists fared better in the courts, however.

Two major decisions—*Solid Waste Agency of Northern Cook County (SWANCC) v. U.S. Army Corps of Engineers* in 2001 and *Rapanos v. United States* in 2006—limited the extent to which the Clean Water Act could be used to protect isolated wetlands across the United States. In response to the confusion created by *SWANCC* and *Rapanos*, in March 2014 the EPA and the Army Corps of Engineers jointly proposed a rule to clarify the definition of the "waters of the United States." The proposed rule, which replaced guidance issued in 2003 and 2008 after the court rulings, clarified that under the Clean Water Act, wetlands with any significant connection to downstream water quality should be protected. According to the proposal, 17 percent of isolated wetlands would automatically receive protection under the Clean Water Act, while the remainder would be subject to case-by-case evaluation.[71] This rule was finalized in February 2018; however, President Trump issued an Executive Order for the EPA administrator Scott Pruitt to review and possibly rescind efforts.[72,73]

OUTCOMES

As a result of industry resistance, increasingly strident attacks by conservative Republicans, and the sheer magnitude of the tasks it has been asked to undertake, the EPA's accomplishments have been neither as dramatic nor as far-reaching as the original air and water pollution statutes demanded. Moreover, a chorus of critics contends that what cleanup has been accomplished has cost far more than necessary because regulations were poorly designed and haphazardly implemented. Nevertheless, the nation has made enormous progress in cleaning up air pollution and has made some gains in addressing water pollution as well.

The EPA reports substantial reductions in air pollution for the six major "criteria" pollutants since the mid-1980s; even as the economy has grown, energy consumption has risen, and vehicle miles traveled have increased. Between 1980 and 2016, concentrations of nitrogen dioxide (NO_2), measured annually, declined 62 percent; sulfur dioxide (SO_2) concentrations decreased 82 percent, carbon monoxide (CO) levels fell 85 percent, and airborne lead concentrations dropped 99 percent. These achievements notwithstanding, in 2016, about 123 million people lived in counties where monitored air was unhealthy at times because of high levels of one or more of the six criteria pollutants.[74]

The nation has also made gains in combating water pollution, although it has not come close to realizing the lofty objectives of the 1972 Clean Water Act. It is difficult to assess overall progress in ameliorating water pollution because several different entities collect data on water quality and each uses a different monitoring design, indicator set, and methods. As a result, the EPA cannot combine their information to answer questions about the quality of the nation's waterways or track changes over time. To address this deficiency, the EPA and its partners implemented a series of aquatic resource surveys that are repeated every five years.[75] A 2012 survey of the nation's lakes, ponds, and reservoirs, the National Lakes Assessment, found that 33 percent of our lakes are in good biological condition. However, 31 percent are in a most disturbed condition.[76] According to the EPA's most recent National Rivers and Streams Assessment, conducted between 2008 and 2009, 55 percent of the nation's river and stream miles do not support healthy populations of aquatic life, with phosphorus and nitrogen pollution and poor habitats the most widespread problems.[77]

Even more important from the perspective of many critics is that nonpoint-source water pollution presents a significant and growing problem that is only beginning to be addressed seriously under the Clean Water Act.[78] Nonpoint sources include farmlands, city storm sewers, construction sites, mines, and heavily logged forests. Runoff from these sources contains silt, pathogens, toxic chemicals, and excess nutrients that can suffocate fish and contaminate groundwater. The EPA's national water quality inventories show that five of the top six water-quality-related sources of river and stream impairment in the United States are nonpoint sources.[79] The act also fails to deal with groundwater, which supplies the drinking water for 34 of the nation's 100 largest cities.[80] Loss and degradation of wetlands contribute to water quality problems as well.[81] That said, the law has resulted in enormous investments in sewage treatment, and as a consequence, many of the most seriously polluted water bodies have been substantially cleaned up.

CONCLUSIONS

As this case makes clear, public attentiveness, especially when coupled with highly visible demonstrations of concern, can produce dramatic changes in politics and policy. Front-page coverage of Earth Day demonstrations in 1970 both enhanced public awareness of and concern about environmental problems and convinced elected officials that environmental issues were highly salient. In response, aspiring leaders competed with one another to gain credit for addressing air and water pollution. Legislators' near-unanimous support for the Clean Air and Clean Water acts suggests that rank-and-file members of Congress also sought recognition for solving the pollution problem or, at a minimum, got on the bandwagon to avoid blame for obstructing such solutions.

The Clean Air and Clean Water acts enacted in the early 1970s departed dramatically from the status quo in both form and stringency. According to the approach adopted in these laws, which has become known derisively as command-and-control but might more neutrally be called prescriptive, uniform emissions standards are imposed on polluters. This approach reflected the framing of the pollution issue: industry had caused the problem, and neither industry nor government bureaucrats could be trusted to address it unless tightly constrained by specific standards and deadlines. The Clean Air and Clean Water acts' ambitious goals reflected the initial urgency of public concern and the immediacy of the legislative response. But the inchoate EPA was destined to fail when it tried to implement the laws as written. The agency encountered hostility from President Nixon, who wanted to weaken implementation of the acts, as well as from its overseers in Congress, who berated it for failing to move more quickly.

Equally challenging was the need to placate interest groups on both sides of the issue. Citizen suit provisions designed to enhance public involvement in the regulatory process resulted in a host of lawsuits by environmentalists trying to expedite the standard-setting process. At the same time, newly mobilized business interests backed by conservative groups used administrative hearings and lawsuits to obstruct implementation of the new laws. Caught in the middle, the EPA tried to enhance its public image—first by cracking down on individual polluters and later by emphasizing the public health aspect of its mission. The agency hoped that by steering a middle course, it could maintain its credibility, as well as its political support. On the one hand, the backlash was effective: in the late 1970s, Congress substantially weakened the requirements of the Clean Air and Clean Water acts. On the other hand, both laws subsequently survived multiple serious challenges, and both they and the EPA continue to enjoy broad public support.

QUESTIONS TO CONSIDER

- Critics charge that the Clean Air and Clean Water acts are classic examples of symbolic politics, in which politicians set goals that are clearly unattainable in order to placate the public. What do you think are the costs and benefits of adopting ambitious and arguably unrealistic legislative goals?
- In retrospect, what are the strengths and weaknesses of the particular approach to pollution adopted in the original Clean Air and Clean Water acts?
- How do you think the creation of the EPA and passage of the Clean Air and Clean Water acts in the early 1970s have affected the environment and our approach to environmental protection in the long run?

NOTES

1. Mary Graham, *The Morning After Earth Day: Practical Environmental Politics* (Washington, D.C.: Brookings Institution, 1999).
2. Frank R. Baumgartner and Bryan D. Jones, *Agendas and Instability in American Politics* (Chicago: University of Chicago Press, 1993), 4.
3. Herbert Kaufman, *The Administrative Behavior of Federal Bureau Chiefs* (Washington, D.C.: Brookings Institution, 1981); Kenneth J. Meier, *Politics and the Bureaucracy* (North Scituate, Mass.: Duxbury Press, 1979).
4. Francis E. Rourke, *Bureaucracy, Politics, and Public Policy*, 3d ed. (Boston: Little, Brown, 1976).
5. Clarence J. Davies III, *The Politics of Pollution* (New York: Pegasus, 1970).
6. Council on Environmental Quality, *Environmental Quality: The First Annual Report of the Council on Environmental Quality* (Washington, D.C.: U.S. Government Printing Office, 1970).
7. John F. Wall and Leonard B. Dworsky, *Problems of Executive Reorganization: The Federal Environmental Protection Agency* (Ithaca, N.Y.: Cornell University Water Resources and Marine Sciences Center, 1971).
8. An inversion is an atmospheric condition in which the air temperature rises with increasing altitude, holding surface air down and preventing the dispersion of pollutants.

9. Gary Bryner, *Blue Skies, Green Politics: The Clean Air Act of 1990* (Washington, D.C.: CQ Press, 1993).

10. Davies, *The Politics of Pollution*.

11. Wall and Dworsky, *Problems of Executive Reorganization*.

12. Davies, *The Politics of Pollution*.

13. This was not the first such fire on the Cuyahoga. In fact, throughout the nineteenth and early twentieth century river fires were common in U.S. cities. Unlike its predecessors, however, this fire garnered national media attention—a reflection of how the times interact with events. See Jonathan H. Adler, "The Fable of the Burning River, 45 Years Later," *The Washington Post*, June 22, 2014.

14. Marc K. Landy, Marc J. Roberts, and Stephen R. Thomas, *The Environmental Protection Agency: Asking the Wrong Questions*, exp. ed. (New York: Oxford University Press, 1994).

15. Samuel P. Hays, *Beauty, Health, and Permanence: Environmental Politics in the United States, 1955–1985* (New York: Cambridge University Press, 1987).

16. Charles T. Rubin, *The Green Crusade: Rethinking the Roots of Environmentalism* (Lanham, Md.: Rowman & Littlefield, 1998). Economists Allen Kneese and Charles Schultze note that on November 16, 1960, the President's Commission on National Goals, composed of eleven distinguished citizens, delivered to President Dwight D. Eisenhower its report, *Goals for Americans*. The report listed fifteen major goals, each of which the commission felt to be an area of national concern for the coming decade. Controlling pollution was not on the list. See Allen V. Kneese and Charles L. Schultze, *Pollution, Prices, and Public Policy* (Washington, D.C.: Brookings Institution, 1975).

17. Graham, *The Morning After Earth Day*.

18. Shabecoff, *A Fierce Green Fire: The American Environmental Movement* (New York: Hill and Wang, 1993), 109–110. Although the environmental groups did not engineer Earth Day, their memberships grew in the 1960s, rising from 124,000 in 1960 to 1,127,000 in 1972. See Robert Mitchell, "From Conservation to Environmental Movement: The Development of the Modern Environmental Lobbies," in *Government and Environmental Politics: Essays on Historical Developments Since World War Two*, ed. Michael Lacey (Baltimore: Johns Hopkins University Press, 1989), 81–113.

19. Kirkpatrick Sale, *The Green Revolution: The American Environmental Movement, 1962–1992* (New York: Hill and Wang, 1993).

20. The following Earth Day anecdotes are assembled from reports in *The New York Times* and *The Washington Post*, April 23, 1970.

21. Mary Etta Cook and Roger H. Davidson, "Deferral Politics: Congressional Decision Making on Environmental Issues in the 1980s," in *Public Policy and the Natural Environment*, ed. Helen M. Ingram and R. Kenneth Godwin (Greenwich, Conn.: JAI Press, 1985), 47–76.

22. Hazel Erskine, "The Polls: Pollution and Its Costs," *Public Opinion Quarterly* 1 (Spring 1972): 120–135.

23. Samuel P. Hays, "The Politics of Environmental Administration," in *The New American State: Bureaucracies and Policies Since World War II*, ed. Louis Galambos (Baltimore: Johns Hopkins University Press, 1987), 23.

24. John C. Whitaker, *Striking a Balance: Environment and Natural Resources Policy in the Nixon-Ford Years* (Washington, D.C.: AEI, 1976).

25. "Transcript of the President's State of the Union Message to the Joint Session of Congress," *The New York Times,* January 23, 1970, 22.

26. Ibid.

27. Council on Environmental Quality, *Environmental Quality.*

28. Ibid.

29. Richard A. Harris and Sidney M. Milkis, *The Politics of Regulatory Change: A Tale of Two Agencies* (New York: Oxford University Press, 1989).

30. Ibid.

31. Congress may not amend an executive reorganization proposal; it must approve or disapprove the entire package. Ordinarily, to stop a reorganization, either chamber must adopt a resolution disapproving it within 60 days of its introduction. But Nixon gave Congress 120 days to decide on the EPA because the reorganization plan was so complex. See "Nixon Sends Congress Plans to Consolidate Environmental Control, Research Agencies," *The Wall Street Journal,* July 10, 1970, 3.

32. As historian Paul Charles Milazzo notes, Muskie was driven not by the will of the people or a love of nature, but by his discovery that the chairmanship of a committee that controlled air and water pollution could be a path to power. The expertise of his staff gave him an institutional base from which to operate as a legislative policy entrepreneur, and he used his power to build a coalition to support a federal water quality program even before most Americans demanded one. See Paul Charles Milazzo, *Unlikely Environmentalists: Congress and Clean Water, 1945–1972* (Lawrence: University of Kansas Press, 2006).

33. U.S. Congress, Senate, *Congressional Record,* 91st Cong., 2d sess., March 4, 1970, S2955.

34. John C. Esposito, *Vanishing Air* (New York: Grossman, 1970), 270, 290–291.

35. Charles O. Jones, *Clean Air: The Policies and Politics of Pollution Control* (Pittsburgh, Pa.: University of Pittsburgh Press, 1975), 192.

36. Alfred Marcus, "Environmental Protection Agency," in *The Politics of Regulation,* ed. James Q. Wilson (New York: Basic Books, 1980), 267–303.

37. Primary standards must "protect the public health" by "an adequate margin of safety." Secondary standards must "protect the public welfare from any known or anticipated adverse effects."

38. An executive order is a presidential directive to an agency that enables the president to shape policy without getting the approval of Congress.

39. John Quarles, *Cleaning Up America: An Insider's View of Environmental Protection* (Boston: Houghton Mifflin, 1976).

40. Alfred Marcus, *Promise and Performance: Choosing and Implementing an Environmental Policy* (Westport, Conn.: Greenwood Press, 1980).

41. The National Environmental Policy Act, which took effect in 1970, requires federal agencies to complete environmental impact statements before embarking on any major project.

42. Milazzo, *Unlikely Environmentalists*, 5.

43. Congress can override a presidential veto with a two-thirds majority in both chambers.

44. Clean Air Act, 42 U.S. Code § 7604-Citizen Suits, Section 304(a)(2).

45. The industry mounted only weak resistance to Muskie's attacks on it. According to journalist Richard Cohen, Muskie later speculated that some industry leaders "could see what was coming" and therefore gave limited cooperation (or got on the bandwagon). The passive attitude of the car manufacturers probably also reflected its strong financial position at the time (imports represented only 13 percent of all U.S. auto sales) and its weak lobbying operation. General Motors, for example, did not even establish a Washington lobbying office until 1969. See Richard Cohen, *Washington at Work: Back Rooms and Clean Air* (New York: Macmillan, 1992).

46. Arnold Howitt, "The Environmental Protection Agency and Transportation Controls," in *Managing Federalism: Studies in Intergovernmental Relations* (Washington, D.C.: CQ Press, 1986), 116.

47. Hays, *Beauty, Health, and Permanence*.

48. Graham, *The Morning After Earth Day*, 31.

49. Quoted in ibid., 53.

50. Howitt, "The Environmental Protection Agency," 125.

51. Marcus, *Promise and Performance*.

52. The number of business-related political action committees (PACs) also increased from 248 in 1974 to 1,100 in 1978. See David Vogel, "The Power of Business in America: A Reappraisal," *British Journal of Political Science* 13 (1983): 19–43.

53. More than 2,000 companies contested EPA standards within the first few years of its operation. See James T. Patterson, *Grand Expectations: The United States, 1945–1974* (New York: Oxford University Press, 1996).

54. Marcus, *Promise and Performance*.

55. Transportation control measures include creating bicycle paths and car-free zones, rationing gas, imposing gas taxes, building mass transit, creating bus lanes, encouraging carpooling, and establishing vehicle inspection and maintenance programs.

56. Quoted in Marcus, *Promise and Performance*, 133.

57. Howitt, "The Environmental Protection Agency."

58. Marcus, *Promise and Performance*.

59. The law required the EPA to grant permits to all industrial and government polluters, including 21,000 municipal sewage treatment facilities.

60. Marcus, *Promise and Performance*.

61. Recall that the original Clean Air Act did not allow the EPA to consider economic and technical factors.

62. Marcus, *Promise and Performance*. Although many of its original provisions were weakened, in some respects the Clean Air Act was strengthened as a result of the 1977 amendments. For instance, the amendments formalized the prevention-of-significant deterioration (PSD) concept, authorizing new standards for areas with good air quality to prevent industry from moving from polluted to clean-air regions. These amendments also allowed the EPA to set Lowest Achievable Emissions Rate standards for new sources in nonattainment areas.

63. Conventional pollutants are solids, biochemical oxygen demand (BOD) pollutants, pH, and fecal coliform.

64. Marcus, *Promise and Performance*.

65. Dick Kirschten, "EPA: A Winner in the Annual Budget Battle," *National Journal*, January 28, 1978, 140–141.

66. Ibid.

67. Paul R. Portney, "EPA and the Evolution of Federal Regulation," in *Public Policies for Environmental Protection*, 2d ed., ed. Paul R. Portney and Robert N. Stavins (Washington, D.C.: Resources for the Future, 2000), 11–30.

68. Marc K. Landy, Marc J. Roberts, and Stephen R. Thomas, *The Environmental Protection Agency: Asking the Wrong Questions*, expanded ed. (New York: Oxford University Press, 1984), 249.

69. Michael Biesecker. "EPA Ends Clean Air Policy Opposed by Fossil Fuel Interests." January 25, 2018. Available at https://www.apnews.com/646836ad590c4230b730fc17cfbcb967/EPA-ends-clean-air-policy-opposed-by-fossil-fuel-interests.

70. John H. Cushman Jr., "Scientists Reject Criteria for Wetlands Bill," *The New York Times*, May 10, 1995.

71. Claudia Copeland, "EPA and the Army Corps' Proposed Rule to Define 'Waters of the United States,'" Congressional Research Service R43455, November 21, 2014.

72. "Final Rule: Definition of 'Waters of the United States'—Addition of Applicability Date to 2015 Clean Water Rule." Available at https://www.epa.gov/wotus-rule/final-rule-definition-waters-united-states-addition-applicability-date-2015-clean-water.

73. Emily Shugerman, "Trump Administration Rolls Back Obama Clean Water Rule." February 1, 2018. Available at https://www.independent.co.uk/news/world/americas/us-politics/trump-clean-water-act-repeal-barack-obama-wotus-scott-pruitt-epa-a8189721.html.

74. U.S. Environmental Protection Agency, "Air Trends." Available at http://www.epa.gov/airtrends/. The agency is continuously revising its estimation methods, so each year's estimates are slightly different from the previous year.

75. U.S. Environmental Protection Agency, "National Aquatic Resource Surveys: An Update," January 2011. Available at http://water.epa.gov/type/watersheds/monitoring/upload/nars-progress.pdf.

76. U.S. Environmental Protection Agency, National Lakes Assessment 2012. Available at https://www.epa.gov/sites/production/files/2016-12/documents/nla_report_dec_2016.pdf.

77. U.S. Environmental Protection Agency, The National Rivers and Streams Assessment 2008–2009. Available at http://water.epa.gov/type/rsl/monitoring/riverssurvey/upload/NRSA200809_FactSheet_Report_508Compliant_130314.pdf.

78. Under Section 303 of the Clean Water Act, states are required to develop lists of impaired waters. For each polluted waterway, the state must determine the total maximum daily load (TMDL) specifying the amount of each pollutant that a body of water can receive and still meet water quality standards and allocating pollutant loadings among point and nonpoint sources. Although the Clean Water Act has required TMDLs since 1972, states only began developing them in the late 1990s, in response to a wave of lawsuits.

79. Thomas C. Brown and Pamela Froemke, "Nationwide Assessment of Nonpoint Source Threats to Water Quality," *BioScience* 62,2 (February 2012): 136–146.

80. Council on Environmental Quality, *Environmental Quality: The Fifteenth Annual Report of the Council on Environmental Quality* (Washington, D.C.: U.S. Government Printing Office, 1984).

81. Since the 1600s the lower forty-eight states have lost half of the nation's 220 million acres of wetlands. Development destroyed some 500,000 acres of wetlands per year in the 1970s, but in recent years both the Fish and Wildlife Service and the National Resource Conservation Service have documented slight gains in wetlands acreage. See Claudia Copeland, "Wetlands: An Overview of Issues," Congressional Research Service, July 12, 2010, RL33483.

CHAPTER THREE

LOVE CANAL

Hazardous Waste and the Politics of Fear

In the summer of 1978, Americans began to hear about a terrifying public health nightmare in Niagara Falls, New York. According to news reports, hundreds of families were being poisoned by a leaking toxic waste dump underneath their homes. Residents plagued by cancer, miscarriages, and birth defects were demanding to be evacuated and relocated. The episode, known simply as "Love Canal," became a national story because "it radicalized apparently ordinary people. [It] severed the bond between citizens and their city, their state, and their country."[1] Love Canal also shaped public attitudes about abandoned toxic dumpsites—about the risks they pose and about government's responsibility for ensuring they are cleaned up. In addition, the incident was the catalyst for the nation's most expensive environmental law: the Comprehensive Environmental Response, Compensation, and Liability Act of 1980, popularly known as the Superfund Act.

The Love Canal case illuminates the role of science and scientific experts in controversies over threats to human health. In such conflicts, experts' emphasis on detachment and objectivity can alienate a public that feels it has been harmed. Experts contend that the public is deeply concerned about environmental threats, such as hazardous waste, that pose relatively small risks, while more serious problems, such as climate change and wetlands loss, get short shrift. The reason is that experts and the public perceive risks differently. For experts, human health risk corresponds closely to statistical estimates of annual fatalities. But ordinary people incorporate a much richer set of considerations—including voluntariness, immediacy, familiarity, and control—into their understanding of risk. The public particularly fears low probability catastrophic and "dread" risks. Moreover, because people's risk perceptions are based on general inferential rules (heuristics) that feel like common sense, they tend to be resistant to change, even in the face of compelling scientific evidence.[2]

Experts often unwittingly contribute to public alarm and confusion. For example, scientists calculate the cancer risk posed by prolonged exposure to small amounts of a chemical by extrapolating from the results of animal bioassays, epidemiological studies, and cellular analyses, all of which yield uncertain results.[3] In doing so, scientists make a variety of assumptions that reflect their values—in particular their beliefs about the levels of risk to which the public ought to be exposed.[4] Because experts' values differ, they often arrive at divergent assessments. The resulting phenomenon of dueling experts breeds mistrust among the public, which lacks the wherewithal to sort out competing technical claims.

Some citizens have tried to enhance their own technical competence by acquiring information about environmental health risks using community-based participatory research, in which communities work collaboratively with scientists to investigate and address issues of local concern. One mode of participatory inquiry is popular epidemiology, a process by which community members gather scientific data and other information and marshal the knowledge of experts to understand the incidence of a disease.[5] Proponents of popular epidemiology believe that tools such as community health surveys bring to the fore environmental data and circumstances that traditional epidemiological studies otherwise would not elicit.[6] More important, they believe that putting information in the hands of citizens empowers them and enables them to transfer problems from the (inaccessible) technical realm to the political arena.

Even when armed with compelling information, however, citizens are likely to encounter tremendous resistance among state and local politicians to address environmental health threats. Those officials feel constrained in responding to public alarm by the need to foster economic development and retain high-income taxpayers.[7] In addition, because industry is mobile, officials tend to be cautious about taking on polluters for fear of alienating companies that employ citizens in the community.[8] Policymakers also may be reluctant to intervene because they want to avoid financial responsibility for problems that promise to be costly to solve.

Publicity can change this political dynamic in several ways. First, media coverage of a risk mobilizes citizens beyond those directly affected because people assume that the more attention the media pay to a risk, the worse it must be. According to psychologist Baruch Fischhoff, "If scientists are studying it and the news reports it, people assume it must be worth their attention."[9] Moreover, because media coverage emphasizes dramatic, newsworthy events, the reports are likely to be alarming and hence effective at activating citizen groups. The media then amplify those groups' political influence by covering their activities. By focusing on human interest anecdotes, journalists create victims, villains, and heroes—the central elements in a compelling causal story. When such stories involve human health threats, particularly to children, they appeal to highly consensual values and so have enormous potential to resonate with the broader public. Media attention galvanizes not only citizens, but also elected officials, who fear the consequences of negative publicity.[10] Critical media coverage may put policymakers on the defensive, forcing them either to justify a problem or act to solve it.[11] At the same time, policymakers use the extent of media coverage to gauge the intensity and nature of public opinion on that issue.[12]

BACKGROUND

In the 1970s a combustible mix of genuine health threats and sensational media coverage came together in the city of Niagara Falls, New York. But the Love Canal story has its roots in the late nineteenth century, when entrepreneur William T. Love received permission from the New York State Legislature to build a canal that would divert the Niagara River away from the falls for about seven miles, dropping nearly 300 feet before it reconnected to the river. The canal was the centerpiece of Love's scheme to construct a vast industrial city fueled by cheap and abundant hydropower.

Love dug a mile-long trench that was ten feet deep and about fifteen feet wide and then built a factory and a few homes alongside it. But his dream collapsed in the mid-1890s when a financial depression caused investors to withdraw their support. In addition, in 1906 the U.S. Congress passed a law barring Love from diverting water from the Niagara because it wanted to preserve the falls. In a final blow, the 1910 advent of alternating current, which allowed power to be transported over long distances, reduced the need to locate an industrial city near its power source. The abandoned canal soon became a popular fishing, swimming, and picnicking spot for residents.

In 1920 the Love Canal was sold at public auction and became a municipal waste-disposal site. Then, in 1942, the Niagara Power and Development Corporation gave the Hooker Chemical and Plastics Corporation permission to dispose of wastes in the abandoned canal, and in 1947 it sold the canal and sixteen acres of surrounding land to Hooker. Between 1942 and 1952 Hooker dumped more than 21,000 tons of toxic chemical wastes at the site, widening or deepening the canal in places to accommodate its needs.[13]

Company officials were aware that the materials they were dumping were dangerous to human health; residents recalled workers at the site rushing to neighboring yards to wash off burns with water from garden hoses. As early as 1943, a letter to the *Niagara Gazette* claimed that the smell was unbearable and that the white cloud that came from the site "killed the grass and trees and burnt the paint off the back of the houses and made the other houses all black."[14] A favorite game of neighborhood children was to pick up phosphorous rocks, throw them against the cement, and watch them explode. In the hot weather, spontaneous fires broke out, and noxious odors wafted through open windows of nearby homes.[15]

By 1952 the canal was nearly full. Hooker and the city of Niagara Falls covered the dumpsite with a protective clay cap and earth, and soon weeds and grasses began to sprout on its surface. In 1953, when city officials were looking for inexpensive land on which to build a school, Hooker obliged them by transferring the sixteen-acre dumpsite to the Board of Education for a token fee of one dollar. At that time, Hooker issued no detailed warnings about the possible hazards posed by the buried chemicals. The company did, however, include in the deed a disclaimer that identified the wastes in a general way and excused it from liability for any injuries or deaths that might occur at the site.[16] School board members toured the area with Hooker representatives and took test borings that showed chemicals in two locations only four feet below the surface. Yet the school board—apparently unconcerned about any potential health threat and despite the misgivings of its own architect—began to build an elementary school and playground at the canal's midsection. When workers started to excavate and discovered chemical pits and buried drums, the board simply moved the school eighty-five feet north and installed a drainage system to divert accumulating water into the Niagara River.

Not long after the 99th Street Elementary School was completed, the school board donated some unused property to the city to build streets and sidewalks. Soon some homebuilders approached the board about trading part of the Love Canal site for parcels the board wanted. At a meeting in November 1957, representatives of Hooker strongly opposed the trade, saying they had made clear at the original transfer that the site was unsuitable for any construction that required basements and sewer lines. Apparently heeding Hooker's warnings, the school board voted against the trade, but developers began to build modest, single-family houses around the borders of the site anyway. During construction, contractors cut channels through the clay walls lining the hidden canal and used topsoil from the canal surface for fill. Because the new houses were on land that was not part of the original transaction between the school board and Hooker, the property owners' deeds did not notify them of chemicals buried in the adjoining land.[17]

In 1960 the school board gave the northern portion of the site to the city, and in 1961, it sold the southern portion at auction. By that time, the city had already installed streets paralleling and crisscrossing the canal. In the late 1960s the state ran the LaSalle Expressway through the southern end of the site, necessitating relocation of a main street and uncovering chemical wastes that Hooker agreed to cart away.[18] By the early 1970s the area around the canal was a working-class neighborhood; there were nearly 100 homes on 97th and 99th streets with backyards abutting a long, empty lot that should have been 98th Street but was really a chemical-filled trench (see Map 3.1).[19]

Throughout the 1950s and 1960s individual residents around the Love Canal site complained to the municipal government about odors and odd afflictions, including rashes and respiratory problems, as well as oily black substances in basements and exposed, rusting barrels in fields around their homes. Members of crews building streets in the area complained of itchy skin and blisters. In 1958 Hooker investigated reports that three or four children had been burned by debris on their former property, but the company did not publicize the presence of the chemicals even after this incident, probably because it feared liability.[20] Municipal records indicate that by 1969, building inspectors had also examined the Love Canal dumpsite area and reported that the conditions were hazardous: as the rusting barrels collapsed, holes appeared on the surface of the field, and chemical residues remained on the field after rainwater had evaporated.[21] Still, the city took no action to investigate the problem further or to remedy it.

Map 3.1 Love Canal Emergency Declaration Area

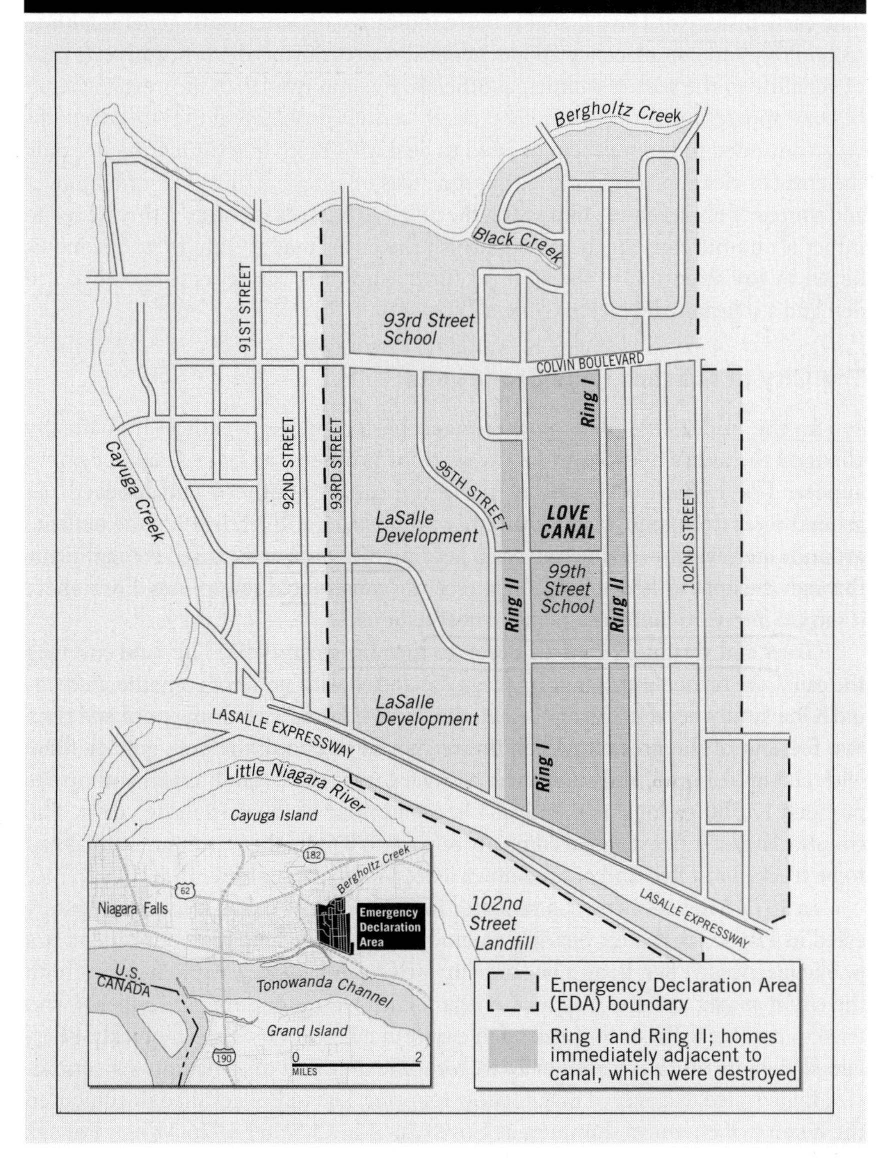

Source: New York State Department of Health, https://www.health.ny.gov/environmental/investigations/love_canal/docs/lc_eda_map.pdf.

THE CASE

The early history of Love Canal is more indicative of Americans' general faith in technology and complacency about chemical wastes during the postwar years than of venality on the part of municipal officials. Even in the 1970s, however, when it became apparent that there were potentially serious problems at the site, local officials continued to ignore them or tried to deal with them quietly, fearing not only the costs of cleanup, but also the consequences of antagonizing a major employer and source of tax revenue. Addressing the possible hazards in the area threatened to inflict economic consequences and tarnish the city's image. Only when the media began to pay attention to the issue, in turn prompting residents to mobilize and demand a solution, did elected officials respond.

The City of Niagara Falls Stonewalls

In the mid-1970s, when a prolonged period of wet weather dramatically changed the area's hydrology, visible signs of problems at Love Canal began to appear. The LaSalle Expressway along the canal's southern end blocked the groundwater from migrating southward to the Niagara River. In 1976 the built-up groundwater overflowed the clay basin holding the waste and carried contaminants through the upper silt layer and along recently constructed sewer lines. From there it seeped into yards and basements of nearby houses.

Trees and shrubs in the area began to turn brown and die. The field covering the canal site turned into a mucky quagmire dotted with pools of contaminated liquid. One family became alarmed when they noticed their swimming pool had risen two feet out of the ground. When they removed the pool, the hole quickly filled with chemical liquid, and soon their backyard was a wasteland. Local authorities pumped 17,500 gallons of chemical-filled water out of the yard in two days. The county's largest waste disposal company refused to handle the wastewater, and it had to be trucked to Ohio and poured down a deep-well disposal site.[22]

In 1977 Michael Brown, a reporter from the *Niagara Gazette*, became interested in Love Canal after hearing an eloquent plea for help from a resident at a public meeting. When Brown began to investigate, it became clear to him that both the city manager and the mayor of Niagara Falls were stonewalling residents who tried to contact them or to speak up in city council meetings. Brown quickly ascertained that city officials had been aware for some time that the situation was serious.

Brown also discovered that a fellow reporter, David Pollack, had documented the history of chemical dumping at Love Canal in October 1976. When Pollack got a private company, Chem-Trol, to analyze the sludge from some Love Canal basements, the company found toxic chemicals and determined that Hooker was their source. Pollack also ascertained that in early 1976 the New York Department of Environmental Conservation (DEC) had begun testing houses around the canal site after tracing high levels of the pesticide mirex in Lake Ontario fish to a dumpsite adjacent to the canal.[23] The DEC's investigation revealed that polychlorinated biphenyls (PCBs) and other toxic materials were flowing from the canal into adjoining sewers.[24] The DEC's study was proceeding slowly, however, because the agency lacked adequate funding, personnel, and equipment. Furthermore, the DEC got little cooperation: Hooker denied all responsibility, and municipal officials—uneasy about antagonizing the city's largest industrial employer and worried about the magnitude of the city's liability—preferred to address the problem discreetly.

In April 1977 the city, with some funding from Hooker, hired the Calspan Corporation to develop a program to reduce the groundwater pollution at Love Canal. Calspan documented the presence of exposed, corroded drums and noxious fumes and notified officials that chemical contamination was extensive. That summer the *Niagara Gazette* published a summary of the Calspan report and urged the city to undertake the cleanup project recommended by its consultants. The city declined, however, and the story was insufficiently dramatic to capture residents' attention. In September, a municipal employee concerned about the city's inaction contacted Rep. John LaFalce, D-N.Y., the district's member of Congress, and urged him to tour the area. Unable to get a response from the city manager, LaFalce asked the Environmental Protection Agency (EPA) to test the air in the basements along 97th and 99th streets. In October, soon after LaFalce's visit, the regional EPA administrator wrote in an internal memorandum saying that, based on what he had seen, "serious thought should be given to the purchase of some or all of the homes affected."[25]

By April 1978, New York Health Commissioner Robert Whalen had become sufficiently concerned that he directed Niagara County's health commissioner to remove exposed chemicals, build a fence around the dumpsite, and begin health studies of area residents.[26] In May, the EPA released the results of its air-sampling studies. At a public meeting at the 99th Street School, agency officials told residents that they had found benzene in the air of their basements. At this point, Michael Brown, shocked by the EPA's reports and disturbed by the unwillingness of local officials to acknowledge the seriousness of the problem, undertook his own investigation.

The Local Media Raise Residents' Concerns

Brown's story in the *Niagara Gazette* on the benzene hazard claimed that there was a "full fledged environmental crisis" under way at Love Canal. With local and county authorities unwilling to investigate further, Brown conducted an informal health survey and found a startling list of residents' ailments—from ear infections and nervous disorders to rashes and headaches. Pets were losing their fur and getting skin lesions and tumors, women seemed to have a disproportionate incidence of cancer, and several children were deaf.[27]

By repeating residents' claims about their health problems, Brown's reporting enhanced their perceptions of the threats posed by the buried chemicals. Until the spring of 1978, residents had been only dimly aware of the potential association between the fumes and chemical wastewater and their health problems. But as articles began to appear in the *Niagara Gazette* and then in the Buffalo papers that May and June, some local people became alarmed. Lois Gibbs, a resident of 101st Street since 1972, made a frightening connection: Gibbs's son, Michael, had begun attending kindergarten at the 99th Street School and had developed epilepsy soon afterward. Gibbs contacted her brother-in-law, a biologist at the State University of New York (SUNY) at Buffalo, and he explained the health problems that could be produced by chemicals dumped in the canal. Gibbs tried to transfer her son to a different school, but school administrators resisted.

Disconcerted, the normally reticent Gibbs started going door to door with a petition demanding that the city address residents' concerns about the school. Talking day after day with neighbors, she discovered that some homeowners were worried that bad news about Love Canal would cause property values to decline and had formed a group to agitate for property tax abatement and mortgage relief. Gibbs also became deeply familiar with the health problems of canal-area residents. To her,

> it seemed as though every home on 99th Street had someone with an illness. One family had a young daughter with arthritis. . . . Another daughter had had a miscarriage. The father, still a fairly young man, had had a heart attack. . . . Then I remembered my own neighbors. One . . . was suffering from severe migraines and had been hospitalized three or four times that year. Her daughter had kidney problems and bleeding. A woman on the other side of us had gastrointestinal problems. A man in the next house down was dying of lung cancer and he didn't even work in the [chemical] industry. The man across the street had just had lung surgery.[28]

Armed with this worrisome anecdotal evidence, Gibbs transformed herself into a highly effective policy entrepreneur. Although this was her first experience with political activism, Gibbs realized that she would need an organization behind her to wield any political clout. Following the strategic advice of her brother-in-law, she founded the Love Canal Homeowners Association (LCHA), which became the most visible and persistent of the citizen groups formed during this period. Gibbs devoted nearly all her waking hours to its activities. Her primary function was to promote a simple causal story—that Hooker had irresponsibly dumped dangerous chemicals that were making residents (and, most important, their children) ill—and link it to her preferred solution: evacuation of and compensation for all the families in the area.[29]

Assessing (and Avoiding) Blame

The first response of government officials to LCHA activism was to try to shift responsibility to other levels of government. "It's a county health problem," said City Manager Donald O'Hara. "But if the state says the city has the authority to move them out and the people want to be moved, then we'll move them."

Dr. Francis J. Clifford, the county health commissioner, responded, "Of course Don O'Hara says the responsibility is with the county. If I were him, I'd say the same thing. The lawyers will fight it out."[30] Everyone hoped to pin the blame on the federal government, with its deep pockets, especially upon hearing of witnesses' claims that the U.S. Army had also dumped wastes into the canal in the early 1950s. But after a brief investigation, the Department of Defense denied those allegations.

In the meantime, the county did little more than install a few inexpensive fans in two homes and erect a flimsy fence that did little to keep children away from the dumpsite. The municipal government was equally dilatory: the city council voted not to spend public money for cleanup because part of the site was owned by a private citizen living in Philadelphia. The city's tax assessor refused to grant any tax abatement on the homes, even though banks would not mortgage them and lawyers refused to title them.[31]

The city was also reluctant to pursue Hooker Chemical because it was an important employer and taxpayer in an area historically dependent on the chemical industry. At that time, Hooker employed about 3,000 workers from the Niagara area. The plant at Niagara Falls was the largest of Hooker's sixty manufacturing operations, and Hooker's corporate headquarters were there as well. Even more important, Hooker was planning to build a $17 million headquarters downtown. Municipal officials were offering Hooker a lucrative package of tax breaks and loans as well as a $13.2 million mortgage on a prime parcel of land.[32]

Hooker maintained from the outset that it had no legal obligations with respect to Love Canal. Once it began getting negative press coverage, the company retaliated with a concerted effort to redefine the problem: it launched a nationwide campaign involving thousands of glossy pamphlets and a traveling two-man "truth" squad to convince the media that the problems at Love Canal were not its fault. Hooker representatives emphasized that the canal had been the best available technology at the time. The company, they said, was merely acting as a good corporate citizen, not admitting guilt, when it contributed money toward the city-sponsored Calspan study and subsequently volunteered to share cleanup costs.

The National Media Expand the Scope of the Conflict

Normally, such a united front by industry and local officials would have squelched attempts at remediation, but citizens' mobilization—combined with sympathetic media coverage of their complaints and demonstrations—enhanced the residents' clout. Early August 1978 marked a turning point in the controversy because the arrival of *New York Times* reporter Donald McNeil on the scene transformed a local issue into a national one. In turn, national press coverage inflamed residents' passion and further escalated tensions.[33] Within a day of McNeil's first report on the situation, reporters from other national newspapers converged on Niagara Falls. The media's coverage of the ensuing confrontations was compelling because it conveyed the image of Hooker as, at worst, a malevolent villain and, at best, an irresponsible one, while depicting residents as ordinary citizens and innocent victims.

Meanwhile, the New York Department of Health (DOH) was collecting blood samples, surveying residents, and testing air samples in homes to ascertain the actual health threats posed by the buried waste. In early July 1978 residents received forms from the state indicating high levels of chemicals in their houses but providing little information about what those levels meant. On August 2 Health Commissioner Whalen announced the results of the DOH's first large-scale health study of the area: researchers had found ten carcinogenic compounds in vapors in the houses, at levels from 250 to 5,000 times as high as those considered safe. They also found that women living at the southern end of the canal suffered an unusually high rate of miscarriages (29.4 percent vs. the normal 15 percent) and that their children were afflicted with an abnormally high rate of birth defects (five children out of twenty-four). Four children in one small section of the neighborhood had severe birth defects, including clubfeet, retardation, and deafness. The people who had lived in the area the longest had the most problems.[34]

Whalen declared a public health emergency and concluded that there was a "great and imminent peril" to the health of the people living in and around Love Canal.[35] He urged people to take precautionary measures to avoid contamination but offered no permanent solutions. He was unwilling to declare the neighborhood uninhabitable but did advise people to avoid spending time in their basements or eating vegetables from their gardens. Finally, he recommended that pregnant women and children under the age of two move out of the area.

On August 3, 1978, about 150 residents—incensed by Whalen's announcement, their anger fueled by the media coverage—met in front of their homes to collect mortgage bills and resolve not to pay them. They also planned to demand aid from the government or Hooker so they could relocate. Said one resident, "If they take my house, I owe $10,000 on it. . . . I couldn't sell it for ten. They won't kick me out for two years, and that's two years' taxes and payments I save to find someplace else."[36] The following evening, the first families began leaving the contaminated site, toting their cribs and suitcases past a scrawled sign that read, "Wanted, safe home for two toddlers."

As the summer wore on, public meetings became increasingly acrimonious. Residents were frustrated because DOH scientists, in their attempts to conduct objective research, were reluctant to counsel fearful residents about the results of their medical tests. Furthermore, government officials refused to act, awaiting definitive study results and, more important, a clear assessment of financial culpability. At one packed meeting in a hot, crowded school auditorium, sobbing young mothers stood up shouting out the ages of their children or the terms of their pregnancies and asked whether they would be moved out of their houses. As one journalist described the scene, "Angry young fathers, still in the sweaty T-shirts they had worn to work in local chemical factories, stood on chairs and demanded to know what would happen to their small children if their wives and infants were moved."[37]

Media coverage of residents' plight was particularly well-timed because Hugh Carey, the governor of New York, was facing an election in November 1978 and so felt compelled to respond to residents' widely publicized pleas. The state set up an office in a local school and provided rent vouchers, medical advice, and moving help for the thirty-seven families with pregnant women or infants under two years of age. Federal officials reacted to the spate of newspaper stories as well: on August 6, 1978, William Wilcox of the Federal Disaster Assistance Administration inspected the site at the request of President Jimmy Carter. The following day, the president declared an emergency at Love Canal and approved federal relief funds for the area.[38]

Shortly after this announcement, Governor Carey toured Love Canal for the first time. During the visit, he pledged to relocate the 239 "inner ring" families and purchase their homes on 97th and 99th streets. Once these families were evacuated, he said, the state planned to raze the houses as well as the 99th Street School and begin a massive construction program to stop the leaching and seal the canal. The remediation plan, laid out in an engineering study sponsored jointly by the city and Hooker, called for ditches to be dug eight to ten feet deep along the periphery of the filled canal and drainage tiles to be laid and covered over. Chemical-laden water leaking out of the old clay canal bed would run down the tiles into a collecting basin, from which it would be pumped and taken for treatment or disposal.

The Outer Ring Residents Are Left Behind

The residents of the "outer ring" between 93rd and 103rd streets were enraged by this solution, feeling that the state was ignoring their predicament. They too had experienced health problems, and studies had detected high levels of chemicals in their houses as well. Under the supervision of Beverly Paigen, a biologist and cancer researcher who was helping residents interpret the information they received from the health department, the LCHA undertook one of the earliest efforts to do popular epidemiology—a telephone survey to ascertain the pattern of illnesses in the historically wet drainage areas around the canal. But when Dr. Paigen analyzed the data and presented her results, complete with qualifying statements about her methodology, the DOH dismissed her findings, saying that the study was meaningless because it was put together by a "bunch of housewives with an interest in the outcome of the study."[39]

Frustrated, Gibbs and the LCHA continued to press the state to relocate the outer-ring families, fearing their political leverage would decline after the gubernatorial election in November. By this point, Gibbs had become adept at raising the salience of her definition of the problem: she orchestrated dozens of public demonstrations and rallies, wrote letters and made phone calls to officials, submitted to interviews on talk shows and news programs, and even testified before a congressional subcommittee. As a result, her version of events dominated national perceptions of the situation.[40]

The state nevertheless continued to resist the idea of relocating the families who lived on the outskirts of the contaminated area. Groups of angry, outer-ring residents began to picket the cleanup site, and on December 11, about fifty people braved twelve-degree temperatures to stop workers' cars attempting to enter the site. The police arrested six demonstrators for obstructing traffic. Deepening the residents' desperation, the state announced it had discovered dioxin at Love Canal. A *New York Times* report dramatized the lethal potential of dioxin, a byproduct of herbicide production, by noting that "three ounces, minced small enough into New York City's water supply, could wipe out the city."[41]

To the disappointment of the LCHA, on January 29, the Federal Disaster Assistance Administration rejected an appeal from the state to reimburse it for the $23 million it had spent removing the inner-ring families and cleaning up the site. This decision increased the financial pressure on state officials and made them even more reluctant to engage in a costly second relocation. Matters became more complicated, however, when a blue-ribbon panel appointed by the governor concluded in early February that the Love Canal site was hazardous, prompting health commissioner David Axelrod to order that pregnant women and families with children under the age of two be removed temporarily from the area. This action only fueled Gibbs's rage: she pointed out that if the place was dangerous for pregnant women and children, it posed a threat to everyone else as well.[42] Although twenty-four families moved, hundreds more remained in limbo. By this time, according to Gibbs, reporters were becoming impatient with the LCHA and were beginning to ask, "If you're so afraid, why don't you just leave?" Residents replied that they were working people who had invested a lifetime of earnings in their homes. Where would they go? they asked. What would they live on?

Experts Versus Citizens

For the families that remained, more uncertainty lay ahead. The $9.2 million cleanup effort encountered frequent delays as people raised safety concerns and public officials quarreled over new health studies. By August 1979, Love Canal residents had endured another long, hot summer made more oppressive by the fumes from the remedial construction site. For weeks people had been calling the LCHA office to say they were experiencing headaches, difficulty breathing, and burning, itching eyes. The DOH scheduled a public meeting for August 21, and residents again anticipated answers to questions about their health and their future.

Instead, what turned out to be the final meeting between residents and the commissioner of public health did little more than cement the antagonism between them. During the meeting, Axelrod acknowledged that his department had found dioxin in the soil but said the remaining residents would have to deal with the risks on their own. With that, what little remained of residents' faith in scientific experts and government officials evaporated. According to Gibbs, the officials "offered no answers, no useful information. The residents' confidence was shaken time and again. They didn't trust the safety plan or the construction plan or the health department officials. People were more frustrated when they left than when they arrived."[43]

At the end of August 1979, the Niagara Falls school board voted to close the 93rd Street Middle School. Shortly thereafter, 110 families moved into hotels, taking advantage of an LCHA-initiated court settlement in which state officials agreed that, until the drainage ditches were complete, they would pay the hotel bills and a $13 per day meal allowance for those suffering medical problems because of fumes from the remedial construction site. The state emphasized that it was not paying for a permanent evacuation.

On November 6, the state finished construction of the multimillion-dollar drainage system at the dumpsite, and officials announced that the families could return to their homes. But a dozen of them refused. Challenging the cutoff of payments and asserting that their homes continued to be unsafe, the group vowed to stay at the motel until they were carried out.[44] As usual, Gibbs was hoping the motel sit-in would draw high-level attention to the situation facing the outer-ring residents. She was well aware that Governor Carey had already signed legislation authorizing the state to spend $5 million to buy up to 550 additional homes in the outer ring, most of them to the west of the canal, including her own house. The buyout had been mired in political squabbles and red tape, however, as local officials continued to resist taking responsibility for the problem.

By the end of 1979, the money for relocating the outer-ring families still was not forthcoming. As negotiations between the state and federal governments dragged on, skeptical residents feared that the public's interest was beginning to wane and that with it would go the pressure on politicians to act. Their fears were confirmed when, on February 24, 1980, the EPA announced that air monitors in two houses several hundred yards from the fence around the canal had detected four suspected carcinogens: benzene, chloroform, trichloroethylene, and tetrachloroethylene. To residents' dismay, that disclosure merited only a few paragraphs in the Metro section of *The New York Times*. Similarly, scant notice was taken of Attorney General Robert Abrams's filing, on April 28, of a lawsuit against the Occidental Petroleum Corporation and two of its subsidiaries, Hooker Chemical Corporation and Hooker Chemicals and Plastics Corporation, charging them with responsibility for the Love Canal disaster. With a presidential campaign in full swing, an influx of Cuban refugees, and the eruption of Mount St. Helens, the focus of the national press had shifted.

Health Studies Breed Panic and More Publicity

Federal officials had, however, unwittingly set in motion the catalyst that would irrevocably change the fortunes of Love Canal residents. In early 1979, the EPA, hoping to bolster its tarnished public reputation for dealing with hazardous waste ineptly, had created the Hazardous Waste Enforcement Task Force, and the task force's first assignment was to collect information about Love Canal. On December 20, the Department of Justice, relying on evidence gathered by the task force, filed a lawsuit on behalf of the EPA against Hooker Chemical, the Niagara County Health Department, the City of Niagara Falls, and the city's board of education. The suit demanded $124.5 million, an end to the discharge of toxins in the Love Canal area, cleanup of the site, and relocation of residents if necessary. The lawsuit was remarkable because it asked the court to hold a private company retroactively liable for wastes it dumped many years ago at a site it no longer controlled.

To establish liability, the Justice Department had to prove that the health damages suffered by residents were linked directly to Hooker's wastes. Evidence had shown that the area was contaminated by toxic chemicals and that residents had unusually high rates of health problems, but to link those damages unequivocally to Hooker required further evidence. The Justice Department and the EPA hired Biogenics of Houston, headed by Dr. Dante Picciano, to investigate chromosomal damage among the residents because such impairment indicates exposure to toxic chemicals. The lawyers intended this study to be purely exploratory: if the results were positive, the lawyers could order a full-blown epidemiological study.

From the outset, Picciano's work was flawed in ways that jeopardized its credibility. First, Picciano was a controversial figure because of a much-publicized falling out with his former employer, Dow Chemical, over his tests of that company's workers. Second, and more critical, the study of Love Canal residents did not include a control group, apparently because the EPA did not wish to spend the time or money for one. Third, researchers selected study participants according to criteria that would maximize the chance of finding defects; that is, they intentionally chose people who had had miscarriages or other health problems. Despite these methodological weaknesses, in spring 1980, the study moved forward, and in early May, Biogenics made the preliminary results available. Picciano reported that he had found chromosomal aberrations in eleven of the thirty-six individuals tested. He concluded that eight of them had a rare condition called "supernumerary acentric fragments," or extra pieces of genetic material.[45] According to Picciano, such abnormalities should occur in only 1 percent of individuals and might forewarn increased risk of miscarriages, stillborn babies, birth defects, or cancer.[46] Nevertheless, he advised prudence in using these data given the absence of a control group.

On May, 17 someone leaked the Biogenics study to *The New York Times*, and media attention quickly returned to Love Canal. The EPA scrambled to control the damage, hastily scheduling a press conference in Niagara, preceded by private consultations with each of the residents in whom Picciano had detected abnormalities. The agency set aside half an hour for each family to meet with a doctor who would explain the results and answer questions. Following the individual sessions, EPA representatives gave local and national press conferences. Even though officials were aware of the need to interpret the study results cautiously, especially as the investigation was exploratory and had not been designed to be scientifically valid, they were unable to control the ensuing media frenzy.

Love Canal residents were stunned by the news, and as White House officials held strategy meetings in Washington, D.C., their dismay grew. Federal officials, recognizing the impact of the media coverage, raced to get the Biogenics study peer-reviewed and decide on a course of action. In the meantime, *The New York Times* reported that Governor Carey and the White House were in an all-out feud over who would foot the bill for a large-scale relocation, which now appeared unavoidable.

At the LCHA offices on Monday, a crowd of anxious residents and the press gathered to await news of the government's next move. When the headline "White House Blocks Love Canal Pullout" appeared in the *Buffalo Evening News*, the crowd became agitated: people began stopping cars to spread the word, and one group set a fire in the shape of the letters "EPA" across the street. A few minutes later, a crowd of 200 blocked traffic on both ends of the street. Gibbs, hoping to defuse residents' ire, summoned the two EPA officials who were still in Niagara to address the crowd. Once the officials arrived, however, Gibbs refused to let them leave, and for several hours residents held the officials hostage in a sweltering LCHA office. Gibbs explained that they were simply protecting the officials from a potentially violent mob. She moved back and forth between the crowd and the hostages, assuring each that the situation was under control. Finally, the FBI threatened to rush the crowd. Three FBI agents, four U.S. marshals, and six members of the Niagara Falls Police Department escorted the EPA officials from the LCHA office without further incident. Although it was resolved quietly, the episode left little doubt about residents' desperation.[47]

Figuring Out Who Will Pay

Ultimately, it was the escalation of panic in Niagara Falls and across the nation—not the health studies, which continued to be inconclusive—that compelled politicians at the federal and state levels to support relocating the remaining Love Canal homeowners. Officials had to work out the particulars, such as whether the move would be permanent or temporary and, most important, who would pay for it. Governor Carey, taking advantage of presidential election-year politics, exerted continuous pressure on the Carter administration to finance a permanent move for residents living adjacent to the canal. He complained that the state had already shelled out $35 million for the initial evacuation and subsequent moves. Every day the press reported a new plea from the state to the federal government for relocation funds. Adding fuel to the fire, on May 21, *The New York Times* reported that a study conducted by Beverly Paigen had found nerve damage in twenty-eight of thirty-five Love Canal residents examined, whereas only two out of twenty in the Niagara Falls control group had similar ailments. Moreover, there was a qualitative difference in the nature of the impairment between the two groups.[48]

That afternoon, following media coverage of the latest study, President Carter declared a second federal emergency at Love Canal and urged the remaining families to move to temporary housing. The second Emergency Declaration Area was bound to the south by the LaSalle Expressway, to the east by uninhabited wooded wetlands, to the north by the Black and Bergholtz creeks, and to the west by 93rd Street (see Map 3.1). The relocation was intended to last up to a year to give the EPA sufficient time to study the results of further medical tests of residents. The Federal Emergency Management Agency (FEMA), the successor to the Federal Disaster Assistance Administration, was to supervise the evacuation and relocation. When asked what finally prompted the federal government to act, EPA spokesperson Barbara Blum replied, "We haven't felt that we've had enough evidence before now."[49]

Five days after the president authorized the temporary relocation of 800 families, only 220 had registered for the move. Aware of their increasing leverage, many of the others said they would not leave until the government bought their houses. Residents also said they would refuse to participate in further health testing programs until an agreement was reached on the purchase of their homes. "We are not going to let the government use us as guinea pigs," said a spokesperson for the LCHA.[50] On May 23 Governor Carey presented a detailed plan for the resettlement and estimated the cost at $25 million, of which the state was committed to paying one-fifth. The tug-of-war between the governor and the president dragged on as the state continued to demand money, while federal officials insisted that federal law did not permit the purchase of homes under an emergency program. In mid-June, Carey proposed that the federal government lend the state $20 million of the $25 million necessary to relocate outer-ring residents.

As government officials struggled to resolve the financial issues, controversy continued to swirl around the health studies that had provoked residents' alarm. On June 14, nearly a month after the original Biogenics study's release, a scientific panel appointed by the EPA debunked its conclusions. Dr. Roy Albert of the New York University Medical Center, who headed the panel, called the results "indeterminate" and "really of no use," primarily because the study lacked contemporary controls.[51] The review panel also challenged Picciano's interpretation of the data based on its own analysis of photocopies of photographs of his chromosome preparations. (Cytogeneticists point out that detecting chromosome damage is subjective, more art than science.[52]) Adding to the confusion, two days later, the EPA released another scientific review, and this one tended to support the Biogenics study. According to the second study group, headed by Dr. Jack Killian of the University of Texas, "Some individuals in the study had aberrations that were beyond the normal limits expected in 36 healthy people."[53]

On June 23, the New York DOH announced that in the early 1960s half the pregnancies of women living on a street bordering Love Canal had ended in miscarriage. A report that compiled health studies performed in 1978, when the state first declared a health emergency, described unusual numbers of miscarriages and birth defects, as well as reduced infant weight. The health effects had peaked in the neighborhood about twenty years earlier, the report said, within a decade of most of the dumping. The journal *Science* rejected the report for publication, however, saying it was statistically unsound.[54]

Throughout the summer, as experts debated the technical merits of the health studies, the press perpetuated the causal story favored by Love Canal residents, portraying them as patriotic, working-class victims of corporate greed who had been abandoned by the government. "I can think of three or four [men] right off hand who say they'll never serve the country again for the simple reason that when they needed us, we were there," said one man. "Now they're turning their backs on us. It kind of makes you feel bad."[55] *The New York Times* ran stories headlined "For One Love Canal Family, the Road to Despair" and "Love Canal Families Are Left with a Legacy of Pain and Anger." Journalists reported that, along with a sense of betrayal, abandonment, and isolation, many in Love Canal were feeling enormous uncertainty and a loss of control over their lives. "They realize that this loss of control stems from long-ago decisions to bury chemicals and then to build homes near that spot, not from decisions they made," said Adeline Levine. "Now control rests in large measure on the decisions of distant political figures."[56]

Touching on issues sure to elicit public sympathy, reporters noted that the disaster had particularly acute consequences for many of the area's children. They interviewed one young pediatrician who had helped to bring about the closing of the 93rd Street School because he had noticed in the mid-1970s that asthma and other respiratory diseases seemed to occur more frequently in Love Canal children than in his other patients. As the crisis unfolded, the children he saw began to manifest psychological problems, which he believed grew out of their intractable illnesses as well as the emotional hothouses in which they lived. "A lot of these kids are really upset that things might grow from their bodies, or that they might die prematurely," Dr. James Dunlop explained.[57]

The Final Evacuation

The media portrait ultimately stirred a national political response: in September 1980, despite the ongoing controversy over the health studies, President Carter signed an agreement to lend the state of New York $7.5 million and provide a grant of another $7.5 million to purchase the homes of the relocated residents. To repay the loan, the newly formed Love Canal Area Revitalization Agency planned to rehabilitate the abandoned area and resell the homes. On October 2 the president signed the bill that made federal support for the Love Canal evacuation and renewal a reality.

Ironically, a week later a five-member panel of scientists that Governor Carey had appointed in June told the press that "inadequate" scientific studies might have exaggerated the seriousness of the health problems caused by toxic wastes. Of the Biogenics study the panel said, "The design, implementation, and release of the EPA chromosome study has not only damaged the credibility of science, but exacerbated any future attempts to determine whether and to what degree the health of the Love Canal residents has been affected." The panel described Paigen's nervous system study as "literally impossible to interpret.... [It] cannot be taken seriously as a piece of sound epidemiological research." After reviewing government and private research from the previous two years, the scientists concluded, "There has been no demonstration of acute health effects linked to exposure to hazardous wastes at the Love Canal site." But the panel added that "chronic effects of hazardous-waste exposure at Love Canal have neither been established nor ruled out as yet in a scientifically rigorous manner."[58]

Residents reacted bitterly to these findings because they feared that the report might allow the government to back out of its agreement to buy area homes. But the panel's eminent chair, Dr. Lewis Thomas of the Memorial Sloan Kettering Cancer Center, responded to residents' denunciations, saying that although claims of health effects were insupportable on scientific grounds, he believed that the anguish caused by the presence of chemicals and the possibility of future findings were reason enough not to live in the area.[59] In the end, the federal government did not renege on the buyout.

OUTCOMES

The crisis at Love Canal affected not only residents of the area but national politics as well, because the episode opened a window for new federal policy. As Daniel Mazmanian and David Morell write, "Love Canal was one of those rare catalytic events, one of those seemingly isolated incidents that so often throws an entire nation into turmoil. This unpredictable turning point in the late 1970s bared both the sensitive chemophobic nerve lurking just below the surface of American public consciousness and the growing lack of trust in both business and government expertise."[60] After Love Canal hit the headlines, the media—abetted by EPA researchers—expanded the scope of the conflict even further by covering similar horror stories across the nation. For example, although it received less publicity than Love Canal, the Valley of the Drums was a 23-acre abandoned industrial waste dump in Bullitt County, Kentucky. In 1979, the EPA documented thousands of deteriorating and leaking drums containing a variety of hazardous substances that were leaking into nearby Wilson Creek, a tributary of the Ohio River.

The Superfund Law

Members of Congress, sensitive to the furor caused by Love Canal, responded quickly with an ambitious new law that established a system for identifying and neutralizing hazardous waste dumpsites. In early December 1980, Congress passed the Comprehensive Environmental Response, Compensation, and Liability Act (CERCLA)—commonly known as the Superfund Act—and on December 11, President Carter signed the bill into law. The EPA had been pressing Congress for a hazardous waste cleanup law for some time, but the specter of an incoming Republican president—and hence the closing of the policy window opened by the Love Canal episode—prompted congressional Democrats to scale back their proposals sufficiently to garner majority support.[61]

CERCLA's structure clearly reflects the impact of Love Canal in defining the problem of abandoned hazardous waste dumps. To avoid the kind of delays experienced at Love Canal, the act authorized the EPA to respond to hazardous substance emergencies and clean up leaking chemical dumpsites if the responsible parties failed to take appropriate action or could not be located. To speed up the process, CERCLA established a $1.6 billion trust fund, the Superfund, financed primarily by a tax on petrochemical feedstocks, organic chemicals, and crude oil imports. (The legislation authorizing the trust fund taxes expired in 1995; although revenues continued to accrue for some years, since 2004 the program has relied solely on appropriations from general revenues.) Congress also responded to the issues of corporate responsibility that were so painfully obvious at Love Canal by instituting retroactive, strict, joint, and several liability.[62] Strict, retroactive liability means that any company that disposed of hazardous waste, even if it did so legally and prior to CERCLA's passage, is automatically liable for the costs of cleanup. Joint and several liability means that one party may be held responsible for the entire cleanup cost, even if it dumped only some of the waste. Such an approach transfers the burden of proof from the victims of toxic pollution to the perpetrators; it also creates a powerful incentive for companies to dispose of waste in a precautionary way or to reduce the amount they generate out of fear of future liability.

Congress passed CERCLA by an overwhelming majority. Although the members of Congress had no real idea of the scope of the problem, they were clearly moved by the anecdotal evidence of human health effects from exposure to toxic waste and, more important, by public fears of such effects.[63] In establishing the Superfund, Congress ignored the wishes of the influential Chemical Manufacturers Association, which had opposed any cleanup legislation. Moreover, the chemical industry failed to get on the bandwagon as passage of the bill became imminent and missed an opportunity to work with Congress and perhaps weaken the law, which is one reason its provisions are so punitive.[64]

Legislators' near unanimity in passing CERCLA reflects not only heightened public attention combined with the industry's failure to cooperate but also the widespread perception that the nation's hazardous waste problem was manageable: initial EPA studies had concluded that between 1,000 and 2,000 sites needed remediation at an estimated total cost of $3.6 million to $4.4 million. By the mid-1980s, however, it was obvious that the EPA had grossly underestimated the number of sites requiring cleanup. By 1986 the agency had identified more than 27,000 abandoned hazardous waste sites across the nation and had assigned almost 1,000 of the most dangerous to its National Priority List. In 1986 Congress passed the Superfund Amendments and Reauthorization Act, which increased funding and tightened cleanup standards. The EPA added about twenty sites annually between 1988 and 2007, fewer than expected; on the other hand, the sites that were added promised to be expensive to clean up, so program expenditures and appropriations—which regularly exceeded $1 billion per year—were declining, even as cleanup costs were increasing. Making matters worse, in 1995 Congress let the taxes on chemical feedstocks expire. As a result, by 2003 the Superfund was drained, and thereafter cleanup was funded annually out of general revenues. By 2015, 386 sites had been deleted from the priority list, while construction had been completed on 1,166 of the 1,321 total sites.[65]

Remediation and Resettlement of Love Canal

With the passage of CERCLA, public and legislative attention turned to other matters, but the cleanup of Love Canal quietly moved forward. By July 1989 a total of 1,030 families had left the area. Residents in all but 2 of the original 239 homes and all but 72 of the 564 outer-ring families chose to move, as did most of the renters in the nearby housing project. Contractors razed the 99th Street Elementary School and the neat rows of houses adjacent to the six-block-long canal and then targeted their remediation efforts at containing the rubble in the forty-acre Love Canal landfill. In addition to creating the landfill, cleanup of the Emergency Declaration Area included decontaminating the area's storm sewers, dredging 3,000 feet of creek bed contaminated by rainwater runoff, returning to the landfill the 11,000 cubic yards of contaminated soil at the 93rd Street Middle School that was transferred there in the 1960s, and excavating lesser amounts of soil from three "hot spots" created when former landowners stole fill from the canal before the problem was detected.[66]

Despite the massive restoration, the stigmatized Love Canal area remained deserted for more than a decade, after which resettlement proceeded slowly. As of December 1991 only about twenty-five families had moved into the neighborhood, even though the federal government had declared the area safe. The only visible reminder of the dump itself was a pasture, isolated by miles of cyclone fence covered with Day-Glo yellow warning signs reading: "Danger—Hazardous Waste Area—Keep Out." Stretching at exact intervals like distance markers on a driving range were brilliant orange pipes used to vent and monitor the landfill.[67]

More than ten years later, in 2004, the EPA finally took the seventy-acre Love Canal cleanup site off the National Priorities List. The site of the former dump was still fenced off; behind the fence surrounding the site, chemical monitors were visible, and two treatment plants hummed. But the EPA reported that neighborhoods to the north and west had been revitalized, with more than 200 boarded-up homes renovated and sold at below-market prices to new owners, as well as ten newly constructed apartment buildings.[68]

In November 2013 the thirty-five-year anniversary of Love Canal prompted a reunion of sorts. Lois Gibbs and others visited the site, renamed Black Creek Village, and bemoaned its resettlement. She noted with dismay that the containment area was fenced in and posted with a sign saying only, "Private Property NO Trespassing."[69] The following spring current and past residents of Niagara Falls filed fifteen new lawsuits (in addition to several that had already been filed) over remediation work at Love Canal.

The plaintiffs claimed that toxic pollutants continued to seep from the dumpsite and were causing birth defects, cancer, and other illnesses. The lawsuits arose out of a 2011 incident in which toxic chemicals allegedly spewed from a city sewer line during the excavation of a fifty-foot section of sanitary pipe. The accident occurred on Colvin Boulevard, near 96th Street, within sight of the fenced-off containment area. The New York Department of Environmental Conservation (DEC) concluded that the incident was isolated, since no other pockets of chemicals were found in any of the other sixteen sites where sewer lines near the Love Canal site were repaired.[70]

Issues surrounding Love Canal continue. In February 2018 a class action lawsuit was filed against the City of Niagara Falls, the Niagara Falls Water Board, and Occidental Petroleum Corporation (the former Hooker Chemical Corporation). The purpose of the suit is to address injuries related to toxic chemicals that are migrating through groundwater or leaking sewage pipes. Moreover, in March 2018, the residents of Wheatfield (where Love Canal waste was stored from 1968 to 2015) wanted damages awarded for residents who lived nearby the toxic dump due to medical concerns similar to those of Love Canal. The outcomes of these cases are pending.[71,72]

CONCLUSIONS

Although it has been cleaned up, Love Canal remains a symbol of public fears about hazardous waste. Ironically, most hazardous waste experts believe those fears are greatly exaggerated. They lay blame for the public's misapprehension of the risks squarely at the feet of the media, which played a critical role in the cycle of activism and political response at Love Canal: through their coverage of scientific studies and residents' ailments, journalists raised local citizens' awareness of and concern about the threats posed by buried chemicals. Once local citizens mobilized, the media's extensive and sympathetic coverage of their complaints, and its framing of the story as a classic David-and-Goliath tale, attracted the sympathy of the national public.

Concerned about the potential liability, as well as costs likely to spiral, local and state officials put off responding to residents' complaints for as long as possible. They hoped instead to address the problem quietly, in anticipation that the furor would die down. Then, once media coverage had raised the visibility of Love Canal, government officials expected that scientific analysis would clarify the dimensions of the problem and suggest an appropriate remedy. Instead, scientific studies—most of which were conducted hastily—only heightened tensions between residents and officials without pinpointing health effects. Ultimately, politicians at both the state and national levels felt compelled to respond—not because the scientific evidence confirmed the problem at Love Canal but because the media had created a strong sense of urgency among residents and the national public.

In hindsight, Governor Carey, Representative LaFalce, and other elected officials acknowledged that, although the first Love Canal evacuation was necessary, the second was probably an overreaction to citizen activism, particularly the "hostage-taking" event, rather than a product of careful evaluation. In the intervening years, some follow-up studies have confirmed the initial findings of adverse health effects at Love Canal, but reputable evaluators have been highly critical of their methodology. Other follow-up studies have shown few effects that can be decisively attributed to buried chemicals. A 1981 study, published in *Science*, concluded that people living near Love Canal had no higher cancer rates than other New York State residents.[73] In late 1982 the EPA and the U.S. Public Health Service released the results of a major assessment, which found that the Love Canal neighborhood was no less safe for residents than any other neighborhood in Niagara Falls. Less than a year later, the federal Centers for Disease Control reported that they too had found Love Canal residents no more likely to suffer chromosomal damage than residents living elsewhere in Niagara Falls.[74] Another follow-up study released by the New York Department of Health in August 2001 failed to find elevated cancer rates among Love Canal residents.[75] A third follow-up study, conducted by the same agency between 1996 and 2005, found that children born at Love Canal were twice as likely as children in other parts of the county to be born with a birth defect. It did not, however, find elevated cancer rates that were statistically significant.[76]

In fact, it is notoriously difficult to prove the existence of residential cancer clusters such as the one suspected at Love Canal. Among hundreds of exhaustive, published investigations of such clusters in the United States through 1999, not one identified an environmental—as opposed to an occupational or medical—cause.[77] Michael Brown, one of the original journalists covering Love Canal for the *Niagara Gazette*, concludes that "perhaps science is simply not up to the task of proving a toxic cause and effect." He points out that "because residents move in and out, because families suffer multiple ailments, . . . because the effects of chemicals when they interact with one another are all but unknown, and because the survey populations are quite limited, attempts to prove a statistically significant effect may be doomed to failure."[78] Although the existence of neighborhood cancer clusters is suspect, however, public horror of them is real. As physician Atul Gawande points out, "Human beings evidently have a deep-seated tendency to see meaning in the ordinary variations that are bound to appear in small samples."[79]

The difficulty of documenting environmental and health effects of chronic exposure to toxic chemicals has rendered common-law liability an ineffectual tool for addressing such problems. As Jonathan Harr's 1995 book *A Civil Action* makes vividly clear, individual citizens who believe they have been harmed by the activities of large, multinational corporations are at a severe disadvantage because legal liability in such cases is difficult to prove.[80] To prevail in a "toxic tort" case, plaintiffs must demonstrate not only that the alleged polluter was responsible for the harmful substance, but also that the substance caused the injuries suffered. Both of these causal linkages can be enormously difficult to prove, and defendants have virtually unlimited resources with which to challenge the individual studies that constitute the building blocks of such arguments.

To avert this problem, the Superfund shifts the burden of proof to polluters by establishing strict liability for hazardous waste dumpsites. The Superfund mechanism also creates incentives for companies to behave in a precautionary way because complying with current disposal rules does not constitute a defense against liability in the future. Critics charge that, in practice, the Superfund throws money at sites that pose negligible risks, unfairly penalizes small businesses that contributed little, if anything, to a site's toxicity, and constitutes little more than a subsidy for lawyers and hazardous waste cleanup firms.[81] In response to these criticisms, throughout the 1990s lawmakers proposed a barrage of legislative reforms (twenty-six in the 102nd Congress, thirty-four in the 103rd, and fifty-seven in the 104th).[82] Congress failed to pass any of these proposals, however, and in 1995 the Clinton administration instituted an administrative reform package that sought to make the distribution of cleanup costs among potentially responsible parties equitable (more in line with their contributions) and reduce the overall cost of remediation by linking cleanup levels to prospective land use at sites that are unlikely to become residential. When President George W. Bush took office in 2001, his EPA administrator, Christine Todd Whitman, expressed no interest in reversing the Clinton-era reforms.[83]

The main problem facing the Superfund program since the early 2000s has been underfunding. Under the George W. Bush administration, funding shortages had several consequences: the EPA put more effort into cost-recovery cases against potentially responsible parties than it had done previously; in response to pressure from state environmental agencies, the agency became more reluctant to list new sites on the National Priority List; and the agency put greater emphasis on short-term remedial actions at abandoned disposal sites. Although Obama's political appointees more effectively enunciated a clear set of enforcement priorities and goals than did squabbling Bush administration officials, they nevertheless faced continuing budgetary constraints. Therefore, they continued to emphasize beneficial reuse of Superfund sites and short-term remedial work. They did, however, place a higher priority on environmental justice concerns than their predecessors.[84]

The Trump administration argues that former EPA administrator Scott Pruitt is strongly committed to clean-up and remediation of superfund sites.[85] Pruitt, and his Superfund Task Force, published a list of sites for immediate attention. "We have made it a priority to get these sites cleaned up faster and in the right way," said Pruitt.[86] However, critics remain skeptical because rushing to remove communities from the National Priorities List could cause long-term consequences.[87]

QUESTIONS TO CONSIDER

- Why did a local story about a toxic waste dump become a national crisis?
- Do you think the government reacted appropriately to the events at Love Canal? Why or why not?
- Ironically, some local officials (and even some residents) resist having nearby abandoned toxic waste sites listed on the National Priority List. Why might that be, and what if anything should the federal government do about it?

NOTES

1. Verlyn Klinkenborg, "Back to Love Canal: Recycled Homes, Rebuilt Dreams," *Harper's Magazine*, March 1991, 72.

2. Paul Slovic, Baruch Fischhoff, and Sarah Lichtenstein, "Rating the Risks," in *Readings in Risk*, ed. Theodore Glickman and Michael Gough (Washington, D.C.: Resources for the Future, 1990), 61–75.

3. Extrapolating from animal tests is difficult because species vary in their responses to chemicals; laboratory animals are often exposed to substances through different routes than are humans in the environment; and testers expose animals to massive doses of a substance. Epidemiological studies involve human subjects and larger samples, but it is notoriously difficult to isolate the effects of a single chemical. Moreover, such studies rely on statistical methods that may fail to detect the kinds of increases in lifetime cancer risks that regulators are concerned about. Finally, cellular analyses do not always reveal a substance's hazardous properties, nor do they quantify the low-dose risk of chemical carcinogens. For a detailed explanation of the shortcomings of these methods, see John D. Graham, Laura C. Green, and Marc J. Roberts, *In Search of Safety: Chemicals and Cancer Risk* (Cambridge, Mass.: Harvard University Press, 1988).

4. Mark E. Rushefsky, *Making Cancer Policy* (Albany: State University of New York Press, 1986).

5. Phil Brown, "Popular Epidemiology and Toxic Waste Contamination: Lay and Professional Ways of Knowing," *Journal of Health and Social Behavior* 33 (September 1992): 267–281.

6. Frank Fischer, *Citizens, Experts, and the Environment* (Durham, N.C.: Duke University Press, 2000).

7. Paul Peterson, *City Limits* (Chicago: University of Chicago Press, 1981); Kee Warner and Harvey Molotch, *Building Rules: How Local Controls Shape Community Environments and Economies* (Boulder, Colo.: Westview Press, 2000).

8. Matthew Crenson, *The Un-Politics of Air Pollution: A Study of Non-Decisionmaking in the Cities* (Baltimore: Johns Hopkins University Press, 1971).

9. Quoted in Daniel Goleman, "Hidden Rules Often Distort Rules of Risk," *The New York Times*, February 1, 1994, C1.

10. Martin Linsky, "Shrinking the Policy Process: The Press and the 1980 Love Canal Relocation," in *Impact: How the Press Affects Federal Policymaking*, ed. Martin Linsky (New York: Norton, 1986), 218–253.

11. David L. Protess et al., "The Impact of Investigative Reporting on Public Opinion and Policy Making," in *Media Power in Politics*, 3d ed., ed. Doris A. Graber (Washington, D.C.: CQ Press, 1994), 346–359.

12. David Pritchard, "The News Media and Public Policy Agendas," in *Public Opinion, the Press, and Public Policy*, ed. J. David Kennamer (Westport, Conn.: Praeger, 1992), 103–112.

13. The main components of the buried wastes were benzene hexachloride (a byproduct of the pesticide lindane), chlorobenzene, dodecyl mercaptan, sulfides, benzyl chloride, and benzoyl chloride. See Allan Mazur, *A Hazardous Inquiry: The Rashomon Effect at Love Canal* (Cambridge, Mass.: Harvard University Press, 1998).

14. Quoted in Adeline Levine, *Love Canal: Science, Politics, and Public Policy* (Lexington, Mass.: Lexington Books, 1982), 11.

15. Michael Brown, *Laying Waste: The Poisoning of America by Toxic Chemicals* (New York: Washington Square Press, 1981).

16. Eric Zeusse, "Love Canal: The Truth Seeps Out," *Reason*, February 1981, 16–33.

17. Levine, *Love Canal*, 11.

18. Linsky, "Shrinking the Policy Process."

19. Andrew Danzo, "The Big Sleazy: Love Canal Ten Years Later," *Washington Monthly*, September 1988, 11–17.

20. Steven R. Weisman, "Hooker Company Knew About Toxic Peril in 1958," *The New York Times*, April 11, 1979, B1.

21. Levine, *Love Canal*.

22. Donald McNeil, "Upstate Waste Site May Endanger Lives," *The New York Times*, August 2, 1978, A1.

23. Mirex was used in the South to control ants and as a flame retardant and plasticizer until the Food and Drug Administration restricted its use.

24. PCBs, which are used to insulate electronic components, are known to kill even microscopic plants and animals.

25. Quoted in Levine, *Love Canal*, 19.

26. Mazur, *A Hazardous Inquiry*.

27. Brown, *Laying Waste*.

28. Lois Marie Gibbs, *Love Canal: My Story* (Albany: State University of New York Press, 1982), 15–16.

29. There was a brief struggle for control of the LCHA during the summer of 1978 between inner-ring resident Thomas Heisner and Gibbs. Heisner articulated residents' economic concerns, while Gibbs's rhetoric focused more on health. Gibbs's views were also more expansive and inclusive than Heisner's; in particular, Gibbs sought to represent *all* residents, including renters. By the end of the summer, Gibbs had prevailed, and Heisner left to form a splinter group. See Elizabeth D. Blum, *Love Canal Revisited* (Lawrence: University of Kansas Press, 2008).

30. Quoted in Donald McNeil, "Upstate Waste Site: Carey Seeks U.S. Aid," *The New York Times*, August 4, 1978, B14.

31. McNeil, "Upstate Waste Site May Endanger Lives."

32. Brown, *Laying Waste.*

33. Mazur, *A Hazardous Inquiry.*

34. Brown, *Laying Waste;* McNeil, "Upstate Waste Site May Endanger Lives."

35. Donald McNeil, "Health Chief Calls Waste Site a Peril," *The New York Times,* August 3, 1978, A1.

36. Quoted in ibid.

37. Donald McNeil, "First Families Leaving Upstate Contamination Site," *The New York Times,* August 5, 1978, 1.

38. Under an emergency declaration, the government can pay only for temporary relocation to save lives, protect public health, and protect property. A disaster declaration, by contrast, is intended to help a community recover after events such as floods or earthquakes in which it is necessary to rebuild houses, schools, highways, and sewer systems.

39. Gibbs, *Love Canal,* 81.

40. Gibbs's story eclipsed not only that of Hooker and local officials, but also that of the African American residents of Griffon Manor, a federal housing project located across the street from Love Canal. Historian Elizabeth Blum argues that significant racial and class tensions pervaded residents' struggle and that, although it died down over time, that tension never wholly disappeared. Blum also notes that Gibbs's tactics were more effective than those employed by the Ecumenical Task Force (ETF), a group of relatively well-educated, middle-class activists who became involved in the Love Canal struggle. Whereas the LCHA relied heavily on confrontational tactics, such as 1960s-era protests and picketing, the ETF preferred a more conciliatory approach. See Blum, *Love Canal Revisited.*

41. Donald McNeil, "3 Chemical Sites Near Love Canal Possible Hazard," *The New York Times,* December 27, 1978, B1.

42. Gibbs, *Love Canal.*

43. Ibid., 59.

44. "Love Canal Families Unwilling to Go Home Facing Motel Eviction," *The New York Times,* November 8, 1979, B4.

45. Human beings have forty-six chromosomes in every cell. As new cells grow and reproduce by division, newly formed chromosomes and their genes are normally exact, complete replications of the originals. Ordinarily, environmental changes such as variations in temperature or barometric pressure, diet, or muscular activity have no effect on the process; moreover, a low level of mutation occurs spontaneously and is difficult to attribute to any specific cause. However, contact with radiation, chemicals, and other environmental hazards may cause abnormal changes in the structure of chromosomes. In some such cases, chromosome material may be missing; more rarely, additional material is detected.

46. It is important to note that everyone has some chromosome damage—probably from viral infections, medical X-rays, or exposure to chemicals and medication. Although

chromosome damage is an important test of exposure to toxic chemicals, not everyone with chromosome damage suffers ill effects. See Gina Kolata, "Chromosome Damage: What It Is, What It Means," *Science* 208 (June 13, 1980): 1240.

47. Josh Barbanel, "Homeowners at Love Canal Hold 2 Officials Until F.B.I. Intervenes," *The New York Times*, May 20, 1980, A1; Josh Barbanel, "Peaceful Vigil Resumed at Love Canal," *The New York Times*, May 21, 1980, B1.

48. Dr. Paigen gave the following explanation: toxic chemicals, two of which—chloroform and trichloroethylene—have traditionally been used as operating room anesthetics, can act on the nervous system in two ways. They can attack the nerve fibers themselves or the fattier myeline sheath that encases the nerve fibers. "Chemicals which are soluble in fat, as the Love Canal chemicals tend to be, tend to concentrate in fatty tissues such as the nervous sheath," she said. Dr. Paigen said the damage showed up in two of the peripheral nervous systems tested. "These are the sensory nerves which control touch and feelings," she said. But she added that the findings probably meant that those who showed peripheral nervous system damage had also suffered some damage to the central nervous system. See Dudley L. Clendinen, "New Study Finds Residents Suffer Nerve Problems," *The New York Times*, May 21, 1980, B7.

49. Quoted in Irving Molotsky, "President Orders Emergency Help for Love Canal," *The New York Times*, May 22, 1980, A1.

50. Quoted in Josh Barbanel, "Many at Love Canal Insist on U.S. Aid Before Moving," *The New York Times*, May 27, 1980, B3.

51. Quoted in John Noble Wilford, "Panel Disputes Chromosome Findings at Love Canal," *The New York Times*, June 14, 1980, 26.

52. Ibid.

53. Quoted in Irving Molotsky, "Love Canal Medical Study Backed," *The New York Times*, June 18, 1980, B4. Although Killian had once supervised Picciano, he dismissed suggestions he might be biased, pointing out that he had no role in the Biogenics study and had participated in the evaluation of it at the request of the EPA.

54. Robin Herman, "Report Cites Miscarriage Rate at Love Canal in 60's," *The New York Times*, June 24, 1980, D16.

55. Quoted in Georgia Dullea, "Love Canal Families Are Left with a Legacy of Pain and Anger," *The New York Times*, May 16, 1980, A18.

56. Quoted in Constance Holden, "Love Canal Residents Under Stress," *Science* 208 (June 13, 1980): 1242–1244.

57. Quoted in Dudley L. Clendinen, "Love Canal Is Extra Tough on Children," *The New York Times*, June 9, 1980, B1.

58. Quoted in Richard J. Meislin, "Carey Panel Discounts 2 Studies of Love Canal Health Problems," *The New York Times*, October 11, 1980, 25.

59. Josh Barbanel, "Love Canal Skeptic Favors Relocation," *The New York Times*, October 12, 1980, 39.

60. Daniel Mazmanian and David Morell, *Beyond Superfailure: America's Toxics Policy for the 1990s* (Boulder, Colo.: Westview Press, 1992), 6.

61. John A. Hird, *Superfund: The Political Economy of Environmental Risk* (Baltimore: Johns Hopkins University Press, 1994).

62. Congress actually deleted the strict, joint, and several liability provision from the law immediately before its passage, but the courts subsequently reinstated the requirement through their interpretation of the statute. See Mazmanian and Morell, *Beyond Superfailure*.

63. Political scientist John Hird writes, "That the prospect of an incoming conservative Republican administration did not doom the Superfund effort entirely in 1980 is a testimony to the political appeal for both Republicans and Democrats of a hazardous waste cleanup program seen as vital by their constituents." See Hird, *Superfund*, 186.

64. Lawrence Mosher, "Environment," *National Journal*, December 30, 1980, 2130.

65. U.S. Environmental Protection Agency, "NPL Site Totals by Status and Milestone," February 9, 2015. Available at http://www.epa.gov/superfund/sites/query/queryhtm/npltotal.htm.

66. Andrew Hoffman, "An Uneasy Rebirth at Love Canal," *Environment*, March 1995, 5–9, 25–30.

67. Ibid.

68. U.S. Environmental Protection Agency, "EPA Removes Love Canal from Superfund List," Press Release, September 30, 2004, #04152. Available at http://www.epa.gov/superfund/accomp/news/lovecanal.htm.

69. Timothy Chipp, "Fight Rages On as Lois Gibbs Returns to Love Canal," *Niagara Gazette*, October 22, 2013.

70. Thomas Prohaska, "15 New Lawsuits Filed Over Love Canal Work," *The Buffalo News*, February 14, 2014; Marlene Kennedy, "Years Later, Love Canal Stew Bubbles Up," *Courthouse News Service*, February 24, 2014.

71. "Napoli Shkolnik PLLC Files a Class Action Lawsuit for Love Canal, New York," February 16, 2018. Available at http://www.marketwired.com/press-release/napoli-shkolnik-pllc-files-a-class-action-lawsuit-for-love-canal-new-york-2245820.htm.

72. Tevlock, Dan. "Wheatfield Landfill Subject of Lawsuit," March 27, 2017. Available at www.investigativepost.org/2017/03/27/wheatfield-landfill-subject-lawsuit/.

73. Dwight T. Janerich et al., "Cancer Incidence in the Love Canal Area," *Science* 212 (June 19, 1981): 1404–1407.

74. Clark W. Heath Jr. et al., "Cytogenetic Findings in Persons Living Near the Love Canal," *Journal of the American Medical Association* 251 (March 16, 1984): 1437–1440.

75. New York State Department of Health, "Love Canal Follow-Up Health Study," August 2001. Available at http://www.health.state.ny.us/nysdoh/lcanal/cancinci.htm.

76. Erika Engelhaupt, "Happy Birthday, Love Canal," *Chemical & Engineering News* 86 (November 17, 2008): 46–53.

77. Atul Gawande, "The Cancer-Cluster Myth," *The New Yorker*, February 8, 1999, 34–37.

78. Michael Brown, "A Toxic Ghost Town," *Atlantic Monthly*, July 1989, 23–28.

79. Gawande, "The Cancer-Cluster Myth."

80. Jonathan Harr, *A Civil Action* (New York: Vintage, 1995).

81. See, for example, Marc K. Landy and Mary Hague, "The Coalition for Waste: Private Interests and Superfund," in *Environmental Politics: Public Costs, Private Rewards*, ed. Michael S. Greve and Fred L. Smith (New York: Praeger, 1992), 67–87.

82. Dianne Rahm, "Controversial Cleanup: Superfund and the Implementation of U.S. Hazardous Waste Policy," *Policy Studies Journal* 26 (1998): 719–734.

83. Robert T. Nakamura and Thomas W. Church, *Taming Regulation* (Washington, D.C.: Brookings Institution Press, 2003).

84. Joel A. Mintz, "EPA Enforcement of CERCLA: Historical Overview and Recent Trends," *Southwestern Law Review* 41 (2012): 645–659.

85. Ibid.

86. EPA Press Office, "EPA Making Strides in Cleaning Up the Nation's Most Contaminated Sites," *EPA News Releases*, January 1, 2018.

87. Bradley Dennis and Juliet Eilperin, "At Superfund Sites, Scott Pruitt Could Flip His Industry-Friendly Script," *The Washington Post*, January 23, 2018.

CHAPTER FOUR

ECOSYSTEM-BASED MANAGEMENT IN THE CHESAPEAKE BAY

The cleanup of the Chesapeake Bay is a massive, thirty-year-long effort, involving six states, the District of Columbia, and the Environmental Protection Agency. The goal has been to reduce pollution levels. However, in 2010, efforts to reduce pollution of the bay had fallen more than 40 percent short of their goals. Rather than concede defeat, state and federal officials pledged to redouble their efforts and set targets for 2025. Some observers remain skeptical, given the growing challenges facing the bay and the lack of progress to date. However, recent longitudinal data suggest sustained attention to the health of the bay has substantially improved the aquatic ecosystem.[1]

Although it is the largest of the nation's 850 estuaries, the Chesapeake is not alone in being sorely degraded. Estuaries, the areas where rivers meet the sea and freshwater and saltwater mix, are essential to the vitality of marine ecosystems. But a host of problems plague the world's great estuaries. Overharvesting has decimated fish, shrimp, and shellfish populations, dramatically altering the food web; fishing with trawls that scour the ocean bottom has left many coastal areas barren (see Chapter 10). Nutrient-laden runoff from farms and septic tanks, as well as wastewater from sewage treatment plants, has caused rampant growth of algae, which prevents sunlight from reaching essential sea grasses; in addition, when algae die and decompose, they consume the oxygen needed by fish and other aquatic species.[2] Coastal development has destroyed wetlands that serve as nurseries for fish; poured silt into the water, where it smothers plant and animal life; and paved over forests and wetlands that once filtered rainfall before it ran off into local waterways. Air pollutants have fallen directly onto bays, as well as onto land, where they run off into waterways—often mobilizing heavy metals along the way. Upstream, dams inhibit the movement of migratory species. Dredging, widening, and straightening channels for navigation disturb sedimentation and water circulation patterns.

As a result of these influences, the condition of estuaries around the world declined markedly during most of the twentieth century. According to the Millennium Ecosystem Assessment, temperate estuaries experienced major damage prior to 1980 from agricultural and industrial development.[3] Although some began to recover as the millennium drew to a close, as of 2018, many major U.S. estuaries remain degraded. They showed symptoms of nutrient overloading, toxic contamination from industrial development, habitat loss and degradation, proliferation of invasive species, and alterations in natural flow regimes. The most disturbing symptom is dead zones, large areas of low oxygen concentrations in which most marine life dies.

Like many estuaries, the Chesapeake faces a welter of problems, but it also benefits from the stewardship of the Chesapeake Bay Program. Established in 1983, the bay program is one of the most venerable instances of ecosystem-based management (EBM), a widely touted "third-generation" approach to addressing environmental problems. EBM entails devising landscape-scale plans in collaboration with stakeholders. It also involves adaptive management: designing interventions with the intent of learning about the system's response and then adjusting management practices in response to new information. In theory, EBM can compensate for many of the failures often attributed to traditional regulatory approaches. Because problems are addressed at a landscape (or watershed) scale, remedies are likely to be both more comprehensive than piecemeal or local solutions and more tailored than uniform national regulations. Landscape-scale planning can also facilitate cooperation among multiple agencies and jurisdictions. Proponents of EBM believe that a collaborative, rather than a top-down, regulatory approach to planning will yield a solution that is a result of buy-in among stakeholders. And, according to its adherents, flexible, adaptive management should result in practices that are responsive to new information and go beyond the minimum required by law.[4]

The Chesapeake Bay case sheds light on how the comprehensive, flexible, and collaborative approach embodied by EBM works in practice. Historically, the bay program has operated as a multi state cooperative, with the EPA exercising a supporting role and the bay states taking different paths, depending on their political culture and proximity to the bay. Many policy scholars believe such flexibility is desirable because it allows each state to devise an approach that is most likely to work in its particular context. History suggests, however, that state-level politicians often resist instituting stringent regulations, fearing that industry or developers will move to neighboring states or that powerful local interests might fund challengers in the next election.

The Chesapeake Bay case also illuminates the utility of nonregulatory approaches in addressing one of the most vexing problems facing the bay: nonpoint-source water pollution.[5] Under the Clean Water Act, the United States has made progress in controlling pollution from point sources, such as sewage treatment plants and factories. But for decades, the Clean Water Act offered little leverage to address polluted runoff from farms or urban and suburban development—two of the major problems facing the Chesapeake. For such diffuse sources, theorists have argued that rules and prescriptions are impractical and that nonregulatory approaches—such as inducements, information, and voluntary measures—are likely to yield more effective and enduring results.[6] Support for nonregulatory approaches has increased dramatically as environmental politics has become more polarized.

Whereas some incentive-based programs have produced environmentally beneficial results (see Chapter 5), there is little evidence to suggest that voluntary (or self-regulatory) programs—which contain no binding performance standards, independent certification, or sanctions—are effective.

Finally, the Chesapeake Bay case enhances our understanding of the role of science in environmental policymaking, particularly the impact of comprehensive assessments of ecosystem functioning, which are a critical component of EBM. Scientists have faced obstacles in trying to gauge not just the impact of individual programs, but also overall progress. The Chesapeake Bay Program features one of the world's most sophisticated hydrologic models. But both the model and the way it has been used have proven controversial: regulators have relied on model results, rather than on monitoring data, to determine the bay's condition, often overstating program achievements as a result. Moreover, money spent on trying to perfect scientists' understanding of the bay is not available for implementation. Increasingly, bay advocates are questioning whether additional study is more important than investing in practices that will bring about change. In short, the role of science in the Chesapeake Bay cleanup, as in other instances of EBM, is far more ambiguous and controversial than a rational model of policymaking would suggest.

BACKGROUND

The 64,000-square-mile Chesapeake Bay watershed stretches from southern New York State to Virginia (see Map 4.1). Beyond its scenic beauty—its 11,000 miles of shoreline feature myriad picturesque inlets and coves—the Chesapeake is among the most productive estuaries in the world. Although it evolved over millions of years, by 2000 B.C., the bay had reached its current configuration: roughly 190 miles long, 30 miles across at its widest point, and an average of 21 feet deep when all of its tributaries are considered. Because it is so shallow—much of it is only 6 feet deep or less—sunlight once nourished the 600,000 acres of underwater grasses that, in turn, provided spawning grounds for the bay's abundant fish and crabs. The grasses also received modest but routine injections of nutrients from its shoreline, as well as from its 150 miles of feeder rivers and streams; those nutrients were then extensively recycled and reused within the estuary. Incredibly, a massive oyster population filtered a volume of water equal to the entire bay in a matter of days.[7]

When the first explorers from Spain and France reached the bay in the 1500s, the Native American population—which stood at about 24,000—had established agricultural villages but done little to deplete the region's natural bounty. In 1607, Captain John Smith established a settlement in Jamestown, Virginia; after circumnavigating the bay the following year, he remarked that "Heaven and earth never agreed better to frame a place for man's habitations."[8] Subsequent settlers concurred, and over the next two centuries, the population of the watershed grew rapidly, as colonists mined the region's abundant forests and fisheries. In 1800, the

Map 4.1 Chesapeake Bay Watershed

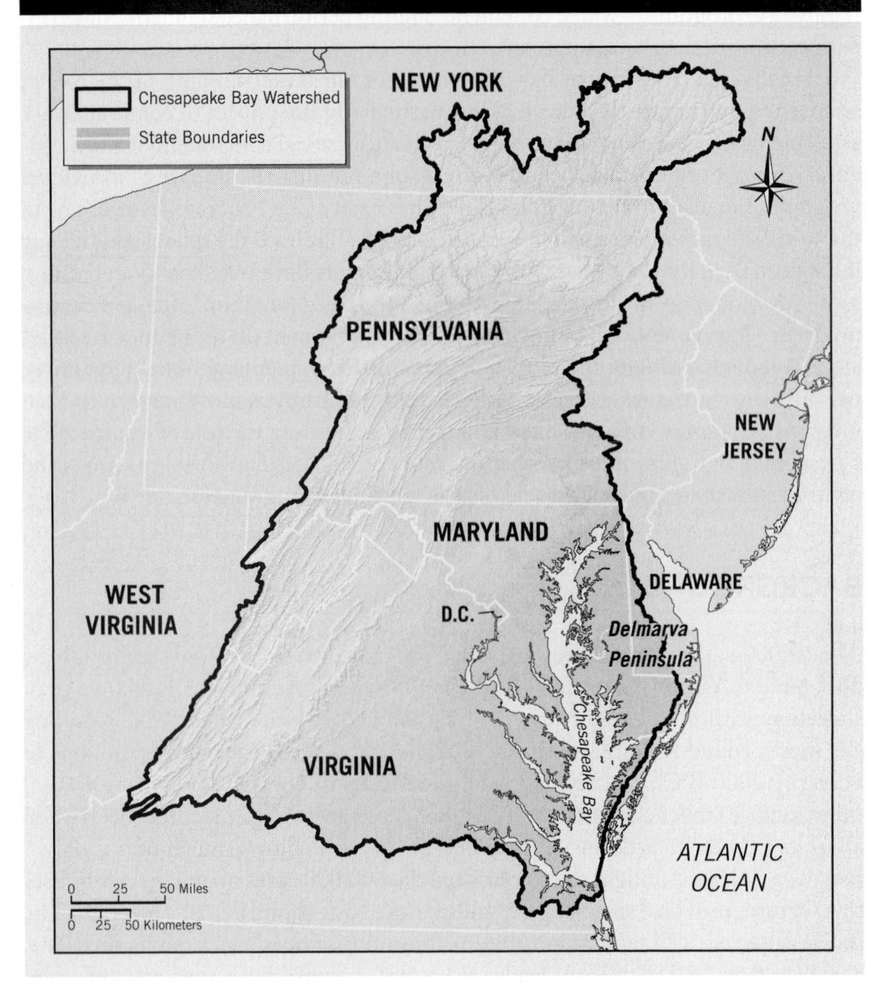

Source: Adapted from Chesapeake Bay Program.

population of the watershed reached 1 million, and by the mid-1800s, half of its forests had been cleared for fuel. At the turn of the twentieth century, the region's population exceeded 3 million, industrial areas were fouling the bay with raw sewage, and even more of the forested area had been harvested for railroad ties. Despite the many blows associated with the rapid settlement of its watershed by Europeans, the bay appeared to be resilient well into the 1950s.

In the 1960s and 1970s, however, the impact of accelerating development, changes in farming practices, and the construction of several large dams began to manifest themselves. Between 1950 and 1980, the population of the bay's six-state drainage basin increased 50 percent, from 8.3 million to 12.8 million. During the same period, the interstate highway system expanded dramatically, and the personal automobile became the dominant transportation mode, creating air pollution and setting the stage for urban sprawl. The amount of land used for residential and commercial purposes nearly tripled, going from 1.5 million acres in 1950 to 4.2 million acres in 1980. The Chesapeake Bay Bridge, which opened Maryland's Eastern Shore to tourists in 1952, sparked the development of marinas and vacation homes. More than 2,800 acres of wetlands around the bay were lost to development during this period.[9]

A group of bay-area businesspeople grew alarmed by these trends, and in 1967 they formed the Chesapeake Bay Foundation, with the goal of raising public awareness of the bay and its declining resources.[10] But it was not until 1972, when Congress passed two laws—the Clean Water Act and the Coastal Zone Management Act (CZMA)—that defenders of the bay gained legal leverage, thanks to a new regulatory framework for controlling pollution of estuaries. The Clean Water Act prohibited dumping of wastes into navigable waters by industry or government without a permit and authorized federal grants for building sewage treatment plants (see Chapter 2). The CZMA provided federal funds for states to develop and implement management plans that would protect, restore, and enhance coastal areas. Although these laws furnished the region's environmentalists with new advocacy tools, the Chesapeake Bay Foundation opted for a more conciliatory approach. Similarly, the Alliance for the Chesapeake Bay—formed in 1972—adopted a nonconfrontational tone in its environmental activism. Like the foundation, the alliance sought to save the bay "through collaboration, partnerships, and consensus," rather than through advocacy and litigation.[11]

Several events in the 1970s and 1980s bolstered environmentalists' efforts to draw attention to the bay's plight. In June 1972 tropical storm Agnes struck, causing record flooding. After the storm, the underwater grasses that form the base of the bay's food chain experienced a precipitous decline. Agnes served as a powerful focusing event. In fact, as political scientist Howard Ernst explains, "Some have argued that the hurricane led to a paradigm shift within the scientific community and later the general public—causing people to abandon the prevailing view of the bay as an extension of the sea and instead to see the Bay as a distinct ecosystem that is dominated by the influence of its watershed."[12] The year after the storm struck, a study by the Army Corps of Engineers (Corps) identified nutrients—particularly nitrogen and phosphorus—as the primary culprits in the bay's degradation. That report, in turn, prompted Sen. Charles Mathias, D-Md., to request (and obtain) $25 million in funding for the EPA to conduct a five-year study of the bay's health that would quantify the impacts of nitrogen and phosphorus and identify their main sources.

The release of the EPA report, "Chesapeake Bay: A Framework for Action," in the fall of 1983, constituted another major focusing event for the bay. The report highlighted several problems that warranted policymakers' attention, but the main concern was that excessive amounts of nitrogen and phosphorous were linked to the growth of algae, whose proliferation led to oxygen depletion and a marked decline in the extent of underwater grasses. Scientists estimated that the amount of water in the bay's channel with little or no dissolved oxygen content (the dead zone) in the summer had increased tenfold since 1950; much of the water deeper than forty feet in the lower bay was anoxic from mid-May through September.[13] The report also identified the three main drivers of the bay's decline: poor agricultural practices, inadequate treatment of sewage, and runoff from urban and suburban development. And it highlighted the consequences of the bay's deteriorating water quality: a precipitous drop in blue crabs, oysters, striped bass, and shad. (Overharvesting these animals also contributed to their decline.) Almost as important as its individual findings was the report's overall holism. Prior to this study, scientists around the bay had studied individual portions of it, believing that problems were localized. By integrating hundreds of individual studies, the EPA report provided a systemic view of the bay and made manifest the relationships among various drivers and different parts of the watershed.[14]

At the urging of Maryland governor Harry Hughes, the EPA report was unveiled in December 1983, at a regional summit involving the states of Maryland and Virginia, the District of Columbia (D.C.), and the EPA. Led by Governor Hughes, the parties decided to form the Chesapeake Bay Program, a voluntary partnership.[15] They signed a three-paragraph Chesapeake Bay Agreement that recognized the historical decline in the bay's living resources and pledged to work cooperatively on reducing the nutrients entering the bay. Working with the bay program was the Chesapeake Bay Commission, a tristate legislative body that had formed in 1980 to advise the general assemblies of Maryland and Virginia and the U.S. Congress on environmental matters related to the bay.

From the outset the assumption was that Maryland, the state with the most direct experience of the bay's problems, would take the lead in the cleanup. And, in fact, Governor Hughes stated that the bay was his top priority and that he anticipated spending $15 million in operating costs and $20 million in capital expenditures on bay restoration in 1984. Virginia, by contrast, expected to spend $6 million in year one, and Pennsylvania projected spending only $1 million.[16] (In 1984 Hughes went to his General Assembly with the Green Book, which contained fourteen points, a $22 million budget, and ten pieces of bay-related legislation.[17]) The EPA signed on to provide only minimal help; in fact, under pressure from the White House Office of Management and Budget, the EPA initially opposed federal legislation to provide $40 million over four years to clean up the bay. EPA Administrator William Ruckelshaus justified the agency's position by saying, "Ultimately it is the citizens of these states—the major beneficiaries of a healthy bay—who must be prepared to assume primary responsibility for protecting their own interests. They must accept a major portion of increased pollution control expenditures."[18]

In early 1984, however, the administration of President Ronald Reagan reversed itself: the president announced in his State of the Union address his intention to "begin the long, necessary effort to clean up a productive recreational area and a special natural resource—the Chesapeake Bay."[19] Motivated by the desire to resuscitate his poor image vis-à-vis the environment prior to the election, Reagan pledged $40 million over four years in the form of matching grants.[20] According to

The Washington Post, the bay cleanup was an obvious choice for this purpose; it was a symbolic project with no obvious detractors, probably because of its proximity to Washington, D.C., and the impact of the widely publicized federal study. Moreover, because of the heavy involvement of the states, the bay program provided an opportunity to tout the administration's "new federalism."[21]

THE CASE

A burst of activity followed the signing of the 1983 agreement and the Reagan administration's financial commitment. First, the bay program itself quickly became a genuinely comprehensive, watershed-scale initiative. In 1984 the EPA, Corps, Fish and Wildlife Service (FWS), U.S. Geological Survey (USGS), and National Oceanic and Atmospheric Administration (NOAA) signed interagency agreements to work together on the bay. Eventually, eleven federal agencies would participate in the bay program. In 1985 Pennsylvania joined the Chesapeake Bay Commission—a crucial step because, although Pennsylvania does not have any Chesapeake waterfront, it comprises a large portion of the watershed for the Susquehanna River, which in turn provides half of the bay's freshwater. At the same time, state and federal officials sought to calm fears that the cleanup would jeopardize states' autonomy. "Cooperation is the key element," Virginia governor Charles Robb emphasized at a press conference in the spring of 1984.[22]

Second, the program began taking concrete steps to facilitate restoration. In 1984 the newly formed bay program created the Chesapeake Bay Water Quality Monitoring Program to track conditions in the bay. In 1985 the Alliance for the Chesapeake Bay initiated a volunteer citizen water-quality monitoring program to complement the government effort. Two years later, the bay program began helping landowners develop plans to reduce nutrients flowing off their land. Although these activities suggested progress, in 1986 a scientific and technical advisory committee issued a report that argued more precise objectives were needed. According to marine ecologist Donald Boesch and his coauthors, the scientific consensus embodied in this report "provided the rationale and credibility" for bold action.[23] In 1987 the bay program partners crafted and signed a new agreement that committed them to 40 percent reductions in the amount of nitrogen and phosphorus entering the bay by 2000. The compact included twenty-nine specific commitments aimed at helping to meet those goals.

The new agreement came in the nick of time: in 1988 Maryland State Senator Bernie Fowler could not wade deeper than ten inches into the Patuxent River and still see the tips of his sneakers. (Fowler recalled that in the 1950s he was able to see his sneakers in water up to his shoulders. The "sneaker index"—the depth at which white sneakers are no longer visible—subsequently became an intuitive measure of the bay's condition, and every second Sunday in June people wade into the Patuxent on Bernie Fowler day.) It soon became apparent, however, that the 40 percent goal politicians had agreed to was not as ambitious as scientists had thought; upon realizing that a large fraction of the nutrients coming into the bay were airborne or runoff from forests, the signatories aimed to reduce only "controllable" nutrients by 40 percent.[24] By excluding airborne nutrients and forest runoff, the bay program set a de facto goal of reducing the total amount of nutrients flowing into the bay annually by about 24 percent for nitrogen (74 million pounds) and 35 percent for phosphorous (8 million pounds).[25] These were far lower targets than what the scientific task force had recommended as a minimum. Moreover, the agreement was entirely voluntary; it contained no penalties or sanctions if the states failed to achieve its objectives.

Despite their modesty, even those goals proved elusive. From 1984 to 2000, the bay program partners spent between $40 million and $50 million a year on bay-related activities.[26] Much of that funding paid for scientific research, particularly on efforts to construct a multimillion-dollar, three-dimensional, hydrologic model of the bay's complex circulatory system. Most of the remainder went toward implementing a host of policies and programs adopted by all three states, D.C., and the EPA in hopes of bringing about improvements. Plainly evident by 2000 was that, although efforts to reduce nutrient pollution (particularly phosphorus) from point sources had been relatively successful, initiatives aimed at nonpoint sources, because they were largely either incentive-based or voluntary, had done little to change the behavior of their targets. Such policies were up against much stronger political and economic incentives to operate in ways that were environmentally destructive. Nevertheless, state officials were reluctant to impose strict mandates because they were loath to antagonize farmers, increase ratepayers' fees, or curb lucrative development.

The 1980s and 1990s: Holding the Line

Between 1983 and 2000, states in the Chesapeake Bay watershed upgraded their sewage treatment plants, urged farmers to adopt best management practices, and enacted local growth-management measures in hopes of stemming the flow of nutrients into the bay. In the meantime, the region's population continued to grow, passing 13 million in 1990 and 15 million eight years later, with most growth occurring in counties adjacent to the bay and its tidal tributaries.[27] Accompanying population growth was a disproportionate increase in both the vehicle miles traveled and the amount of developed land in the watershed. Automobile use, based on miles traveled, rose at four times the rate of population growth between 1970 and 1997.[28] Between 1975 and 1999, 1.12 million acres of agricultural land in the watershed were lost; in the 1990s the rate of land conversion more than doubled over the previous decade.[29] Meanwhile, animal agriculture became far more concentrated, adding to the burden of manure being spread on the region's shrinking cropland acreage. Not surprisingly, these trends overwhelmed efforts to improve the bay's condition.

Modeling, Monitoring, and Understanding the Bay. To provide a scientific foundation for its restoration activities, the Chesapeake Bay Program assembled a scientific and technical advisory committee comprising experts from universities, research institutes, and federal agencies. The program also forged links to a network of well-established researchers throughout the watershed. In particular, members of the Chesapeake Research Consortium, an association of six institutions that had a long-standing research engagement with issues affecting the bay (Johns Hopkins University, the University of Maryland, the Smithsonian Institution, the Virginia Institute of Marine Science, Old Dominion University, and Pennsylvania State University), furnished valuable studies of how the bay works.

Between 1980 and 2000 the network of scientists working in the region generated a remarkable volume of research that deepened the scientific understanding of the bay's functioning. Figuring out the precise causes and consequences of the bay's problems was tricky because of the complexity of the system: the bay regularly experiences multiple circulation patterns and a range of salinity levels, so its behavior is highly variable from one area to the next, from season to season, and from year to year. Adding to the complexity are weather variations: although drought can raise salinity levels in parts of the bay, creating problems for underwater grasses, overall conditions tend to improve in dry years because groundwater and runoff transport fewer nutrients to the bay. In wet years, by contrast, large volumes of sediment, phosphorus, and nitrogen—often in pulses—typically cause conditions to worsen. Despite these challenges, during the 1980s and 1990s scientists succeeded in quantifying the contributions of various sectors to the bay's nutrient load. They estimated that roughly one-quarter of the bay's nutrients came from point sources, primarily the major sewage treatment plants in the watershed—that is, those processing more than 500,000 gallons of wastewater a day. More than one-half of the nitrogen and phosphorus came from nonpoint sources, with agriculture being the main culprit and shoreline development the second-largest contributor. Atmospheric sources were responsible for the remainder.[30]

While scientists worked to quantify the relative contributions of various sectors, engineers built watershed-scale computer models that policymakers could use to both choose policy tools and gauge their effectiveness. In the 1980s the Corps and the EPA (in collaboration with other agencies) developed two models: the Chesapeake Bay Watershed Model, which estimates past and current nutrient loadings, and the Water Quality Model, which estimates the effects of the loads generated by the Watershed Model on bay water quality. Widely considered the world's most sophisticated estuary model, the Water Quality Model simulates the growth of algae, the vertical and horizontal movement of water and nutrients, the amount of oxygen available in different parts of the bay, and numerous other variables.[31] The EPA also deployed a third model, the Regional Acid Deposition Model, to illuminate the extent and impact of airborne nitrogen deposition on the bay region. Together, this suite of models allowed bay program officials to simulate changes in the bay ecosystem as a result of modified agricultural management, land-use practices, sewage treatment, and air pollution controls.

Another scientific activity undertaken during the 1980s was monitoring the actual conditions in the bay. The bay program established the Chesapeake

Bay Monitoring Program, a cooperative effort that involves Maryland, Virginia, Pennsylvania, D.C., several federal agencies, ten research institutions, and more than thirty scientists. Between them, these entities monitored nineteen physical, chemical, and biological characteristics of the bay, collecting data twenty times per year in the bay's main stem and tributaries. Among the features monitored are freshwater inputs, nutrients and sediment, chemical contaminants, plankton and primary production, benthic species, finfish and shellfish, underwater grasses, water temperature, salinity, and dissolved oxygen. In addition to water-quality indicators, bay program partners monitor land-use changes using remote-sensing technology, produce data on fifteen types of land cover, and generate information on wetlands loss.

Despite the breadth of the bay program's scientific and technical apparatus, in the 1990s critics charged there was a chasm between scientists generating basic scientific insights and the engineers developing the models used to make management decisions. From the outset, modeling dominated the bay program. Research scientists pointed out that, although models are only as good as their assumptions, it was difficult for scientists to provide input to the modeling effort.[32] As a result, the models tended to lag behind the scientific understanding; for example, for many years, the engineers (and resource managers) assumed nitrogen was not a key factor influencing the bay's condition, despite scientific studies indicating it was a limiting nutrient.[33]

Critics also charged that adaptive management lagged in the Chesapeake Bay Program. In 1994 political scientist Timothy Hennessey argued that the bay program was engaging in adaptive management because program elements and institutional structures had undergone significant changes in light of new information.[34] But Donald Boesch disagreed, pointing out that the term *adaptive management* refers specifically to learning about how ecosystems respond to intervention from management experiments. And although a lot of money was being spent on monitoring, there had been very little analysis of the data that were being collected—much less effort to adjust models or management practices in light of those data.

Treating Sewage. Although the science suggested they were not as large a source of nutrients in the bay as agriculture, sewage treatment plants were the most straightforward objects of pollution-reduction efforts. In dense urban areas such as Baltimore and Washington, D.C., sewage treatment plants were the main contributor of nutrients; more important, they were relatively easy targets because they were considered point sources and so required permits under the Clean Water Act's National Pollution Discharge Elimination System (NPDES); and finally, federal loans and grants were available to help defray the cost of upgrading them. In the 1980s most wastewater treatment plants used secondary treatment, which employs biological processes to catch the dissolved organic matter that escapes primary treatment. But to remove significant amounts of nitrogen, they would have to install biological nutrient removal (BNR), a tertiary-treatment technology that relies on naturally occurring bacteria to convert bioavailable nitrogen into inert nitrogen gas that is released into the atmosphere. BNR can lower nitrogen effluence in wastewater from 18 milligrams (mg) per liter to as little as 3 mg per liter.[35]

Maryland and D.C. voluntarily created incentives that prompted some improvements in their sewage treatment plants, but progress was slower in Virginia. Maryland provided a 50 percent cost share to local governments that agreed to upgrade their plants to 8 mg per liter of nitrogen with BNR. As a result, by 1998 twenty-three of the state's sixty-five wastewater treatment plants had BNR, and construction or design was under way on another twenty. The Blue Plains Wastewater Treatment Plant, which serves millions of people in the Washington, D.C., metropolitan area, also made great strides in controlling phosphorus during the 1980s: in 1961 it was adding more than 6 million pounds of phosphorus each year to the Potomac; by 1991 it was down to 750,000 pounds—about as low as technologically feasible at the time. Then, in the fall of 1996, Blue Plains began treating half of its flow with BNR, promptly reducing nitrogen in that portion of its treated waste by at least 40 percent.[36] Blue Plains anticipated treating all of its waste with BNR by 2000. In Virginia, by contrast, state and local governments squabbled over the state's role and on whether to focus on the contributions of sewage or agriculture. It was not until the late 1990s that Virginia instituted a cost-share program. As a consequence, only two of thirty-one sewage treatment plants in the Virginia portion of the Potomac and Shenandoah river basins had BNR in place by the late 1990s.[37]

Even more challenging than improving sewage treatment plants was upgrading the many septic systems in the watershed. Most new development between 1980 and 2000 occurred in rural areas, where individual septic systems were used to treat wastewater. By the end of the twentieth century, about 33 million pounds of nitrogen were entering the bay each year from septic systems.[38] Improved septic-system technology could have reduced dissolved nitrogen losses to groundwater by about half, but because installing such systems was both expensive and voluntary, few homeowners chose to do so. Rural residents also balked at efforts to regulate septic tanks, while also resisting the suggestion of a public sewer system, which they feared would entice new development.

As it turned out, among the cheapest and most effective measures taken during the 1980s and 1990s to reduce nutrients in sewage was a phosphate detergent ban, adopted first by Maryland and then in 1986 by Washington, D.C., in 1988 by Virginia, and in 1989 by Pennsylvania. The bans meant that millions of pounds of phosphorus never entered the waste stream in the first place, cutting by one-third the amount of phosphorus going into sewage treatment plants. The amount of phosphorus leaving sewage treatment plants declined even more: Maryland reported that the ban reduced by 1,741 pounds per day the amount of phosphorus flowing into the bay from its sewage treatment plants, a 45 percent decrease; after Virginia imposed its ban, the phosphorus level in wastewater discharged from sewage treatment plants dropped 50 percent from 1987 levels.[39]

Reducing Agricultural Pollution. Like septic systems, agricultural pollution posed a daunting challenge during the 1980s and 1990s: although the hazards associated with the region's agricultural operations were increasing steadily, states were reluctant to impose restrictions on those operations. Farmers working the 87,000 agricultural operations constituted less than 4 percent of the bay region's labor force, but they worked a quarter of the watershed's 41 million acres.[40] By the early 1990s, they were applying nearly 700 million pounds of commercial fertilizer annually, while also spreading thousands of tons of manure.[41] Excess nutrients from commercial fertilizer and manure percolate through the soil into groundwater or runoff during heavy rains. In addition, an acre of cropland dumps 400 pounds of sediment per year into the bay.[42]

According to the Chesapeake Bay Program, poultry manure was the largest source of excess nitrogen and phosphorus reaching the bay from the lower Eastern Shore of Maryland and Virginia in the 1980s and 1990s.[43] Chicken manure is richer in nutrients than waste from livestock, and there is more of it; as a result, manure from livestock accounts for less than one-tenth as much nutrient pollution as chickens. Also contributing to the problem were poultry slaughterhouses: by the late 1990s the Delmarva Peninsula (see Map 4.1) was home to twelve slaughterhouses that killed 2 million chickens each day and used more than 12 million gallons of water to wash away fat, feathers, and other chicken waste.[44] Although the wastewater from slaughterhouses was treated before being discharged into local waterways, it nevertheless contained high levels of nutrients.

Exacerbating the problems associated with poultry production in the Chesapeake Bay was the rapid concentration in the industry: between 1954 and 1987, the size of the average chicken farm in Pennsylvania grew 100-fold, from 1,000 to 100,000 chickens; similar increases occurred in Maryland and Virginia (and in cow and hog farming).[45] By 1998 there were 5,816 broiler houses with an annual production of 606 million birds on the Delmarva Peninsula. Those chickens annually produced 750,000 tons of litter—far more than could be used locally as fertilizer.[46] Even as animal numbers increased, the amount of land used to grow crops declined, so more manure was being spread on less acreage.

Among the best management practices (BMPs) recommended to farmers who were growing crops were practicing conservation tillage (minimal plowing), applying nutrients under the soil surface, planting cover crops, building ponds to silt out the soil before it reaches creeks, constructing terraces and grass-planted drainage ditches to filter soil from water, protecting streams with fencing, and installing manure-holding structures.[47] The bay program estimated that fully implementing BMPs could prevent as much as 100 million pounds of nitrogen from entering the bay each year.[48] But cash-strapped crop farmers resisted changing their practices voluntarily because applying excess fertilizer served as insurance against low yields, while any negative environmental consequences were distant and indirect.

At the same time, concentrated animal feeding operations (CAFOs) successfully repelled regulations on their industry under the Clean Water Act by either the federal government or the states.[49] Impeding efforts to reduce the waste coming from poultry houses was the success by wealthy poultry companies in fending off responsibility for controls. In the chicken trade, the poultry companies supply chicken growers with chicks and feed and then reclaim the birds when they are ready to be slaughtered. But the contract growers, who operate on thinner margins, are responsible for the manure produced in the interim. Environmentalists and many regulators argued that only the large poultry companies had the sophistication and the resources to develop alternative uses for the chicken manure—for example, by converting it into fertilizer pellets or burning it to make energy. Moreover, it was far simpler to hold five large companies accountable than 2,700 chicken farmers. But the companies rejected this idea. Chicken production is "an intensely competitive business," explained James A. Perdue, president of Perdue Farms, Inc. "If our costs go up, we simply can't survive."[50] They argued that the market, not state regulators, should determine the fate of the chicken manure.

Bowing to pressure from agricultural interests, the states devised lax agricultural regulations that were complemented by minimal enforcement, as public officials preferred to work with the industry rather than adopt an adversarial posture. Maryland was the most aggressive of the bay states in regulating agriculture, and its experiences with the poultry industry are instructive. In 1989 Maryland established a Nutrient Management Program, a voluntary nutrient-reduction program designed to help agricultural operators reduce the nutrients running off their fields. That program received meager funds and was a low priority at the state level; as a result, few farmers participated. Nevertheless, agricultural interests successfully resisted pressure for mandatory nutrient management plans throughout the 1990s, until a focusing event drew attention to the inadequacy of existing nutrient-management efforts: in 1997 a toxic outbreak of the microbe *Pfiesteria piscicida*, which thrives in nutrient-rich water, killed millions of fish and caused individuals to fall ill. Galvanized by the outbreak, in 1998 Maryland's General Assembly passed and Governor Parris Glendening signed the Water Quality Improvement Act, which required all but the smallest Maryland farmers to develop nutrient management plans by the end of 2001 and then implement those plans or face substantial penalties. Importantly, the law also linked the permits for national poultry processors to the proper disposal of such waste.

In response to the new law, Maryland poultry producers formed a political action committee to support candidates sympathetic to the industry and mobilized to resist efforts to regulate their activities. In July 1999 a Delmarva trade group kicked off a public relations campaign to counter the image of chicken companies as polluters. "Much of the work that we have done in 1998, and will do in 1999, involves protecting you from government intrusion," Kenneth M. Bounds, president of Delmarva Poultry Industry, Inc., told farmers, bankers, grain salespeople, and others in the chicken business at a banquet in April 1999. "Our industry is under attack, and everyone must rally to its defense. Our critics are armed with fear and misinformation. We must overcome them."[51]

By the 2001 deadline, only 20 percent of Maryland's 1.7 million acres of farmland were governed by nutrient management plans; only 2,152 of more than 7,000 farms had even submitted such plans; and nearly 3,000 farms had filed for delays, while the rest had not filed any forms at all.[52] Moreover, the plans talked about amounts but not where, when, or how nutrients would be applied. In any case, the state had no system in place for monitoring implementation of the plans, so it was impossible to discern their effectiveness.[53] In 2007 Governor Martin O'Malley again proposed regulations on poultry operations but applied them only to 200 of the 800 largest companies, which produced about half the state's poultry litter. (Those rules took effect in early 2009.)

The bay states were similarly reluctant to limit releases of nitrogen and phosphorus in wastewater from poultry slaughterhouses, setting inadequate limits and imposing weak enforcement measures. After Maryland complained to the EPA about Virginia's lax oversight of Tyson's slaughterhouse in Temperanceville, the largest source of pollution among Delmarva poultry plants, Virginia sought ways to avoid taking action against the industry. The ruling philosophy under then-governor George Allen involved working with business. "We like to use persuasion," he said. "It's a positive approach to it, as opposed to a dictatorial approach."[54] Moreover, although the poultry companies routinely sent wastewater that exceeded their permit limits to municipal treatment plants, the state rarely penalized them. For local officials, the quandary was that, although treating the plants' polluted wastewater cost taxpayers money, without the industries the tax base would be far smaller.[55]

Although poultry was the number one contributor, another major source of agricultural pollution was Pennsylvania's dairy industry, which dumped more nutrients into the bay via the Susquehanna than Maryland or Virginia.[56] In the early 1990s there were some 100,000 dairy cows in Lancaster County, which was among the top milk-producing counties in the nation; each cow produced about 120 pounds of manure every day.[57] As of 1991 Pennsylvania was relying on voluntary programs to curb farm runoff; with little incentive to comply, only 144 of the thousands of farmers in the county had signed on. In 1993 Pennsylvania passed the Nutrient Management Act, a first step toward mandatory nutrient management plans for high-density animal operations. The law required the state conservation commission to draft regulations within two years, farmers to develop nutrient management plans within one year of the regulations' issuance, and implementation of those plans, with state assistance, within three years of submission.

But actual implementation of the law was slow: it took the conservation commission four years to draw up regulations. The rules they ultimately issued covered only 5 percent to 10 percent of Pennsylvania farms, and they included loopholes that further weakened the law. Complicating regulators' task was Lancaster County's large population of Amish, who use only manure for fertilizer.[58] Citizen groups such as the Little Conestoga Watershed Alliance worked with farmers to fence cows out of streams; they also encouraged farmers to create buffer zones between streams and farm fields and to not spread manure so often. But many farmers, whose operations were already only marginally profitable, worried about making changes that might reduce their income.

Curbing Shoreline Development. Runoff from urban and suburban areas constitutes a third major source of nutrient pollution in the Chesapeake. During the 1980s and 1990s, increasing amounts of lawn fertilizers, pet waste, and other contaminants in stormwater runoff added to the bay's nutrient loading. Air pollution from the ever-increasing number of automobiles contributed as well. But like controlling agricultural pollution, curbing development along the 11,684-mile shoreline of the bay and its tributaries was a controversial proposition.

As with agriculture, Maryland moved most aggressively among the bay states to regulate land use. The Maryland Critical Area Law, passed in 1984, aimed to regulate development through land-use policies; specifically, it designated a 1,000-foot collar of land surrounding the bay and its tidal waters as the "critical area." It allowed intensive building on only 5 percent of undeveloped shoreline; on the remaining 95 percent of undeveloped land, it allowed construction of one house per twenty acres. It established a 100-foot protective buffer for bayside development.[59] It created a twenty-five-member Chesapeake Bay Critical Area Commission to establish criteria for implementing the law at county and municipal levels. By 1986 the legislature had approved the commission's recommendations that

1. All counties and municipalities within the critical area must develop local protection programs, subject to commission approval.
2. Proposals must include local zoning and development plans for minimizing the adverse effects of growth.
3. Each local area must also categorize the land within the critical area as either an intensely developed area (IDA), a limited development area (LDA), or a resource conservation area (RCA).[60]
4. All localities must reinforce and bolster state sediment and stormwater control programs, preserve and enhance forestlands in the critical area, and implement soil-conservation and water-quality plans in agricultural areas.[61]

By 1989 all but two counties of the sixteen counties and forty-four municipalities affected by the law had local programs in place.

But once again, implementation was something else altogether. Thousands of subdivided parcels of less than twenty acres within the RCAs—many of them hurriedly created during the extended grace period before the criteria went into effect—were "grandfathered."[62] Furthermore, as the Chesapeake Bay Foundation predicted in 1988, developers quickly found ways to capitalize on vagueness in the law. For example, the law allowed only an average of one house every twenty acres in the critical area. Developers realized they could cluster houses at the shoreline and leave the rest of the land undeveloped and still meet the letter—if not the spirit—of the law.[63]

After several years of debate, Virginia imposed its own set of restrictions on development, but neither the laws nor their implementation did much to reduce impacts on the bay. In 1988 Virginia's General Assembly passed the Chesapeake Bay Preservation Act, which affected the eighty-four independent cities, counties, and towns that border on tidal waters and their tributaries—a region known as Tidewater Virginia. The law created the Chesapeake Bay Local Assistance Board to oversee the Chesapeake Bay Local Assistance Department, which administers the act. The law also created two conservation designations: Resource Management Areas (RMAs), which contain wetlands, and Resource Protection Areas (RPAs), which are especially sensitive. At a minimum, RPAs were supposed to contain all tidal wetlands, nontidal wetlands contiguous with tidal wetlands and tributary streams, tidal and nontidal shorelines, and coastal sand dunes. Importantly, the act required a vegetated buffer zone no less than 100 feet adjacent to any RPA and both sides of any stream.

Like Maryland, however, Virginia provided exemptions for approved agricultural and forestry practices—a huge loophole.[64] Other compromises reached during negotiations over the law further weakened it. In particular, half of the bay watershed—the area west of Interstate 95—was not included in the program; on lots recorded before October 1, 1989, localities could reduce the buffer to as little as 50 feet; and the newly created local assistance department received few tools to enforce local compliance. An analysis of the law's implementation by journalists Jennifer Peter and Scott Harper revealed that developers were easily skirting the law and most local governments were openly flouting it.[65] Cities routinely approved applications to build within the 100-foot buffer, often without public comment; some projects were allowed to build within 50 feet of the water's edge.[66] City officials declined to enforce the law because they worried about antagonizing developers and losing high-income taxpayers, and they were reluctant to risk infringing on landowners' property rights. In December 2001 the Virginia legislature amended the law to tighten the shoreline development rules, but many loopholes remained, and enforcement duties continued to fall to local governments, some of whom made clear they would resist the new rules.[67]

Assessing and Reporting Progress. In the mid-1990s a variety of observers celebrated the bottom-up, predominantly voluntary approach of the bay program, which they believed was yielding results. Kari Hudson, writing in *American City & County*, praised the state of Maryland for "departing from the traditional mandates and regulations" and instead working collaboratively with local officials and citizens to generate and vet ideas, a structure that had ostensibly generated more sensible ideas than state officials working alone might have produced and bred greater commitment to the cleanup. Hudson also praised the citizen monitoring program, which gave locals a stake in the restoration effort.[68]

In fact, however, much of the improvement that occurred between 1987 and 2000 was attributable to bans on phosphorous detergents and to upgraded sewage treatment plants, neither of which required stewardship, voluntarism, or local participation. Another piece of data that belied the enthusiasm for voluntarism was a 1994 poll conducted by the University of Maryland's Survey Research Center, which found that despite a decade of educational efforts, most people did not believe that the actions of individuals were responsible for the bay's problems. Respondents also said (inaccurately) that farmers were the *least* responsible for pollution; instead, they blamed business and industry for the bay's woes.[69]

More important, as 2000 approached, it was increasingly clear that conditions in the bay itself were not improving. By the mid-1990s, oysters had been ravaged by overharvesting, disease, and pollution; harvests were down to 2 million pounds annually, from a high of 118 million pounds in the 1880s.[70] In 1996 the melting of heavy winter snow followed by Hurricane Fran led to record-high water flows into the bay, which resulted in massive nutrient loading. Studies found no improvement in the abundance of underwater grasses and no decline in oxygen depletion since the 1970s.[71]

Despite these troubling symptoms, the bay program reported in the late 1990s that states were on track to reach the 40 percent reduction goal for phosphorus and to come close for nitrogen—a claim the media triumphantly reiterated. The bay program's optimism was largely a result of its heavy reliance on computer models to estimate the amount of nutrients entering the bay. In fact, monitoring by the USGS based on samples from the bay's tributaries and main stem suggested the picture was considerably worse than the models suggested. According to the monitoring results, between 1985 and 1999, there was no significant reduction in the total nitrogen and total phosphorus loads at most of the thirty-one sites tested; only two sites showed a significant decrease in nitrogen, and only five showed significant phosphorus reductions.[72] The Chesapeake Bay Commission's 2001 report on the state of the bay acknowledged that, despite modeling results showing a 15 percent reduction in the amount of nitrogen entering the bay between 1985 and 2000, "water-quality monitoring data from the bay's largest tributaries revealed no discernible trends in nutrient loads." Eventually, the bay program was forced to concede that it had failed by a considerable margin, falling short by 2.3 million pounds per year for "controllable" phosphorus and 24 million pounds per year for "controllable" nitrogen.[73]

2000–2014: New Agreements

Rather than giving up in the face of discouraging results, in 2000, with the threat of federal regulation looming, the bay partners recommitted themselves to cleaning up the Chesapeake. To demonstrate their seriousness, they signed Chesapeake 2000, a twelve-page compact containing five major goals and dozens of specific actions. In addition, three more states joined the bay program as "headwater partners": Delaware, New York, and West Virginia. Even with the cooperation of the additional states, however, meeting the new goals promised to be difficult. About sixteen times more nitrogen and seven times more phosphorus was entering the bay in 2000 than had done so prior to the arrival of British colonists, and upgraded Corps modeling showed that a 50 percent reduction in nitrogen would be necessary to avert unacceptable ecological consequences.[74] Furthermore, the decade of the 2000s was marked by unusual climate variability in the bay watershed. Heavy rains throughout the year following several years of drought washed large amounts of nutrients and silt into rivers and streams that feed the bay. Hurricane Isabel, which struck in September 2003, dumped millions of gallons of raw sewage and thousands of tons of sediment into the bay, as hundreds of septic systems failed and wells were breached. The hurricane's destructive force was magnified by the one-foot increase in relative sea level that had occurred over the twentieth century, a consequence of climate change and the gradual subsidence of land.[75]

The Threat of Total Maximum Daily Loads and Chesapeake 2000. Section 303(d) of the 1972 Clean Water Act requires the EPA and the states to enforce total maximum daily loads (TMDLs) for pollutants entering the nation's waterways. For point sources, the state can enforce a TMDL through permits; for nonpoint sources, the state must show it has adequately funded programs to reduce runoff.[76] The state must write a TMDL for all segments of any waterway that fails to meet its water-quality standards. If a state does not write a TMDL for each impaired segment, the EPA must write one itself.

To help the states comply with these settlements, the EPA began working on an overall TMDL for the Chesapeake Bay. It put that process on hold, however, when on June 28, 2000, the governors of Maryland, Virginia, and Pennsylvania signed Chesapeake 2000, which contained a set of commitments the states hoped would remove the bay from the federal government's list of threatened waterways and thereby avert the need for a TMDL (and the associated possibility of sanctions). Chesapeake 2000 established ten milestones against which the program would judge itself; among those, water quality was always foremost. But others included increasing the native oyster population tenfold, based on a 1994 baseline, by 2010; reducing the rate of harmful sprawl development by 30 percent by 2012; restoring 114,000 acres of underwater grasses; and achieving and maintaining the 40 percent nutrient-reduction goal agreed to in 1987. As before, the goals were all voluntary.

New Initiatives to Address Pollution. In an effort to meet the aspirations of Chesapeake 2000, Maryland, Virginia, Pennsylvania, and D.C. continued devising ways to reduce nutrient pollution from wastewater treatment plants. In the spring of 2004 Maryland passed a "flush tax," a $2.50 monthly surcharge on sewage and septic users. The measure was expected to raise $60 million per year that would be used to upgrade wastewater treatment plants. Virginia borrowed $250 million for improving its sewage treatment plants, while Pennsylvania voters approved a $250 million bond issue for the same purpose. In January 2005 all six of the watershed states agreed to limit the amount of nitrogen and phosphorus discharged from wastewater treatment facilities under a new EPA policy that allowed permits to contain nutrient limits. By 2009 the states had reissued permits for 305 significant wastewater treatment plants, covering 74 percent of the flow from the 485 such facilities in the region.[77]

The bay states also took additional steps to reduce pollution from agriculture. For example, state regulators in Pennsylvania proposed tough new rules to control phosphorus in farm runoff. In addition, Pennsylvania submitted to the EPA a new strategy for planting cover crops to keep soil from eroding during the winter. But these measures were relatively primitive compared to those of Maryland and Virginia: the EPA estimated that Pennsylvania was only 23 percent of the way to its nitrogen-reduction goals, compared to Maryland's 57 percent and Virginia's 35 percent.[78] In an effort to improve its performance, in 2007 Pennsylvania began experimenting with a new approach to reducing nutrient runoff from farms: a system of tradable pollution credits similar to the one implemented in Virginia in 2006. A developer who wanted to build a new sewage treatment plant would have to buy enough credits each year from a farmer employing BMPs to offset any pollution from the plant. The developer of a 100-house subdivision would need about 700 credits annually.[79] A small company named Red Barn started a credit-trading business and was trying to coax farmers into the program by emphasizing the money-making opportunities. Many of the area's Amish farmers were loath to participate, however, and those who did found it difficult to get credits certified.

But much of the focus among the bay states in the early 2000s was on urban-suburban runoff. In part this was because, in late 1999, the EPA issued rules requiring municipalities to obtain permits for their stormwater runoff. It was also because urban-suburban runoff was the only source of bay pollution that was actually rising. The main culprits were lawns, which were fast becoming the watershed's largest "crop," and rapidly expanding paved areas: between 1990 and 2000, the population in the Chesapeake Bay watershed had increased 8 percent, but the amount of impervious cover increased 41 percent, or a whopping 250,000 acres.[80] Analyses by the USGS showed that in the late 1990s, as a result of suburban development, pollution began creeping up in eight of the Chesapeake's nine biggest tributaries after a slow decline in the 1980s and 1990s.[81]

State environmental agencies concluded in 2005 that it would take about $12 billion to address stormwater runoff problems in Maryland, Virginia, and D.C. In hopes of reducing those costs, municipal governments were experimenting with a variety of green infrastructure (or low-impact development) approaches, including rain barrels and cisterns, rain gardens, sunken road medians, roof drain infiltrators, green roofs, swales, and other tools to slow rainwater and allow it to percolate through the soil rather than sending it directly into culverts and streams. Stafford County, Virginia, required builders to apply for a waiver if they declined to install rain gardens or other water filters. Maryland's Anne Arundel County required developers renovating old properties, builders, and owners of single-family homes to install stormwater systems on site if the paved surface around the building exceeded 5,000 square feet. And Arlington, Virginia, undertook an ambitious project to retrofit a school and community-center complex with three-story-high cisterns to collect rainwater that could be used to irrigate the lawn; contractors also dug the parking lot to a depth of twelve feet, filled it with gravel and dirt, and topped it with asphalt tiles that are separated just enough to let rainwater seep through.[82] Such experiments sometimes required repairs because they were done improperly to begin with. But the biggest obstacle was public perception. "When water comes in [to a rain garden], it does stand for a certain time period," explained Wade Hugh, chief of watershed management for Prince William County. "People will have to get over the perception that they have a drainage problem."[83]

Assessing and Reporting Progress. New initiatives notwithstanding, monitoring data gathered in the mid- and late 2000s suggested that climate variability and sprawl development were overwhelming any nutrient reductions achieved. The 2004 aerial survey of underwater grasses in the bay revealed that about one-third had died during 2003, when heavy rain sent large amounts of nitrogen into the bay; as a result, the total area covered declined from about 90,000 acres in 2002 to about 65,000 acres in 2003.[84] The 2005 survey showed that grasses had bounced back somewhat in 2004 to cover 78,260 acres—about double the acreage they covered when the survey began in 1984, though less than half the 185,000 acres biologists were aiming for by 2010.[85] But the acreage covered by underwater grasses declined again in 2006 to 59,090 acres; a dry spring increased the bay's salinity, stressing the grasses, and heavy rains in early June muddied the upper and middle bay, blocking critical light.[86] There was bad news on the dissolved oxygen front as well: scientists were dismayed to report in early August 2005 that they had detected one of the largest volumes of oxygen-depleted water ever recorded in the Chesapeake Bay.[87]

Making matters worse, reports emerged beginning in late 2003 that administrators in charge of the bay program had repeatedly concealed information suggesting that their efforts were failing, downplayed negative trends, and even issued reports overstating their progress.[88] Following a pattern first raised by *The Washington Post* in late 2003 and 2004 and subsequently documented by the Government Accountability Office (GAO), program officials had continued to rely on computer models that inflated the benefits of bay protection measures and gave a distorted picture of conditions in the bay; although monitoring data were sparse, the available information made it clear that the models' version of the bay was too rosy.[89] For example, the Chesapeake Bay Foundation's sixth annual state-of-the-bay report, released in late 2003, was titled "The Bay's Health Remains Dangerously Out of Balance and Is Getting Worse." By contrast, EPA officials had offered a more hopeful assessment, suggesting that the bay program was moving in the right direction, albeit not as fast as it could be, and that the trend for the previous twenty years was one of improvement. Eventually, however, bay program officials conceded that they had relied on the models, even though it was clear they overstated the program's benefits, in order to preserve the flow of federal and state money to the project. "To protect appropriations you were getting, you had to show progress," explained Bill Matuszeski, who headed the bay program from 1991 to 2001.[90]

After insisting for years that the 2010 goals were within reach, in January 2007 Rich Batiuk, associate director of the bay program, informed a meeting of the Chesapeake Bay Commission that the cleanup of the bay was well behind schedule and would likely miss the 2010 deadline by a wide margin.[91] Batiuk highlighted some of the program's successes: a substantial regrowth of sea grasses in the northern bay, reductions in the nitrogen and phosphorus released by sewage treatment plants, and the comeback of the bay's rockfish population that began in the 1980s. But overall, he conceded, the cleanup was way off the pace set in 2000: crab populations were still below historic levels, oxygen readings were at just 29 percent of the goal set for 2010, and oysters were at a paltry 7 percent. Underwater grasses, considered a relative success, had reached only 42 percent of the level sought by 2010. The main culprit, said Batiuk, was rapid population growth: an additional 800,000 people moved into the watershed between 2000 and 2005, negating improvements that were made in reducing pollution.

Some critics suggested the disease afflicting the bay program was too much analysis and planning and too little action; while understanding of the bay's problems had improved enormously, it had proven far more difficult to actually fix them. "We have done a truly tremendous job of defining the problem, and we have done a truly tremendous job of defining the solution," said J. Charles (Chuck) Fox, a former EPA official. "But we have not yet succeeded in actually implementing the solution."[92] This sentiment echoed views expressed by others closely associated with the bay program. "We know what we have to do," said former Chesapeake Bay Program director Matuszeski. "What we don't want to do is pay for it."[93] Roy Hoagland, vice president for environmental protection and restoration with the Chesapeake Bay Foundation, agreed. "We have the science and the solutions," he said. "All we need is the implementation and the dollars."[94] In the face of sharp criticism, in 2007 the EPA restarted work on a TMDL for the entire bay as well as plans to allocate pollution among the hundreds of sewage treatment plants, stormsewer systems, and farm fields that contribute to the bay's pollution.

Environmentalists Rethink Their Tactics. Even as the EPA was conceding that its efforts had proven inadequate, advocates of bay restoration—scientists, environmentalists, and watermen and waterwomen were becoming restive with the collaborative, voluntary approach embodied by the Chesapeake Bay Program. In 2003 William C. Baker, president of the Chesapeake Bay Foundation, questioned the culture of consensus and cooperation that was the hallmark of the bay cleanup but made it difficult for participants to acknowledge the estuary's decline. In a dramatic turnaround, in June 2003 the foundation called for regulatory oversight of the restoration, with firm deadlines and penalties if the state and federal governments did not meet their commitments. "There simply must be a binding, legal framework if the agreed-upon goals are to be met," said Baker.[95] In late August the foundation condemned the voluntary effort among the states, saying that the collaborative approach needed to be replaced by a governing body with the power to create and enforce laws and the authority to levy taxes in the six watershed states and D.C. to pay for cleanup efforts. "Today, we are stuck with the deafening silence of a leadership vacuum," Baker said. "All the scientific data suggests that the bay is not improving. The governance structure simply is not working."[96]

The bay foundation's more aggressive stance did not sit well with many of its allies in government; Baker's complaints prompted a flurry of letters decrying its newfound assertiveness. Nevertheless, in 2004 the foundation announced that it would engage in a litigation program designed to hold polluters and regulators accountable—a striking shift from the foundation's normal way of operating.[97] The foundation's 2004 report complained that the "bay's health has languished at about one-quarter of its potential for the last six years." The restoration, the foundation asserted, was "stalled," and the bay itself was "a system dangerously out of balance." In late 2004 the foundation filed a lawsuit accusing the EPA of dragging its feet on denying permits for sewage treatment plants that did not control their nitrogen releases. The Maryland Watermen's Association also mulled over a class action suit against polluters. And offshoots of the Waterkeepers Alliance, an environmental group known for its confrontational tactics, set up shop on several Chesapeake Bay tributaries.[98]

In October 2008 a group of environmentalists, fishery workers, and former politicians filed another lawsuit against the EPA over its failed Chesapeake Bay cleanup—a response to EPA officials' admission over the summer that the program would not meet 2010 targets. Two months later, a group of scientists who had been studying the bay for years met and came up with a statement that began, "We have concluded that after 25 years of efforts, the formal Bay Program and the restoration efforts under the voluntary, collaborative approach currently in place have not worked."[99] The scientists joined forces with environmentalists and former Maryland officials to call for a major shift in the approach to restoration; they even set up a website that laid out an alternative plan.[100] They argued that tough regulations, based on the principle that clean water is a right of all citizens, should replace voluntary measures, which had not worked. "People are not buying in. They are not doing the right thing," said ecologist Walter Boynton. "At this stage, we need to use a much heavier hand."[101] Boynton and others were motivated by the results of long-term studies that showed worsening conditions; for example, after being polluted by the Patuxent River for years, the bay was actually sending contamination back upstream into the river.

The Obama Administration and TMDLs Revisited. In the winter of 2008–2009, a new version of the bay program's Chesapeake Bay Watershed Model that incorporated improvements in the scientific understanding of both BMPs and meteorology confirmed that all of the cleanup strategies to date had yielded results that fell far short of program goals. The ongoing bad news about the Chesapeake spurred environmental officials to begin wondering whether it was time to lower expectations for the restoration. For several weeks, EPA officials gathered data to determine whether the program's goal of restoring the bay to health was "an impossible stretch." Among the findings of this effort was that, although the bay program had pushed farmers to erect fences to keep cows away from riverbanks, as of February 2009 such fences existed on just 11.7 percent of the riverbank farmland in the watershed. Program officials projected a best-case compliance rate of 32 percent. Similarly, a contractor working for the bay program estimated that if the program tried to get landowners to upgrade their septic tanks, only one-third of them were likely to comply.[102]

But Chuck Fox, who had become the senior adviser to the EPA administrator for the Chesapeake Bay, quashed the research related to lowering expectations on the grounds that it was diverting energy from the challenge of cleaning up the bay. In fact, instead of conceding defeat, the Obama administration opted to redouble its efforts. On May 12, 2009, President Obama declared the Chesapeake Bay a national treasure. He also issued Executive Order 13508, Chesapeake Bay Restoration and Protection, which created a Federal Leadership Committee headed by EPA Administrator Lisa Jackson to develop a coordinated, long-term strategy for protecting and restoring the estuary. To demonstrate its commitment, the administration proposed a record $35.1 million for the Chesapeake Bay Program in its budget request for fiscal year 2010.[103]

For their part, bay program partners vowed to recommit as well. At its annual meeting the executive council agreed to postpone until 2025 the actions needed to reach the goals set in 2000. In hopes of quieting critics, the council established two-year pollution-reduction milestones—although they again failed to provide sanctions for failing to meet them. To meet its two-year milestones, Maryland pledged to cut 3.75 million pounds of nitrogen and 193,000 pounds of phosphorus runoff by increasing cover crops, expanding farm conservation programs, improving septic systems, and upgrading several sewage treatment plants and power plants by the end of 2011.[104] In addition, to implement its Stormwater Management Act of 2007, Maryland advanced a stringent set of regulations requiring environmental site design "to the maximum extent feasible" on all new development and redevelopment sites. Virginia pledged to reduce nitrogen emissions by 3.39 million pounds and phosphorus by 470,000 pounds by 2011. Among the actions it planned to take in the next two years were planting 119,000 acres of cover crops, reforesting 12,500 acres, putting 9,000 acres under stormwater management, conserving 10,000 acres of forest buffers, getting 258,000 new acres under nutrient management, and improving wastewater treatment. Washington, D.C., intended to cut 159,000 pounds of nitrogen (its phosphorus goal was attained). To facilitate this reduction, the city faced a dramatic rise in residential monthly sewer bills, from $30 per month in 2008 to $110 per month in 2025. The D.C. Council proposed to change the fee for stormwater runoff from $7 per year to a fee calculated based on the amount of paved area on a property. New buildings would be required to have features that reduce runoff; subsidies and incentives would be introduced to help owners of existing buildings reduce their runoff. (In April 2010 the GAO declared that the federal government was exempt from the fee, leaving city residents on the hook to fund the $2.6 billion effort to upgrade the city's combined sewer system.) The other watershed states made substantial commitments as well.[105]

Nevertheless, as Howard Ernst pointed out, Obama's executive order provided no new funding or statutory authority and placed no new legal responsibilities on states or localities, and the new nitrogen and phosphorus targets adopted by the states were insufficient to attain the levels scientists had indicated were necessary.[106] Moreover, states' efforts to achieve even these nutrient reductions faced familiar political obstacles. Although Maryland's development rules were approved in 2009, in March 2010 Maryland officials agreed to make developer-friendly adjustments to head off attempts to weaken the law even further. Similarly, in October 2009 Virginia's Soil and Water Conservation Board approved regulations reducing the amount of phosphorus allowed in stormwater discharges from construction sites from 0.45 pounds to 0.28 pounds per acre per year within the Chesapeake Bay watershed but then immediately suspended them after an intense backlash from homebuilders and local officials.[107]

Making matters worse, states in the region were cutting expenditures on pollution control in the face of budget constraints. For instance, the Maryland legislature shifted $25 million out of its Chesapeake Bay 2010 Trust Fund, which funnels money into efforts to stop nonpoint-source pollution, leaving the program with less than $10 million.[108] This was on top of already struggling enforcement: between 2003 and 2008, the number of inspectors at Maryland's Department of the Environment fell from 156 to 132, while the number of sites requiring inspection grew 20 percent; as a result, there were 132 inspectors covering 205,000 sites, a 1-to-1,500 ratio, which was significantly worse than the 1-to-1,090 ratio in 2003.[109]

Fortunately, to some extent, cuts by states were offset by increases in federal spending. The 2008 farm bill provided $188 million to reduce nutrient and sediment pollution in the bay watershed. The federal stimulus bill also provided $878 million to the EPA's Clean Water State Revolving Loan Fund for the six states that comprise the bay watershed.[110] Congress boosted funding for the EPA's bay program office by more than $20 million in 2010, with most of the increase directed toward states to help implement their programs. And the bay program was slated to receive an unprecedented $63 million in the president's fiscal year 2011 budget.

Unmoved by the states' pledges and the injections of federal money, environmentalists were fed up with the nonregulatory approach employed by the bay program for more than twenty-five years. In September 2009 they presented the administration with more than 19,000 signed letters and postcards asking the federal government to set stricter pollution-control rules for the bay. Their message, in a nutshell, was, "We need to penalize bad actors."[111] The EPA seemed to get the point. In late December the agency announced it would pursue a new, get-tough approach in which it would punish states that did not meet their goals, or did not set sufficiently high goals, for cutting pollution flowing into the bay.

On May 12, 2010, the Obama administration released a "sweeping vision" for a revived Chesapeake Bay and its watershed, with a commitment to complete the job by 2025. The report contained scores of actions and deadlines, including conservation practices to be implemented on 4 million acres of farmland, permanent protection of another 2 million acres (in addition to the 7.8 million acres already protected, constituting 19 percent of the watershed), and restoration of the black duck population to 100,000 wintering birds (compared to about 37,000 birds in 2010). To ensure the new strategy stayed on track, agencies planned to develop two-year milestones, and an independent evaluator was expected to review actions taken and suggest improvements. The administration projected that fifty-five of the bay's ninety-two segments would meet water quality standards by 2025 (versus three in 2010), and 70 percent of its streams would be in "fair, good, or excellent" condition (in 2010 just 45 percent met those criteria).[112]

To facilitate meeting the strategy's goals, the EPA promised to draft new rules for CAFOs by June 30, 2012, and finalize them by 2014. The new rules would cover smaller operations than are currently regulated, make it easier to designate new CAFOs, and set more stringent requirements for the use of manure and other wastes. (As of June 2018, the EPA had just wrapped up receiving public comments.)[113] In addition, by September 2011 the EPA intended to initiate national stormwater rules with specific provisions for the Chesapeake Bay that will further control runoff from development. And by 2020 it planned to implement Clean Air Act programs to reduce emissions from power plants, factories, ships, and motor vehicles, as well as air pollution from agriculture. Finally, the agency pledged to improve enforcement of existing air and water pollution permits.

In addition to nutrient-control measures, the administration's strategy envisioned a network of land and water habitats to support the region's wildlife and insulate them against the impacts of climate change. It pledged to restore 30,000 acres of tidal and nontidal wetlands and enhance the function of an additional 150,000 acres of degraded wetlands by 2025—a relatively modest goal considering that as of 2010 an estimated 1 million acres of tidal and nontidal wetlands in the watershed were available for restoration or enhancement. It called for expanding riparian buffers to cover 181,440 miles, or 63 percent of the watershed's available stream banks and shorelines. (In 2010 about 58 percent of the 288,000 total riparian miles had buffers.) It proposed removing dams or constructing fish passages to open 1,000 additional stream miles by 2025. This included stepped-up efforts to combat invasive species, while aiming to restore native oyster habitat and populations to twenty tributaries by 2025.

The EPA followed up its draft restoration plan on July 1, 2010, with a draft TMDL, telling the states how many pounds of nutrients the bay would need to eliminate as part of its new "pollution diet": about 63 million pounds of nitrogen and 4.1 million pounds of phosphorus between 2010 and 2025. EPA computer models suggested the Chesapeake would tolerate a maximum annual load of 187.4 million pounds of nitrogen (later revised to 185.9 million pounds) and 12.5 million pounds of phosphorus if it is to sustain habitat suitable for crabs, waterfowl, and bottom-dwelling species. Marking a shift from the past, the new TMDL included the amount of nitrogen that could be allowed to land on the bay from the air: 15.7 million pounds.

On January 18, 2011, the EPA issued the final TMDL for the Chesapeake Bay, the largest and most complex ever written. The American Farm Bureau filed suit in federal court to block the TMDL, saying the costs of the cleanup would devastate farms and possibly drive them out of the region. It also argued that the science underpinning the TMDL was flawed.[114] At its annual meeting that year, the federation issued a call to arms against the bay cleanup plan, concerned that it was a harbinger of more far-reaching requirements for other watersheds, including the Mississippi River.[115] The federation's complaints notwithstanding, in mid-September 2013, U.S. District Court Judge Sylvia H. Rambo upheld the TMDL.

In September 2012 state and federal leaders proclaimed that polluters were finally on track to meet cleanup goals, having met most of their 2009–2011 milestones. In June 2014 the EPA reported similar results: the partnership as a whole exceeded the 2013 milestone targets for nitrogen and phosphorus. Most of those reductions came as a result of improvements in wastewater treatment plants and lower-than-anticipated wastewater flows, however; the region would have to accelerate agricultural and stormwater controls to meet pollution reduction targets set for 2017.[116]

Despite some encouraging results in terms of reducing nutrient emissions, water quality monitoring continued to reveal disturbing trends. A new technique employed by the USGS showed that while the bay region had seen long-term declines in nutrients in many areas since 1985, those trends had leveled off in recent years, and improvements for phosphorus had largely halted.[117] Observers offered several reasons for the discrepancy between reported achievements and water quality outcomes. One possibility was that officials were overestimating the effectiveness of actions taken—either because the actions taken did not work as promised or because their performance did not hold up over time. To address this possibility, in December 2013 the bay program's Principal Staff Committee agreed to a set of "verification principles" that would guide the development of a consistent system for tracking the implementation of pollution-control practices throughout the watershed.

OUTCOMES

On June 16, 2014, a new Chesapeake Bay Watershed Agreement[118] was signed setting new goals to track progress in order to hold partners accountable for their work. Signatories included representatives from the entire watershed, from the headwaters to bay communities. The goal was to provide continued collaboration among partners to maintain a commitment for a shared vision.[119]

The Chesapeake Bay Foundation has also been tracking pollution reduction goals to determine the likelihood of bay jurisdictions to meet their agreement to fully implement plans by 2025. Midpoint assessments were evaluated in 2017. Specifically, targets are being met to achieve pollution-reduction goals for phosphorus and sediment. Yet, the most significant shortfalls remain with polluted runoff.[120]

More promising is research led by Jonathan Lefcheck of the Bigelow Laboratory for Ocean Sciences in 2018. Lefcheck and colleagues partnered with the Chesapeake Bay Program to examine longitudinal data from 1984 (e.g., submerged water vegetation and water quality). Their analysis suggests good news: anti-pollution efforts have decreased nitrogen and phosphorus levels.[121]

These strides could be derailed by the Trump administration. For example, President Trump has attempted to cut the federal funding for bay programming. However, Congress has fully restored the budget to continue cleanup efforts.[122]

CONCLUSIONS

With nearly $6 billion spent over a quarter-century to reduce nutrients, the signs in the late 2000s were that the impacts of relentless growth were overwhelming pollution-control efforts in the Chesapeake Bay.[123] As Donald Boesch and Erica Goldman explain, accumulated changes in the Chesapeake Bay watershed had caused a change in state, "from a relatively clear-water ecosystem, characterized by abundant plant life in the shallows, to a turbid ecosystem dominated by abundant microscopic plants in the water column and stressful low-oxygen conditions during the summer."[124] What is more, the bay's degraded state seemed to have become resilient. Making matters even worse, the combination of a rising sea level, erosion, climbing ocean water temperatures, and increasingly violent storms associated with climate change threatened to decimate the bay's plant and animal life in the coming decades, setting back whatever progress had been made.[125] Meanwhile, renewed efforts at restoration appeared likely to be dogged by the baseline problem: as Bernie Fowler mused in the early 1990s,

> When people like me grow old and die off, we risk leaving a whole generation that has no idea what this river really was. No memory banks in those computers at EPA can recall the ten barrels of crabs one person used to catch out there, and all the hardheads, and the thrill of the oyster fleet coming in at sunset, the shuckers in the oyster house all singing harmony while they worked. If we can't make some headway soon, these children will never, never have the hope and the dream of bringing the water back, because they just won't have any idea how enriching it used to be.[126]

The bay's ongoing deterioration prompted some introspection among activists about the trade-off between scientific research and action. Initially, scientific research was essential to catalyzing public awareness and a political response; advocates were able to translate a massive synthesis of scientific studies into a simple and compelling story about the bay's plight. As the years wore on, however, critics began to challenge the bay program's preoccupation with studies and ever-more elaborate models, which seemed to provide the responsible parties with an excuse for delaying action. They called for additional resources to be invested not in research but in the implementation of policies that would change behavior. Recent news about submerged aquatic vegetation shows promise in the recovery of the bay, but many other unsettling reports question if ecosystems can indeed recover.

The condition of the bay also led some skeptics to question the value of a comprehensive approach. Mike Hirshfield, chief scientist and strategy officer of the environmental group Oceana, describes the bay program as comprehensively piecemeal: it is a laundry list of items to attack, he says, but there is no interaction among the various components. Hirshfield suggests that more progress might have been made had the bay program focused more narrowly on agriculture, the single biggest source of nutrients, rather than trying to be holistic.[127] Others challenged the utility of the whole-system models employed by the bay program, suggesting that they can lead to conclusions that are dramatically wrong.[128] In fact, according to Linda Pilkey-Jarvis and Orrin Pilkey, the more complex the model, the less accurate it becomes.[129] Yet, as Donald Boesch and his colleagues observe, "Because they yield clear numerical results with which to gauge progress, . . . models have a seductive appeal to policymakers and managers, an appeal that risks false confidence and misconception."[130]

Finally, the experience of the bay program provided support for the view that nonregulatory approaches are insufficient to bring about fundamental changes in industry, municipal, or household behavior. Voluntary programs, which are notable for their lack of sanctions, seem particularly dubious; the flexibility inherent in a voluntary approach, while it allows for tailored responses to pollution, also means there is no accountability when conditions worsen or fail to improve. So it is hardly surprising that a series of studies have found that companies enrolled in voluntary environmental programs do no better, and sometimes perform worse, than nonparticipants.[131] In his 2009 book, *Fight for the Bay*, Howard Ernst suggested that the bay program mimic the Lake Tahoe Regional Authority, which was seeking to restore a once-pristine mountain lake on the California-Nevada border. "They monitor, they file lawsuits," he said. "It's not pretty, it's not nice, but it works."[132]

QUESTIONS TO CONSIDER

- The Chesapeake Bay Program has repeatedly set ambitious goals for cleanup that the region has never come close to meeting. What are the pros and cons of establishing lofty aspirations?
- What is your assessment of the value of voluntary environmental programs? Under what conditions might they be an appropriate alternative to more prescriptive regulatory approaches, such as mandates and technology standards?
- Critics have decried the use of complex models in environmental decision making, but supporters regard them as essential tools. What, in your view, is the appropriate role of models in environmental policymaking, and how might their potential drawbacks be minimized?

NOTES

1. Dietrich, Tamara. "Deep Dive Shows Damaged Chesapeake Bay Can Heal. " *The Washington Post*, March 19, 2018.
2. More precisely, microscopic bacteria consume the algae that fall to the bottom. Because of their high metabolism, these bacteria rapidly draw oxygen from the water. In deep water, a barrier known as the pycnocline—which separates heavier, saltier water from lighter freshwater—prevents new oxygen from reaching the bottom. As a result, water below the pycnocline becomes hypoxic or anoxic, creating a "dead zone" with little or no oxygen for fish and other aquatic species. See Karl Blankenship, "Westerly Breezes Take the Wind Out of Bay Cleanup's Sails," *Bay Journal*, May 2010. Scientists have identified more than 400 dead zones around the world, affecting nearly 95,000 square miles of coastal waters. See Howard R. Ernst, *Fight for the Bay: Why a Dark Green Environmental Awakening Is Needed to Save the Chesapeake Bay* (Lanham, Md.: Rowman & Littlefield, 2009).
3. Millennium Ecosystem Assessment, *Ecosystems & Human Well-Being: Synthesis Report* (Washington, D.C.: Island Press, 2005). Based on data from 110 estuaries around the world, the World Wildlife Federation and the Zoological Society of London concluded that there was a pronounced decline in estuarine bird and fish species between 1980 and the mid-1990s, at which point the rate of loss slowed, and in 2001 recovery began. The deterioration they measured was largely a result of massive declines in tropical species, which offset gains in temperate species during the latter decades of the twentieth century. See Stefanie Deinet et al., "The Living Planet Index for Global Estuarine

Systems," Technical Report, World Wildlife Federation and the Zoological Society of London, 2010. Available at http://assets.panda.org/downloads/lpi_estuaries_project_report_final.pdf.

4. Judith A. Layzer, *Natural Experiments: Ecosystem-Based Management and the Environment* (Cambridge, Mass.: The MIT Press, 2008).

5. This chapter focuses on efforts to reduce nutrient pollution from point and nonpoint sources. There are other important initiatives under way as well, including management of crab and oyster harvests, oyster restoration, toxics reduction, and others. For more on these efforts, see Tom Horton, *Turning the Tide* (Washington, D.C.: Island Press, 2003).

6. Dewitt John, *Civic Environmentalism: Alternatives to Regulation in States and Communities* (Washington, D.C.: CQ Press, 1994); Marion R. Chertow and Daniel C. Esty, *Thinking Ecologically: The Next Generation of Environmental Policy* (New Haven, Conn.: Yale University Press, 1997).

7. Oysters, clams, and other bivalves filter water through their gills, removing plankton and sediment from as many as fifty gallons of water per day. They also consume large quantities of nitrogen. See Betsy Carpenter, "More People, More Pollution," *U.S. News & World Report*, September 12, 1994, 63–65.

8. Peter McGrath, "An American Treasure at Risk," *Newsweek*, 12 December 1983, 68.

9. Figures are from the Chesapeake Bay Program (http://www.chesapeakebay.net/) and Howard R. Ernst, *Chesapeake Bay Blues: Science, Politics, and the Struggle to Save the Bay* (New York: Rowman & Littlefield, 2003).

10. According to the Chesapeake Bay Foundation's website, the businesspeople met with Rogers C. B. Morton, a member of Congress from Maryland's Eastern Shore. Morton told them that they could not expect government to fix all the bay's problems and encouraged them to form a private-sector organization that could "build public concern, then encourage government and private citizens to deal with these problems together." See Chesapeake Bay Foundation, "Our History." Available at http://www.cbf.org/Page.aspx?pid=392. The foundation's origins explain its largely nonconfrontational approach to advocacy.

11. Alliance for the Chesapeake Bay. Available at http://allianceforthebay.org/?page_id-133.

12. Ernst, *Chesapeake Bay Blues,* 58. There was an interstate conference on the bay in 1933, and the concept of treating the bay as a single resource emerged at that time as well.

13. "Chesapeake Bay Water Succumbing to Wastes," *Engineering News-Record*, October 6, 1983; Tom Kenworthy, "Oxygen-Devoid Area of Bay Reported Up Tenfold Since 1950," *The Washington Post*, December 30, 1983, B5.

14. Donald F. Boesch, president, University of Maryland Center for Environmental Science, personal communication, 2003.

15. This Chesapeake Bay Program superseded the first one, which was established by the EPA in 1976 to conduct the comprehensive study of the bay.

16. Angus Phillips, "Bay Supporters Meet in Fairfax on Cleaning Up the Chesapeake," *The Washington Post*, December 7, 1983, C4.

17. J. Charles Fox, vice president, Chesapeake Bay Foundation, personal communication, 2003. The Green Book contained a long list of initiatives, including tripling the state share of sewage treatment construction costs, expanding enforcement of effluent permits and de-chlorinating all sewage effluent in two years, starting a five-year planting effort to restore bay grasses, hatching 10 million rockfish (striped bass) and shad per year, and doubling the oyster catch through expanded seed programs. Hughes also talked about establishing a Critical Areas Commission to work with local governments on land-use plans for property adjoining the bay and tributaries, establishing a youth conservation corps and a $5-per-day sport fishing license, doubling financial aid to the Chesapeake Bay Foundation's education program, and providing cost-share funds and technical assistance to farmers seeking to keep sediment and fertilizers on their land. See Angus Phillips, "States Agree to Work on Bay Clean-Up," *The Washington Post,* December 10, 1983, B1.

18. Quoted in Phillips, "States Agree to Work on Bay Clean-Up," B1.

19. Angus Phillips, "$40 Million Bay Proposal Delights Cleanup Group," *The Washington Post,* January 26, 1994, C1. As Howard Ernst points out in *Fight for the Bay,* Reagan's was just the latest in a long line of presidential promises to save the bay.

20. Legislation ultimately funded the cleanup at $52 million, and additional money was provided in separate bills. See Alison Muscatine and Sandra Sugawara, "Chesapeake Bay Emerges as Reagan's Environmental Cause," *The Washington Post,* October 7, 1984, E40.

21. Ibid. Reagan's "new federalism" involved devolving more authority for public programs to the states, which were presumably better equipped than federal bureaucrats to solve local problems.

22. Quoted in Tom Sherwood, "Robb, Ruckelshaus Seek to Calm Roiled Waters of Bay Cleanup," *The Washington Post,* May 10, 1984, C4.

23. Donald F. Boesch, Russell B. Brinsfield, and Robert E. Magnien, "Chesapeake Bay Eutrophication: Scientific Understanding, Ecosystem Restoration, and Challenges for Agriculture," *Journal of Environmental Quality* 30 (2001): 303–320.

24. The program justified this shift, which occurred in 1992, because the airshed of the Chesapeake Bay, at 350,000 square miles, is far larger than the watershed; many of the airborne emissions falling on the watershed come from West Virginia, New York, and Delaware, none of which were signatories to the bay agreement. From the perspective of the bay, however, the shift was extremely significant, since a 40 percent reduction was already the minimum based on runs of the primitive, two-dimensional model the bay program was working with at the time. In fact, the figure was based not on break points in the watershed model—that model suggested a linear relationship between nutrients and dissolved oxygen levels—but on break points in the marginal cost curve. Although many scientists, including those at the Chesapeake Bay Foundation, believed the 40 percent figure was too low and the idea of reducing only "controllable" nutrients by 40 percent ridiculous, they nevertheless decided to see whether the program could deliver even that level of improvement.

25. Boesch et al., "Chesapeake Bay Eutrophication."

26. Todd Shields, "Fish Kills Seen as 'Alarm Bell' for Chesapeake, Tributaries," *The Washington Post,* August 17, 1997, B1.

27. Donald F. Boesch and Jack Greer, *Chesapeake Futures: Choices for the 21st Century*, Chesapeake Bay Program Science and Technical Advisory Committee Publication No. 03-001, 2003.

28. Charles Babington, "Bay Governors Differ on Waste Limits," *The Washington Post*, December 9, 1998, 11.

29. Boesch and Greer, *Chesapeake Futures*. Boesch and Greer also point out that the average size of new single-family homes went from 1,500 square feet in 1970 to 2,265 square feet in 2000. The average lot size increased substantially during this period as well.

30. According to the bay program's 1999 *State of the Bay* report, 25 percent of phosphorous came from point sources, 66 percent from nonpoint sources, and 9 percent from the atmosphere; with respect to nitrogen, 22 percent came from point sources, 57 percent from nonpoint sources, and 21 percent from the atmosphere. These proportions are periodically adjusted to reflect new information.

31. Karl Blankenship, "After Review, Bay Program Moves to Improve Water Quality Model," *Bay Journal*, January/February 2000.

32. Kevin Sellner, director, Chesapeake Research Consortium, personal communication, 2003.

33. Boesch, personal communication. A limiting nutrient is one that limits a plant's growth. Such nutrients cause pollution when they enter a water body in large amounts because they enable algae to grow.

34. Timothy M. Hennessey, "Governance and Adaptive Management for Estuarine Ecosystems: The Case of Chesapeake Bay," *Coastal Management* 22 (1994): 119–145.

35. The average BNR system in the bay watershed results in nitrogen effluence of about 6 milligrams per liter. See Ernst, *Chesapeake Bay Blues*. Because BNR relies on biological activity, it is more effective in the summer, when temperatures are warmer and rates of biological activity are faster. See Karl Blankenship, "BNR Treatment Exceeds Expectations at Blue Plains," *Bay Journal*, December 1996.

36. Blankenship, "BNR Treatment Exceeds Expectations."

37. Eric Lipton, "Va. Raises Commitment to Chesapeake Cleanup," *The Washington Post*, August 3, 1998, B1.

38. Boesch and Greer, *Chesapeake Futures*. In a typical septic system, sewage flows into an underground tank. Solids settle in the tank, while liquids are allowed to leach into the soil and eventually into groundwater. Tanks must be pumped periodically to remove waste that has not been broken down by bacteria. About 5 percent of septic systems in Maryland and Virginia fail each year. See Stephen C. Fehr and Peter Pae, "Aging Septic Tanks Worry D.C. Suburbs," *The Washington Post*, May 18, 1997, 1.

39. D'Vera Cohn, "Bans on Phosphates Said to Aid Bay Cleanup," *The Washington Post*, January 23, 1989, B4.

40. Boesch and Greer, *Chesapeake Futures*. About half of these operations primarily raised crops, while 42 percent focused on animals.

41. According to the Chesapeake Bay Program, of the nitrogen loads from agriculture, 43 percent comes from manure, while 57 percent comes from commercial fertilizer. A little more than half (51 percent) of the phosphorus from agriculture comes from commercial fertilizer, while just under half (49 percent) comes from manure.

42. Carpenter, "More People, More Pollution."

43. According to the EPA, poultry on the lower shore sent more than four times as much nitrogen into the bay as the biggest nonagricultural source—leaky septic tanks and runoff from developed areas—and more than three times as much phosphorus as the second-largest nonfarm source, sewage treatment plants. See Peter S. Goodman, "An Unsavory Byproduct," *The Washington Post*, August 1, 1999, 1.

44. Goodman, "An Unsavory Byproduct"; Peter S. Goodman, "Permitting a Pattern of Pollution," *The Washington Post*, August 2, 1999, 1. Delmarva is the local name for the peninsula of Delaware, Maryland, and Virginia.

45. Ernst, *Chesapeake Bay Blues*.

46. Goodman, "An Unsavory Byproduct."

47. In the 1990s farmers began experimenting with another option: adding phytase, an enzyme, to feed pellets to reduce the nitrogen and phosphorus in animal manure. In addition, some farmers in New York State were experimenting with grass feeding and rotational grazing, which not only reduced nutrient runoff, but saved money on feed, fertilizer, chemicals, and veterinary operations. A third option was to install anaerobic digesters that turned manure into energy, leaving nitrogen and phosphorus in forms that are more flexible and can be transported more economically.

48. Ernst, *Chesapeake Bay Blues*.

49. CAFOs are defined by statute as lots where a large number of animals are kept for more than forty-five days in a twelve-month period and where crops or vegetation are not grown. Although CAFOs must get permits under the NPDES program, neither the EPA nor the states pursued this avenue in the 1980s and 1990s. In 2001 the EPA estimated that at least 13,000 CAFOs were required to have permits, but the federal government and states had issued only 2,520 permits. The GAO found that until the mid-1990s the EPA directed few resources toward the permit program for CAFOs, instead giving higher priority to other polluters; as a result, EPA regulations allowed some 60 percent of CAFOs to avoid regulations altogether. See Michele M. Merkel, "EPA and State Failures to Regulate CAFOs Under Federal Environmental Laws," Environmental Integrity Project, Washington, D.C. Available at http://www.environmentalintegrity.org/pdf/publications/EPA_State_Failures_Regulate_CAFO.pdf; Government Accountability Office, *Livestock Agriculture: Increased EPA Oversight Will Improve Environmental Program for Concentrated Animal Feeding Operations*, GAO-03-285 (January 2003).

50. Quoted in Peter S. Goodman, "Who Pays for What Is Thrown Away?" *The Washington Post*, August 3, 1999, 1.

51. Quoted in Goodman, "An Unsavory Byproduct."

52. Ernst, *Chesapeake Bay Blues*.

53. Annemarie Herbst, "Regulating Farm Nutrient Runoff: Maryland's Experience with the Water Quality Improvement Act." Unpublished masters thesis (Cambridge, Mass.: Massachusetts Institute of Technology, Department of Urban Studies and Planning, 2005). According to Herbst, the central reason for the law's failure to yield results was the decision to acquiesce to farmers and place responsibility for implementation of the law in Maryland's farmer-friendly Department of Agriculture, rather than the Department of Environment.

54. Quoted in Goodman, "Permitting a Pattern of Pollution."

55. Ibid.

56. The Susquehanna contributes half the freshwater reaching the bay. It deposits one-half of the phosphorous and three-quarters of the nitrogen in the upper part of the bay, above the Potomac; it is the source of about 40 percent of the nitrogen and 20 percent of the phosphorus in the bay as a whole. See Tasha Eichenseher, "Susquehanna Tops List of Endangered Waterways," *Greenwire*, April 13, 2005.

57. David Finkel, "Chesapeake Bay Blues," *The Washington Post Magazine*, July 7, 1991, W9.

58. David A. Fahrenthold, "Pennsylvania Pollution Muddies Bay Cleanup," *The Washington Post*, May 16, 2004, 1.

59. Reflecting the political clout of agriculture, however, the Critical Area Law reduced the width of the buffer to twenty-five feet for agricultural land.

60. Each designation came with its own regulations. For land designated as intensely developed, new development must be clustered in previously developed areas, and impervious surface area must be kept to 15 percent or less of the site. For land designated as limited development, new development must protect natural areas and maintain slopes, and impervious surface areas must be 15 percent or less of the site. And for RCAs, only residential development is allowed and is limited to one dwelling per twenty acres.

61. "The Maryland Initiative: Lesson for the Nation," *EPA Journal* 15 (September/October 1989): 29–34.

62. The grandfathering clause, along with the extended grace period, were adopted to defuse objections to the law; a third provision, allowing each jurisdiction a floating growth zone, served the same purpose. See Patrick W. Beaton and Marcus Pollock, "Economic Impact of Growth Management Policies Surrounding the Chesapeake Bay," *Land Economics* 68 (1992). Poor record keeping prevented state and county officials from knowing which lots were grandfathered, so it was extremely difficult to limit development in the critical area. See Anita Huslin, "Watching, Conserving Chesapeake Waterways," *The Washington Post*, May 8, 2003, T12.

63. Eugene L. Meyer, "Erosions of Protections for Bay Shoreline Feared," *The Washington Post*, December 27, 1988, C1.

64. The law allowed agricultural operations to plow their fields within twenty-five feet of a stream if they completed pesticide and nutrient management plans, although the state did not have the staff to ensure those plans were being implemented. Timber companies were exempted if they agreed to follow state forestry guidelines and stay at least twenty-five feet from creeks and streams. But only 7 percent of logging

operations abided by those guidelines, according to a state study released in 2000. And local officials contended that timbering was used as a regulatory shield to clear land for new development. See Jennifer Peter and Scott Harper, "Flawed Law Fails the Bay," *The Virginian-Pilot*, July 29, 2001, 1.

65. Peter and Harper, "Flawed Law Fails the Bay"; Jennifer Peter and Scott Harper, "Exceptions Rule in Local Cities," *The Virginian-Pilot*, July 30, 2001, 1.

66. For example, according to the *Virginian-Pilot* analysis, Virginia Beach approved 806 of 868 exemptions to the 100-foot buffer rule; Chesapeake had granted all but one application to cut the buffer to 50 feet and 95 percent of the 223 requests to intrude even farther. In total, South Hampton Roads cities—except for Suffolk—had granted more than 1,000 requests to bypass the law between its passage in 1988 and 2000; in neighboring Isle of Wight County, almost every exception to the law had been approved. See Peter and Harper, "Flawed Law Fails the Bay"; Peter and Harper, "Exceptions Rule in Local Cities."

67. Scott Harper, "Bay Development Rules Tightened," *The Virginian-Pilot*, December 11, 2001, 1; Scott Harper, "Two Cities Reject Bay Buffer Rule," *The Virginian-Pilot*, March 2, 2002, 1.

68. Kari Hudson, "Restoring the Chesapeake: Bottom-Up Approach Is Winning Pollution Battle," *American City & County*, June 1, 1995. Available at http://americancityandcounty.com/mag/government_restoring_chesapeake_bottomup.

69. D'Vera Cohn, "Poll Shows People Disagree with Governments Over Who Pollutes the Bay," *The Washington Post*, May 6, 1994, D3.

70. Carpenter, "More People, More Pollution."

71. Ernst, *Chesapeake Bay Blues*.

72. Ibid.

73. Chesapeake Bay Commission, "Seeking Solutions: Chesapeake Bay Commission, Annual Report 2001," p. 34. Available at http://www.chesbay.us/Publications/CBC%20annual%20report%202001.pdf

74. Mike Hirshfield, senior vice president. Oceana, personal communication, 2003; Donald F. Boesch and Erica B. Goldman, "Chesapeake Bay, USA," in *Ecosystem-Based Management for the Oceans*, ed. Karen MacLeod and Heather Leslie (Washington, D.C.: Island Press, 2009), 268–293.

75. Anita Huslin, "Bay's Rise May Add to Impact of Storms," *The Washington Post*, September 29, 2003, B1.

76. Former EPA official J. Charles Fox contends that the distinction between point sources and nonpoint sources is artificial. "In a strict reading of the Clean Water Act—in twenty years I've never seen a nonpoint source," he says. Most of what have been defined as nonpoint sources, in other words, could easily be redefined as point sources and required to have permits. The 1987 Clean Water Act Amendments, for example, clarified that municipal stormwater is a point source, and cities must have permits for it. Fox, personal communication.

77. U.S. Environmental Protection Agency, "Progress in Reducing Pollution From Wastewater Facilities," n.d. Available at http://www.epa.gov/reg3wapd/pdf/pdf_npdes/WastewaterProgress.pdf.

78. Fahrenthold, "Pennsylvania Pollution Muddies Bay Cleanup," B3.

79. With the help of the World Resources Institute, a Washington, D.C.-based environmental organization, the state calculated the value of credits based on the impact of a new management practice and the distance of the farm from the bay. A typical credit was worth between $2 and $9 for a reduction of about 1.6 pounds of pollutants. See Felicity Barringer, "A Plan to Curb Farm-to-Watershed Pollution of Chesapeake Bay," *The New York Times*, April 13, 2007, 10.

80. Sopan Joshi, "A Mandatory Sewage Plan in Search of Federal Funding," *The Washington Post*, September 26, 2008, B1. A study issued by the Chesapeake Stormwater Network in March of 2010 found that lawns had become the biggest single "crop" in the watershed, and lawn fertilizer was one of the principal contaminants in urban-suburban runoff. See Robert McCartney, "Redefining the Beautiful Lawn When It Comes to Bay's Health," *The Washington Post*, April 25, 2010, C1. Another study, released in July 2008, suggested that cities were responsible for more water-quality problems than previously believed and that climate change was causing intense rainfall after dry spells, washing nitrogen pollution from paved areas into neighboring waterways. The study, published in the journal *Environmental Science and Technology*, found a link between booming development in the watershed and high nitrate concentrations in half of Maryland's 9,000 streams.

81. David A. Fahrenthold, "Pollution Is Rising in Tributaries of Bay, Data Show," *The Washington Post*, December 5, 2007, B1.

82. Lisa Rein, "As Pressure Increases, So Do Ways to Curb Polluted Runoff," *The Washington Post*, May 23, 2005, 1.

83. Quoted in ibid.

84. David A. Fahrenthold, "Pollution Kills a Third of Bay Grasses," *The Washington Post*, May 19, 2004, B1.

85. "Survey Shows Increase in Vital Underwater Grasses," *The Washington Post*, May 26, 2006, B3.

86. Lucy Kafanov, "Submerged Gasses Decline by 25 Percent—EPA," *E&E News PM*, March 30, 2007.

87. Eric Rich, "Bay Scientists Chagrined at Being Right," *The Washington Post*, August 4, 2005, T3.

88. Peter Whoriskey, "Bay Pollution Progress Overstated," *The Washington Post*, July 18, 2004, 1; Peter Whoriskey, "Oxygen Levels in Bay Disputed," *The Washington Post*, July 23, 2004, B1; Elizabeth Williamson, "GAO Denounces Bay Cleanup Efforts," *The Washington Post*, November 16, 2005, B1.

89. The bay program initially said that the flow of nitrogen and phosphorus from rivers into the bay had declined nearly 40 percent since 1985. A revised computer model

suggested that phosphorus pollution had dropped about 28 percent since 1985, while nitrogen pollution had dropped about 18 percent. But according to the USGS, the observed concentrations of those nutrients flowing into the bay from rivers had changed little; eight of the nine rivers entering the bay showed no trend or increased concentrations of phosphorus since the late 1980s, while seven of the nine major rivers showed no trend or increased concentrations of nitrogen. See Whoriskey, "Bay Pollution Progress Overstated." A review led by Tom Simpson of the University of Maryland found that many of the equations on which the EPA based its estimates were based on small-scale experiments or the educated guesses of experts. In fact, eighteen of the thirty-six measures considered by the model had less of an effect than they had been credited with. As a result, the model consistently overstated the effectiveness of remedial measures. See David A. Fahrenthold, "Cleanup Estimate for Bay Lacking," *The Washington Post,* December 24, 2007, B1. Similarly, research by two University of Maryland scientists found no improvement in the volume of oxygen-depleted water in the bay, contrary to reports from the Chesapeake Bay Program that there had been indications of an improving trend since 1985. See Whoriskey, "Oxygen Levels in Bay Disputed."

90. Quoted in David A. Fahrenthold, "Broken Promises on the Bay," *The Washington Post,* December 27, 2008, 1.

91. David A. Fahrenthold, "A Revitalized Chesapeake May Be Decades Away," *The Washington Post,* January 5, 2007, 1.

92. Quoted in David A. Fahrenthold, "What Would It Take to Clean Up the Bay by 2010?" *The Washington Post,* January 29, 2007, 1.

93. David A. Fahrenthold, "Bay Program Ready to Study Less, Work More," *The Washington Post,* September 26, 2006, B1.

94. Williamson, "GAO Denounces Bay Cleanup Efforts."

95. Anita Huslin, "Environmental Activists Issue New Call to Save Bay," *The Washington Post,* June 17, 2003, B1.

96. Quoted in Tim Craig, "Bay Cleanup Criticized," *The Washington Post,* August 22, 2003, B8.

97. Christian Davenport, "Bay Advocate Turns to Court," *The Washington Post,* July 22, 2004, T10.

98. David A. Fahrenthold, "Advocates for Bay Churn Waters," *The Washington Post,* September 5, 2004, C1. Waterkeepers are full-time, paid public advocates for particular rivers; communities support them because they provide citizens with the tools to challenge well-heeled industrial and development interests.

99. Quoted in Ernst, *Fight for the Bay.*

100. See "Chesapeake Bay Action Plan." Available at http://www.bayactionplan.com/.

101. Quoted in David A. Fahrenthold, "Scientists Urge More Aggressive Cleanup," *The Washington Post,* December 9, 2008, B2.

102. David A. Fahrenthold, "Evaluation of Chesapeake Goals Killed," *The Washington Post*, May 4, 2009, B1.

103. Allison Winter, "U.S., States Announce New Goals, Oversight," *Greenwire*, May 12, 2009.

104. Ibid.

105. Pennsylvania agreed to reduce nitrogen by 7.3 million pounds and phosphorus by 300,000 pounds by 2011 through a variety of actions; Delaware pledged to cut nitrogen by 292,072 pounds; New York intended to reduce nitrogen 870,500 pounds and phosphorus 86,700 pounds; and West Virginia planned to reduce nitrogen 42,254 pounds and phosphorus 3,364 pounds. The total reductions pledged by the jurisdictions amounted to 15.8 million pounds of nitrogen and 1.1 million pounds of phosphorus. See Karl Blankenship, "Executive Order, 2-Year Deadlines Boost Efforts to Clean Up Bay," *Bay Journal*, June 2009.

106. Ernst, *Fight for the Bay*.

107. David A. Fahrenthold, "Rules on Storm Water Eased for Some Builders," *The Washington Post*, March 10, 2010, B6.

108. David A. Fahrenthold, "Environmental Protections Take Hit in Fiscal Crunch," *The Washington Post*, January 26, 2009, B1.

109. David A. Fahrenthold, "Md. Runs Short of Pollution Inspectors," *The Washington Post*, September 23, 2008, B1. In early 2010 the Chesapeake Bay Riverkeepers petitioned the EPA to remove Maryland's authority to issue discharge permits, claiming the Maryland Department of the Environment was not enforcing compliance with the permits. See Rona Kobell, "Riverkeepers Ask EPA to Take Away MDE's Authority to Issue Permits," *Bay Journal*, January 2010.

110. Karl Blankenship, "Bay Only 38% of the Way Toward Meeting Water, Habitat Goals," *Bay Journal*, April 2009.

111. David A. Fahrenthold, "EPA Asked to Toughen Pollution-Prevention Laws," *The Washington Post*, September 2, 2009, B4.

112. Karl Blankenship, "New Federal Bay Strategy Promises Unprecedented Effort," *Bay Journal*, June 2010.

113. The EPA's CAFO program under the Clean Water Act is a morass, with rules repeatedly challenged in court and the agency retreating. For a complete history of actions taken since 2001, see http://water.epa.gov/polwaste/npdes/afo/CAFO-Regulations.cfm.

114. The federation released a report in late 2010 comparing the results of the EPA's Watershed Model to the results of a model developed by the U.S. Department of Agriculture. It argued that because there were substantial differences between the two, the EPA should hold off on issuing its TMDL. A panel convened by the bay program's Scientific and Technical Advisory Committee rejected the review, conducted by a Michigan-based consulting firm, calling it "flawed" and characterized by "poor scientific

merit." Quoted in Karl Blankenship, "Agribusiness's Report Critical of Bay Model Called Flawed, Misleading," *Bay Journal*, October 2011.

115. Darryl Fears, "Farm Bureau Targets EPA Pollutant Limits for Chesapeake Bay," *The Washington Post*, February 28, 2011.

116. Karl Blankenship, "Overall, Bay Region Exceeded 2012–13 Nutrient, Sediment Reduction Goals," *Bay Journal*, May 10, 2014.

117. Karl Blankenship, "Water Quality Monitoring Shows Some Long-Term Improvements Fading Away," *Bay Journal*, February 1, 2015.

118. Chesapeake Watershed Agreement. (2014). Available at https://www.chesapeakebay.net/documents/FINAL_Ches_Bay_Watershed_Agreement.withsignatures-HIres.pdf

119. Chesapeake Bay Program. 2014, June 16. Chesapeake Watershed Agreement. Available at http://chesapeakebay.net

120. Blueprint Progress: Tracking Milestones. 2019. http://www.cbf.org/how-we-save-the-bay/chesapeake-clean-water-blueprint/blueprint-progress-tracking.html

121. Dietrich, "Deep Dive Shows Damaged Chesapeake Bay Can Heal".

122. Ibid.

123. David Fahrenthold, "Broken Promises on the Bay," *The Washington Post*, December 27, 2008.

124. Boesch and Goldman, "Chesapeake Bay, USA," 271.

125. A report on climate change by the Chesapeake Bay Program's Scientific and Technical Advisory Committee suggested that a rising sea level would exacerbate flooding and submerge the estuary's wetlands; the combination of warmer waters and rising carbon dioxide levels would encourage algae growth; and a hotter bay would make it harder for underwater eelgrass—which provides habitat for blue crabs and other marine life—to survive. On the other hand, warmer water may favor the blue crab, which evolved in tropical conditions, but will diminish the amount of eelgrass, a favored habitat. See Lauren Morello, "A Changing Climate Will Roil and Raise Chesapeake Bay," *Greenwire*, June 24, 2009. Already, many bay communities are feeling the effects of rising sea levels. See David A. Fahrenthold, "Losing Battles Against the Bay," *The Washington Post*, October 25, 2010, 10.

126. Tom Horton, "Hanging in the Balance," *National Geographic* (June 1993): 4–35.

127. Hirshfield, personal communication.

128. Sam Luoma, senior scientist, USGS, personal communication.

129. Linda Pilkey-Jarvis and Orrin H. Pilkey, "Useless Arithmetic: Ten Points to Ponder When Using Mathematical Models in Environmental Decision Making," *Public Administration Review* 68 (May/June 2008): 469–479.

130. Boesch et al., "Chesapeake Bay Eutrophication," 311.

131. Peter deLeon and Jorge E. Rivera, "Voluntary Environmental Programs: A Symposium," *Policy Studies Journal* 35 (November 2007): 685–688; Dinah A. Koehler, "The Effectiveness of Voluntary Environmental Program—A Policy at a Crossroads," *Policy Studies Journal* 35 (November 2007): 689–722; Nicole Darnall and Stephen Sides, "Assessing the Performance of Voluntary Environmental Programs: Does Certification Matter?" *Policy Studies Journal* 36 (February 2008): 95–117.

132. Quoted in Angus Phillips, "Old Ideas Are Polluting the Chesapeake Bay," *The Washington Post*, December 14, 2008, D2.

CHAPTER FIVE

MARKET-BASED SOLUTIONS

Acid Rain and the Clean Air Act Amendments of 1990

The Clean Air Act of 1970 was the federal government's first serious step toward reducing the nation's air pollution (see Chapter 2). In 1977 Congress amended the Clean Air Act and scheduled it for reauthorization in 1981. But President Ronald Reagan opposed any action on the bill, and legislators were deeply divided, particularly on the issue of acid rain.[1] As a result, the act languished in Congress throughout the 1980s. Finally, after a decade of stalemate, Congress and the administration of President George H. W. Bush agreed on a clean air plan that included as its signature provision an innovative allowance-trading approach to reducing the emissions that cause acid rain. The regulations required by the Clean Air Act Amendments of 1990 took effect in 1995 and aimed by the year 2000 to cut in half emissions of sulfur dioxide (SO_2), acid rain's main precursor. By 2010 SO_2 emissions were well below the cap, and reductions had been achieved at a far lower cost than projected.

This case illustrates the importance of regional concerns in Congress and how acid rain has increasingly become an issue for developing countries such as China. The regional emphasis of Congress was particularly pronounced in this case because acid rain is a transboundary problem: those who bear the environmental burden of its effects are in a different region from those who cause it. As previous cases have made clear, the organization and procedures of Congress provide its members with many ways of resisting the imposition of direct, visible economic costs on their constituents: authorizing subcommittees can delete legislative provisions during markup; committee chairs can prevent a bill from reaching the floor, as can a majority of committee members; and legislative leaders, such as the Speaker of the House or the Senate majority leader, can refuse to schedule a vote on a bill. Even if a bill does reach the floor, its opponents may attach enough hostile amendments to cripple it or, in the Senate, a minority can filibuster it.

In a legislative contest over environmental protection, science can play a critical, even if indirect, role. Opponents of environmentally protective policies typically emphasize scientific uncertainty and call for more research in hopes of delaying change. But proponents of those policies rely on the available scientific evidence, however uncertain, to buttress their claims of environmental harm and the need for prompt government action. They are more likely to be successful if science furnishes a simple and direct relationship between cause and effect, sympathetic victims, loathsome villains, and an imminent crisis; the backing of credible, legitimate scientific knowledge brokers can further bolster the case for a protective policy.[2] If environmentalists manage to persuade the public that a problem is serious and warrants remediation, a legislative leader is more likely not only to adopt the issue, but also to succeed in recruiting allies. A widely accepted story provides a handy explanation that rank-and-file members can give constituents.[3]

International pressure can also affect legislative dynamics. Because SO_2 emitted in the United States precipitates in Canada, the Canadian government, as well as Canadian scientists and environmentalists, pushed to get acid rain onto the U.S. policy agenda and then actively tried to influence policy. According to one observer, the issue "overshadowed almost all other elements of the bilateral relationship" between the United States and Canada.[4] Although they had little direct effect on the legislative process, the Canadians provided a steady stream of reliable information to, and publicity for, the proregulation side, which in turn helped to shape public and elite views about the need for controls.

Finally, policy entrepreneurs can facilitate the passage of new legislation by furnishing leaders with a novel solution that transforms the political landscape by broadening supportive coalitions or breaking up opposing ones. In this case, the allowance-trading mechanism proposed for acid rain helped break the legislative logjam by appealing to some Republican members of Congress, who believed such market-based approaches were superior to conventional approaches in two respects. First, in theory, market-based policy instruments allow polluters who can clean up cheaply to do more and those for whom cleanup is costly to do less, which in theory reduces overall pollution to the desired level at a minimum total cost—an efficiency rationale. Second, market-based tools ostensibly create incentives for businesses to develop and adopt new technologies and to continue reducing pollution beyond the target level because they can save money by doing so.[5] In addition to attracting supporters, the allowance-trading mechanism can divide opponents by promising benefits to polluters that can clean up cheaply (or have already done so) and then sell their pollution allowances. If proponents of policy change can devise an attractive, coalition-building solution, they are more likely to be able to capitalize on the political momentum unleashed when a policy window opens.

BACKGROUND

Scientists began documenting the effects of acid rain more than 100 years ago. Robert Angus Smith, a nineteenth-century English chemist, was the first to detect the occurrence of acidic precipitation. He coined the term *acid rain* and conducted extensive tests of its properties, but his research was largely ignored. Similarly, work by ecologist Eville Gorham on the causes of acidic precipitation and its consequences for aquatic systems in the 1950s and 1960s "was met by a thundering silence from both the scientific community and the public at large."[6] But in the late 1960s, Swedish soil expert Svante Odén rekindled scientific interest in acid rain. Integrating knowledge from the scientific disciplines of limnology, agricultural science, and atmospheric chemistry, Odén developed a coherent analysis of the behavior of acidic deposition over time and across regions. He hypothesized that acidic precipitation would change surface water chemistry, lead to declines in fish populations, reduce forest growth, increase plant disease, and accelerate materials damage.[7]

Odén's research laid the groundwork for conclusions unveiled at the 1972 United Nations Conference on the Human Environment in Stockholm. At the conference, a group of Swedish scientists presented a case study demonstrating that acidic precipitation attributable to SO_2 emissions from human-made sources—primarily coal-fired industrial processes and utilities—was having adverse ecological and human health effects.

Growing scientific concern about acid rain was a primary factor in getting the problem on the U.S. political agenda. In 1975, the Forest Service sponsored an international symposium on acid rain and forest ecosystems, and shortly thereafter, Professor Ellis Cowling, a Canadian expert, testified at a congressional hearing on the need for more funding to study the phenomenon. In 1977 President Jimmy Carter's Council on Environmental Quality suggested that the United States needed a comprehensive national program to address acid rain. This argument gained momentum in the executive branch, and in 1978 Carter began to take steps toward that goal by establishing the Bilateral Research Consultation Group on the Long-Range Transport of Air Pollution to conduct a joint investigation with Canada. Calling acidic precipitation "a global environmental problem of the greatest importance,"[8] the following year the president asked Congress to expand funding for research and to investigate possible control measures under the Clean Air Act.

Fueling environmentalists' interest in acid rain policy was a series of reports in 1978, 1979, and 1980 by the International Joint Commission (IJC), which had been created to address U.S.–Canada water quality issues in the Great Lakes. Those reports suggested that, if not controlled, acid rain could render 50,000 lakes in the United States and Canada lifeless by 1995, destroy the productivity of vast areas of forest, and contaminate the drinking water supplies of millions of people.[9] In addition, the Bilateral Research Consultation Group issued two reports, in October 1979 and November 1980, which together comprised the first comprehensive statement of scientific knowledge about acid rain and its likely effects in eastern North America. One of the group's major findings was that at least 50 percent of Canada's acid deposition originated in the United States, whereas only 15 percent of acid rain in the United States came from Canada.[10]

The concern among environmentalists and high-level Canadian officials generated by these reports created pressure on the United States to do more than simply study acid rain. So in 1980 President Carter signed the U.S.–Canada memorandum of intent to negotiate an agreement on transboundary air pollution. He also approved the Acid Precipitation Act of 1980, which established the Interagency

Task Force on Acid Precipitation and charged it with planning and implementing the National Acid Precipitation Assessment Program, a comprehensive research program to clarify the causes and effects of acid rain.

Impatient for a more substantial policy response, Canadian and U.S. advocates of acid rain controls appealed directly to the EPA to regulate acid rain-causing emissions under the Clean Air Act. Although Carter's EPA administrator Douglas Costle conceded that government intervention was warranted, he also expressed concern about the political risks, noting that "in an election year, it is best to keep your head down" rather than embark on a new environmental regulation program.[11] In the administration's final days, however, Costle acknowledged that U.S. emissions were contributing significantly to acid rain over sensitive areas of Canada, and he laid the foundation for the next EPA administrator to invoke Clean Air Act Section 115, under which the EPA can require states to reduce the impact of air pollution on foreign countries.

THE CASE

In the early 1980s the topic of acid rain moved from the relative obscurity of international scientific inquiry into the domestic political spotlight, as the U.S. media propagated the problem definition articulated by scientific knowledge brokers and environmentalists. During the 1970s few Americans considered acid rain a policy problem. Yet in 1983 a Harris poll found that nearly two-thirds of those questioned were aware of acid rain and favored strict controls on SO_2 emissions.[12] By this time acid rain had also become a serious foreign-policy issue between the United States and Canada, and environmentalists had begun pressuring members of Congress to introduce acid rain controls as part of the Clean Air Act reauthorization. But President Reagan and powerful members of Congress opposed such measures, and for nearly a decade, they used their political resources to thwart them.

The Emerging Scientific Consensus on Acid Rain

By the early 1980s scientists knew that the precursors to acid rain, SO_2 and nitrogen oxides (NO_x), were released by a variety of natural mechanisms, such as volcanic eruptions, lightning, forest fires, microbial activity in soils, and biogenic processes, but that industrial activity that relied on burning fossil fuels was spewing these substances into the air in quantities that dwarfed nature's output. They estimated that in 1980 the United States emitted about 26 million tons of SO_2 and 21 million tons of NO_x; that the thirty-one states east of the Mississippi River emitted about 80 percent of that SO_2 (22 million tons) and two-thirds of the NO_x (14 million tons); that nearly three-quarters of the SO_2 emitted east of the Mississippi came from power plants; and that forty enormous coal-fired plants clustered in the Midwest and the Ohio and Tennessee valleys—a single one of which emitted about 200,000 tons of SO_2 a year—were responsible for almost half of the region's SO_2 emissions.[13]

Furthermore, scientists were beginning to recognize that, because these enormous plants had smokestacks as high as the tallest skyscraper, their SO_2 emissions traveled hundreds or even thousands of miles downwind, turning to acid and falling to earth along the way.[14] Therefore, more than 50 percent of the acid sulfate deposited on the Adirondacks of New York came from Midwestern sources; about 20 percent came from the large metal smelters in Ontario, Canada; and less than 10 percent originated in the Northeast. Although the sources of NO_x were more diverse than those of SO_2, scientists estimated that more than half of all NO_x emissions also came from the smokestacks of power plants and industrial sources. Automobiles, as well as other forms of transportation, were the second major source of NO_x.

Scientists were also gaining confidence in their understanding of the consequences of acid rain, particularly for aquatic systems. They knew that healthy lakes normally have a pH of around 5.6 or above; that when a lake's pH drops to 5.0, its biological processes begin to suffer; and that at pH 4.5 or below, a lake is generally incapable of supporting much life. Because precipitation in the eastern United States had an average pH of 4.3, that region's lakes and streams—whose soils lacked alkaline buffering—had become highly acidic.[15] Lakes at high altitudes appeared to be particularly sensitive to acid rain because, surrounded only by rocky outcroppings, they tend to be poorly buffered. Furthermore, researchers discovered that when water runs off snow or ice or during snowmelt, high-altitude lakes receive a large pulse of acid capable of killing fish and other aquatic life outright.[16]

Scientific comprehension of the impacts of acidification on terrestrial ecosystems, particularly evergreen forests, was more uncertain, and researchers had only a sketchy understanding of the complex interaction between acid rain and other pollutants. Since the 1960s scientists had observed a massive decline in many forests in parts of Europe and the eastern United States, particularly in high-elevation coniferous forests. For example, in sites above 850 meters in New York's Adirondacks, the Green Mountains of Vermont, and the White Mountains of New Hampshire, more than half of the red spruce had died. At lower elevations, researchers had documented injury to hardwoods and softwoods. This loss of forest vitality could not be attributed to insects, disease, or direct poisoning because it was occurring in stands of different ages with different histories of disturbance or disease. Pollution was the only known common factor affecting all of them. In spite of the troubling dimensions of forest decline, however, scientists had been unable to establish a firm causal link between acid rain and forest damage—although all the hypotheses to explain failing tree health in the United States and Europe implicated acid rain, if not as a lethal agent then as a major stressor. Laboratory and field studies showed that acidic deposition could damage leaves, roots, and microorganisms that form beneficial, symbiotic associations with roots; impair reproduction and survival of seedlings; leach nutrients such as calcium and magnesium from soils; dissolve metals such as aluminum in the soil at levels potentially toxic to plants; and decrease a plant's resistance to other forms of stress, including pollution, climate, insects, and pathogens.[17]

Scientists were also learning about acid rain's adverse effects on soils. In particular, they had ascertained that, although the forests in the Northeast and Canada grow on naturally acidic soils, acid rain can still damage these soils. The explanation for this is that most of the important ions in soil are positively charged—hydrogen (acidity); calcium and magnesium (nutrients); and aluminum, lead, mercury, and cadmium (heavy metals)—and are thus bound to the negatively charged surface of large, immobile soil particles. But acid deposition depletes the ability of soil particles to bind positively charged ions and thus unlocks the acidity, nutrients, and toxic metals.[18] Even in lakes that had not begun to acidify, the concentration of calcium and magnesium in the water was increasing with time, suggesting that these nutrients were being leached from surrounding soils.

Beyond its ecological impacts, scientists suspected that acid rain was damaging human health, directly or indirectly. They had determined that, once acid rain dissolves mercury, lead, cadmium, and other toxic metals in the environment, it transports them into drinking water supplies and thereby into the food chain. In addition, some reports suggested that downwind derivatives of SO_2, known as acid aerosols, pose respiratory health threats to children and asthmatics in the eastern United States. Finally, studies showed that particles of acid sulfate scatter light and reduce visibility, creating a haze that was most pronounced in summer but was present in all seasons. In the eastern United States, visibility was about twenty-five to forty miles absent pollution; it declined to as little as one mile during pollution episodes.

Crafting and Disseminating a Compelling Story About Acid Rain

As scientists became more concerned about acid rain, some prominent experts began translating the available research into a simple and compelling causal story in hopes of alerting the public and prompting policymakers to adopt policies that would mitigate the phenomenon. In the process, these experts propelled into common currency the term *acid rain*—a shorthand that conveyed the phenomenon's "unnaturalness," while masking its complexity. They portrayed acid rain as a serious problem with potentially irreversible effects. And they suggested the only way acid rain could be fixed was by substantially reducing emissions of SO_2 and, to a lesser extent, NO_x.

Because scientific knowledge brokers served as journalists' primary source of information, early coverage of the issue in the popular press echoed their precautionary tone and reiterated a simple story that linked the emissions of dirty, coal-fired utilities in the Midwest to the death of pristine lakes and forests in the Northeast. Media reports suggested that although a potential ecological catastrophe loomed, it could be averted by drastic cuts in emissions. One of the earliest articles, a 1974 *Time* magazine piece titled "The Acid Threat," explained that although scientists did not fully understand the effects of increased acidity in rain or snow, they suspected it of causing forest damage, killing fish, and corroding buildings, bridges, and monuments. Similarly, a 1978 article in the *Saturday Review* titled

"Forecast: Poisonous Rain, Acid Rain" charged that dangerously acidic rains were falling nationwide and poisoning rivers and lakes and killing fish. In 1979 *U.S. News & World Report* introduced the issue by saying, "Rain that is increasingly acidic is beginning to devastate lakes, soil and vegetation throughout North America—and alarming scientists who fear the damage may be irreversible."[19] And a 1980 *Time* magazine piece concluded, "As ecologists point out, doing nothing about acid rain now could mean nightmarish environmental costs in the future."[20]

Similarly, a 1977 story in *The New York Times* began with an alarming statement: "The brook trout and spotted salamanders of Adirondack Mountain lakes are becoming vanishing species because of deadly rain and snow containing corrosive acids that kill them."[21] The *Chicago Tribune* also ran a series of worrisome articles on acid rain, prompted by the release of the IJC assessments. In the summer of 1979, science journalist Casey Bukro wrote that "Canadian and American scientists [had] discovered something that appears to be a new kind of ecological doomsday machine" in which "[s]ulfur fumes from coal burning and nitrogen gases from auto emissions are carried thousands of miles by the wind and fall back to Earth as sulfuric and nitric acid rain or snow. Acid rains kill fish in lakes and streams and damage the fertility of the soil. The damage could be irreversible."[22] Following Bukro's articles, a *Chicago Tribune* editorial concluded portentously, "So far it is only fish that are being wiped out—fish and the other creatures for whom fish are an indispensable link in the food chain. But when any life form crashes, there is a disturbing message for human beings. Something is going wrong."[23] Even the *Los Angeles Times* reported on the environmental threat posed by acid rain in the eastern United States; one article began, "Those sweet showers and pure fluffy snowfalls turn out not to be so sweet and pure after all. In fact, some rain and snow are downright filthy."[24]

Resistance to Acid Rain Controls in the Reagan Administration

The emerging scientific consensus and unsettling media coverage notwithstanding, President Reagan made his opposition to acid rain controls clear in 1981. Neither he nor the members of his administration believed the benefits of mitigating acid rain outweighed the costs of regulating generators of SO_2 and NO_x. David Stockman, head of the OMB during Reagan's first term, expressed the administration's cornucopian view when he wondered, "How much are the fish worth in those 170 lakes that account for four percent of the lake area of New York? And does it make sense to spend billions of dollars controlling emissions from sources in Ohio and elsewhere if you're talking about a very marginal volume of dollar value, either in recreational terms or in commercial terms?"[25] But administration officials were rarely so forthright about their values; instead, they continued to contest the science and cite scientific uncertainty as a rationale for delay in formulating a policy to reduce SO_2 and NO_x emissions.

In June 1981 the National Academy of Sciences (NAS) released a comprehensive study of the effects of acid rain and other consequences of fossil fuel combustion. The report's authors concluded that "continued emissions of sulfur and nitrogen oxides at current or accelerated rates, in the face of clear evidence of serious hazards to human health and to the biosphere, will be extremely risky from a long-term economic standpoint as well as from the standpoint of biosphere protection."[26] The NAS called for a 50 percent reduction in the acidity of rain falling in the Northeast. The Reagan administration, while not actually disputing the NAS's scientific findings, labeled the report as "lacking in objectivity." In October 1981 the EPA issued a press release reiterating the administration's position that "scientific uncertainties in the causes and effects of acid rain demand that we proceed cautiously and avoid premature action."[27]

In 1982 the EPA released its own "Critical Assessment" of acid rain, a long-awaited, 1,200-page document that was the combined effort of fifty-four scientists from universities and research institutes around the country. The report's findings clearly indicated a link between human-made emissions in the Midwest and dead and dying lakes and forests, materials damage, and human health effects in the eastern and southeastern United States and southeastern Canada. But political appointees at EPA downplayed the report, refusing to draw inferences from it or employ the models proposed in it.

Then, in February 1983, the United States and Canada released the results of the bilateral study initiated by President Carter in 1980. The two countries had hoped to reach agreement, but they deadlocked over a major section of the report dealing with the effects and significance of acid rain. In an unusual move that reflected high-level opposition to acid rain controls in the United States, the two nations issued separately worded conclusions:

> The Canadians concluded that reducing acid rain "would reduce further damage" to sensitive lakes and streams. The U.S. version omitted the word "damage" and substituted "chemical and biological alterations." The Canadians concluded that "loss of genetic stock would not be reversible." The U.S. version omitted the sentence altogether. The Canadians proposed reducing the amount of acid sulfate deposited by precipitation to less than 20 kilograms of acid sulfate per hectare per year—roughly a 50 percent reduction—in order to protect "all but the most sensitive aquatic ecosystems in Canada." The United States declined to recommend any reductions.[28]

The Reagan White House decided not to have the U.S.–Canada report peer reviewed by the NAS as originally planned and instead handpicked a scientific panel to review it. To the administration's dismay, however, its panel strongly recommended action, saying, "It is in the nature of the acid deposition problem that actions have to be taken despite the incomplete knowledge. . . . If we take the conservative point of view that we must wait until the scientific knowledge is definitive, the accumulated deposition and damaged environment may reach the point of irreversibility."[29]

Although the panel delivered its interim conclusions well in advance of House Health and Environment Subcommittee hearings on acid rain, the White House did not circulate them. Instead, administration officials testified before the subcommittee that no action should be taken until further study was done.

It appeared that administration obstruction might end when William Ruckelshaus, a respected environmentalist and the original head of the EPA, assumed leadership of the beleaguered agency in spring 1983.[30] Although Ruckelshaus described acid rain as one of the "cosmic issues" confronting the nation and named it a priority, it quickly became apparent that the administration had no intention of pursuing acid rain regulation. By August 1984 the White House still had not released the report prepared by its expert panel, leading some members of Congress to question whether the administration was suppressing it. Gene Likens, director of the Institute of Ecosystem Studies of the New York Botanical Gardens and a member of the panel, expressed his irritation in an interview with *The Boston Globe:*

> I have been concerned for some time that the Administration is saying that more research is needed because of scientific uncertainty about acid rain. I think that is an excuse that is not correct. You can go on exploring scientific uncertainty forever. We must do something about acid rain now. Clearly in this case, it was an economic consideration that was most important to the Reagan Administration because of the cost of fixing the problem.[31]

When a spokesperson for the White House attributed the delay to panel chair William Nierenberg, director of the Scripps Institute of Oceanography, he retorted, "Let me say that somebody in the White House ought to print the damn thing. I'm sick and tired of it. What we said in effect was, as practicing scientists, you can't wait for all the scientific evidence."[32]

The administration not only refused to release the panel's precautionary conclusions, but went so far as to alter important passages of a 1984 EPA report titled "Environmental Progress and Challenges: An EPA Perspective," moderating statements about the urgency of environmental problems. The OMB deleted a sentence referring to the findings of an NAS report on the adverse effects of acid rain and replaced a sentence on the EPA's intent to establish an acid rain control program with one that simply called for further study of the problem.[33]

Congressional Divisions in the Reagan Years

While the White House dawdled, a coalition of environmental lobbyists urged legislators to pass comprehensive Clean Air Act revisions that included provisions to reduce acid rain, since the existing legislation did not give the EPA sufficient authority to address that problem. Richard Ayres, cofounder of the Natural Resources Defense Council, was the leading policy entrepreneur for the Clean Air Coalition, an umbrella organization of environmental and other lobbies trying to link SO_2 and NO_x control policies to the acid rain problem. The coalition counted among its allies the Canadian Coalition on Acid Rain, the first registered Canadian lobby in the United States to work for a nongovernmental, nonbusiness citizens' organization and a vocal participant in the U.S. policy debate.[34] At the same time, however, a unified coalition of utilities, industrial polluters, the eastern and Midwestern high-sulfur coal industry, the United Mine Workers (UMW), and public officials from eastern and Midwestern coal-producing states made its opposition to acid rain controls known to Congress.

In the face of intense lobbying, members of the House and Senate split over three central issues, all of which involved the regional allocation of costs and benefits. First, how much and how quickly should polluters have to reduce SO_2 emissions?[35] Proposals generally called for reductions of between 8 million and 12 million tons of SO_2 emissions from 1980 levels, the bulk of which would have to be made by the Midwestern states whose utilities burned large quantities of high-sulfur coal.[36] Second, what means should polluters be able to use to reduce SO_2 emissions? Allowing utilities to switch from high-sulfur to low-sulfur coal would be the most cost-effective solution for some, but members from high-sulfur coal-producing states such as Illinois, Kentucky, Pennsylvania, and West Virginia feared major job losses if fuel switching became widespread. On the other hand, requiring all generators to install costly scrubbers was unacceptable to western states already using low-sulfur coal. The third and most contentious issue was who would pay for the proposed reduction in SO_2 emissions. One financing model employed the "polluter pays" principle; the second included subsidies for the regions bearing the greatest share of the cleanup costs. Midwestern representatives opposed "polluter pays," but western and southern states disliked cost-sharing because they neither caused nor suffered from the effects of acid rain.

Congressional leaders sympathetic to opponents of regulation capitalized on a membership sharply divided along regional lines to foil efforts at acid rain legislation. The Senate Energy Committee—led by senators from coal-mining states, such as Wendell Ford, D-Ky., and Richard Lugar, R-Ind.—opposed acid rain controls. By contrast, the relatively liberal Senate Environment and Public Works Committee was populated by members from eastern states affected by acid rain. Robert Stafford, R-Vt., and George Mitchell, D-Maine, introduced bills to curtail emissions of acid rain precursors with a variety of financing provisions, and the Environment Committee repeatedly endorsed these proposals.[37] But in each session Majority Leader Robert Byrd, D-W.Va., representing a state that produces high-sulfur coal, was able to block consideration of those bills on the Senate floor.

Similar regional divisions stymied efforts in the House. The Energy and Commerce Committee annually debated but failed to approve bills to spread the costs of emissions reductions across electricity users nationally. On several occasions, the committee's powerful chair, John Dingell, D-Mich., used his parliamentary power to scuttle clean air legislation because he feared it would include more stringent automobile emissions standards. In 1984 the chair of the Energy Committee's Health and Environment Subcommittee, Henry Waxman, a liberal California Democrat, teamed up with Gerry Sikorski, D-Minn., and 125 others to sponsor a major acid rain bill. The Waxman–Sikorski measure established a goal of a 10-million-ton reduction in SO_2 and a 4-million-ton reduction in NO_x and distributed the cost nationally by creating a fund financed by a one mill (one-tenth of a cent) federal tax per kilowatt hour of electricity used. But insurmountable political divisions among the constituencies of the Health and Environment Subcommittee killed the measure: environmentalists, supported by recreation interests from New England and New York and by the Canadian government, demanded more stringent emissions reductions than the bill provided, while coal miners and utilities refused to support any action at all.

Acid Rain's Rising Salience

While Congress debated, scientists continued to amass evidence of acid rain's impacts. By the mid-1980s more than 3,000 scientific articles on acid rain had been published, nearly all of them depicting the phenomenon as a serious problem.[38] There were skeptics in the scientific community. Volker Mohnen, an atmospheric scientist, was vocal in his dissent from the scientific consensus, insisting that changes in Midwestern emissions were unlikely to produce proportional changes in Northeastern acid rain.[39] Soil scientists Edward Krug and Charles Frink rejected the hypothesis that acid rain was the cause of lake acification, suggesting instead that the culprit was land-use changes.[40] But the overwhelming majority of scientists studying the issue, even as they acknowledged the complexity of the relationship between emissions and ecological damage, continued to agree on the qualitative explanation for acid rain's causes and consequences and to argue that enough was known about acid rain to select an emissions reduction strategy. Reports issued by the NAS and the congressional Office of Technology Assessment reflected the scientific consensus on the topic.

Abundant media coverage of the mainstream scientific view helped to raise the salience of acid rain. Reporting on acid rain by the major newspapers—*The New York Times*, *The Washington Post*, *The Boston Globe*, the *Chicago Tribune*, and the *Los Angeles Times*—remained steady throughout the 1980s, peaking at 216 articles in 1984 and ranging from 129 to 144 articles per year through 1988. As important as the number of stories, the coverage largely adopted a precautionary tone. Between 1974 and 1981, 72 percent of the articles on acid rain conveyed the idea that it was harmful. Even after the antiregulation forces mobilized, coverage remained largely favorable to environmentalists: 51 percent of the articles published between 1982 and 1985 depicted acid rain as dangerous, while 44 percent were neutral. And 62 percent of those issued between 1986 and 1988 had a proregulation slant, while 31 percent were neutral.[41] Surveys suggested that constant media coverage of the issue had raised public concern. According to a 1986 review of existing polls, by the mid-1980s the North American public was aware of acid rain as an environmental problem and, although their understanding of it was limited, regarded it as a serious problem. Most believed that concrete actions should be taken to address acid rain and were willing to pay for abatement measures.[42]

Prospects for Controls Improve in the Bush Administration

During the Reagan years, interests that opposed addressing acid rain were able to rest assured that they had friends in high places, both in Congress and the White House. As the Reagan administration drew to a close, however, several factors combined to give hope to proponents of acid rain controls. First, in 1987 the House got a rare chance to vote on clean air legislation: Representative Dingell, who had been bottling up Clean Air Act revisions in committee, sponsored legislation to extend the act's deadlines into 1989 as a way to take the urgency out of producing a new air pollution bill before the 1988 elections.[43] But to Dingell's chagrin, his amendment lost, 162 to 247. Instead, the House backed a proenvironmental amendment that pushed the deadlines only to August 1988, thereby keeping the heat on Congress to write a new clean air law. Waxman and others interpreted the vote to mean that they had the support to pass clean air legislation if only they could get it out of committee and onto the floor.[44] In the Senate the prospects for Clean Air Act reauthorization improved as well when Senator Byrd stepped down as majority leader and was succeeded by George Mitchell, a determined advocate of acid rain controls.

Perhaps most important, presidential candidate George H. W. Bush made reauthorizing the Clean Air Act a centerpiece of his election campaign. Bush and his advisers hoped to capitalize on a proenvironmental backlash against the Reagan years: memberships in environmental groups had exploded during the 1980s, and public opinion polls revealed that a large majority of Americans considered themselves environmentalists.[45] Furthermore, acid rain control was especially important to Canadian Prime Minister Brian Mulroney, a confidante of Bush.[46] Reflecting these influences, in a speech in Detroit, candidate Bush repudiated the Reagan administration approach, saying "the time for study alone has passed." He went on to offer a detailed plan "to cut millions of tons of sulfur dioxide emissions by the year 2000."[47] Although his position on some issues ultimately disappointed environmentalists, Bush did follow through on his campaign promise to offer revisions to the Clean Air Act. Soon after taking office, he assembled a team of advisers to craft a White House version of what would become the Clean Air Act Amendments of 1990.

An Innovative Allowance-Trading Plan. The Bush team spent most of its time working on the acid rain details of the Clean Air Act reauthorization bill, and these provisions proved to be the most innovative part of the president's proposal. Led by C. Boyden Gray, the White House counsel and a policy entrepreneur for market-based regulatory approaches, the group selected a system of allowance trading combined with a nationwide cap on total SO_2 emissions. Designed primarily by Dan Dudek of the Environmental Defense Fund, the system aimed to reduce SO_2 emissions by 10 million tons by the year 2000. (Apparently, the administration selected the figure of 10 million tons in part because it was a double-digit reduction that represented a 50 percent cut, both of which signaled seriousness.)[48] This kind of incentive-based mechanism, which had been softening up in the policy community for well over a decade, rested on a simple premise: the federal government should not dictate levels of pollution control required of individual companies, as it had done in the more conventional regulations of the 1970 Clean Air Act. Rather, it should limit total emissions and distribute tradable allowances to all polluters, thereby creating incentives for companies to reduce their own pollution at lowest cost. Utilities could meet their emissions allowance using any of a variety of methods, or they could purchase allowances from other plants and maintain the same level of emissions. Those who were able to clean up cheaply could sell their excess allowances to those that could not.

Politically, the allowance-trading system faced opposition on several fronts. Many EPA staffers were skeptical of the new approach, and utility executives did not trust the government to administer a trading program properly. Some environmentalists were suspicious as well, seeing emissions trading as conferring a "right to pollute" on companies that could now buy their way out of cleaning up their mess.[49] On the other hand, allowance trading had some advantages over more conventional approaches. It satisfied President Bush's preference for market-based (as opposed to prescriptive) regulatory solutions. Moreover, the plan facilitated the creation of a broad coalition of supporters that included the clean coal-burning western states (whose utilities would not be forced to install scrubbers or subsidize acid rain controls), eastern and western producers of low-sulfur coal (for which demand was likely to accelerate), Republican advocates of less government interference in the market (for whom a trading system satisfied both efficiency and individual liberty concerns), and the northeastern states affected by acid deposition. Within this alliance, of course, different groups had concerns about the details of the plan. In particular, representatives of western states worried that their utilities, which were already relatively clean, would not be able to obtain allowances that could facilitate economic growth in their states and that most of the allowances would be concentrated in the Midwest.

Those who stood to lose the most under the White House plan and consequently mobilized to oppose it were the UMW, with its strong presence in high-sulfur coal-mining states, and the Midwestern utilities that traditionally relied on high-sulfur coal. The UMW argued that the bill would force major coal purchasers to shift to low-sulfur coal, jeopardizing thousands of coal-mining jobs. Midwesterners contended that the Bush plan was skewed against them: nine states responsible for 51 percent of the nation's SO_2 emissions were going to have to accomplish 90 percent of the reduction and bear 73 percent to 80 percent of the cost—in the form of higher electricity bills—in the initial phase of the plan.

The Bush Plan in the Senate. In short, when Bush presented his plan to Congress, the battle lines were already drawn. In the Senate, Majority Leader Mitchell realized that most of the resistance to clean air legislation would come from outside the Environment and Public Works Committee because the regional balance of the committee clearly favored supporters of acid rain controls: of its sixteen members, five were from New England, two were from New York and New

Jersey, and four others were from the Rocky Mountain states that produce low-sulfur coal. Furthermore, Sen. Max Baucus, D-Mont., had assumed the chair of the Environmental Protection Subcommittee upon Mitchell's ascension to majority leader, and he had consistently supported clean air legislation—in part because Montana produces and uses low-sulfur coal, which many utilities would turn to under a market-based system. The ranking minority member of the subcommittee, John Chafee, R-R.I., also had a strong environmental record; moreover, his state had low per capita SO_2 emissions and therefore had little to lose from stringent acid rain controls. So the committee was able to mark up and approve (15–1) the president's clean air bill with relative ease. The controversy arose later, in the full Senate.

It took ten full weeks from its introduction on January 23, 1990, for the Senate to pass the Clean Air Act Amendments. During that time, the chamber worked continuously on the bill, virtually to the exclusion of all other business. The lengthy process was unusual for a complex and controversial proposal in at least one respect: there was no extended talk or procedural delay on the Senate floor; rather, Senator Mitchell presided over most of the debate behind closed doors in his suite. Commented Sen. Tom Daschle, D-S.D., "In my [12] years in Congress, I have never seen a member dedicate the attention and devotion to an issue as George Mitchell has done on clean air. He has used every ounce of his energy to cajole senators and come up with imaginative solutions."[50] As majority leader, Mitchell was both a persistent advocate of clean air legislation and a pragmatist: he wanted a clean air bill, and he knew that he would have to strike deals to get one. As a result, he often antagonized environmentalists, who hoped for greater regulatory stringency and less compromise than Mitchell was willing to tolerate. For example, despite objections from some major environmental groups, Mitchell and Baucus embraced the emissions-trading system devised by the Bush team in exchange for White House acceptance of changes in other parts of the bill.

Although there was clear momentum in the Senate in favor of reaching agreement on the clean air bill, the Appalachian and Midwestern senators, led by Byrd, were not ready to give up their fight to disperse the region's acid rain cleanup costs. With extensive help from his aide, Rusty Mathews, Byrd crafted a formula that won the approval of coal-state senators: he proposed to give Midwestern power plants bonus emissions allowances to encourage them to adopt scrubbers rather than low-sulfur coal. An informal head count revealed that the Senate would not support Byrd's plan, however, and he was forced to scramble to salvage something from the acid rain debate. Mathews came up with another alternative: giving the heaviest polluters extra allowances in Phase I to help them buy their way out of the cleanup effort. Eventually, Byrd and his cosponsor, Sen. Christopher Bond, R-Mo., were able to persuade Mitchell and others to accept this modification, known as the Byrd–Bond amendment, in exchange for their support on the acid rain provisions of the bill, and on April 3 the Senate passed its version of the Clean Air Act Amendments by a vote of 89 to 11.

Acid Rain Controls in the House. The decade-long rivalry between Waxman and Dingell continued to dominate the clean air debate in the House Energy and Commerce Committee. Dingell had chaired the full committee since 1979, and Waxman had chaired the Health and Environment Subcommittee since 1981. To augment his influence and counteract Waxman, Dingell had worked assiduously with House Speaker Jim Wright, D-Texas, to recruit sympathetic members to Energy and Commerce. By 1990, one-third of the committee's members were from the industrial heartland—the Midwest had nine seats and the Appalachian coal states another seven—and only two were from New England. Nevertheless, when the Bush administration presented its proposal, Dingell took the advice of some important allies in the House (most notably Majority Whip Tony Coelho, D-Calif.) and from his wife, Debbie, a General Motors executive, that delay and defensive tactics would no longer succeed, and once a tipping point was reached, he would have to get on the bandwagon if he wanted to avoid blame for inaction. According to one Democratic aide, "Dingell always knew that he could delay action for a time, but that he then would have to fight for the best deal he could get."[51]

Dingell had assigned Waxman's Environment Subcommittee responsibility for marking up most of the clean air package but had strategically referred the two most controversial titles of the bill, acid rain and alternative fuels, to the Subcommittee on Energy and Power chaired by Philip Sharp, D-Ind., who had long opposed acid rain legislation in hopes of protecting Indiana's old and dirty utilities. In November 1989 Sharp told reporters he planned to dramatize his concerns about the administration's proposal by stalling the acid rain provisions in his subcommittee. Two of Sharp's Midwestern allies on the committee, Edward Madigan, R-Ill., and Terry Bruce, D-Ill., publicly criticized Dingell for failing to deliver on a promise to help solve their cost-sharing problems. In an effort to patch things up, Dingell unveiled a plan to enroll support among the "cleans"—predominantly western states—for a cost-sharing program that would relieve some of the burden on the Midwestern "dirties."[52]

The protracted recalcitrance of the utility lobby and its supporters in the House had not earned them many friends, however, and the Midwestern contingent was unable to persuade members from other states to accept their cost-sharing proposals. The utilities also had increased their vulnerability to tough controls by stubbornly insisting throughout the 1980s that there was no acid rain problem and refusing to provide members of Congress and their aides with information on their operations. Making matters worse for the utilities, the administration's plan had fractured the antiregulation coalition: Midwestern representatives were split between those who favored giving utilities flexibility in reducing their emissions and those who wanted to protect high-sulfur coal miners and therefore preferred a scrubber mandate. The utilities themselves were divided, depending on their energy source and the extent to which they had already engaged in pollution control in response to stringent state laws that were themselves the result of lobbying by environmentalists.[53] In particular, the heavily polluting utilities and those that had already begun to clean up—such as those in Minnesota, New York, and Wisconsin—could not resolve their differences.

To facilitate negotiations among members of this fractious group, nearly all of the committee meetings were conducted behind closed doors and often at night. Like Mitchell in the Senate, Dingell hoped to spare rank-and-file lawmakers the constant pressure to serve constituent demands by freeing them from the speechmaking and position-taking associated with public hearings on controversial legislation.[54] The "dirties" and "cleans" finally agreed on a compromise in which Midwestern utilities got more time to reduce emissions, additional emissions allowances, and funds to speed development of technology to reduce coal emissions. Members from the "clean" states extracted some concessions as well: they obtained greater flexibility for their utilities under the trading system, and they got all the allowances the Midwesterners initially had been reluctant to give them. The Energy and Commerce Committee completed its drafting of the clean air legislation on April 15, 1990, and the full House passed its version of the bill on May 23 by a vote of 401 to 21.

Reconciling the House and Senate Bills. Following Senate and House passage of their separate versions of the clean air bill, all that remained was to reconcile the differences in conference committee. The House conferees controlled most aspects of the conference committee bill, but the Senate prevailed on the acid rain provisions.[55] Despite their failures in the House and Senate chambers, Midwestern representatives still hoped to extract some relief from the conference committee. Representative Sharp and his colleagues worked during the final days to salvage aid to their region, requesting additional allowances to ease the pain for the dirty coal-fired utilities that would be hardest hit by the new rules. The Midwesterners among the House conferees settled on a proposal to give some additional allowances to Illinois, Indiana, and Ohio utilities each year until 1999. After 2000, ten Midwestern states would get a small number of extra allowances each year.[56]

When they presented the plan to a group of pivotal House and Senate conferees, however, the reaction was unenthusiastic. Rep. Dennis Eckart, D-Ohio, then took the plan to Senator Mitchell, who agreed to support it in exchange for a few concessions. In the end, said Eckart, Mitchell broke the impasse by convincing the skeptics to pacify the Midwest. In the process, the Midwesterners lost on some other demands, including a provision in the House bill that would have protected high-sulfur coal mining jobs by requiring utilities to install pollution control devices rather than switching to low-sulfur coal.[57]

In general, despite the extra allowances granted to some Midwestern utilities, the acid rain provisions represented a big loss for the Midwest and the oldest, dirtiest electricity generators. The final bill, according to one veteran Washington, D.C., lobbyist, "proved the axiom that divided industries do badly on Capitol Hill."[58] That said, House conferees did manage to extract from the Bush administration a concession on displaced worker assistance, even though the president had threatened to veto the bill if it included unemployment benefits to workers who lost jobs as a result of the law. Once the conferees had resolved or agreed to abandon the remaining issues, the final package had to pass the House and Senate. Reflecting the momentum behind the law and the skills of the legislative leaders who had shepherded it through the legislative process, the six days of debate that followed were perfunctory. The House passed the clean air conference report 401 to 25 on October 26, and the Senate passed it 89 to 10 the following day. President Bush signed the Clean Air Act Amendments into law on November 15, 1990.

Implementing the Acid Rain Provisions

Title IV of the 1990 Clean Air Act Amendments required a nationwide reduction in SO_2 emissions by 10 million tons—roughly 40 percent from 1980 levels—by 2000. To facilitate this reduction, Title IV established a two-phase SO_2 allowance-trading system under which the EPA issued every utility a certain number of allowances, each of which permitted its holder to emit one ton of SO_2 in a particular year or any subsequent year.[59] A utility that generated more emissions than it received allowances for could reduce its emissions through pollution control efforts or could purchase allowances from another utility. A utility that reduced its emissions below the number of allowances it held could trade, bank, or sell its excess allowances. Title IV also mandated a reduction in NO_x emissions of 2 million tons by 2000. To achieve this reduction, the law required the EPA to establish emissions limitations for two types of utility boilers by mid-1992 and for all other types by 1997. Finally, Title IV required all regulated utilities to install equipment to monitor SO_2 and NO_x emissions continuously to ensure compliance.[60]

EPA Rulemaking Under Title IV. For opponents of acid rain regulations, the implementation of Title IV was the last remaining opportunity to subvert the goal of reducing SO_2 and NO_x emissions. They were gratified when, less than a year and a half after signing the much-heralded Clean Air Act Amendments, President Bush instituted the Council on Competitiveness headed by Vice President Dan Quayle. The council was driven, in Quayle's words, by "the desire to minimize regulations and to make regulations as unburdensome as possible while meeting the requirements of the statutes."[61] Bush wanted to retreat from the most onerous commitments of the act because, with a reelection campaign imminent, he needed to accommodate his allies in the Republican Party and in the business community. Efforts to derail the acid rain and other programs were largely unsuccessful, however, and on January 15, 1991, the EPA issued a report delineating the dozens of deadlines the agency would have to meet to comply with the new law. By summer 1991 the agency had resolved most of the major SO_2 rulemaking issues; the following December it published its proposed rules for notice and comment; and by early 1993 it had promulgated its final SO_2 regulations.[62]

Vigorous and supportive congressional oversight helped to ensure that the regulatory process stayed on track. Because committee Democrats wanted to take credit for the act as the 1992 presidential and congressional races neared, high-ranking members of the authorizing committees monitored Clean Air Act implementation zealously. In April 1992 Waxman wrote an angry *New York Times* editorial chiding the Bush administration for failing to issue the regulations necessary to carry out the act. Even Dingell joined Waxman in this effort. Having thrown his support behind the law, Dingell was not going to let the White House undermine it, so he made sure the EPA had the wherewithal to write rules under the Clean Air Act Amendments. One manifestation of congressional support was the agency's ability to increase its acid rain staff in fiscal year 1992 from fifteen to sixty-three.

In addition to congressional oversight, the statute itself encouraged compliance because it contained clear guidelines and strict penalties if the agency failed to put SO_2 and NO_x regulations in place, a feature that reflected Congress's determination to make the act self-enforcing. Waxman explained that "the specificity in the 1990 Amendments reflects the concern that without detailed directives, industry intervention might frustrate efforts to put pollution control steps in place. . . . History shows that even where EPA seeks to take strong action, the White House will often intervene at industry's behest to block regulatory action."[63]

A comparison between the statutory provisions for NO_x and SO_2 reveals the importance of statutory design for effective implementation. Although the EPA required utility boilers to meet new NO_x emissions requirements, there was no emissions cap, so overall emissions continued to rise. Moreover, the NO_x provisions did not contain a hammer: NO_x limitations did not apply unless rules were actually promulgated, so there was considerable incentive for regulated sources to delay and obstruct the rulemaking process. In fact, the NO_x rule was delayed by a court challenge and was not finalized until early 1995. (NO_x reductions began in 1996 and were increased in 2000.) According to one EPA analyst, "Poor statutory construction [for NO_x regulation] has cost the environment in terms of delayed protection, has cost the industry in terms of burner reconfiguration expenditures, and has cost the public, who ultimately pays the capital, administrative, and litigation expenses."[64]

The SO_2 Allowance-Trading System. Between 1991 and 1992 the EPA issued all of the Phase I permits for SO_2, certified the continuous monitoring systems that utilities had installed and tested, and developed an emissions tracking system to process emissions data for all sources. The first phase of the SO_2 trading system required 263 units in 110 coal-fired utility plants (operated by sixty utilities) in twenty-one eastern and Midwestern states to reduce their SO_2 emissions by a total of approximately 3.5 million tons per year beginning in 1995.[65] The EPA allocated allowances to utilities each year based on a legislative formula.[66] The agency provided some units with bonus allowances, the subject of so much congressional haggling; for example, high-polluting power plants in Illinois, Indiana, and Ohio got additional allowances during each year of Phase I that they could sell to help generate revenues to offset cleanup costs. To appease representatives of "clean" states, the EPA also allowed plants to receive extra allowances if they were part of a utility system that had reduced its coal use by at least 20 percent between 1980 and 1985 and relied on coal for less than half of the total electricity it generated. Additional provisions gave clean states' utilities bonus allowances to facilitate economic growth.

Phase II, which began in 2000, tightened emissions limits on the large plants regulated in Phase I (by cutting the allocation of allowances to those units by more than half) and set restrictions on smaller, cleaner coal-, gas-, and oil-fired plants. Approximately 1,420 generating units were brought under the regulatory umbrella in Phase II, during which the EPA issued 8.95 million allowances for one ton of SO_2 emissions each to utilities annually.[67] To avoid penalizing them for improvements made in the 1980s, Title IV permitted the cleanest plants to increase their emissions between 1990 and 2000 by roughly 20 percent. Thereafter, they were not allowed to exceed their 2000 emission levels. The EPA did not allocate allowances to utilities that began operating after 1995; instead, it required them to buy into the system by purchasing allowances. Doing so ensured that overall emissions reductions would not be eroded over time.

Once it set up the system, the EPA's ongoing role in allowance trading was simply to receive and record allowance transfers and to ensure at the end of each year that a utility's emissions did not exceed the number of allowances held. Each generator had up to thirty days after the end of the year to deliver to the EPA valid allowances equal to its emissions during the year. At that time, the EPA cancelled the allowances needed to cover emissions. If a generator failed to produce the necessary allowances, the EPA could require it to pay a $2,000-per-ton excess emissions fee and offset its excess emissions in the following year. Because the excess emissions fee was substantially higher than the cost of complying with the law by purchasing allowances, designers of the system expected the market to enforce much of the emissions reductions.

OUTCOMES

At first, utility executives refused to trade allowances. So allowance broker John Henry called C. Boyden Gray at the White House and asked whether the administration could order the Tennessee Valley Authority (TVA), a federally owned electricity provider, to start buying allowances to compensate for excess emissions from its coal-fired power plants. In May 1992 the TVA did the first deal, paying $250 per ton for the allowances it needed.[68] The following spring the Chicago Board of Trade held the first of three SO_2 allowance auctions. That auction reaped $21 million and attracted more participants than most observers had expected.[69] The EPA managed to sell the allowances it was offering at prices ranging from $130 to $450, with an average price of $250 per ton. Over the next ten years, allowance prices fluctuated but remained below $200; in 2004 allowance prices began to rise, spiking briefly at $1,600 per ton in December 2005 before falling back below $200.[70] (Anticipation of the Bush administration's Clean Air Interstate Rule, or CAIR, described below, combined with the interruption in natural gas supplies caused by Hurricane Katrina, caused the rise in allowance prices. Falling natural gas prices and the installation of scrubbers in response to CAIR caused allowance prices to fall again.[71]) Trading, although initially sluggish, soon picked up, reaching about 9 million trades between economically unrelated entities by 2007.[72]

The Economic Impacts of Title IV

To some observers, the generally lower-than-expected allowance prices suggested that utilities were investing more heavily in emissions reduction than projected and anticipated having more allowances than they needed when emissions limits were tightened. Others worried that such low prices might cause utilities that had planned to install pollution control devices to reconsider because it might prove cheaper simply to buy more allowances. Indeed, one utility that bid in the first auction, the Illinois Power Company, stopped construction on a $350 million scrubber and began making private deals to stockpile permits that it intended to use between 1995 and 2000.[73] But a 1996 Resources for the Future report suggested that, although most were not trading allowances, utilities had in fact cut pollution.[74] (The study found that the bulk of EPA-cited trades were record-keeping changes in which utilities consolidated ownership.)

Utilities were cutting their emissions because it had become substantially cheaper to do so. By allowing utilities to choose how to achieve compliance, Title IV created a marketplace in which pollution control alternatives competed directly with one another, prompting price cuts and innovation among coal marketers, railroads, and scrubber manufacturers. Fuel switching and blending with low-sulfur coal was the compliance option of choice for more than half of all utilities. But the most important factor accounting for lower compliance costs, according to analyst Dallas Burtraw, was the nearly 50 percent reduction in transportation costs for moving low-sulfur coal from the Powder River Basin to the East—a consequence of deregulating the railroads in the 1980s.[75] (Lower transportation costs, in turn, enabled experimentation with fuel switching and blending.) As a result of these savings, emissions reductions during the first five years of the program were achieved at substantially lower cost relative to alternative regulatory approaches, such as technology or performance standards.[76] The EPA anticipated that by 2010 the cost of the program would be $1 billion to $2 billion a year, only one-quarter of the originally forecast cost.[77]

The Ecological Impacts of Title IV

Assessments of Title IV revealed some impressive and unexpected accomplishments beyond low compliance costs. According to the EPA, by 2008 SO_2 emissions from sources covered by Title IV were 7.6 million tons, down from 15.7 million tons in 1990, when the act was passed (see Figure 5.1). (Total SO_2 emissions from all sources declined from nearly 26 million tons in 1980 to 11.4

million tons in 2008.[78]) Moreover, the USGS found that in 1995, sixty-two sites in the mid-Atlantic and Ohio River Valley regions experienced, on average, a 13.8 percent decline in sulfur compounds and an 8 percent drop in hydrogen ions. The authors attributed this "greater-than-anticipated" decline to the implementation of Phase I of the Clean Air Act Amendments.[79] Similarly, reports from the Adirondacks in 2000, 2002, and 2003 showed declining sulfur levels in many of the regions' lakes.[80]

Assessments of the act's ecological impacts were less sanguine, however; within a few years of the program's enactment, it was clear to scientists that the reductions mandated in Title IV would not be sufficient to stem the ecological damage done by acid rain. A report released in April 1996, based on more than three decades of data from New Hampshire's Hubbard Brook Forest, suggested that the forest was not bouncing back as quickly as expected in response to emissions reductions.[81] "Our view and that of soil scientists had been that soils were so well buffered that acid rain didn't affect them in any serious way," said ecologist Gene Likens, commenting on the report.[82] Even though the deposition of acid had abated, its effects lingered. Vegetation in the Hubbard Brook Experimental Forest had nearly stopped growing since 1987, and the pH of many streams in the Northeast remained below normal. Researchers speculated, and subsequent evidence confirmed, that years of acid deposition had depleted the soil's supply of alkaline chemicals and that the soil was recovering only gradually.[83] A second likely cause of the slow comeback of Hubbard Brook was that, because NO_x emissions had not declined significantly (the NO_x program had only just taken effect), nitrates continued to acidify the rain falling in the Northeast.

The Hubbard Brook findings were consistent with the observations of plant pathologist Walter Shortle of the Forest Service. In fall 1995 Shortle's group had reported in the journal *Nature* that, because acid rain was no longer sufficiently neutralized by calcium and magnesium, it was releasing aluminum ions from minerals into the soil, where they are toxic to plants.[84] Another study, conducted in the Calhoun Experimental Forest in South Carolina, confirmed that acid rain dissolves forest nutrients much faster than was previously believed.[85] And in October 1999 the USGS concurred that, while sulfur levels in rain and streams were declining, the alkalinity of stream water had not recovered.[86]

A 1999 National Acid Precipitation Assessment Program report summarizing the research on acid rain affirmed that, although the 1990 Clean Air Act Amendments had made significant improvements, acid precipitation was more complex and intractable than was suspected when that law was passed. The authors noted that high-elevation forests in the Colorado Front Range, West Virginia's Allegheny Mountains, the Great Smoky Mountains of Tennessee, and the San Gabriel Mountains of southern California were saturated or near saturated with nitrogen; the Chesapeake Bay was suffering from excess nitrogen, some of which was from air pollution; high-elevation lakes and streams in the Sierra Nevada, the Cascades, and the Rocky Mountains appeared to be on the verge of chronically high acidity; and waterways in the Adirondacks were becoming more acidic, even as sulfur

Figure 5.1 SO_2 Emissions From Acid Rain Program Sources, 1980–2008

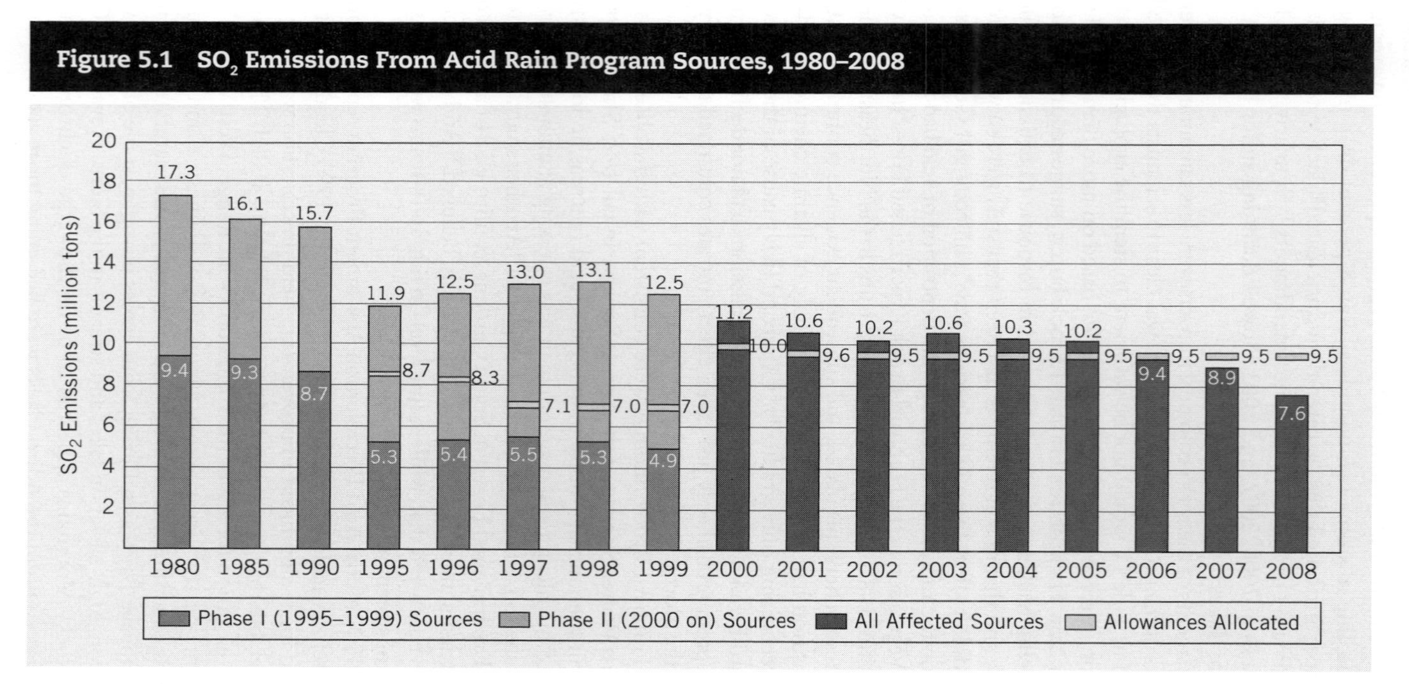

Source: U.S. Environmental Protection Agency, "Allowance Markets Assessments: A Closer Look at the Two Biggest Prices Changes in the Federal SO_2 and NOx Allowance Markets," White Paper, April 23, 2009. Available at http://www.epa.gov/airmarkt/resource/docs/marketassessmnt.pdf.

deposits declined.[87] The report concluded that if deposits of sulfur and nitrogen were not reduced, these sensitive forests, lakes, and soils would continue to deteriorate. According to Jack Cosby, a University of Virginia environmental scientist, "It's been the near consensus of scientists that the Clean Air Act amendments haven't gone far enough."[88]

A study conducted by ten leading researchers and published in the March 2001 issue of *Bioscience* further verified that acid rain's effects continued to be felt. Although acid rain had decreased 38 percent since controls were implemented, 41 percent of lakes in the Adirondacks and 15 percent of lakes in New England were either chronically or periodically acidic, largely because the buffering capacity of surrounding soils had been depleted.[89] The authors also noted that research since 1990 had clarified how acid deposition was harming the region's forests: acid fog and rain depletes calcium, an important plant nutrient, from spruce needles, leaving them susceptible to freezing; acid rain also mobilizes aluminum, which interferes with maple tree roots' ability to take up nutrients. According to Gene Likens, "The science on the issue is clear. Current emission control policies are not sufficient to recover sensitive watersheds in New England."[90] The article claimed that an 80 percent reduction in SO_2 would be necessary to restore the region's soils, lakes, and streams within twenty-five years.

The effects of acid rain were also becoming apparent in the Southeast. Impacts there took longer to manifest themselves because southern soils are generally deeper than northern soils and so had absorbed more acid. Once the soils were saturated, however, acid levels in nearby waters began to skyrocket. Even in the West, acid rain was attracting attention, as scientists studying spruce trees in the Rockies found that trees downwind of populous areas showed high levels of nitrogen and low ratios of magnesium in their needles.[91] And researchers were beginning to observe additional consequences of acid rain: ornithologists at Cornell University charged that acid rain was implicated in the decline of the wood thrush and other northeastern bird species. They argued that by depleting the soil of calcium, acid rain caused the number of earthworms, millipedes, slugs, and snails to plummet, leaving females and chicks calcium deficient.[92]

New Research Spurs Calls for More Controls

Partly in response to new findings about the enduring ecological impacts of acid rain, but primarily in reaction to concerns about the local health effects of ground-level ozone (smog), in the late 1990s pressure began mounting for stricter emissions controls, particularly on NO_x. In November 1997 the EPA called for new state air-pollution plans to cut the NO_x emissions of utilities and other large sources in the states east of the Mississippi by 85 percent. Over the protests of Midwestern governors and power plant executives, the EPA in 1998 ordered twenty-two states in the East and Midwest to reduce emissions of NO_x by an average of 28 percent, mainly during the summer when it contributes to smog.[93] The following year Sen. Daniel Patrick Moynihan, D-N.Y., proposed new legislation to cut SO_2

by 50 percent more than required by the 1990 amendments, reduce NO_x 70 percent by 2005, and establish a trading scheme for NO_x similar to the one created for SO_2. But the bill attracted minimal support in Congress, where some members thought the government had done enough to address the issue.[94] In 2000 New York officials and environmental groups seized on a General Accounting Office (GAO, later the Government Accountability Office) report showing that nitrogen levels were continuing to rise in Adirondack waterways as ammunition for their case that Congress needed to tighten acid rain controls.[95] But officials representing Midwestern utilities remained adamant that further regulations were unnecessary, pointing out that it was too early to evaluate the impacts of existing NO_x reductions, which had taken effect in 1996 and were tightened in 2000. Furthermore, although most scientists supported policies to reduce NO_x, some were dubious about proposals that focused only on utility emissions, pointing out that much of the nitrogen in the atmosphere comes from motor vehicle emissions.[96]

While Congress wrestled with the question of whether and how to reduce power plant emissions, the Clinton administration pursued tighter controls on another front with greater success. In September 1999 New York Attorney General Eliot Spitzer announced his intention to sue seventeen southern and Midwestern power plants, as well as several utilities in New York State, to force them to reduce their SO_2 and NO_x emissions. Spitzer employed a novel argument under the New Source Review (NSR) rule of the Clean Air Act: that power companies had expanded and upgraded their old plants sufficiently to be considered new plants for regulatory purposes.[97] In November Connecticut Attorney General Richard Blumenthal said he would follow Spitzer's lead and sue sixteen coal-fired power plants. Around the same time, following a two-year investigation by the EPA, the Department of Justice filed lawsuits against seven utility companies across the country. By the end of 2000, the litigation appeared to be bearing fruit: in November 2000 the Virginia Electric Power Company signed a landmark $1.2 billion settlement in which it agreed to cut SO_2 and NO_x emissions from its eight coal-burning plants by 70 percent, or 252,000 tons per year, within twelve years.[98] A month later the Cinergy Corporation of Cincinnati reached a settlement in which it agreed to spend $1.4 billion over twelve years to cut SO_2 from its ten coal-burning plants in Kentucky.[99]

In 2001, however, President George W. Bush disavowed the Clinton administration's litigation strategy and instead began exploring ways to relax the NSR rule to give utilities more flexibility to enhance their plants without adding pollution controls. After the administration published five revised NSR rules in the *Federal Register* on December 31, 2002, nine northeastern states immediately filed suit. The lawsuit challenged four of the new provisions, each of which would allow plants to increase their emissions without penalty. In their press release, the attorneys general framed the issue in stark terms, portraying corporate polluters as the villains and public health (rather than lakes and forests) as their main concern. Undaunted, in August 2003 the administration finalized an additional NSR rule, which said that pollution upgrades would be required only if companies spent more than 20 percent of the replacement cost of the entire plant—an extremely generous standard. Once again the state attorneys general filed suit. In 2006 the Court of Appeals for the D.C. Circuit unanimously vacated the 20 percent rule, and the Supreme Court declined to hear the case. Then, after a series of conflicting district- and appeals-court rulings on the other rule changes, in *Environmental Defense et al. v. Duke Energy* (2007), the Supreme Court dealt the Bush administration's NSR interpretation a fatal blow. Shortly thereafter, utilities began settling cases initiated by the Clinton administration but long delayed by the Bush administration's wavering. That trend culminated in October 2007, when American Electric Power—the nation's largest coal-burning utility—agreed to install $4.6 billion worth of pollution-control devices at sixteen power plants in the Midwest and mid-Atlantic.[100]

On a parallel track, the Bush administration promoted its "Clear Skies" legislative initiative, which sought first and foremost to reduce smog and soot levels but, as a second-order effect, would also mitigate acid rain. First introduced in early 2002, Clear Skies would have established a 2.1-million-ton cap on NO_x by 2009 and a 1.7-million-ton cap by 2018. For SO_2, the plan called for a 4.5-million-ton cap by 2010 and a 3-million-ton cap by 2018. Clear Skies failed repeatedly to get out of the Senate Environment and Public Works Committee, however, because a small number of Republicans agreed with environmentalists that it would achieve less than full enforcement of the existing Clean Air Act, which generally requires any region that does not meet federal air standards to reach those goals within five years, and because it failed to address carbon-dioxide emissions (see Chapter 12). Stymied in Congress, EPA Administrator Mike Leavitt in December 2003 unveiled the Clean Air Interstate Rule (CAIR), which established by administrative fiat a regime similar to that in Clear Skies but applicable only to the twenty-eight eastern states and Washington, D.C. The administration claimed that the three separate emissions-trading schemes established under CAIR, which became final in March 2005, would yield a 73 percent reduction in SO_2 emissions and a 61 percent drop in NO_x emissions by 2015.[101]

Most environmentalists applauded the new rule, but in July 2005 a handful of environmental groups filed a lawsuit challenging language in the rule's preamble that would limit the government's ability to reduce emissions in the future. At the same time, industry filed a dozen lawsuits arguing that the EPA had overstepped its authority in making the rule.[102] In July 2008 the Court of Appeals for the D.C. Circuit threw out CAIR, contending that "No amount of tinkering with the rule or revising of the explanations will transform CAIR, as written, into an acceptable rule."[103] According to the court, one of CAIR's "fundamental flaws" was that it involved banking SO_2 allowances used under the acid rain program. (After the court rejected CAIR, SO_2 allowance prices dropped precipitously.) In December, however, the court reversed itself, allowing CAIR to remain in force until the EPA crafted a replacement.

The administration of President Barack Obama moved quickly on the issue of SO_2 and NO_x emissions, and in July 2011 it finalized a rule to replace CAIR. The

Cross-State Air Pollution Rule (CSAPR), which was slated to take effect in 2012, featured tighter emissions caps than CAIR. The rule also averted another of the "fundamental flaws" in its predecessor: it limited interstate allowance trading to ensure that each state would not be prevented from complying with federal air-quality rules because they were downwind of power plants using the allowances. The CSAPR gave each state an emissions cap that takes downwind effects into account. But critics worried that the new rule would undermine the national market for SO_2 allowances, which were trading at a mere $5 per ton—a level that gave utilities little incentive to add new emissions controls. (Sure enough, by 2012 Title IV was, for all intents and purposes, defunct; the EPA had effectively eliminated interstate allowance trading in favor of more geographically limited trading that focused on the health benefits of emissions reductions on specific downwind states.[104]) In any case, in late December 2011, in response to a suit filed by fourteen upwind states and a coalition of industry interests, the U.S. Court of Appeals for the D.C. Circuit stayed CSAPR pending judicial review, and in August 2012 the court vacated the rule. In late April 2014 the Supreme Court reversed the circuit court's decision, and in 2016, the EPA updated the CSAPR for 2018 NAAQS. Starting May 2017, the CSAPR will reduce summertime NO_x emissions from 22 eastern states to reduce ground-level exposure for millions.[105] But in the meantime, CAIR remained in place and emissions reductions continued: according to the EPA, by 2015 annual SO_2 emissions had fallen 81 percent since 1970, while annual NO_x emissions had declined 54 percent.[106]

CONCLUSIONS

The acid rain case provides a vivid example of the interaction between science and politics in environmental policymaking. Critics of the 1990 Clean Air Act Amendments point out that Congress had virtually completed work on the bill before the multimillion-dollar National Acid Precipitation Assessment Program report, commissioned by President Carter ten years earlier, had even been released; therefore, they conclude, science could have had no real impact on the legislation. But the assessment program had released drafts of the report, and its major findings were already well established. Just as opponents of acid rain controls justified their position on the grounds of scientific uncertainty, proponents of acid rain controls translated scientific claims generated throughout the 1980s into a compelling political story that raised the issue's salience. By generating support among voters, advocates created incentives for leaders to tackle the issue. Hoping to capitalize on widespread public concern in the United States and Canada, in 1988 presidential candidate George H. W. Bush promised to address the problem of acid rain.

When Bush was elected and George Mitchell became majority leader of the Senate, policy entrepreneurs like C. Boyden Gray saw an opportunity to link the environmental problem to a market-based solution. Such an approach changed the political dynamics of the issue because it was palatable to many Republicans and westerners. The emissions-trading scheme also helped to break up the formidable coalition of utilities that had been monolithic in their opposition to regulations, a strategy that worked in part because environmentalists had lobbied successfully for the state-level acid rain laws that had prompted some power plants to clean up.

After a decade of delay, the Clean Air Act Amendments passed by overwhelming majorities in both chambers, making manifest both the power of well-placed opponents to maintain the status quo against a popular alternative and the importance of legislative leaders in facilitating a successful policy challenge. President Bush's imprimatur on the new bill contrasted sharply with the Reagan administration's obstruction and was critical to attracting Republican support. Congressional leaders, particularly Mitchell and Dingell, brokered deals that were crucial to building a majority coalition. Their willingness to hold meetings behind closed doors, away from the scrutiny of lobbyists, was particularly crucial.

As it turned out, the allowance-trading program brought about greater reductions in SO_2 emissions at lower costs than expected—illustrating the pitfalls of cost projections, which typically fail to account for the market's dynamic response to regulation. The allowance-trading program generated other benefits as well: emissions reductions occurred relatively quickly, as a result of allowance banking; the EPA made no exemptions or exceptions from, or relaxations of, the program's requirements; and the hot spots that were feared did not materialize.[107] On the other hand, a careful historical review finds that the trading system did not induce significant technological innovation; in fact, most of the performance and capital cost improvements in the dominant SO_2 control technology occurred before the 1990 law was enacted.[108] Furthermore, it is noteworthy that Germany dramatically decreased its SO_2 emissions—by an even greater proportion than the United States, in less time, and in a cost-effective fashion—despite relying on prescriptive regulation. As one analyst remarks, the German experience undermines the claim made by many economists that market-based instruments are always more efficient and effective than "command-and-control" regulations.[109]

Despite the gains associated with Title IV and the cap-and-trade programs that have superseded it, subsequent scientific research suggests that further measures will be needed to mitigate the problems associated with acid rain in the United States and in developing countries like China. A study released in 2013 by Gene Likens found that two-thirds of the ninety-seven streams and rivers his team studied in the East had grown more alkaline—regardless of whether they were downstream of agricultural land, forestland, or urban land. Upon further investigation, the researchers found that rural rivers had become more alkaline as well. The likely cause was acid rain, which had been eating away large chunks of rock, particularly limestone. The runoff had produced carbonates that flowed into rivers. These findings were important because alkalinity stimulates the growth of certain types of algae, and too much algae sucks the oxygen out of water. Moreover, if alkaline water mixes with sewage, it converts the ammonia in sewage into a more toxic form.[110]

Due to changing weather patterns, the westerlies are moving the air pollution across the Pacific from China to the United States. Steve Davis, a UC-Irvine Earth scientist, argues that the United States is a contributor to China's increasing pollution problem because we have outsourced most of our manufacturing. According to Davis, "Rain doesn't easily wash it out of the atmosphere, so it persists across long distances."[111] As a result, Chinese manufacturing has spiked the country's levels of acid rain, which is traveling back to the United States.

QUESTIONS TO CONSIDER

- Critics often cite acid rain regulations as evidence of misplaced environmental priorities, saying the 1990 National Acid Precipitation Assessment Program report made clear that acid rain does not pose a serious threat. Do you agree with this critique? Why, or why not?
- More generally, has science played a sufficient role in the development of U.S. acid rain policy? If not, why not, and how might its impact be increased?
- Why do proponents believe emissions-trading programs are more effective, economically and ecologically, than uniform emission limitations? Does the evidence support these claims? Can you think of any drawbacks to the emissions-trading approach?
- How can developing countries like China strike a balance between finding emission-efficient ways of consumption while providing energy for their populace?

NOTES

1. Acid rain is a common term that describes all forms of acidic precipitation, as well as dry deposition of acid materials. It occurs when sulfur and nitrogen combine with oxygen in the atmosphere to produce sulfur dioxide (SO_2) and nitrogen oxides (NO_x). Within hours or days, these pollutants spontaneously oxidize in the air to form sulfate and acid nitrate, commonly known as sulfuric and nitric acids. These acids usually remain in the atmosphere for weeks and may travel hundreds of miles before settling on or near the earth in rain, snow, mist, or hail, or in dry form. Scientists measure the relative acidity of a solution on a logarithmic potential hydrogen (pH) scale. While all types of precipitation are somewhat acidic—"pure" rainfall has a pH of about 5.6—in industrial regions, the pH of rain is often around 4.0 and can drop as low as 2.6. Because the pH scale is logarithmic, a full pH unit drop represents a tenfold increase in acidity.
2. Judith A. Layzer, "Sense and Credibility: The Role of Science in Environmental Policymaking" (Ph.D. diss., Massachusetts Institute of Technology, 1999).
3. John W. Kingdon, *Congressmen's Voting Decisions*, 3d ed. (Ann Arbor: University of Michigan Press, 1989).

4. Alan Schwartz, quoted in Leslie R. Alm, *Crossing Borders, Crossing Boundaries: The Role of Scientists in the U.S. Acid Rain Debate* (Westport, Conn.: Praeger, 2000), 8.

5. Robert N. Stavins, "Market-Based Environmental Policies," in *Public Policies for Environmental Protection*, 2d ed., ed. Paul R. Portney and Robert N. Stavins (Washington, D.C.: Resources for the Future, 2000), 31–76.

6. Ellis B. Cowling, "Acid Precipitation in Historical Perspective," *Environmental Science & Technology* 16 (1982): 111A–122A.

7. Ibid. During the same period that Odén developed his theory, Ontario government researchers were documenting lake acidification and fish loss in a wide area surrounding Sudbury and attributing those effects to SO_2 emissions from the Sudbury smelters. This work was suppressed by the Ontario government, however. See Don Munton, "Dispelling the Myths of the Acid Rain Story," *Environment* 40 (July–August 1998): 4–7, 27–34.

8. Jimmy Carter, quoted in James L. Regen and Robert W. Rycroft, *The Acid Rain Controversy* (Pittsburgh: University of Pittsburgh Press, 1988), 118.

9. Paul Kinscherff and Pierce Homer, "The International Joint Commission: The Role It Might Play," in *Acid Rain and Friendly Neighbors*, rev. ed., ed. Jurgen Schmandt, Judith Clarkson, and Hilliard Roderick (Durham, N.C.: Duke University Press, 1988), 190–216.

10. Marshall E. Wilcher, "The Acid Rain Debate in North America: 'Where You Stand Depends on Where You Sit,'" *The Environmentalist* 6 (1986): 289–298.

11. Quoted in Philip Shabecoff, "Northeast and Coal Area at Odds over Acid Rain," *The New York Times*, April 19, 1980, 22.

12. Regens and Rycroft, *The Acid Rain Controversy*.

13. Roy Gould, *Going Sour: The Science and Politics of Acid Rain* (Boston: Birkhauser, 1985).

14. Ironically, the tall smokestacks installed by utilities and factories in the 1970s to disperse pollution were largely responsible for the acidification of rain in the northeastern United States. See Gene E. Likens, Richard F. Wright, James N. Galloway, and Thomas J. Butler, "Acid Rain," *Scientific American* 24 (October 1979), 43–51.

15. Only 5 percent to 10 percent of the water in a lake comes from rain that has fallen directly on the lake; most of the water is runoff from the surrounding watershed and has been neutralized by the alkaline soil.

16. Acidity can kill fish in several ways: by interfering with their salt balance, by causing reproductive abnormalities, by leaching aluminum into the lake at levels toxic to fish gills, or by killing the organisms on which they feed. See Michael R. Deland, "Acid Rain," *Environmental Science & Technology* 14 (June 1980): 657; Likens et al., "Acid Rain."

17. Gould, *Going Sour;* Office of Technology Assessment, *Acid Rain and Transported Air Pollutants: Implications for Public Policy* (Washington, D.C.: Office of Technology Assessment, 1984); U.S. GAO, *An Analysis of Issues Concerning Acid Rain*, GAO/RCED-85-13 (December 1984), 3.

18. Gould, *Going Sour.*

19. "The Growing Furor over Acid Rain," *U.S. News & World Report,* November 19, 1979, 66.

20. "Acid from the Skies," *Time,* March 17, 1980.

21. Harold Faber, "Deadly Rain Imperils 2 Adirondack Species," *The New York Times,* March 28, 1977.

22. Casey Bukro, "'Acid Rains' Have Scientists Worried," *Chicago Tribune,* July 15, 1979, 5.

23. *Chicago Tribune,* "A New Worry—Acid Rain," July 20, 1979, section 3, p. 2. Available at http://archives.chicagotribune.com/1979/07/20/page/34/article/a-new-worry-acid-rain.

24. Barbara Moffett, "Acid Rain Poses Potential Danger," *Los Angeles Times,* May 15, 1980, 6.

25. Quoted in Regens and Rycroft, *The Acid Rain Controversy,* 85.

26. National Research Council, *Atmosphere-Biosphere Interactions: Toward a Better Understanding of the Ecological Consequences of Fossil Fuel Combustion* (Washington, D.C.: National Academy Press, 1981).

27. Quoted in Gould, *Going Sour,* 30.

28. Ibid., 32.

29. Quoted in ibid., 33.

30. The EPA was beleaguered because the Reagan administration had appointed high-level administrators hostile to environmental protection who aggressively cut its budget and reorganized its personnel. In March 1983, following a slew of negative media coverage of the agency and a congressional investigation into malfeasance by top-level officials, Administrator Anne Gorsuch, as well as twenty of her colleagues, resigned in disgrace. See Judith A. Layzer, *Open for Business: Conservatives' Opposition to Environmental Regulations* (Cambridge, Mass.: MIT Press, 2012).

31. Quoted in Michael Kranish, "Acid Rain Report Said Suppressed," *The Boston Globe,* August 18, 1984, 1. For an extended discussion of the panel's work and subsequent delay in releasing its report, see Naomi Oreskes and Erik M. Conway, *Merchants of Doubt: How a Handful of Scientists Obscured the Truth on Issues from Tobacco Smoke to Global Warming* (New York: Bloomsbury Press, 2010).

32. Quoted in ibid.

33. Philip Shabecoff, "Toward a Clean and Budgeted Environment," *The New York Times,* October 2, 1984, 28.

34. Alm, *Crossing Borders, Crossing Boundaries.*

35. Congress focused on SO_2 emissions because scientists believed SO_2 was the primary culprit in acid rain (accounting for more than two-thirds of the acidity) and because it was the simplest—both technically and politically—to address.

36. The heaviest sulfur-emitting states in the proposed acid rain control region were Florida, Illinois, Indiana, Kentucky, Missouri, Ohio, Pennsylvania, Tennessee, West Virginia, and Wisconsin. These states accounted for two-thirds of the SO_2 emitted by

the thirty-one states east of the Mississippi. See Steven L. Rhodes, "Superfunding Acid Rain Controls: Who Will Bear the Costs?" *Environment,* July–August, 1984, 25–32.

37. Two of the more prominent proposals were reported out of the Environment Committee in 1982 and 1984 (S. 3041). Both required a 10-million-ton reduction in SO_2 emissions in the thirty-one states east of the Mississippi. Both allowed fuel switching. In 1985 Senator Stafford introduced the Acid Rain Control Act, which divorced the issue from the Clean Air Act revisions and contained provisions similar to those in S. 3041.

38. Sharon Begley, "Acid Rain's 'Fingerprints,'" *Newsweek,* August 11, 1986, 53.

39. Volker Mohnen, "Acid Rain," Hearings Before the Subcommittee on Natural Resources, Agriculture Research and Environment of the Committee on Science and Technology, U.S. House of Representatives, 97th Cong., 1st sess., September 18–19, November 19, December 9 (Washington, D.C.: U.S. Government Printing Office, 1981).

40. Edward C. Krug and Charles R. Frink, "Acid Rain on Acid Soil: A New Perspective," *Science* 221 (August 5, 1983): 520–525.

41. Aaron Wildavsky, *But Is It True? A Citizen's Guide to Environmental Health and Safety Issues* (Cambridge, Mass.: Harvard University Press, 1995).

42. Keith Neuman, "Trends in Public Opinion on Acid Rain: A Comprehensive Review of Existing Data," *Water, Air & Soil Pollution* 31 (December 1986): 1047–1059.

43. George Hager, "The 'White House' Effect Opens a Long-Locked Political Door," *Congressional Quarterly Weekly Report,* January 20, 1990, 139–144.

44. Richard Cohen, *Washington at Work: Back Rooms and Clean Air,* 2d ed. (Boston: Allyn and Bacon, 1995).

45. For example, Cambridge Reports found the percentage of people who said the amount of government regulation and involvement in environmental protection was "too little" increased from 35 percent in 1982 to 53 percent in 1988. A series of polls conducted by *The New York Times*/CBS News asked a national sample of Americans whether they agreed with the following statement: "Protecting the environment is so important that requirements and standards cannot be too high, and continuing environmental improvements must be made regardless of cost." In 1981, 45 percent said they agreed with this statement. By the end of Reagan's first term, this proportion had increased to 58 percent; in 1988, shortly before Reagan left office, it had climbed to 65 percent. Between 1980 and 1989 membership in the major national environmental groups grew between 4 percent (Audubon) to 67 percent (Wilderness Society) annually; the Sierra Club nearly tripled in size over the course of the decade. See Robert Cameron Mitchell, "Public Opinion and the Green Lobby: Poised for the 1990s," in *Environmental Policy in the 1990s,* ed. Norman J. Vig and Michael E. Kraft (Washington, D.C.: CQ Press, 1990), 81–99.

46. Marc K. Landy, Marc J. Roberts, and Stephen R. Thomas, *The Environmental Protection Agency: Asking the Wrong Questions from Nixon to Clinton,* exp. ed. (New York: Oxford University Press, 1994).

47. Quoted in Norman J. Vig, "Presidential Leadership and the Environment: From Reagan and Bush to Clinton," in *Environmental Policy in the 1990s,* 2nd ed., 80–81.

48. Gabriel Chen, Robert Stavins, and Robert Stowe, "The SO_2 Allowance Trading System and the Clean Air Act Amendments of 1990: Reflections on Twenty Years of Policy Innovation," Harvard Kennedy School Faculty Research Working Paper Series, RWP12-003, January 2012. It is noteworthy that scientists had long been advocating a roughly 50 percent reduction in emissions as a starting point. The target ultimately chosen by the White House represented a 50 percent reduction in emissions from the power sector from the 1980 level of about 17.5 million tons. By contrast, the White House team had focused on identifying the point at which the marginal cost of abatement would begin to rise sharply, which they estimated at between 7 million and 8 million tons.
49. Richard Conniff, "The Political History of Cap and Trade," *Smithsonian*, August 2009.
50. Quoted in Cohen, *Washington at Work*, 96.
51. Ibid., 82.
52. The clean states wanted assurances that they would get credit for earlier emissions reductions and would be allowed to expand their utility capacity if needed without busting tight emissions caps. They also did not want to foot the bill for other states' cleanup.
53. Michael Oppenheimer, personal communication, 2003.
54. Cohen, *Washington at Work*.
55. Alyson Pytte, "Clean Air Conferees Agree on Industrial Emissions," *Congressional Quarterly Weekly Report*, October 20, 1990, 3496–3498.
56. In Phase I, Illinois, Indiana, and Ohio utilities each got 200,000 extra allowances per year; in Phase II, ten Midwestern states got 50,000 extra allowances each year.
57. Alyson Pytte, "A Decade's Acrimony Lifted in Glow of Clean Air," *Congressional Quarterly Weekly Report*, October 27, 1990, 3587–3592.
58. Quoted in Cohen, *Washington at Work*, 128.
59. These allowances are like checking account deposits; they exist only as records in the EPA's computer-based tracking system, which contains accounts for all affected generating units and for any other parties that want to hold allowances.
60. This last element, continuous, rigorous emissions monitoring, was an idea generated by a career EPA official named Brian McLean. His monitoring system enabled a simple system whereby the EPA could simply check to see whether a power plant's emissions matched up with its allowances at the end of each year. See Conniff, "The Political History of Cap and Trade."
61. Quoted in Cohen, *Washington at Work*, 211.
62. Brian McLean, "Lessons Learned Implementing Title IV of the Clean Air Act," 95-RA120.04 (Washington, D.C.: U.S. Environmental Protection Agency, 1995).
63. Quoted in Landy et al., *The Environmental Protection Agency*, 290.
64. McLean, "Lessons Learned," 9.
65. According to the EPA, another 182 units joined Phase I of the program as substitution or compensation units, bringing the total of Phase I affected units to 445. See U.S.

Environmental Protection Agency, "Clean Air Markets." Available at http://www.epa.gov/airmarkets/progsregs/arp/basic.html#phases.

66. For example, in Phase I, an individual unit's allocation was the product of a 2.5-pound SO_2 per million Btu emission rate multiplied by the unit's average fuel consumption between 1985 and 1987.

67. Dallas Burtraw and Sarah Jo Szambelan, "U.S. Emissions Trading Markets for SO_2 and NO_x," RFF Discussion Paper, RFF DP 09-40, October 2009. Available at http://www.rff.org/documents/RFF-DP-09-40.pdf.

68. Conniff, "The Political History of Cap and Trade."

69. Barnaby J. Feder, "Sold: The Rights to Air Pollution," *The New York Times,* March 30, 1993, B1.

70. Details available at http://www.epa.gov/capandtrade/allowance-trading.html.

71. U.S. Environmental Protection Agency, "Allowance Markets Assessments: A Closer Look at the Two Biggest Prices Changes in the Federal SO_2 and NO_x Allowance Markets," White Paper, April 23, 2009. Available at http://www.epa.gov/airmarkt/resource/docs/marketassessmnt.pdf.

72. Ibid.

73. Feder, "Sold."

74. Dallas Burtraw, "Trading Emissions to Clean the Air: Exchanges Few but Savings Many," *Resources* (Winter 1996): 3–6.

75. Jeff Johnson, "Utilities Cut Pollution, Save Money, but Few Trade Allowances," *Environmental Science & Technology* 29(12): 1995. 547A. DOI 10.1021/es00012a734

76. Nathaniel O. Keohane, "Cost Savings from Allowance Trading in the 1990 Clean Air Act: Estimates from a Choice-Based Model," in *Moving to Markets in Environmental Regulation,* ed. Jody Freeman and Charles O. Kolstad (New York: Oxford University Press, 2007), 194–229. Studies of the costs of Title IV generally find a 43 percent to 55 percent savings compared to a uniform emissions standard. See Burtraw and Szambelan, "U.S. Emissions Trading Markets for SO_2 and NO_x."

77. Zachary Coile, "'Cap-and-Trade' Model Eyed for Cutting Greenhouse Gases," *San Francisco Chronicle,* December 3, 2007, A1.

78. U.S. Environmental Protection Agency, "Acid Rain and Related Programs: 2008 Emission, Compliance, and Market Data." Available at http://www.epa.gov/airmarkets/documents/progressreports/ARP_2008_ECM_Analyses.pdf.

79. James A. Lynch, Van C. Bowersox, and Jeffrey W. Grimm, "Trends in Precipitation Chemistry in the United States, 1983–1994: An Analysis of the Effects in 1995 of Phase I of the Clean Air Act Amendments of 1990, Title IV," USGS Report 96-0346 (Washington, D.C.: U.S. Geological Survey, 1996).

80. U.S. GAO, *Acid Rain: Emissions Trends and Effects in the Eastern United States,* GAO/RCED-00-47 (March 2000); "Acid Rain Compounds Decrease in N.Y. Lakes," *Greenwire,* September 20, 2002; "Lakes Less Acidic—Study," *Greenwire,* April 15, 2003.

81. G. E. Likens, C. T. Driscoll, and D. C. Buso, "Long-Term Effects of Acid Rain: Response and Recovery of a Forest Ecosystem," *Science* 272 (April 12, 1996): 244–246.

82. Quoted in Jocelyn Kaiser, "Acid Rain's Dirty Business: Stealing Minerals from Soil," *Science* 272 (April 12, 1996): 198.

83. William K. Stevens, "The Forest that Stopped Growing: Trail Is Traced to Acid Rain," *The New York Times*, April 16, 1996, C4. A series of studies by Dr. Lars Hedin and six colleagues, published in 1994 in the journal *Nature*, found that reductions in the release of alkaline particles offset the cuts in sulfates by 28 percent to 100 percent. See William K. Stevens, "Acid Rain Efforts Found to Undercut Themselves," *The New York Times*, January 27, 1994, 14.

84. Gregory B. Lawrence, Mark B. David, and Walter C. Shortle, "A New Mechanism for Calcium Loss in Forest-Floor Soils," *Nature* 378 (November 9, 1995): 162–165.

85. Daniel Markewitz et al., "Three Decades of Observed Soil Acidification in the Calhoun Experimental Forest: Has Acid Rain Made a Difference?" *Soil Science Society of America Journal* 62 (October 1998): 1428–1439.

86. "Good News–Bad News Story for Recovery of Streams from Acid Rain in Northeastern U.S.," USGS News Release, October 4, 1999.

87. James Dao, "Federal Study May Give New York Allies in Acid Rain Fight," *The New York Times*, April 5, 1999, B1.

88. Quoted in ibid.

89. Charles T. Driscoll et al., "Acidic Deposition in the Northeastern United States: Sources and Inputs, Ecosystem Effects, and Management Strategies," *Bioscience* 51 (March 2001): 180–198; Kirk Johnson, "Harmful Effects of Acid Rain Are Far-Flung, a Study Finds," *The New York Times*, March 26, 2001, B1.

90. Quoted in Tim Breen, "Acid Rain: Study Finds Little Improvement in Northeast," *Greenwire*, March 26, 2001.

91. Kevin Krajick, "Long-Term Data Show Lingering Effects from Acid Rain," *Science* 292 (April 13, 2001): 196–199.

92. "Acid Rain in Northeast May Cause Population Decline—Report," *Greenwire*, August 13, 2002.

93. This was the NO_x Budget Program, which was in place from 1999 to 2002; it was replaced from 2003 to 2008 by a slightly different NO_x cap-and-trade program. See John H. Cushman Jr., "U.S. Orders Cleaner Air in 22 States," *The New York Times*, September 25, 1998; Sam Napolitano, Gabrielle Stevens, Jeremy Schreifels, and Kevin Culligan, "The NO_x Budget Trading Program: A Collaborative, Innovative Approach to Solving a Regional Air Pollution Problem," *Electricity Journal* 20,9 (November 2007): 65–76.

94. Dao, "Federal Study."

95. James Dao, "Acid Rain Law Found to Fail in Adirondacks," *The New York Times*, March 27, 2000, 1.

96. Ibid. Because stationary sources reduced their NO_x emissions in the 1980s, by 1998 on- and off-road vehicles and engines accounted for 53 percent of NO_x emissions.

97. The 1970 Clean Air Act grandfathered old plants but imposed strict emissions standards on new plants. The 1977 Clean Air Act Amendments established the NSR permitting process, which requires existing sources that make substantial modifications to install updated pollution controls. To avoid this requirement, many utilities began making significant upgrades but characterizing them as routine maintenance.

98. Richard Perez-Pena, "Power Plants in South to Cut Emissions Faulted in Northeast Smog," *The New York Times*, November 16, 2000, 1.

99. Randal C. Archibold, "Tentative Deal on Acid Rain Is Reached," *The New York Times*, December 22, 2000, 2.

100. Matthew L. Wald and Stephanie Saul, "Big Utility Says It Will Settle 8-Year-Old Pollution Suit," *The New York Times*, October 9, 2007.

101. The three CAIR cap-and-trade programs were the SO_2 annual trading program, which began in 2010, and the NO_x annual trading program and NO_x ozone-season trading program, both of which began in 2009.

102. Darren Samuelsohn, "Industry Files 12 Lawsuits Against EPA's CAIR Rule," *Greenwire*, July 12, 2005.

103. Quoted in Del Quentin Wilber and Marc Kaufman, "Judges Toss EPA Rule to Reduce Smog, Soot," *The Washington Post*, July 12, 2008, 3.

104. Chen, Stavins, and Stowe, "The SO_2 Allowance-Trading System."

105. Final Cross-State Air Pollution Rule. Clean Air Markets. U.S. Environmental Protection Agency. Available at https://www.epa.gov/airmarkets/final-cross-state-air-pollution-rule-update.

106. U.S. Environmental Protection Agency, "Clean Air Interstate Rule, Acid Rain Program, and Former NO_x Budget Trading Program: 2012 Progress Report, SO_2 and NO_x Emissions, Compliance, and Market Analyses." Available at https://www.epa.gov/sites/production/files/2017-09/documents/2014_full_report.pdf.

107. A. Danny Ellerman, "Are Cap-and-Trade Programs More Environmentally Effective Than Conventional Regulation?" in *Moving to Markets in Environmental Regulation*, 48–62.

108. Margaret R. Taylor, Edward S. Rubin, and David A. Houshell, "Regulation as the Mother of Invention: The Case of SO_2 Control," *Law & Policy* 27 (April 2005): 348–378.

109. Frank Wätzold, "SO_2 Emissions in Germany: Regulations to Fight *Waldsterben*," in *Choosing Environmental Policy: Comparing Instruments and Outcomes in the United States and Europe*, ed. Winston Harrington, Richard D. Morgenstern, and Thomas Sterner (Washington, D.C.: Resources for the Future, 2004), 23–40.

110. Christopher Joyce, "Rivers on Rolaids: How Acid Rain Is Changing Waterways," *Morning Edition*, September 13, 2013.

111. University of California Irvine, "Made in China for Us," Available at https://www.eurekalert.org/pub_releases/2014-01/uoc--mic011514.php.

PART TWO

HISTORY, CHANGING VALUES, AND NATURAL RESOURCE MANAGEMENT

CHAPTER SIX

OIL VERSUS WILDERNESS IN THE ARCTIC NATIONAL WILDLIFE REFUGE

The controversy over drilling for oil on the Coastal Plain of Alaska's Arctic National Wildlife Refuge (ANWR, pronounced "anwar") began shortly after the passage of the Alaska National Interest Lands Conservation Act (ANILCA) in 1980. After nearly four decades of intense debate, a provision in the 2017 Tax Cuts and Jobs Act (TCJA) opened the Coastal Plain to oil drilling. The ongoing conflict pits those who want to explore for oil and gas in the refuge against individuals who want to see the area off-limits to all development. Ordinarily, decisions about oil and gas development on public lands are made by federal agencies, but an unusual statutory provision in ANILCA requires legislative approval for oil exploration in ANWR, so combatants have waged this battle almost entirely in Congress.

Over the years, the contest has featured fierce public relations campaigns by advocates on both sides, as well as a variety of parliamentary maneuvers by legislators. Nearly two decades after the issue first arose, oil and gas development in the refuge was the major environmental issue in the 2000 presidential campaign, and the centerpiece of President George W. Bush's national energy policy. In spring 2005 the Senate voted by a narrow margin to allow oil exploration in ANWR as part of the budget process. Ultimately, however, the Bush administration's drive to open the refuge failed, leaving ANWR's fate to future debate. Following multiple failed attempts to designate portions of the refuge as wilderness, ANWR again took center stage with the passing of the TCJA of 2017. In congruence with President Donald J. Trump's energy and budget agendas, the controversial bill contained a provision mandating lease sales within the Coastal Plain, and thus opening the refuge to oil drilling and energy development. For some, President Trump ended an epic, four-decade battle. However, the question becomes how drilling opponents intend to contest the provision both in Congress and the court.

The ANWR dispute epitomizes the divide between proponents of natural resource extraction and advocates of wilderness preservation. Such controversies are particularly intractable because, unlike disputes over human health, which involve relatively consensual values, debates about developing wilderness tap into ecological values, about which differences are often so wide as to be unbridgeable. For wilderness advocates, compromise is unthinkable: any development in a pristine area constitutes a complete loss, a total violation of the spiritual and aesthetic qualities of the place. For development proponents, such an attitude reflects a lack of concern about human economic needs. Both sides have tried to advance their values using tactics appropriate for the legislative arena, where statute requires that the battle over ANWR be fought. In this case, however, wilderness advocates have managed to expand the scope of the conflict by using symbols, metaphors, and images to transform an apparently local battle into a national contest. Forced to respond, development advocates have enlisted symbols, metaphors, and images of their own to appeal to widely held values such as economic growth, national security, and local control.

The ANWR case also features an unusual twist on typical natural resource policymaking: wilderness advocates have the advantage because *not* drilling is the status quo, and it is always much easier to block policy change than to bring it about. The legislative complexity of energy policymaking has enhanced the advantage held by preservationists. At least two committees in each chamber, and as many as seven in the House, claim jurisdiction over energy decisions. Traditionally, a single committee has jurisdiction over a bill; since the 1970s, however, multiple referral—or referral of a bill to more than one committee—has become more common, particularly in the House. Multiple referral, which can be simultaneous or sequential, complicates the legislative process because a bill must clear all of the committees of reference before it can go to the floor.[1] Legislating energy policy is also daunting because so many organizations mobilize in response to the myriad provisions that threaten to impose costs or promise to deliver benefits. The involvement of multiple interests creates the possibility of antagonizing one or more of them, something legislators prefer to avoid. To get around this obstacle, drilling proponents have repeatedly tried a low-visibility tactic: attaching refuge-opening riders to budget bills, which have the additional benefit of being immune to Senate filibusters.[2]

BACKGROUND

The ANWR debate is the most recent manifestation of a century-old conflict over both Alaska's resource-based economy and the state's relationship to the federal government. The battle over the trans-Alaska pipeline, in particular, set the stage for the passage of ANILCA in 1980, which, in turn, created the legal and institutional context for the protracted struggle over ANWR. Shortly after ANILCA's passage, two distinct coalitions formed—one to advocate wilderness designation for the newly established refuge and the other to promote oil and gas development in the area.

A History of Resource Exploitation in Alaska

Alaska has long been a battleground between development interests and environmentalists. From the time the United States acquired the Alaska territory from Russia in 1867, American adventurers seeking to make their fortunes have exploited the area's natural attributes. The first terrestrial resource to attract attention was gold. Discovery of the Klondike gold fields in 1896, when the territory's white population was little more than 10,000, precipitated an influx of prospectors and others hoping to profit from the boom. Development there from the late 1800s to the mid-1900s conformed to the boom-and-bust cycle typical of natural resource-based economies. Prior to Alaska attaining statehood in the late 1950s, the federal government tried to mitigate this trend by setting aside millions of acres of land to conserve natural resources—including coal, oil, gas, and timber—to ensure their orderly development.

Environmentalists also have long-standing interests in Alaska's natural endowments. As early as the 1930s, forester and outspoken wilderness advocate Bob Marshall began pleading for the preservation of Alaskan wilderness.[3] While the U.S. Navy was searching for oil and gas, the National Park Service began to investigate Alaska's recreational potential, and in 1954—after surveying the Eastern Brooks Range—it recommended that the federal government preserve the northeastern corner of the state for its wildlife, wilderness, recreational, scientific, and cultural values.

Fearing further inroads by environmentalists, boosters such as Robert Atwood—editor and publisher of the *Anchorage Daily Times*—and Walter Hickel—millionaire real estate developer, contractor, and hotelier—advocated statehood to support greater economic development. In 1957 a pro-statehood pamphlet declared, "If Alaska becomes a state, other people, like ourselves, will be induced to settle here without entangling federal red tape and enjoy an opportunity to develop this great land."[4] On January 3, 1959, after more than a decade of debate, Congress made Alaska a state, and for the next twenty years, the state and its native tribes jockeyed to select lands for acquisition. In 1966 Stewart Udall, secretary of the Department of the Interior, froze applications for title to Alaska's unappropriated land until Congress could settle outstanding native claims. The same year, however, Hickel was elected governor. With his swearing-in, a long period of development got under way in the state, best symbolized by the rush to develop the Prudhoe Bay oil field that began with Atlantic Richfield's strike in the winter of 1967–1968.[5]

The Trans-Alaska Pipeline Controversy

In 1969 President Richard Nixon appointed Hickel to head the interior department, a position from which he was well placed to act as a policy entrepreneur for his pet project, the trans-Alaska pipeline. On February 10, 1969, the "big three" companies that had been developing oil reserves on Alaska's North Slope came together into a loose consortium called the Trans-Alaska Pipeline System (TAPS) and announced plans to build a 798-mile-long pipeline from the oil fields at Prudhoe Bay to the port of Valdez. This announcement coincided with a surge in national environmental consciousness, however, making a clash inevitable. At first, opponents were able to delay the project by citing technical impediments, administrative and legal requirements, and native claims. The history of the oil industry's destructive practices in the Arctic also fueled public skepticism about TAPS. Resistance crystallized in 1970, and national environmental groups filed three lawsuits to stop the pipeline, citing the Mineral Leasing Act and the National Environmental Policy Act (NEPA).[6] But in January 1971 an interior department staff report recommended building the pipeline, even though it would cause environmental damage.

By this time, President Nixon had replaced Hickel with Rogers C. B. Morton, who was dubious about the pipeline and wanted more information. Interior department hearings held in February 1971 "became a showcase of unyielding positions" in which environmentalists warned that the pipeline would destroy pristine wilderness, while oil interests claimed that Alaskan oil was economically essential and that the environmental risks were minimal.[7] Environmental activists subsequently formed the Alaska Public Interest Coalition to coordinate and intensify the public campaign against the pipeline.[8] Complicating matters for the coalition, however, was the December 1971 passage of the Alaska Native Claims Settlement Act (ANCSA), which had been strongly supported by oil and gas interests. ANCSA transferred to native Alaskans ownership of 44 million acres and proffered a settlement of nearly $1 billion, half of which was to be paid from oil production royalties, thereby giving native Alaskans a powerful financial stake in oil development. At the same time, the act prohibited the state of Alaska or its native tribes from selecting lands within a proposed oil pipeline corridor as part of that 44 million acres to ensure that neither the state nor a native tribe could use its property rights to block the project.

In March 1972 the interior department released its final environmental impact statement (EIS) on the pipeline. The document stressed the need to develop Alaskan oil to avoid increasing dependence on foreign sources.[9] Environmentalists immediately challenged the EIS in court, arguing that it was based on unreliable data and failed to consider transportation alternatives. Later that year, Secretary Morton declared the pipeline to be in the national interest, but public sentiment remained strongly against it until the summer of 1973, when forecasts of an energy crisis opened a policy window for development advocates. The oil industry capitalized on the specter of an oil shortage to generate legislative interest in the pipeline, and in July the Senate narrowly passed an amendment offered by Mike Gravel, D-Alaska, which declared that the interior department's EIS fulfilled all the requirements of NEPA, thus releasing the project from further legal delay.[10] Just before the end of its 1973 session, Congress passed the Trans-Alaska Pipeline Authorization Act (361–14 in the House and 80–5 in the Senate), and on November 16 President Nixon signed it into law.[11]

The Alaska National Interest Lands Conservation Act (ANILCA)

Although development interests prevailed in the pipeline controversy, environmentalists in the 1960s and 1970s also succeeded in staking some claims to Alaska. On December 9, 1960, interior secretary Fred Seaton set aside 8.9 million acres as the Arctic National Wildlife Range. Then, acting under Section 17d(2) of ANCSA, which authorized the secretary to set aside "national interest" lands to be considered for national park, wilderness, wildlife refuge, or national forest status, Seaton added 3.7 million acres to the range. In the late 1970s environmentalists began campaigning to get formal protection for the lands Seaton had set aside. Their campaign ended in 1980 when Congress enacted ANILCA, which placed 104 million acres of public domain land under federal protection.[12] The act reclassified the original 8.9-million-acre Arctic National Wildlife Range as the Arctic National Wildlife Refuge, adding to the refuge all lands, waters, interests, and submerged lands under federal ownership at the time Alaska became a state. In addition, ANILCA expanded the original refuge by adding 9.1 million acres of adjoining public lands, extending it west to the pipeline and south to the Yukon Flats National Wildlife Refuge.

ANILCA made significant compromises, however, such as excluding from federal protection areas with significant development potential, thereby setting the stage for another showdown over resource development in Alaska. The provision of particular consequence was Section 1002, which instructed the interior department to study the mineral and wildlife resources of the refuge's 1.5-million-acre Coastal Plain to determine its suitability for oil and gas exploration and development. Although the House of Representatives had favored wilderness designation for this area, often called the "ecological heart" of the refuge, the Senate had resisted in the absence of a thorough assessment of its oil and gas potential. As a compromise, Section 1002 prohibited leasing, development, and production of oil and gas in the refuge unless authorized by Congress.[13]

At first glance, oil and gas development might appear out of place in a wildlife refuge; after all, the statutory purpose of the refuge system, established in 1966, is "to provide, preserve, restore, and manage a national network of lands and waters sufficient in size, diversity, and location to meet society's needs for areas where the widest possible spectrum of benefits associated with wildlife and wildlands is enhanced and made available."[14] Furthermore, ANILCA specifies that ANWR is to be managed in the following way:

> (i) to conserve fish and wildlife populations and habitats in their natural diversity including, but not limited to, the Porcupine caribou herd (including the participation in coordinated ecological studies and management of this herd and the Western Arctic caribou herd), polar bears, grizzly bears, muskox, Dall sheep, wolves, wolverines, snow geese, peregrine falcons, and other migratory birds and arctic char and grayling; (ii) to fulfill the international treaty obligations of the United States with respect to fish and wildlife and their habitats; (iii) to provide, in a manner consistent with purposes set forth in subparagraphs (i) and (ii), the opportunity for continued subsistence uses by local residents; and (iv) to ensure, to the maximum extent practicable and in a manner consistent with the purposes set forth in subparagraph (i), water quality and necessary water quantity within the refuge.[15]

A provision of Section 4(d) of the act that establishes the refuge system, however, authorizes the use of a refuge for any purpose, including mineral leasing, as long as that use is "compatible with the major purpose for which the area was established."[16] And, in fact, private companies hold mining leases as well as oil and gas leases on many of the nation's wildlife refuges: as of 2003 oil and gas exploration and development were occurring in about 25 percent of them (155 of 575), while 4,406 oil and gas wells—1,806 of which were active—were sprinkled throughout 105 refuges.[17] Although there is little systematic documentation of the environmental impact of resource development in wildlife refuges, an investigation by the U.S. General Accounting Office (GAO, later the Government Accountability Office) in 2003 found that those effects varied from negligible to substantial and from temporary to long term.[18] With funding of only $2 per acre, compared to $13 per acre for the Park Service and $7 per acre for the Forest Service (in 2003), the Fish and Wildlife Service has had limited resources with which to resist or mitigate the effects of such incursions.[19]

THE CASE

Vast and extremely remote, ANWR is the most northerly unit, and the second largest, in the National Wildlife Refuge System. ANWR encompasses about 19.6 million acres of land in northeast Alaska, an area almost as large as New England. It is bordered to the west by the trans-Alaska pipeline corridor, to the south by the Venetie–Arctic Village lands and Yukon Flats National Wildlife Refuge, to the east by Canada, and to the north by the Beaufort Sea. Fairbanks, the nearest city, is about 180 miles south of the refuge boundary by air. Two native villages, Kaktovik on Barter Island and Arctic Village on the south slope of the Brooks Range, abut the refuge (see Map 6.1).

The refuge is geographically diverse. It contains a full range of boreal forest, mountain, and northern slope landscapes and habitats, as well as the four tallest peaks and the largest number of glaciers in the Brooks Range. The refuge's northern edge descends to the Beaufort Sea and a series of barrier islands and lagoons, and the valley slopes are dotted with lakes, sloughs, and wetlands. In the southeastern portion, the Porcupine River area, groves of stunted black spruce grade into tall, dense spruce forests. In addition to portions of the main calving ground for the 130,000-member Porcupine caribou herd and critical habitat for the endangered peregrine falcon, ANWR is home to snow geese, tundra swans, wolves, wolverines, arctic foxes, lynx, marten, and moose. The rivers, lakes, and lagoons contain arctic grayling, lake trout, whitefish, northern pike, turbot, arctic cod, and several varieties of salmon. The waters offshore harbor summering bowhead whales, and the coastal lagoons provide year-round habitat for polar bears and ringed and bearded seals.

Although a quintessential wilderness seldom visited by humans, ANWR is an unusual poster child for the preservationist cause: it is not among the most

Map 6.1 The Arctic National Wildlife Refuge

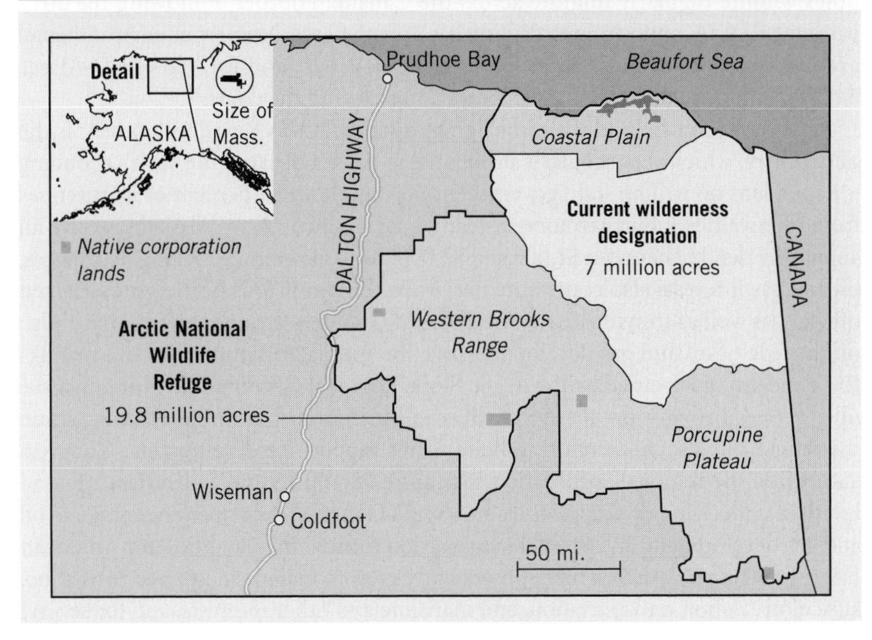

Source: U.S. Fish and Wildlife Service. David Butler/Globe Staff

spectacular of America's wild places. Yet advocates of preservation have created a powerful image of the refuge as the nation's last remaining true wilderness, a symbol of the American frontier. Their opponents have also crafted a compelling case, however, one in which oil development in the Coastal Plain will enhance the country's economic and military security while posing only a marginal risk to an uninspiring landscape.

The ANWR Coalitions

Spearheading the movement to preserve ANWR's Coastal Plain is the Alaska Coalition (now coordinated by the Alaska Wilderness League), a consortium of Alaskan, national, and international environmental groups, as well as sporting, labor, and religious organizations. Among the most influential supporters of the refuge are the Sierra Club, the Wilderness Society, the Natural Resources Defense Council, the National Wildlife Federation, Defenders of Wildlife, the Audubon Society, the Northern Alaska Environmental Center, and the Trustees for Alaska. Backing the environmentalists and their allies are the Gwich'in Eskimos, a subgroup of the Athabascan tribe that for thousands of years has relied on the Porcupine caribou herd for its survival. The government of Canada, another player on this side, also opposes opening the refuge to development because the Porcupine caribou and other wildlife regularly migrate across the Canadian border. Following the 2017 passage of TCJA, opposition to drilling has proven fierce. Numerous environmental groups, as well as Senators Maria Cantwell, D-Wash., and Ed Markey, D-Mass., have vowed to challenge the bill in both Congress and the court.[20]

A major force behind the campaign to open up ANWR to development is the oil industry, which is particularly influential in Alaska because the state's economy is dependent on it. The state government derives about 85 percent of its revenues from oil royalties.[21] Furthermore, instead of paying income tax, Alaskans receive an annual dividend of between $1,000 and $2,000 from oil revenues. Taking no chances, oil and gas interests also contribute heavily to the campaigns of the state's elected officials, as well as to sympathetic members of Congress around the country.[22] Also on the side of oil and gas development are the Inupiat Eskimos, who live in Kaktovik and are represented by the Arctic Slope Regional Corporation. Not only have oil revenues brought prosperity to the small community of Inupiat living within the refuge, but also, in an effort to build native support for development, Congress authorized the regional corporation to acquire 92,160 acres of subsurface mineral interests underlying the Inupiat's claim, should Congress ever open the refuge to oil and gas development. In 1987 the corporation formed the Coalition for American Energy Security (CAES), which subsequently grew to include motorized recreation, agriculture, labor, transportation, and maritime and other interests, and for several years it was the primary lobbying arm on behalf of ANWR development. In the early 1990s Arctic Power—a much more substantial lobbying organization heavily funded by the major oil companies—replaced CAES. After the passage of TCJA in 2017, Senator Lisa Murkowski, R-Alaska), and oil industry allies planned follow up legislation - the Artic Policy Act and Shipping and Environmental Artic Leadership Act. As of 2019, both pieces of legislation are still pending.[23]

The Battle to Frame the Problem, 1986–1992

From the outset, the parties in the ANWR debate framed the problem in contrasting ways, describing scientific and economic risks and uncertainties in terms that were consistent with their values. Proponents of drilling argued that the nation needed to develop its domestic oil reserves to preserve national security and boost the economy, and they asserted that the supply available in ANWR was significant enough to justify the minimal risk of environmental harm. Opponents of development argued that the amount of oil that ANWR was likely to hold would satisfy just a tiny fraction of total domestic demand and therefore would only marginally reduce the nation's dependence on foreign oil. Moreover, they said, development would certainly create air and water pollution, generate toxic wastes, disturb wildlife habitat, and, most important, destroy the wilderness character of the landscape. By the late 1980s, when the conflict was in full swing, both sides had developed elaborate arguments about the national interest in ANWR and had marshaled and interpreted scientific and economic evidence to support their claims and rebut those of their opponents.

The Potential Benefits of Oil and Gas Development. The Department of Interior has developed the authoritative models predicting the amount of oil that lies under the refuge's Coastal Plain.[24] The first of these, the 1987 Arctic National Wildlife Refuge, Alaska, Coastal Plain Report, known as the "1002 Report," concluded that there was a 19 percent chance of finding economically recoverable oil reserves. The report judged it possible to recover between 600 million and 9.2 billion barrels of oil, with a mean of 3.2 billion barrels. If that average were correct, the Coastal Plain would be the third largest oil field in U.S. history. According to the report's mean case scenario, ANWR would produce 147,000 barrels per day by the year 2000, peak at 659,000 barrels per day in 2005, fall over the next decade to 400,000 barrels per day, and continue to decline for the remaining thirty- to fifty-year life of the field. At its peak, ANWR's production would equal 4 percent of U.S. consumption, 8 percent of domestic production, and about 9 percent of imports, while increasing the gross national product (GNP) by $74.9 billion and reducing the annual trade deficit by approximately $6 billion.[25]

Wielding these official estimates, proponents of development advanced the argument that drilling in the Coastal Plain would enhance national security. They cited the country's growing dependence on imported oil, noting that of the 17.3 million barrels per day consumed in 1990, an average of 7.6 million barrels per day were imported, with 2.1 million of those coming from the politically unstable Middle East. In 1990 the Energy Department predicted that, without changes in the nation's energy policy, imports would account for 65 percent of the U.S. oil supply by 2010. With chilling language, drilling advocates stressed America's vulnerability to reliance on foreign oil, saying, "Remember that two million barrels of oil a day coming down that pipeline from Alaska's Prudhoe Bay played a major role in breaking OPEC's grip. And recall how that grip felt, with its lines at the gasoline pumps and aggravated inflation."[26] Sen. Ted Stevens, R-Alaska, warned that "without the Arctic Wildlife Range [*sic*] resources, this nation faces the threat of a precarious future in which OPEC nations would be able to hold another and perhaps more serious oil embargo over our heads."[27]

Environmentalists used the same information but framed it differently to argue that the amount of recoverable oil that might lie within the refuge was unlikely to enhance the nation's energy security; they noted that it would most likely amount to "less than 200 days worth of oil at current consumption rates" and "less than 3 percent of our daily demand."[28] Environmentalists also contended that the nation's need for additional oil supplies was more apparent than real. They pointed out that the rise in U.S. petroleum imports in the late 1980s was accompanied by a decline in conservation, which was not surprising because the price of oil was low and federal incentives to conserve had atrophied. Environmentalists had official data to support their claims as well: the Congressional Research Service estimated that energy conservation research and development expenditures in real dollars fell from $322 million in 1981 to $129 million in 1989. Funding for renewable energy research and development also dropped dramatically, from $708 million in 1981 to $110 million in 1990, before rising to $157 million in 1991.[29] Reflecting the low oil prices and lack of conservation incentives, automotive fuel efficiency fell in the late 1980s for the first time since 1973. Daniel Lashof, senior scientist for the Union of Concerned Scientists, concluded, "There is simply no way we can produce our way to energy security. [Oil from ANWR] cannot significantly affect the world market prices or the world markets for oil. The only way we can insulate our economy from price shocks is by improving our efficiency and reducing our total consumption of oil."[30]

Drilling advocates had a subsidiary argument about the benefits of drilling in the refuge, however: they claimed that opening ANWR to development would ease the federal deficit and create jobs. In 1990 the prodevelopment CAES suggested that if the interior department's high-end estimates of the refuge's oil potential were correct, the GNP could rise $50.4 billion by 2005 and the trade deficit could decline by billions of dollars. Citing a 1990 study by Wharton Econometrics Forecasting Associates, CAES contended that developing a 9.2-billion-barrel oil field (the high-end estimate, on which it again chose to focus) would create 735,000 jobs by 2005.[31] CAES produced literature that reinforced its claims. At the height of the controversy, the organization was distributing a glossy, eight-page monthly newsletter with articles such as "Leasing ANWR Could Ease Budget Woes" and "Rising Oil Prices Push Up U.S. Trade Deficit," as well as a postcard-sized map of the United States showing the dollar amounts generated in each state by existing oil development in the North Slope area.

Environmentalists responded to the deficit argument in turn, pointing out that independent analysts, using different oil price projections and revised tax and financial assumptions, had generated quite different total expected present values of developing ANWR. In 1990 those values ranged from $0.32 billion to $1.39 billion, compared to the interior department's $2.98 billion.[32] Furthermore, they pointed out, GNP figures disguised the allocation of oil revenues, most of which would go to oil companies. To highlight the self-interested motives of the prodrilling coalition, environmentalists noted that the oil industry reaped more than $41 billion in profit from North Slope oil development and transportation between 1969 and 1987, according to a study done for the Alaska Department of Revenue.[33]

Antidrilling advocates also rebutted the Wharton study's job creation figures, citing a Congressional Research Service analysis that characterized the estimates as "generous" and based on the "most optimistic of underlying scenarios."[34] They added that developing photovoltaics, wind, and other alternative energy sources would yield new jobs as well. Such jobs, they contended, would have the advantage of being local and stable, not subject to the boom-and-bust cycles that in the past had ravaged Texas, Oklahoma, and Alaska. Finally, environmentalists reiterated that conservation measures could have a much bigger impact on domestic oil consumption, and therefore on the trade deficit, than increased domestic oil production.

For example, they said, an increase in fuel economy standards to 40 miles per gallon for new cars and 30 miles per gallon for light trucks could save more than 20 billion barrels of oil by 2020—more than six times the total amount the government believed lay under ANWR's Coastal Plain.

The Environmental Costs of Oil and Gas Development. Not only did environmentalists contest estimates of the economic and security value of ANWR reserves, but they also raised the specter of certain and catastrophic ecological impacts of oil development in the refuge. Most important, they emphasized, because ANWR's Coastal Plain is so windy, it provides relief from insects and safe calving grounds for the Porcupine caribou herd, the second largest caribou herd in Alaska and the seventh largest in the world. They pointed out that most biologists believe the cumulative disturbance resulting from North Slope oil field activity had prevented many Central Arctic herd caribou from feeding optimally, resulting in reduced summer weight gain, decreased pregnancy rates, and increased mortality. Biologists forecast similar effects on the Porcupine caribou herd in the event of Coastal Plain development.

Antidrilling advocates also highlighted biologists' concern about the effect of oil exploration on the approximately 2,000 polar bears that roam the land from northwestern Alaska to the Northwest Territories in Canada. Studies showed that pregnant polar bears habitually use the Coastal Plain and the refuge lands to the east as sites for onshore denning during the winter. Although only a relatively small number of female bears use the Coastal Plain for dens, the area is nonetheless important: researchers found that the productivity of land dens in the Beaufort Sea polar bear population from 1981 to 1989 was significantly higher than that of dens on the offshore ice pack. Furthermore, biologists believed that the Beaufort Sea polar bear population could sustain little, if any, additional mortality of females; the number of animals dying annually was already equal to the number born each year.[35]

Not surprisingly, the oil industry and its allies rejected environmentalists' assertion that oil development would disturb wildlife, emphasizing the uncertainty surrounding scientific estimates of likely wildlife impacts. A BP Exploration publication portrayed environmentalists as unscientific and misanthropic, saying, "Claims of serious adverse effects on fish and wildlife populations are simply not supported by scientific data and may in fact represent an objection to the presence of humans in a wild area."[36] Prodrilling advocates got a boost when interior secretary Donald Hodel wrote an editorial in *The New York Times* saying, "We do not have to choose between an adequate energy supply on the one hand and a secure environment on the other. We can have both."[37]

The two sides also disputed the size of the footprint that would be made by oil development on the Coastal Plain. CAES pointed out that oil operations on the Coastal Plain would occupy only nineteen square miles, an area roughly the size of Dulles Airport outside of Washington, D.C. Environmentalists responded that, although the Prudhoe Bay complex covered a mere 9,000 acres, that acreage was spread across more than 800 square miles. The Office of Technology Assessment concurred with environmentalists, saying,

> Although the industry argues—correctly—that the actual coverage of the surface is likely to be less than one percent of the coastal plain, the physical coverage would be spread out like a spiderweb, and some further physical effects, like infiltration of road dust and changes in drainage patterns, will spread out from the land actually covered.[38]

Environmentalists added that the oil industry had a poor record of both estimating likely environmental impacts and repairing damage. To support this charge they cited a Fish and Wildlife Service report comparing the actual and predicted impacts of the trans-Alaska pipeline and the Prudhoe Bay oil fields. The report concluded that the actual area of gravel fill and extraction was 60 percent greater than expected; gravel requirements surpassed predictions by 400 percent; and road miles exceeded the planned number by 30 percent.[39]

Finally, environmentalists argued that, although it does support a host of wildlife, ANWR's Coastal Plain is a fragile ecosystem that is more vulnerable than temperate ecosystems to damage from industrial activity. To support this claim, they pointed out that ANWR is an Arctic desert that receives an average of only seven inches of moisture a year. Its subsoil, permafrost, remains frozen year round, and only the top few inches to three feet of ground thaw in the summer. They noted that the combination of low temperatures, a short growing season, and restricted nutrients limits the area's biological productivity. Because plants grow slowly, any physical disturbance—even driving vehicles across the terrain in winter—can scar the land for decades, and regrowth could take centuries. They added that, like other Arctic areas, ANWR contains large concentrations of animal populations but relatively low diversity. Mammal populations, in particular, gather in large numbers, thereby heightening their vulnerability to catastrophe. They warned that arctic plants are more sensitive to pollutants than are species that grow in warmer climates; toxic substances persist for a longer time in cold environments than in more temperate climates; and the severe weather and complex ice dynamics of the Arctic complicate both environmental protection and accident cleanup.[40]

In addition to proffering different causal stories and numbers in the debate, the two sides adopted contrasting symbols that revealed the chasm separating their values. Environmentalists characterized the refuge as the last great American wilderness, a vast expanse of untrammeled landscape that provides a home for increasingly rare Arctic wildlife. They likened opening the refuge to "destroy[ing] America's Serengeti" for a few weeks' supply of oil.[41] Development proponents described the area as a barren, marshy wilderness in the summer and a frozen desert the rest of the year. "ANWR's Coastal Plain is no Yellowstone," scoffed a *Wall Street Journal* editorial.[42]

The Political Contest, 1986–1992

These competing arguments had a profound impact on legislative politics. By nationalizing the issue, environmentalists enabled non-Alaskan legislators to claim credit with their environmentalist constituents for protecting the refuge. Development interests responded by trying to persuade uncommitted legislators that opening the refuge would provide their constituents with tangible benefits, including jobs, cheap oil, and national security. Both sides tried to capitalize on focusing events to gain the upper hand. Protection advocates had the advantage for several reasons. First, not drilling is the status quo, and it is easier to block a policy change in Congress than to enact one. Second, ANWR falls under the jurisdiction of multiple committees in Congress, which also makes obstructing policy change easier. Third, the sheer complexity of national energy policy makes it contentious and therefore unattractive to most legislators, who prefer to avoid provoking opposition.

ANWR Gets on the Congressional Agenda. In late 1986 Trustees for Alaska and other environmental groups learned that the interior department planned to circulate for public review the EIS it had prepared for oil leasing in ANWR at the same time it submitted its recommendation to Congress. Representing the Reagan administration's position, William Horn, the department's assistant secretary for fish and wildlife, justified releasing the draft report to the public and Congress simultaneously on the grounds that the refuge might contain "a supergiant oil field that does not exist anywhere else in the United States."[43] The environmentally oriented Alaska Coalition promptly filed a lawsuit against the department, arguing successfully that NEPA requires public participation in the preparation of an EIS prior to its release to Congress. Compelled by the court, the Fish and Wildlife Service held public hearings in January 1987.

On June 1, after incorporating public comments and responses, Secretary Hodel submitted a final report on the oil and gas potential of the Coastal Plain, the projected impacts of development, and a recommendation that Congress authorize oil and gas leasing in Area 1002. In the report's introduction, Hodel said, "The 1002 area is the nation's best single opportunity to increase significantly domestic oil production." He embraced arguments made by the development coalition in support of his conclusions, saying production from the area would "reduce U.S. vulnerability to disruptions in the world oil market and could contribute to our national security."[44] After projecting only minimal impacts on caribou and musk-oxen, the report assured readers that oil development and wildlife could coexist.

The Natural Resources Defense Council and the Steering Committee for the Gwich'in tribe immediately filed lawsuits against the interior department, alleging that the report failed to properly evaluate potential environmental damage. The Gwich'in suit also claimed that the interior department had overestimated the revenues to be gained from drilling by using unrealistically inflated assumptions about oil prices. The plaintiffs requested that a new EIS be prepared that was based on realistic revenue projections and addressed the survival of Gwich'in culture.

Congressional Action, 1987–1989. Concurrently with the legal battles over the adequacy of the 1002 EIS, Congress began debating the merits of developing the refuge. In 1987 sponsors introduced seven bills concerning ANWR, five of which would have opened up the 1.5-million-acre Coastal Plain to oil and gas exploration, development, and production. In the House, two committees—Merchant Marine and Fisheries and Insular and Interior Affairs—held extensive but inconclusive hearings.

In the Senate, William Roth, R-Del., introduced S 1804, an ANWR wilderness designation bill, while Frank Murkowski, R-Alaska, introduced S 1217, a bill to open up the refuge. Two Senate committees, Energy and Natural Resources and Environment and Public Works, held hearings but failed to report bills to the floor. S 1217 was stopped at least in part because of a series of scandals that broke in 1987. First, in July, the press revealed that the interior department had begun dividing the Coastal Plain among Alaskan native corporations for leasing to oil companies without congressional approval. Shortly thereafter, environmentalists discovered that the department had excluded the conclusion of one of its own biologists that oil development would result in a 20 percent to 40 percent decline in the Porcupine caribou population.[45] What the department claimed was an editing error environmentalists attributed to political censorship. Either way, the controversy was sufficient to derail the bill's progress.

The following year, the House Subcommittee on Fisheries and Wildlife Conservation and the Environment, chaired by Don Young, R-Alaska, approved (28–13) HR 3601, which favored leasing. The subcommittee considered but did not adopt several environmentally oriented amendments. Instead, the bill attempted to soften the impact of drilling by requiring oil and gas development to impose "no significant adverse effect" on the environment; establishing a three-mile-wide coastal "insect relief zone" for caribou; and allowing only "essential facilities" within 1.5 to 3 miles of the coast. The House took no further action on the legislation, however, because the Interior Committee, under the proenvironment Rep. Morris "Mo" Udall, D-Ariz., claimed equal jurisdiction over the matter and failed to report out a bill. The Senate also debated the wilderness and oil leasing issue, and advocates of protection on the Environment Committee were able to thwart supporters of drilling on the Energy Committee.[46]

With the inauguration of George H. W. Bush as president in 1989, environmentalists hoped their prospects might improve because as a candidate, Bush had tried to distance himself from President Reagan's antienvironmental record. In January the Alaska Coalition sent a petition signed by 100 environmental groups and local civic associations to the new president and his designated interior secretary, Manuel Lujan Jr. (a former member of Congress who had introduced legislation to open the refuge), requesting that they protect the refuge from development.[47]

But the letter, although it affected the administration's language, did not prompt a shift in its position: in February President Bush called for "cautious development" of ANWR. Backed by the Bush administration, the Senate Energy and Natural Resources Committee approved an ANWR oil and gas leasing bill, S 684, on March 16, 1989, by a vote of twelve to seven.[48]

The House Merchant Marine and Interior committees also held ANWR hearings, and a week after the Senate committee's report, Rep. Walter Jones, D-N.C., introduced a companion bill (HR 1600) to open up the refuge.[49] But House deliberations were interrupted in March by what was for environmentalists a serendipitous focusing event: the *Exxon Valdez* oil spill, which inundated Alaska's Prince William Sound with nearly 11 million gallons of oil. Environmentalists capitalized on the devastating impact of the spill and subsequent news coverage of oil-soaked wildlife. Jones reluctantly conceded the issue, citing hysteria. "We hoped before the oil spill to come out with a bill in July," he said, "but due to the emotional crisis this spill has created, we think it is best to put it on the back burner for the time being until the emotionalism has subsided." Even the prodevelopment Bush White House acknowledged that "it will be some time before [ANWR legislation] is considered."[50] Within Alaska, where support for drilling had run high, the oil industry and its supporters were chastened: the Democratic governor, Steve Cowper, expected the spill to "cause a permanent change in the political chemistry in Alaska."[51]

At least in the short run, the *Exxon Valdez* spill helped environmentalists derail industry's arguments that reserves could be developed without threatening environmental harm. Lodwrick Cook, the chair and chief executive of ARCO, confessed that "ARCO and the [oil] industry have lost a certain amount of credibility because of this spill, and we're going to have to work toward recapturing that."[52] On the advice of Secretary Lujan, the oil industry launched a major public relations campaign. It was confident that approval of an oil-spill preparedness bill, combined with the passage of time, would defuse public ire over the *Valdez* fiasco.[53] In fact, within a year, visible signs of damage to Prince William Sound had disappeared, and supporters of oil exploration in ANWR prepared to revisit the issue.

Congressional Action, 1990–1992. In the summer of 1990 international events opened another policy window—this time for drilling advocates. The threat of war in the Persian Gulf following the invasion of Kuwait by Iraqi forces sparked renewed debate over U.S. dependence on foreign oil and bolstered support for drilling in ANWR. Rep. Billy Tauzin, D-La., expressed his contempt for environmental values when he said, "It's amazing to me how we could put caribou above human lives—our sons and daughters."[54] Although the Bush administration insisted that the subsequent U.S. invasion (in January 1991) did not constitute a "war for oil," proponents of domestic oil and gas development seized on the event to revive drilling proposals. By September the oil industry and its backers in the Bush administration were again pressing Congress to open the Coastal Plain to drilling. As one House aide remarked, "The Middle East crisis wiped the *Exxon*

Valdez off the ANWR map as quickly as the *Exxon Valdez* wiped ANWR off the legislative map."[55]

Pressing its advantage, in April 1991 the Bush administration's development-friendly interior department released a revised estimate of the probability of finding oil in ANWR, raising it from 19 percent to 46 percent—a level almost unheard of in the industry—and estimating peak production of 870,000 barrels per day by the year 2000. Based on 1989 data, which included more accurate geological studies and test wells drilled on the periphery of the refuge, the Bureau of Land Management now estimated that the Coastal Plain most likely contained between 697 million and 11.7 billion barrels of oil.[56]

In the 102nd Congress, prodrilling senators Bennett Johnston, D-La., and Malcolm Wallop, R-Wyo., introduced S 1120, the Bush administration's "National Energy Security Act of 1991," Title IX of which permitted development of ANWR's Coastal Plain. In an effort to win over several Democratic senators who had denounced the plan because it was not accompanied by higher automotive fuel economy standards, Johnston's bill included a provision directing the secretary of transportation to raise the corporate average fuel economy (CAFE) standards substantially, although it did not specify how much. In a further effort to appease proenvironment legislators, Johnston proposed that federal proceeds from ANWR oil be used to fund energy conservation and research. Although Johnston touted the bill as a compromise, the bill's preamble made its purpose clear: "to reduce the Nation's dependence on imported oil" and "to provide for the energy security of the Nation."

Again, however, antidrilling advocates were able to take advantage of unforeseen developments to bolster their depiction of the oil industry as greedy profiteers. In January 1991 *The New York Times* reported that in the last quarter of 1990, oil industry profits had jumped dramatically as oil prices skyrocketed in anticipation of the Gulf War. Net income for Exxon, the world's largest oil company, had more than tripled, and Mobil's profits had climbed 45.6 percent; Texaco's, 35 percent; and Amoco's, 68.6 percent. "In fact," journalist Thomas Hayes reported, "these quarterly earnings might have been much larger, but each of the top five lowered their reported profit with bookkeeping tactics that oil giants often employ legally to pare taxes in periods when profits—and public resentment—are on the rise."[57] Furthermore, instead of an oil shortage, the world was experiencing a glut, as Saudi Arabia increased its output and demand slumped.

Although these developments weakened prodrilling arguments, the Senate Energy Committee nevertheless was persuaded by the interior department's revised estimates of the probability of finding oil in the refuge and the promise of jobs in the deepening recession, and in late May 1991 the committee approved S 1120 by a vote of seventeen to three. The sheer comprehensiveness of the energy bill undermined its prospects in the chamber as a whole, however, as its myriad provisions prompted the mobilization of a host of interests. Energy consumer groups joined environmentalists in trying to block measures to restructure the utility industry, fearing they would raise electricity rates. The auto industry, which disliked the CAFE standards, as well as small utilities disadvantaged by some other provisions, opposed the bill altogether. The prodrilling coalition gained the support of large electric utilities that stood to benefit from the bill, as well as the nuclear power industry lobby. Coalitions taking a variety of positions waged an all-out war for public opinion and the votes of uncommitted members of Congress, resulting in a cacophony of competing messages.

Further hampering the bill's chances of passing, in October 1991 the Senate Environment and Public Works Committee approved legislation that would prohibit oil and gas drilling in ANWR, and six Democratic senators pledged to filibuster S 1120 if it came up for a vote on the floor.[58] On November 1 Johnston called for a vote on cloture, which would have ended the filibuster, but failed by ten votes to get the necessary three-fifths majority. The bill's complexity had simply created too much highly charged opposition for senators to risk the political fallout that accompanies a controversial decision. The Senate eventually passed a national energy bill in 1992, by a vote of 94 to 4, but it contained neither ANWR nor automotive fuel-efficiency provisions.

The House also considered bills authorizing oil and gas leasing in the refuge in 1991 and 1992. In 1991 the House postponed consideration of an ANWR drilling bill after allegations surfaced that Alyeska, operator of the Trans-Alaska Pipeline System, had coercively silenced its critics. Rep. Mo Udall reintroduced his ANWR wilderness bill, but it did not make it to the floor for a vote. Finally, on May 2, 1992, the House passed (381–37) an energy bill that, like its Senate counterpart, lacked provisions on ANWR or fuel-efficiency standards. And on October 24, 1992, the president signed the ANWR-free Energy Policy Act into law.

ANWR Proposals on the Quiet, 1993–1999

The election of Bill Clinton to the presidency in 1992 temporarily punctured the hopes of drilling advocates, as the newly installed administration made clear its staunch opposition to oil development in the refuge. Based on more conservative geologic and economic assumptions than those made by its predecessor, the Clinton administration's interior department in June 1995 reduced its estimate of economically recoverable oil in the refuge to between 148 million and 5.15 billion barrels. The department's 1995 Coastal Plain Resource Assessment also affirmed environmentalists' claims that "there would be major environmental impacts from oil and gas development on the coastal plain" and documented even greater dependence of the Porcupine caribou herd on the 1002 area than had earlier reports.[59]

Nevertheless, in 1995, after two years of relative quiescence, congressional sponsors resumed their efforts to open up the refuge, hoping that the ascension of the Republican-controlled 104th Congress would enhance the prospects for a prodrilling policy. A chief difference between 1995 and preceding years was that two ardent proponents of drilling (both Alaskans) had acquired positions of power on the very committees that previously had obstructed efforts to open the refuge. Murkowski became chair of the Senate Environment and Natural Resources Committee, and Young took over the House Natural Resources Committee (previously the House Interior Committee), which he promptly renamed the House Resources Committee. Both men immediately announced their intention to attach ANWR drilling provisions to the omnibus budget reconciliation bill in the fall.[60] Such a strategy was promising because small riders attached to major bills tend to attract less public notice than individual pieces of legislation and, perhaps more important, the budget reconciliation bill cannot be filibustered. Moreover, the president was less likely to veto omnibus legislation than a less significant bill.

By the middle of the legislative session, Sen. Joseph Lieberman, D-Conn., an ANWR defender, was expressing his concern that a majority in both houses favored opening the refuge. Sure enough, the fiscal 1996 budget resolution (H. Con. Res. 67), approved by the House and adopted by the Senate in May, assumed $1.3 billion in revenue from ANWR oil leases. Furthermore, the day before passing the resolution, the Senate voted fifty-six to forty-four to kill an amendment that would have prohibited exploration in the Coastal Plain. But environmentalists had an ally in the White House. Clinton's interior secretary, Bruce Babbitt, voiced the administration's resistance to the congressional proposal and chastised prodrilling legislators as greedy. "In effect, we are being asked to jeopardize an irreplaceable piece of our national heritage over a three-year difference in budget projections by the people in green eyeshades," Babbitt said.[61] Citing the ANWR provision as one of the chief reasons, President Clinton vetoed the 1996 budget bill. After a week-long standoff during which the federal government shut down altogether, the provision's sponsors backed down and removed the ANWR rider.

Undaunted, drilling advocates continued their campaign to open the refuge. In 1997 *The Washington Post* reported that Arctic Power, a group formed in 1992 to promote drilling in the refuge, had contracted with a Washington, D.C.-based lobbying group (Decision Management Inc.) to target Democratic senators John Breaux and Mary Landrieu of Louisiana with information about the economic ties between Alaska and their state.[62] Drilling proponents also forged an alliance with the Teamsters Union based on their claim that opening the refuge would produce hundreds of thousands of jobs, even though several analyses refuted that assertion.[63] In support of their national security argument, drilling advocates touted an Energy Department finding that, while domestic oil production had been declining, the percentage of U.S. oil that was imported rose from 27 percent in 1985 to nearly 50 percent in 1997. In 1998 revised USGS estimates also bolstered drilling advocates' claims: based on new geological information, technological advances, and changes in the economics of North Slope drilling, agency experts projected there was a 95 percent chance that more than 5.7 billion barrels of oil would be technically recoverable and a 5 percent chance of extracting more than 16 billion barrels from the refuge. At a price of $20 per barrel, the mean estimate of the amount of economically recoverable oil was 3.2 billion barrels.[64] A fact sheet from the House

Resources Committee embraced the new figures and said, "The fact is, ANWR will help balance the budget, create jobs, increase domestic production, reduce oil import dependence and the trade deficit."[65]

Despite concerted efforts by congressional proponents, the Clinton administration's unwavering opposition stymied efforts to open the refuge for the remainder of the decade. Although sponsors introduced bills in the 105th Congress (1997–1998), neither chamber debated the issue. In the 106th Congress (1999–2000), sponsors again introduced both wilderness designation and energy development bills. In addition, assumptions about revenues from ANWR were originally included in the fiscal year 2001 budget resolution as reported by the Senate Budget Committee. The House-Senate conference committee rejected the language, however, and it was excluded from the final budget passed in April 2000.

The 2000 Presidential Campaign and President Bush's Energy Policy

In the summer of 2000 the debate over ANWR's fate again captured the national spotlight when presidential candidate George W. Bush expressed his support for oil development there. The issue subsequently became a main point of contention between Bush and Vice President Al Gore, the Democratic candidate. Gore adopted the rhetoric of environmentalists about the superior benefits of conservation and the importance of preserving wilderness, while Bush cited the nation's need for energy independence and the slowing economy's vulnerability to higher energy prices. When Bush won the election, drilling advocates again felt certain their opportunity had arrived.

Hoping to boost support for drilling, a spokesperson for ARCO—which would be a major beneficiary of permission to drill in the refuge—testified at a House Resources Committee hearing that the company had learned a lot about environmentally friendly oil development over the years. "We can explore without leaving footprints," he said, "and the footprint required for new developments is a tenth of what it once was."[66] BP Amoco, another leading player in Alaska, contended that technological advances enabled it to extract oil with minimal environmental impact: thanks to cutting-edge technology for steering drill bits, the newest wells occupy a much smaller area than older wells; pipelines are built higher, to let caribou pass beneath them, and feature more elbows, which reduce the amount of oil spilled in an accident; and often, rather than constructing roads, companies airlift workers to the site.[67]

To enhance its credibility, BP spent much of the 1990s trying to transform its image from black to green by acknowledging the threat of climate change and embracing alternative fuels. (Taking no chances, the company also donated tens of thousands of dollars to the prodevelopment lobbying group Arctic Power, as well as to Republican politicians, including Bush.[68]) The company's British CEO, John Browne, wrote in a memo to employees that BP's values "may be manifested in different ways, but they have much in common: a respect for the individual and the diversity of mankind, a responsibility to protect the natural environment."[69] Environmentalists remained skeptical, however: the World Wildlife Fund's Francis Grant-Suttie said that "on the PR level they have been successful at differentiating themselves from others, but by virtue of what they're doing on the coastal plain, you can see it's sheer rhetoric."[70] In fact, BP's operations in Alaska were extremely shoddy, as a series of accidents and spills in the 2000s revealed. The company was routinely neglecting maintenance as its Alaskan oil fields continued to yield oil well beyond the anticipated life of the equipment used to extract and transport it.[71]

For President Bush, getting access to ANWR was an important symbol of his administration's priorities; during his first month in office Bush said, "I campaigned hard on the notion of having environmentally sensitive exploration at ANWR, and I think we can do so."[72] Claiming that the nation faced an energy crisis, at the end of January President Bush created a task force headed by Vice President Dick Cheney to devise ways to reduce America's "reliance upon foreign oil" and to "encourage the development of pipelines and power-generating capacity in the country."[73] Drilling advocates were confident, and opponents worried, that a sympathetic president and his allies in Congress would be able to capitalize on concerns about a spike in energy prices and California's energy deregulation fiasco to win passage of a drilling bill.[74]

The Bush administration's rhetoric set in motion yet another ferocious lobbying campaign by Arctic Power, which hired the law firm Patton Boggs, the public relations firm Qorvis, and media consultant Alex Castellanos to create a series of radio and television ads. In late March Arctic Power announced the formation of a new coalition, the Energy Stewardship Alliance. At the same time, the Audubon Society launched its own ad campaign, and Audubon and Defenders of Wildlife began mobilizing citizens to send e-mails and faxes to Congress.

The arguments employed by both sides had changed little over the years. Proponents of drilling emphasized increasing economic security and reducing dependence on foreign sources. They cited the high end of USGS oil estimates (16 billion barrels, an estimate that included state waters and native lands in addition to the ANWR 1002 area), neglecting to note the low odds associated with that number, and they claimed that technological improvements would dramatically minimize the footprint and environmental impact of oil and gas development. They disparaged the aesthetic qualities of the refuge, making sure to fly uncommitted legislators over the area in winter to emphasize its barrenness. Opponents responded by citing the low end of the range of USGS oil estimates (3.2 billion barrels) and reminding the public that such an amount would hardly put a dent in U.S. reliance on foreign energy sources. They rejected drilling advocates' arguments about environmentally sound exploration, noting that serious hazardous waste spills were continuing to occur at Prudhoe Bay. They argued that conservation, not exploration, was the solution to the nation's energy woes. For example, in a *New York Times* editorial, Thomas Friedman cited the Natural Resources Defense Council's estimate that increasing the average fuel efficiency of new cars, SUVs, and light trucks from 24 miles per gallon to 39 miles per gallon in a decade would save 51 billion barrels of oil, more than fifteen times the likely yield from ANWR. And they touted the spiritual importance of untrammeled wilderness: likening the refuge to a cathedral, Friedman compared drilling to online trading in church on a Palm Pilot, saying, "It violates the very ethic of the place."[75]

The two sides also continued their fierce debate over the impact of drilling on the refuge's wildlife. Late winter snows in 2001 caused one of the worst years for the Porcupine caribou herd in the thirty years scientists had been observing them. Thousands of newborn calves died because their mothers gave birth before reaching the Coastal Plain. Antidrilling advocates pointed out that industrial development would have a similar effect by keeping calving mothers off the Coastal Plain. But drilling proponents retorted that this occurrence was a natural pattern that had little to do with whether oil exploration should be allowed. They pointed out that the Central Arctic herd continued to flourish despite Prudhoe Bay development (dismissing the caveat that most of the Central Arctic herd had moved away from its historic calving grounds, something that might not be possible in ANWR).[76] In spring 2002 the USGS released a study warning that caribou "may be particularly sensitive" to oil exploration in ANWR. But a week later, after interior secretary Gale Norton asked scientists to plug two "more likely" scenarios into its model, the agency issued a more sanguine two-page follow-up. In response to the ensuing uproar, one biologist who worked on the study acknowledged, "The truth is, we just don't know what the impact would be."[77]

Journalists often reinforced environmentalists' message by using poetic language to describe the refuge. For example, *Los Angeles Times* reporter Kim Murphy rhapsodized, "The Alaskan Arctic is a place where the dawning summer is transcendent, an Elysian landscape trembling with tentative new life after a dark winter of vast frosts and howling winds."[78] Charles Seabrook wrote in *The Atlanta Journal-Constitution*, "It is a raw, elemental land, a place of exquisite beauty and enormous geometry—one of the Earth's last great wild places. The hand of man is hardly known here. The cutting edges of light, water and wind predominate. Grizzly bears roam the wide-open spaces, and great herds of migrating caribou appear in the summer and fall." By contrast, he observed, "Prudhoe lights up the tundra for miles with powerful industrial lights. Steam erupts from eight-story buildings. More than 500 roads link 170 drilling sites along the coast."[79]

In a blow to the Bush administration's plans, the House Budget Committee released a budget for 2002 that did not include anticipated oil revenue from drilling in the refuge, saying that it would provoke too much controversy. The chair of the Senate Budget Committee indicated he was likely to follow suit. Although the administration continued to emphasize increasing the supply of fossil fuels rather than conserving and developing alternative fuels, by the end of March, it seemed to be backing away from ANWR drilling proposals.[80] And by the summer of 2001, despite Vice President Cheney's road trip to promote an energy plan that focused on developing domestic supplies, the context of the issue had once again changed. Oil prices were falling, supplies had stabilized, and the sense of crisis—so critical to passing major legislation—had vanished. As a result, although the House passed energy bills containing ANWR drilling provisions in 2001 and 2002, opponents in the Senate were able to block them.

In 2003 and 2004 the debate seesawed, as both sides seized on expert reports and fortuitous events to bolster their claims about drilling in the refuge. Just prior to the 2003 Senate energy bill debate, the National Academy of Sciences released its two-year study of the impacts of oil development in Prudhoe Bay. The panel, which included several oil company experts and Alaskan organizations, concluded that, even though oil companies had greatly improved their practices in the Arctic, decades of drilling on the North Slope had produced a steady accumulation of environmental damage that would probably increase as exploration spread and would heal only slowly, if at all. The report concluded that industrial development had "reduced opportunities for solitude and [had] compromised wild-land and scenic values over large areas" and that "[c]ontinued expansion will exacerbate existing effects and create new ones." The panel stopped short of making a recommendation, saying, "Whether the benefits derived from oil and gas activities justify acceptance of the foreseeable and undesirable cumulative effects is an issue for society as a whole to debate and judge."[81] Environmentalists felt the report supported their position, but in August, a massive blackout across the Midwest and Northeast briefly buoyed Republican lawmakers' hopes of turning public opinion in favor of expanded energy development. Environmentalists were vindicated again, however, when a GAO report released in October found that the Fish and Wildlife Service's record of protecting wildlife refuges from the environmentally damaging effects of oil and gas drilling had been spotty and that refuge managers lacked the resources and training to properly oversee oil and gas activities.[82]

During this period, aware that they lacked the votes to stop a filibuster, Senate leaders tried instead to amend the budget resolution, which requires only fifty-one votes to pass. In hopes of wooing Senate holdouts, they traded local benefits for votes. For example, in 2003 they offered Norm Coleman, R-Minn.—normally an opponent of drilling—up to $800 million in federal loan guarantees to build a power plant in one of his state's most economically depressed regions.[83] Although this approach failed, prospects for drilling brightened dramatically in the fall of 2004, when President Bush was reelected, and Republican gains in the Senate appeared to net drilling advocates three new seats.[84] Environmentalists again geared up for a major grassroots lobbying effort, but this time their efforts were insufficient: in March 2005, by a vote of fifty-one to forty-nine, the Senate approved a budget that included instructions allowing the Energy and Natural Resources Committee to lift drilling prohibitions in order to collect $2.4 billion in revenues from ANWR leasing over five years.[85]

The House budget, which passed the same week by a vote of 218 to 214, did not include revenues from drilling in ANWR, and environmentalists hoped to persuade members to prevent those revenues from being included in the conference report. But in late April House and Senate conferees agreed on a budget that *did* include revenues from ANWR drilling, and both chambers approved the deal. Importantly, the budget resolution did not mention ANWR specifically, so it shielded members from accountability on the issue. By contrast, the reconciliation process would compel the authorizing committees in each chamber to approve explicit drilling language, depriving legislators of any political cover for their vote. Environmentalists did their best to dissuade members of the House Resources Committee from including such language, and when Hurricane Katrina struck in late August, they hoped to capitalize on the delay to shore up support for their position. But many observers feared that high gas prices, which rose even further in Katrina's aftermath, would make the political climate more hospitable to drilling in ANWR and other environmentally sensitive areas.

By mid-autumn 2005 approval of ANWR drilling appeared virtually certain, even though a 2005 Gallup poll found that only 42 percent of respondents favored opening the refuge, while 53 percent opposed it, and in early November the Senate passed its budget bill.[86] But to environmentalists' delight, leaders in the House stripped the ANWR provision from the companion bill after a small group of moderate Republicans threatened to withhold their support for the budget if the item were included. One of the dissenting Republicans, Charlie Bass, R-N.H., wrote a letter to Rules Committee Chair David Drier that was signed by twenty-four Republican colleagues. In it he said, "Including the drilling provisions in the Deficit Reduction Act would undermine the protection of public spaces by valuing the worth of the potential resources contained within these lands over their conservation value. . . . Rather than reversing decades of protection for this publicly held trust, focusing greater attention on renewable energy sources, alternate fuels, and more efficient systems and household appliances would net more energy savings."[87] In late December Sen. Ted Stevens, R-Alaska, made a last-ditch attempt to open the refuge by tucking a provision to allow ANWR drilling in a must-pass $435-billion military spending bill. But Sen. Maria Cantwell, D-Wash., led a successful filibuster that prevented the provision from being included in the final bill. Ultimately, 2005 ended with ANWR narrowly but decisively protected.

OBAMA LEGACY

In 2007 and 2008, oil prices shot up dramatically—rising $1.40 per gallon over eighteen months and closing in on $4.00 per gallon—largely as a result of political instability in Nigeria and the Middle East.[88] President Bush seized on the steep rise in prices to ask Congress to lift the decades-old offshore drilling ban and to open up the Arctic refuge, attaching these solutions to the ostensible problem of constricted energy supplies.[89] Domestic drilling proceeded to become a major issue during the 2008 presidential campaign, with Republicans gleefully adopting the slogan, "Drill, Baby, Drill!" By August then Democratic presidential candidate Barack Obama had agreed to consider offshore oil drilling if it were part of a comprehensive energy and climate change plan (see Chapter 12).

Obama's willingness to support domestic drilling did not extend to ANWR, however, so his election to the presidency and the ascension of large Democratic majorities in the House and Senate rendered legislative action to develop the refuge a remote possibility, while enhancing prospects for new wilderness designations in the refuge. Nevertheless, members of the Alaskan delegation expressed outrage when, in spring 2010, the Fish and Wildlife Service began reviewing new areas within the refuge for wilderness designation as it sought to update its twenty-two-year-old Comprehensive Conservation Plan. The Alaskans suggested that interior secretary Ken Salazar terminate the review and instead use the department's resources to conduct studies of the refuge's oil and gas potential.[90] Alaska Governor Sean Parnell weighed in as well with his opposition to a wilderness review or any other process that would complicate the prospects for oil and gas development in the refuge. The Alaskan delegation's position was severely undermined, however, when—in the midst of the ANWR planning exercise—a massive blowout at an offshore oil rig in the Gulf of Mexico raised new questions about oil-drilling technology and America's dependence on oil. The April 22, 2010, explosion, which killed eleven workers, triggered a spill that continued for three months before BP—the company that had been operating the rig—was finally able to cap the well (see Chapter 11). In the intervening time, about 210 million gallons of oil spewed into the gulf, fouling Louisiana's wetlands and wreaking untold ecological damage.[91]

In early 2015, to the dismay of the Alaskan congressional delegation, President Obama announced that he would ask Congress to designate 12 million of the refuge's 19 million acres as wilderness, thereby giving it the highest level of protection available. If enacted, this would be the largest wilderness designation since Congress passed the Wilderness Act in 1964. In justifying his request, Obama explained, "Alaska's National Wildlife Refuge is an incredible place—pristine, undisturbed. It supports caribou and polar bears, all manner of marine life, countless species of birds and fish, and for centuries, it supported many Alaska Native communities. But it's very fragile."[92] But Lisa Murkowski, the new chair of the Senate Energy and Natural Resources Committee and a Republican from Alaska, vowed to squelch the plan. "What's coming is a stunning attack on our sovereignty and our ability to develop a strong economy that allows us, our children and our grandchildren to thrive," she warned. She added, "It's clear this administration does not care about us and sees us as nothing but a territory. . . . I cannot understand why this administration is willing to negotiate with Iran, but not Alaska. But we will not be run over like this. We will fight back with every resource at our disposal."[93] In response, Rep. Jared Huffman, D-Calif., proposed to the 114th Congress that the 1.5 million-acre Coastal Plain be designated as wilderness. Although the bill never reached the floor for debate, they saw new life within an amendment to the Sportsmen's Heritage and Recreational Enhancement (SHARE) Act in February 2016. Despite bipartisan support, the amendment lost in the House vote, 176–226.[94]

OUTCOMES

After failing to open ANWR for thirty-seven years, drilling proponents found new hope in 2016. As a focal point of President Trump's energy and budget agendas, ANWR once again entered the spotlight as drilling advocates recognized new opportunities with a Republican-controlled White House and Congress. President Trump's Office of Management and Budget (OMB) Director, Mick Mulvaney, stated in 2017 that opening ANWR "is entirely consistent with, and in fact a central part of, the president's desire to be not only energy independent, but energy dominant."[95]

Additionally, a four-page provision within the TCJA of 2017 mandated two lease sales within the Coastal Plain (the first within four years and the second within seven years) of at least 400,000 acres. According to the provision, surface development was capped at 2,000 acres. Expecting to receive $2.2 billion in bids, the federal government would also receive approximately seventeen percent in royalties to be split 50–50 with the state of Alaska.[96]

Despite surface development limitations, opponents stressed that any human development would potentially disturb, either directly or indirectly, tens of thousands of acres.[97] Moreover, environmentalists argued that the desired $1.1 billion in federal revenue would fall short, pointing to the recent auction in nearby National Petroleum Reserve–Alaska, which only brought in $1.1 million from seven bids.[98]

With little to no debate in the House and Senate, President Trump signed TCJA into law on December 22, 2017. Wearing Incredible Hulk earrings to celebrate, Sen. Murkowski rejoiced in victory after winning a battle her state and family had fought since 1980, stating, "Alaskans can now look forward to our best opportunity to refill the Trans-Alaska Pipeline System, thousands of jobs that will pay better wages, and potentially $60 billion in royalties for our state alone."[99] Following the TCJA passage, Murkowski expressed intent to seek an additional standalone bill on ANWR energy development and environmental protection.[100]

Numerous environmental groups, such as the Alaska Wilderness League, Sierra Club, and Natural Resource Defense Council, as well as Senators Maria Cantwell, D-Wash., and Ed Markey, D-Mass., have vowed to fight to protect ANWR. Low oil prices coupled with the large expense of drilling in an icy, remote region devoid of infrastructure deter oil companies seeking immediate involvement. Moreover, delays and roadblocks from legal challenges will continue to make companies wary of development in ANWR.

CONCLUSIONS

This case starkly illuminates the polarization between advocates of natural resource development and proponents of wilderness protection. Recent public opinion data indicate the majority (53%) of Americans oppose drilling in ANWR, while only

35% support such action.[101] Advocates of drilling in ANWR see little value in saving a place that few people will visit; for them, preserving the refuge is tantamount to placing the welfare of animals above the security and comfort of human beings. True cornucopians, they believe that technological innovation will solve any environmental problems drilling might cause.[102] By contrast, those who value wilderness take satisfaction in simply knowing that it exists. As one biologist said, "There are a lot of people who will never go to the refuge, but I get some peace of mind knowing that there is an area of naturalness on such a scale."[103] Wilderness advocates are adamant in their opposition to any kind of development in ANWR; for them, compromise is defeat. According to Tim Mahoney, chair of the Alaska Coalition, for example, "To say that we can have oil and caribou is akin to saying that we can dam the Grand Canyon and still have rocks."[104] Assurances that oil companies now have the technological capability to minimize the environmental impact of drilling makes little impression on wilderness advocates, and repeated spills and accidents undermine their credibility. As Deborah Williams, executive director of the Alaska Conservation Foundation, concludes, "The one thing you can't get away from is that in the end, even with all this technology, you've got a massive industrial complex."[105]

Given the chasm separating the two sides, it is clear that the enormous amounts of information generated and rhetoric employed during the nearly two decades of debate over ANWR have not been aimed at changing the minds of the opposition. Instead, advocates have targeted the public and uncommitted members of Congress, with the goal of shaping legislators' perceptions of the issue's salience and the likely electoral consequences of their vote. Drilling advocates have tried to capitalize on focusing events—the Gulf War in the 1980s and rising energy prices in the 1990s and again between 2007 and 2008—to cast the issue of ANWR drilling as a source of economic and military security, just as they used the energy crisis of 1973 to get approval for the Trans-Alaska pipeline. They have also tried to take advantage of policy windows, such as the turnover in congressional leadership in the mid-1990s, and the elections of George W. Bush in 2000 and 2004 and President Donald J. Trump in 2016.

Until late 2017, the environmentalists' ability to fend off development interests' efforts for nearly thirty-seven years is a tribute to their ability to generate national support, which they did by using the refuge's "charismatic megafauna"—particularly caribou and polar bears—to reinforce the symbolism of the last remaining wilderness. In addition, environmentalists have exploited the fact that not drilling was the status quo: they have taken advantage of multiple jurisdictions (in the House), filibusters (in the Senate), and the threat of a presidential veto to thwart ANWR exploration bills. And they have benefited from the complexity of energy policy, which prompts diverse interests to mobilize. Interest-group activity in turn has stymied congressional decision making because legislators fear the electoral repercussions of antagonizing attentive constituents. Even some of the major oil companies have found the level of controversy daunting; by 2004 BP, Conoco-Phillips, and ChevronTexaco had retreated from publicly supporting legislative efforts to open up the refuge.[106]

On the other hand, Americans continue to consume oil in prodigious quantities, despite warnings dating back to the 1970s that our oil addiction renders us vulnerable to oil shocks and international disturbances. Nearly 70 percent of the 21 million barrels of oil the United States consumes every day goes for transportation, most of that by individual drivers. Although the efficiency of the typical American car jumped from 13.8 miles per gallon in 1974 to 27.5 miles per gallon in 1989 after Congress imposed CAFE standards, efficiency stabilized and then began to fall in the 1990s, as oil prices declined and Congress refused to raise the CAFE standards or impose gas taxes. Even as fuel efficiency rose to 44 miles per gallon in Europe, it fell back to 24 miles per gallon in the United States, largely thanks to the popularity of SUVs and light trucks. Not until 2007, with oil at $82 a barrel (equivalent to $95 a barrel in 2015 dollars), did Congress finally approve a major increase in the CAFE standard, to 35 miles per gallon by 2020.[107] (President Obama subsequently accelerated that schedule administratively and President Trump is determined to roll back.)

By 2018, however, the price of oil had plummeted again, to $66 a barrel, mostly as a result of the fracking boom that enabled oil and gas companies to exploit fossil fuels trapped in shale (see Chapter 14). Low oil prices temporarily reduced the pressure to exploit Alaskan reserves. But as long as Americans (and others around the world) continue to live oil-dependent lifestyles, both ANWR and the nation's coastline will be tempting targets for new development. In the meantime, we need to watch the Millennial and Centennial generations—their addiction to oil appears to be less than their parents, with preferences toward alternative sources of energy.

QUESTIONS TO CONSIDER

- Why do you think the conflict over ANWR has been so intractable?
- Did opponents of drilling in ANWR neglect any arguments or tactics that might have helped them fend off the powerful proponents of developing oil and gas in the refuge? If so, what sorts of arguments might they have made, or what kinds of tactics should they have tried?
- Some critics argue that environmentalists have done a disservice to their cause by focusing so heavily on ANWR. What are the pros and cons of expending lobbying resources on blocking oil exploration in the refuge?

NOTES

1. For more detail on the impact of multiple referral on the legislative process, see Garry Young and Joseph Cooper, "Multiple Referral and the Transformation of House Decision Making," in *Congress Reconsidered*, 5th ed., ed. Lawrence C. Dodd and Bruce J. Oppenheimer (Washington, D.C.: CQ Press, 1993), 211–234.
2. A filibuster is a tactic that is available only in the Senate and involves employing "every parliamentary maneuver and dilatory motion to delay, modify, or defeat legislation."

Walter Oleszek, *Congressional Procedures and the Policy Process*, 4th ed. (Washington, D.C.: CQ Press, 1996), 249.

3. Roderick Nash, *Wilderness and the American Mind*, 3d ed. (New Haven, Conn.: Yale University Press, 1982).

4. Quoted in Peter A. Coates, *The Trans-Alaska Pipeline Controversy: Technology, Conservation, and the Frontier* (Bethlehem, Pa.: Lehigh University Press, 1991), 84.

5. John Strohmeyer, *Extreme Conditions: Big Oil and the Transformation of Alaska* (New York: Simon & Schuster, 1993).

6. The first suit charged that TAPS was asking excessive rights of way under the Mineral Leasing Act of 1920. A second suit joined one filed by native villagers that did not want the pipeline to cross their property. The third claimed that the interior department had not submitted an environmental impact statement, as required by NEPA. See Strohmeyer, *Extreme Conditions.*

7. Ibid., 83.

8. The Alaska Public Interest Coalition comprised a diverse set of interests, including the Sierra Club, the Wilderness Society, the National Wildlife Federation, the National Rifle Association, Zero Population Growth, Common Cause, the United Auto Workers, and others.

9. NEPA requires federal agencies to prepare an EIS for any major project. L. J. Clifton and B. J. Gallaway, "History of Trans-Alaska Pipeline System." Available at http://tapseis.anl.gov/documents/docs/Section_13_May2.pdf.

10. Coates, *The Trans-Alaska Pipeline Controversy.* Congress had already amended the Mineral Leasing Act of 1920 to render environmentalists' other legal challenge to the pipeline moot.

11. Strohmeyer, *Extreme Conditions.*

12. Public domain land is federally owned land that has not been designated for a specific purpose and is still eligible for withdrawal under federal land laws.

13. This provision is unusual; ordinarily, the Fish and Wildlife Service can allow oil development in a wildlife refuge without congressional approval.

14. U.S. Department of the Interior, *Arctic National Wildlife Refuge, Alaska, Final Comprehensive Conservation Plan, Environmental Impact Statement, Wilderness Review, Wild River Plans* (Washington, D.C.: U.S. Government Printing Office, 1988), 12.

15. P.L. 96-487; 94 Stat. 2371.

16. P.L. 89-669; 80 Stat. 926.

17. U.S. GAO, *Opportunities to Improve the Management and Oversight of Oil and Gas Activities on Federal Lands*, GAO-03-517 (August 2003).

18. Ibid.

19. Ted Williams, "Seeking Refuge," *Audubon*, May–June 1996, 34–45, 90–94.

20. Lunney, K. (2017, December 20). "Drilling Opponents Take Fight to Courts, Public." *Greenwire*. Available at https://www.eenews.net/greenwire/stories/1060069571/search?keyword=arctic+national+wildlife+refuge.

21. As part of its statehood deal, Alaska received title to 90 percent of the revenues from oil development within its borders.

22. According to the Center for Responsive Politics, between 1989 and 2010 oil and gas interests gave more than $266 million to congressional candidates. The industry also spent more than $1 billion on lobbying between 1998 and 2010. See http://www.opensecrets.org/industries/indus.php?ind=E01.

23. Sobczyk, N. (2018, January 17). "Little Drilling Expected Before 2030; Murkowski Seeks New Bill." *E&E News*. Available at https://www.eenews.net/stories/1060071129.

24. Estimates of the recoverable amount of oil vary depending on the price of oil and the costs of extraction. Projections of the former depend on a host of variables; the latter depend on technological developments, geography, and environmental regulations.

25. U.S. Department of the Interior, *Arctic National Wildlife Refuge, Alaska, Coastal Plain Resource Assessment: Report and Recommendation to the Congress of the United States and Final Legislative Environmental Impact Statement* (Washington, D.C.: U.S. Fish and Wildlife Service, 1987).

26. Peter Nulty, "Is Exxon's Muck-up at Valdez a Reason to Bar Drilling in One of the Industry's Hottest Prospects?" *Fortune*, May 28, 1989, 47–49.

27. Quoted in Philip Shabecoff, "U.S. Proposing Drilling for Oil in Arctic Refuge," *The New York Times*, November 25, 1986, A1.

28. Lisa Speer et al., *Tracking Arctic Oil* (Washington, D.C.: Natural Resources Defense Council, 1991), 30.

29. "National Energy Policy," *Congressional Digest*, May 1991, 130–160.

30. Quoted in ibid., 153.

31. Isabelle Tapia, "Time for Action on Coastal Plain Is Now," Newsletter, Coalition for American Energy Security, March 1991.

32. Speer et al., *Tracking Arctic Oil*.

33. Ibid.

34. Bernard Gelb, "ANWR Development: Analyzing Its Economic Impact," Congressional Research Service, February 12, 1992.

35. U.S. House of Representatives, Committee on Merchant Marine and Fisheries, "ANWR Briefing Book," April 1991.

36. BP Exploration (Alaska) Inc., *Major Environmental Issues*, 3d ed. (Anchorage: BP Exploration, 1991).

37. Donald P. Hodel, "The Need to Seek Oil in Alaska's Arctic Refuge," *The New York Times*, June 14, 1987, Sec. IV, 25.

38. U.S. Office of Technology Assessment, *Oil Production in the Arctic National Wildlife Refuge: The Technology and the Alaskan Context* (Washington, D.C.: Office of Technology Assessment, 1989), OTA-E-394, 13.

39. U.S. Fish and Wildlife Service, *Comparison of Actual and Predicted Impacts of the Trans-Alaska Pipeline Systems and Prudhoe Bay Oilfields on the North Slope of Alaska* (Fairbanks, Alaska: U.S. Fish and Wildlife Service, 1987).

40. Gail Osherenko and Oran Young, *The Age of the Arctic: Hot Conflicts and Cold Realities* (New York: Cambridge University Press, 1989).

41. Shabecoff, "U.S. Proposing Drilling."

42. "Energy Realism," *The Wall Street Journal*, May 30, 1991, A1.

43. Quoted in Shabecoff, "U.S. Proposing Drilling."

44. U.S. Department of the Interior, *Arctic National Wildlife Refuge*, vii, 186.

45. Joseph A. Davis, "Alaskan Wildlife Refuge Becomes a Battleground," *Congressional Quarterly Weekly Report*, August 22, 1987, 1939–1943.

46. Joseph A. Davis and Mike Wills, "Prognosis Is Poor for Arctic Oil-Drilling Bill," *Congressional Quarterly Weekly Report*, May 21, 1988, 1387.

47. Philip Shabecoff, "Bush Is Asked to Ban Oil Drilling in Arctic Refuge," *The New York Times*, January 25, 1989, A16.

48. Joseph A. Davis, "Arctic-Drilling Plan Clears Committee," *Congressional Quarterly Weekly Report*, March 18, 1989, 578.

49. The House considered three other bills to open the refuge and one to designate the area as wilderness in 1989.

50. Quoted in Philip Shabecoff, "Reaction to Alaska Spill Derails Bill to Allow Oil Drilling in Refuge," *The New York Times*, April 12, 1989, A17.

51. Quoted in Richard Mauer, "Oil's Political Power in Alaska May Ebb with Spill at Valdez," *The New York Times*, May 14, 1989, A1.

52. Quoted in Richard W. Stevenson, "Why Exxon's Woes Worry ARCO," *The New York Times*, May 14, 1989, Sec. 3, 1.

53. Philip Shabecoff, "Oil Industry Gets Warning on Image," *The New York Times*, April 4, 1989, B8.

54. Quoted in Phil Kuntz, "ANWR May Be Latest Hostage of Middle East Oil Crisis," *Congressional Quarterly Weekly Report*, September 8, 1990, 2827–2828.

55. Ibid.

56. U.S. Department of the Interior, Bureau of Land Management, Overview of the 1991 Arctic National Wildlife Refuge Recoverable Petroleum Resource Update, Washington, D.C., April 8, 1991. For comparison, experts estimated that the Prudhoe Bay oil field, North America's largest, contained about 14 billion barrels of recoverable oil.

57. Thomas Hayes, "Oil's Inconvenient Bonanza," *The New York Times*, January 27, 1991, A4.

58. Christine Lawrence, "Environmental Panel Sets Up Floor Fight over ANWR," *Congressional Quarterly Weekly Report*, October 19, 1991, 3023.

59. U.S. Department of the Interior, *Arctic National Wildlife Refuge, Alaska, Coastal Plain Resource Assessment: Report and Recommendation to the Congress of the United States* (Washington, D.C.: U.S. Government Printing Office, 1995).

60. The budget reconciliation bill reconciles tax and spending policies with deficit reduction goals.

61. Quoted in Allan Freedman, "Supporters of Drilling See an Opening," *Congressional Quarterly Weekly Report*, August 12, 1995, 2440–2441.

62. Bill McAllister, "Special Interests: Lobbying Washington," *The Washington Post*, October 30, 1997, 21.

63. See, for example, Bernard Gelb, "ANWR Development: Economic Impacts," Congressional Research Service, October 1, 2001; Dean Baker, "Hot Air over the Arctic: An Assessment of the WEFA Study of the Economic Impact of Oil Drilling in the Arctic National Wildlife Refuge," Center for Economic Policy and Research, September 5, 2001.

64. U.S. Department of the Interior, U.S. Geological Survey, "Arctic National Wildlife Refuge, 1002 Area, Petroleum Assessment, 1998, Including Economic Analysis," USGS Fact Sheet FS-028-01, April 2001; U.S. Geological Survey, news release, May 17, 1998.

65. On file with the author. For a more recent version of the ANWR fact sheet, see http://naturalresources.house.gov/anwr/.

66. Quoted in Andrew Revkin, "Hunting for Oil: New Precision, Less Pollution," *The New York Times*, January 30, 2000, D1.

67. Neela Banerjee, "Can BP's Black Gold Ever Flow Green?" *The New York Times*, November 12, 2000, Sec. 3, 1; Revkin, "Hunting for Oil." In 2000 BP acquired ARCO, giving it an even larger stake in the region's oil.

68. In 2000 BP contributed $50,000 to Arctic Power, $34,421 to candidate Bush (compared to $4,250 to candidate Gore), and $613,870 to the Republican Party. See Banerjee, "Can BP's Black Gold Ever Flow Green?"

69. Quoted in ibid.

70. Ibid.

71. Journalist Abrahm Lustgarten documents the extent of BP's negligence in Alaska in his book *Run to Failure: BP and the Making of the Deepwater Horizon Disaster* (New York: W.W. Norton & Company, 2012).

72. Quoted in Joseph Kahn and David E. Sanger, "President Offers Plan to Promote Oil Exploration," *The New York Times*, January 30, 2001, A1.

73. Ibid.

74. Andrew Revkin, "Clashing Opinions at a Meeting on Alaska Drilling," *The New York Times*, January 10, 2001, A1.

75. Thomas Friedman, "Drilling in the Cathedral," *The New York Times*, March 2, 2001, A23.

76. Kim Murphy, "Caribou's Plight Intersects Oil Debate," *Los Angeles Times*, July 5, 2001, 1.

77. Quoted in Michael Grunwald, "Warnings on Drilling Reversed," *The Washington Post*, April 7, 2002, 1.

78. Murphy, "Caribou's Plight."

79. Charles Seabrook, "Alaska," *The Atlanta Journal-Constitution*, July 22, 2001, 1.

80. Katharine Q. Seelye, "Facing Obstacles on Plan for Drilling for Arctic Oil, Bush Says He'll Look Elsewhere," *The New York Times*, March 30, 2001, A13.

81. Quoted in Andrew C. Revkin, "Experts Conclude Oil Drilling Has Hurt Alaska's North Slope," *The New York Times*, March 5, 2003, A1; Andrew C. Revkin, "Can Wilderness and Oil Mix? Yes and No, Panel Says," *The New York Times*, March 11, 2003, F2.

82. U.S. GAO, *Improvement Needed in the Management and Oversight of Oil and Gas Activities on Federal Land*, GAO-04-192T (October 2003).

83. Dan Morgan and Peter Behr, "Energy Bill Add-ons Make It Hard to Say No," *The Washington Post*, September 28, 2003, A10.

84. Helen Dewar, "GOP Gains Boost Chance of Alaska Drilling," *The Washington Post*, December 16, 2004, A02.

85. As part of the budget process, each chamber's budget resolution can include reconciliation instructions that require authorizing committees with jurisdiction over particular spending and revenue policies to make legislative changes in those programs to ensure a specific budgetary goal is reached. Although the budget committees write these instructions based on specific assumptions about which policies will be changed, the authorizing committees have complete discretion over the specific changes and must only meet the spending or revenue targets given in the budget resolution. Once the budget resolution is approved by both chambers, the relevant authorizing committees report their legislation to the budget committees, which then combine those recommendations into omnibus packages—largely an administrative task, as the Budget Act prohibits budget committees from making substantive changes in the legislation. The House and Senate vote separately on their respective omnibus packages, which are then sent to conference committee to be unified. Both chambers must approve the final version of the reconciliation bill before it goes to the president for signature.

86. Ben Geman, "ANWR Vote Spurs Debate over Environmentalists' Clout," *Greenwire*, March 24, 2005. On the other hand, the poll also showed that people's views were not strongly held—a result that probably led some legislators to conclude they had leeway on the issue.

87. Quoted in Catherine Dodge and Laura Litvan, "Some House Republicans Want Alaska Drilling Out of Budget Plan," *Bloomberg News and Commentary*, November 8, 2005. Available at http://www.bloomberg.com/apps/news?pid=newsarchive&sid=afZ9s79t6dLg. Since all 202 House Democrats were expected to vote against the package, the 231 Republicans could afford to lose no more than 13 to get the 218 votes needed for passage.

88. Sheryl Gay Stolberg and David M. Herszenhorn, "Bush Says Pain From Economy Defies Easy Fix," *The New York Times*, April 30, 2008; Jad Mouawad, "Oil Prices Are Up and Politicians Are Angry," *The New York Times*, May 11, 2008.

89. The congressional moratorium on offshore oil drilling was enacted in 1982 and has been renewed every year since. It prohibits oil and gas leasing on most of the outer continental shelf, between 3 and 200 miles offshore. In addition, since 1990 it has been supplemented by George H. W. Bush's executive order, which directed the interior department not to conduct offshore leasing or preleasing activity in areas covered by the legislative ban until 2000. In 1998 President Clinton extended the offshore leasing prohibition until 2012. President George W. Bush considered lifting it but did not. In March 2010 President Obama opened up much of the East Coast to oil and gas drilling, in hopes of facilitating agreement on a comprehensive energy and climate change bill.

90. Patrick Reis, "Alaska Delegation Seeks to Block New ANWR Wilderness," *Environment and Energy Daily*, May 5, 2010.

91. Seth Borenstein, "Study Shows Latest Government Spill Estimate Right," *U.S. News & World Report*, September 24, 2010.

92. Quoted in Juliet Eilperin, "Obama Administration to Propose New Wilderness Protections in the Arctic Refuge—Alaska Republicans Declare War," *The Washington Post*, January 26, 2015.

93. Quoted in Coral Davenport, "Obama Will Move to Protect Vast Arctic Habitat in Alaska," *The New York Times*, January 25, 2015.

94. Jared Huffman. (2016). House takes historic vote on Rep. Huffman's amendment to protect Arctic Refuge, receives bipartisan support [Press release]. Available at https://huffman.house.gov/media-center/press-releases/house-takes-historic-vote-on-rep-huffman-s-amendment-to-protect-arctic.

95. Martinson, E. (2017, May 23). "Opening ANWR to drilling Is Priority in Trump's Proposed Budget." *Anchorage Daily News*. Available at https://www.adn.com/politics/2017/05/23/opening-anwr-to-oil-drilling-is-priority-in-trumps-proposed-budget/.

96. Tax Cuts and Jobs Act of 2017. H.R. 1. Title II. (2017)

97. Cornwall, W. (2017). "Battle over Drilling in arctic Refuge Reignites." *Science* 358:6366, 978–979. Available at http://science.sciencemag.org/content/358/6366/978.full.

98. Sobczyc, N. (2018, January 17). "Little Drilling Expected Before 2030; Murkowski Seeks New Bill." *E&E News*. Available at https://www.eenews.net/stories/1060071129.

99. Hobson, M. (2017, December 22). "In Alaska, State Leaders Take a Long-Awaited Victory Lap. Energywire." *E&E News*. Available at https://www.eenews.net/energywire/stories/1060069717/search?keyword=arctic+national+wildlife+refuge.

100. Sobczyc, "Little drilling expected before 2030."

101. Rasmussen Reports. (2017). "Most Oppose ANWR Drilling." [Data set]. Pulse Opinion Research, LLC. Available at http://www.rasmussenreports.com/public_content/politics/general_politics/december_2017/most_oppose_anwr_drilling

102. For a classic exposition of this argument, see Gale A. Norton, "Call of the Mild," *The New York Times*, March 14, 2005, 21.

103. Quoted in Banerjee, "Can BP's Black Gold Ever Flow Green?"

104. Quoted in Davis, "Alaskan Wildlife Refuge Becomes a Battleground."

105. Quoted in Revkin, "Hunting for Oil."

106. Arctic Power and oil executives acknowledge that the prospect of trying to drill in the refuge is not particularly attractive in business terms because of the threat of litigation and policy reversal. Further complicating matters for the oil industry, rising Arctic temperatures in the last three decades have cut the frozen season, the only period oil prospecting convoys are allowed to cross the tundra, from 200 days to 100 days. See Brad Knickerbocker, "Clash over Policies on Energy Pollution," *The Christian Science Monitor*, February 10, 2005; Neela Banerjee, "BP Pulls Out of Campaign to Open Up Alaskan Area," *The New York Times*, November 26, 2002, C4; Neela Banerjee, "Oil Industry Hesitates over Moving into Arctic Refuge," *The New York Times*, March 10, 2002, Sec. 1, 39; Andrew C. Revkin, "Alaska Thaws, Complicating the Hunt for Oil," *The New York Times*, January 13, 2004, F1.

107. Nelson D. Schwartz, "American Energy Policy, Asleep at the Spigot," *The New York Times*, July 6, 2008, BU1.

CHAPTER SEVEN

FEDERAL GRAZING POLICY

Some Things Never Change

In 2018, President Donald J. Trump pardoned the Oregon cattle ranchers who occupied the Malheur National Wildlife Refuge in an armed standoff with federal agents over the fees they refused to pay for grazing on federal land. This brought new life to a century-long debate. Ranchers with permits to graze sheep and livestock have transformed much of the arid public rangelands, which once featured a diverse array of ecosystems, into desert. Yet the nation's two largest land managers, the Forest Service and the Bureau of Land Management (BLM), have been reluctant to implement policies to restore the range. When these agencies have tried to institute environmentally protective practices, ranchers' congressional supporters have stymied their efforts.

Federal grazing policy exemplifies the extent to which policy history constrains current debates and decisions. The legacies of past policies influence the present in several ways. First, the ideas embedded in its authorizing statute(s), along with subsequent interpretations by the courts, shape an agency's view of its mandate; an agency is likely to resist adopting practices that seem contrary to its founding mission.[1] Second, as sociologist Theda Skocpol observes, past policies "affect the social identities, goals, and capabilities of interest groups that subsequently struggle or ally in politics."[2] Interest groups privileged by past policies in turn create the context for current decisions and constrain an agency's choices among alternatives.[3] Third, the pattern of congressional reactions to administrative decisions, which reflects the interests of powerful, well-organized constituents, creates a sense of what is appropriate and legitimate behavior among agency employees. This is possible because, in general, members of Congress concerned about local development interests try to protect them by gaining positions on the authorizing committees and appropriations subcommittees with jurisdiction over the issues most relevant to those interests.[4] From that vantage point, lawmakers can stave off legislative changes perceived as damaging to their constituents in a variety of ways. Authorizing committees set the congressional agenda because they have gatekeeping power; in addition, they craft legislation and get a second crack at the bills after the conference committee does its work. Authorizing committees also exercise control over agencies under their jurisdiction through informal oversight and carefully structured administrative procedures.[5] Appropriations subcommittees guide an agency's policy implementation by determining whether it has sufficient resources—both in terms of budget and capable staff—for the tasks it has been asked to do. In fact, political scientist Herbert Kaufman contends that appropriations subcommittees, particularly in the House, are among the most zealous superintendents of every bureau.[6]

In practice, the extent to which a single interest can dominate policymaking on an issue varies, depending on the intensity of congressional interest and the number of committees with jurisdiction.[7] Internally simple, autonomous, and unified Congress-agency "subsystems"—such as the grazing policy subsystem—are especially difficult for reformers to infiltrate. To break a minority's grip on such a subsystem, challengers must raise public concern sufficiently to give congressional leaders outside the subsystem incentives to expend their political capital on disrupting the status quo. They must also make a case salient enough to persuade uncommitted legislators to get on the bandwagon. Doing so typically involves redefining a problem to highlight an aspect the public heretofore had not considered, such as an activity's environmental costs. A successful campaign can shift not only the public's attention but also that of election-conscious legislators to the issue.[8] Reframing an issue is not simple, however; challengers need a compelling political story, complete with villains and victims and the threatened loss of something the public values. Advocates of grazing policy reform have struggled to come up with a story that generates widespread public concern. Faced with the potent symbols wielded by supporters of the status quo, reformers have had little success in their efforts to disrupt the grazing policy subsystem and therefore to make federal grazing policy more environmentally protective. Instead, the greatest shifts have come as a result of piecemeal, voluntary initiatives initiated by stakeholder collaboratives across the Rocky Mountain West.

BACKGROUND

"There is perhaps no darker chapter nor greater tragedy in the history of land occupancy and use in the United States than the story of the western range," lamented associate Forest Service Chief Earl H. Clapp in 1936 in *The Western Range*.[9] Although the public domain rangeland was in deplorable condition when federal managers assumed control in the early 1900s, policies to regulate its use in the first two-thirds of the twentieth century did little to restore it.[10] The Forest Service and the BLM established grazing lease and permit systems during this period, but federal land managers faced enormous resistance when they tried to restrict grazing. Legal scholar Charles Wilkinson describes the essence of range policy during the twentieth century as "a series of attempts to resuscitate the range from the condition it reached in the late 1800s . . . [that] proceeded in the face of ranchers who continue to assert their 'right' to graze herds without regulation."[11] Because the original grazing policy subsystem was so tightly controlled, ranchers' congressional allies were consistently able to thwart efforts to restore the range, establishing a pattern that proved exceedingly difficult for contemporary environmentalists to break.

Introducing Grazing in the West

The introduction of cattle onto the western range began during the century following the Revolutionary War, as the federal government sought to dispose of its landholdings west of the Mississippi River. Large portions of the West were too dry for homesteaders to cultivate, but the vast, grassy rangelands beckoned to cattle- and sheep-raisers. Because homesteading laws forbade any single claimant to acquire more than 160 acres, an area considered uneconomical for stock raising in the arid West, ranchers devised creative methods to gain control of larger expanses of prairie and grasslands.[12] One commonly used approach was to find and claim a 160-acre plot with access to water and timber (perhaps supplementing this acreage with dummy claims bought from other homesteaders or family members); build a base ranch there with a residence, hay pastures, corrals, and barns; and then establish informal control over a much larger area by illegally fencing the surrounding public domain. In the spring, summer, and fall, ranchers would turn their stock loose to graze on tens of thousands of acres of public lands (see Map 7.1).[13]

By the 1870s and 1880s stockmen had spread across the West. Between 1865 and 1885 the cattle population skyrocketed from an estimated 3 million to 4 million, mostly in Texas, to about 26 million, along with 20 million sheep, blanketing the area from the Great Plains westward.[14] Unregulated use of the western range by livestock wrought ecological havoc: grazing depleted or degraded more than 700 million acres of grassland during this early period, and massive cattle die-offs occurred periodically in the late 1800s and early 1900s. Although stockmen blamed these disasters on severe weather, the cattle were clearly vulnerable to natural fluctuations because overgrazing had so debased the range.

The Origins of Grazing Regulation

Federal regulation of grazing began in 1906, when the first Forest Service director, Gifford Pinchot, announced his intention to require stockmen whose cattle or sheep grazed on national forests to obtain a permit and pay a fee. The proposed charge was $0.05 per animal unit month (AUM—one unit equals one horse or cow, or five sheep or goats). Although this charge was less than one-third of the forage's market value, stockmen rebelled, insisting that the agency could not tax them through administrative fiat. Eventually, however, the stockmen acquiesced to Forest Service regulation, recognizing that restricting access would protect their own interests in the range. Permittees soon began to exploit their "permit value" by adding it to the sale price of their ranches.[15]

Not surprisingly, the combination of low fees and lax congressional homesteading policies increased the demand for national forest range. At the same time, the

Map 7.1 The Western Range

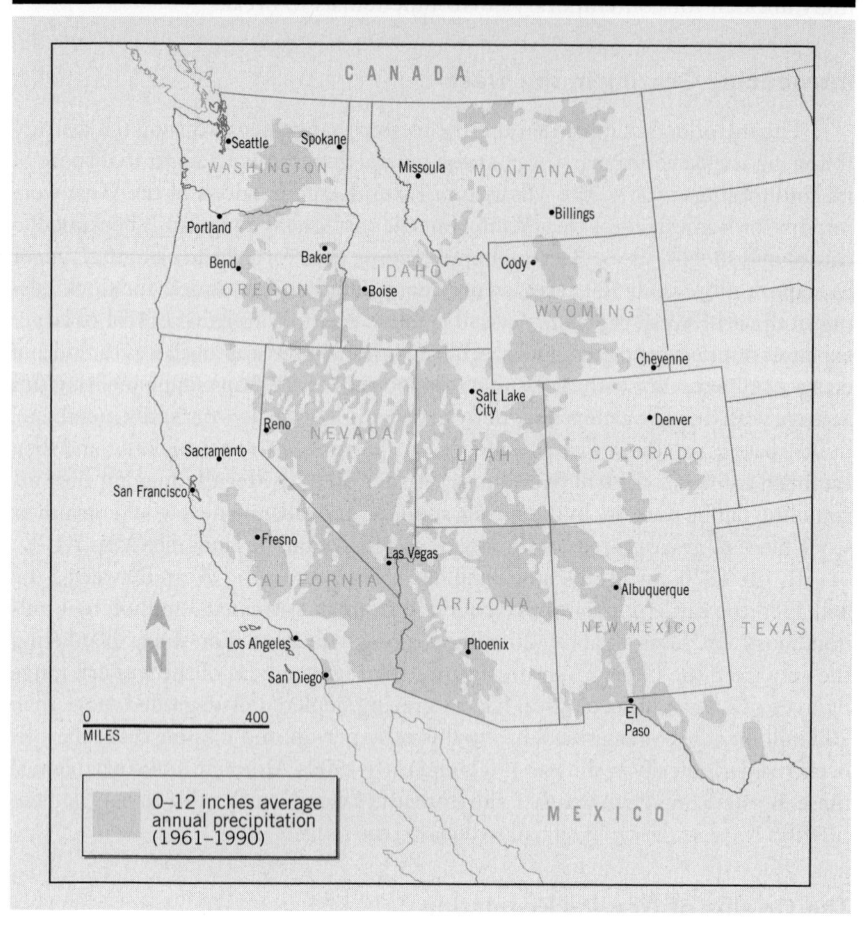

Source: © Linda Marston, 1997. Originally published in: Debra L. Donahue, *The Western Range Revisited* (Norman: University of Oklahoma Press, 1999), 8.

onset of World War I prompted the Forest Service to issue temporary grazing permits and allow stock numbers far greater than the carrying capacity of the range. As a result, between 1908 and 1920, stock raising in the national forests rose from 14 million AUMs to an all-time high of 20 million AUMs.[16] In 1919 the House Appropriations Committee began insisting on more substantial grazing fee increases, mostly in hopes of raising revenue to offset war debts, but Sen. Robert Stanfield, R-Ore.—a rancher and permittee himself—orchestrated a full-scale revolt. In hopes of deterring the appropriators, he held hearings on the fee system and the general administration of rangelands and traveled throughout the West stirring up ranchers' complaints.

Stanfield's "stage managed senatorial attack on Forest Service policy" infuriated forestry experts and conservationists, who promptly retaliated.[17] Representatives of the Society of American Foresters and the American Forestry Association toured the country and criticized the livestock industry in editorials and press releases. The eastern press also struck back: *The New York Times* charged that western senators were resisting the "march of civilization" and called for their assaults to be checked; *The Saturday Evening Post* ran an article by the current Forest Service Chief William Greeley condemning the demand by stockmen to secure special privileges for a few users.[18] Despite the spirited defense by conservationists, however, the Forest Service ultimately retreated from its proposal to increase fees.

While Forest Service permittees were resisting efforts to raise grazing fees, those who ran their livestock on the remaining public domain continued to fend off attempts to regulate *those* lands as well. But by the mid-1930s a series of droughts and the precipitous fall in livestock prices during the Great Depression had caused a crisis in the ranching industry, and some leading ranchers began to believe that a government leasing program on the public domain could stabilize the industry. They opposed turning the program over to the Forest Service, however, afraid the agency would charge them the economic value of the forage and reduce the number of cattle permitted on the land. To placate the stockmen, President Franklin D. Roosevelt's interior secretary, Harold Ickes, promised ranchers favorable terms if they supported keeping the public domain under his department's jurisdiction.[19]

In 1934, with the backing of the largest public domain ranchers, Congress passed the Taylor Grazing Act, named for its sponsor Edward Taylor, D-Colo., a Forest Service critic. The act created the Division of Grazing within the interior department and established a system of grazing permits for public domain land akin to the one administered by the Forest Service. To avoid the vitriolic criticism ranchers had leveled at the Forest Service over the years, the new division set grazing fees at $0.05 per AUM, even though by this time Forest Service fees were three times that amount. And to avoid the perception that the grazing division was a distant bureaucratic organization centered in Washington, Secretary Ickes established a decentralized administration that drew its chief officers not, as the Forest Service did, from a pool of trained professionals but from men with "practical experience" who had been residents of public lands states for at least a year. Finally, the Taylor Grazing Act and its early amendments required the secretary to define regulations for grazing districts "in cooperation with local associations of stockmen." These grazing advisory boards quickly became the dominant force in administering the system, particularly because the division was chronically underfunded and understaffed.[20]

Grazing Policy Controversies, 1940 Through the Late 1960s

By the time the Taylor Grazing Act was passed, the public domain grasslands of the West were already severely depleted. When federal land managers in the new division tried to regulate grazing, however, ranchers' congressional allies intervened. In 1944, when Director Clarence Forsling proposed tripling the grazing fee from $0.05 to $0.15 per AUM, the congressional response was "immediate and harsh."[21] Sen. Pat McCarran, D-Nev., head of the Public Lands Committee, held a series of inflammatory hearings in the West, and shortly thereafter Congress essentially dismantled the division, which by then had been renamed the Grazing Service.[22] In 1946 President Harry S. Truman combined the remnants of the service with the General Land Office to form the Bureau of Land Management. Thus, the BLM began its life with no statutory mission, yet with responsibility for administering 2,000 unrelated and often conflicting laws enacted over the previous century. It had only eighty-six people to oversee more than 150 million acres of land, and its initial budget was so deficient that the grazing advisory boards paid part of some administrators' salaries.[23]

The Forest Service also ran afoul of the stock industry in the mid-1940s. In 1945, when Forest Service grazing permits were due for a decennial review, the agency planned to reduce the number of livestock on the range, shorten the grazing season, and exclude stock altogether in some areas in an effort to rejuvenate the land.[24] Rep. Frank Barrett, R-Wyo., who felt such a move warranted a punitive response, gained authorization to investigate the agency's grazing policies. Emulating McCarran and Stanfield before him, Barrett held hearings throughout the West soliciting criticism of the Forest Service.

Conservationists again were outraged at what they called "the great land grab." Bernard DeVoto, a noted western historian, used his *Harper's Magazine* column to dissent, writing in one essay,

> A few groups of Western interests, so small numerically as to constitute a minute fraction of the West, are hellbent on destroying the West. They are stronger than they otherwise would be because they are skillfully manipulating in their support sentiments that have always been powerful in the West—the home rule which means basically that we want federal help without federal regulation, the "individualism" that has always made the small Western operator a handy tool of the big one, and the wild myth that stockgrowers constitute an aristocracy in which all Westerners somehow share.[25]

The Atlantic Monthly rushed to defend the land management agencies with an article by Arthur Carhart that claimed the Barrett hearings were "rigged" and designed to "throw fear into the U.S. Forest Service," as evidenced by the "transparent manipulations of the meetings, the bias displayed by the chairman . . . the very odor of the meetings."[26] Articles critical of the livestock industry also appeared in *The Nation, Colliers, Reader's Digest*, and hundreds of daily newspapers around the country.

But again, although conservationists managed to stir up the eastern establishment sufficiently to block rancher-friendly legislation, the incendiary hearings caused the land management agencies to retreat from their reform efforts. To improve relations with the stockmen, for example, the Forest Service agreed to cut fees in return for range improvements and to hold hearings on reducing cattle numbers at the request of the affected rancher. As scholars Samuel Dana and Sally Fairfax note,

> The events of 1945–1950 . . . amply demonstrated that congressional supporters of the reactionary cattle operators were quite prepared to destroy an agency that did not meet their peculiar set of goals—through budget cuts, legislative enactment, and simple harassment. Obviously, the ability of a McCarran or a Barrett to hold "hearings" year after year and to tie up the time of BLM and Forest Service officials testifying, gathering data, and defending themselves is a tremendous weapon that members of Congress used to bring recalcitrant officials into line.[27]

In the 1950s, under President Dwight D. Eisenhower, the Forest Service became even more deferential to ranchers, providing its permittees fencing, stock driveways, rodent control, poisonous and noxious plant control, revegetation of grass and shrubs, water development, corrals and loading facilities, and brush control. In the opinion of Forest Service critic William Voigt, "No changes in Forest Service policy with respect to grazing have been more critical than those which . . . shifted so much of the burden of rehabilitating the damaged forest ranges from individual permittees to the taxpayer at large."[28] In addition, Forest Service and BLM grazing fees, although they rose occasionally during this period, fell even further below the real market value of the forage.

Little changed in the 1960s. In 1961 President John F. Kennedy raised the stakes when he delivered a message on natural resources that highlighted the issue of public lands user fees. The only concrete result, however, was more studies. Then, touting an interdepartmental report showing that fees were well below their economic value, in 1969 President Lyndon Johnson's Bureau of the Budget announced what would have been the most progressive grazing policy shift in history: both the BLM and the Forest Service were to raise their fees to a market-value $1.23 per AUM and then index fees to the rates on private lands, phasing in the increase over a ten-year period. Grazing fees did rise during the following years but not as much as Johnson had wanted. The stockmen—with the help of their congressional allies—were once again able to stave off a substantial portion of the planned increase.[29]

THE CASE

With the advent of the environmental era in the late 1960s and early 1970s came a spate of new environmental laws that affected public lands management, most notably the National Environmental Policy Act (NEPA), the Clean Air and Clean Water acts, and the Endangered Species Act. At the same time, conservation biologists and range ecologists were beginning to provide the scientific basis for a challenge to the existing range management regime. Encouraged by these developments, environmentalists tried to use their newfound political clout to challenge ranchers' dominance over federal grazing policy. But each push for reform encountered the legacy of past policies: agencies intimidated by entitled ranchers, who were backed by their well-positioned congressional allies. Unable to redefine the grazing issue in a way that captured national public attention, environmentalists could not mount a successful legislative or administrative challenge and so during subsequent decades made only marginal gains.

Environmentalists' Arguments

The environmentalists' case was (and remains) that livestock overgrazing was destroying the western range and that federal land management practices were exacerbating rather than ameliorating this trend. Some environmentalists wanted to remove cattle from the western public lands altogether; others argued that at a minimum, ranchers' prerogatives ought to be severely curtailed and the range restored to ecological health.

The Effect of Grazing on the Range. Environmentalists charged that when ranchers grazed extremely large numbers of livestock, the animals inflicted serious, long-term, and sometimes permanent damage to the range. Environmentalists cited several reasons for the damage. First, cattle eat the most palatable and digestible plants before eating anything else. Such selective grazing, combined with the limited tolerance of some plant species for grazing, can prompt shifts in the composition of plant communities.[30] Second, cows are heavy consumers of water, which is in short supply west of the 98th meridian. Beyond that line, precipitation drops below twenty inches a year, the lower limit for many nonirrigated crops, and even more important, it falls irregularly, leading to frequent droughts. Third, herds of cattle compact the soil, making rainwater run off, which causes erosion, gullying, and channel cutting.

Environmentalists added that cattle were particularly destructive to the West's precious riparian zones, the lush, vegetated areas surrounding rivers and streams. According to Barry Reiswig, manager of the Sheldon Wildlife Refuge in Nevada and Oregon, "You'll hear lots of [range conservationists], even people on my staff, crowing about how 99 percent of the range is in good or excellent condition, but they ignore the fact that the one percent that is trashed—the riparian zones—is really the only part important for most wildlife species."[31] Riparian zones are important ecologically because they sustain a much greater quantity of species than the adjoining land, providing food, water, shade, and cover for fish and wildlife. They also benefit humans by removing sediment from the water table, acting as sponges holding water and stabilizing both stream flow and the water table, and dissipating flood waters. Cattle degrade riparian zones by eating tree seedlings, particularly the cottonwoods, aspen, and willow on which species such as bald eagles and great blue herons rely. Cattle trampling on and grazing streamside vegetation also cause stream banks to slough off, channels to widen, and streams to become shallower, so they support fewer and smaller fish.[32]

To support their allegations about overgrazing, environmentalists cited evidence that more than half of private and public rangeland was in fair or poor condition.[33] They also referred to data collected by the U.S. General Accounting Office (GAO, later the Government Accountability Office) confirming that most of the West's riparian areas were in poor condition, and a 1990 EPA report stated, "Extensive field observations in the late 1980s suggest riparian areas throughout much of the West are in the worst condition in history."[34] Environmentalists highlighted the findings of Texas Tech soil scientist Harold Dregne that, by the mid-1970s, 98 percent of the arid lands in the western United States—some 464 million acres of privately and publicly owned rangeland—had undergone some degree of desertification.[35] Although some traditionally trained range scientists disputed environmentalists' claims about the ecological impact of livestock grazing, particularly on rangeland plants, most conservation biologists agreed that past overgrazing had eroded soil, destroyed watersheds, and extinguished native grasses and other vegetation on which wildlife feed.[36]

The Failure of Range Management. Environmentalists also charged that federal grazing policy had failed to restore, and in some cases had further depleted, a range already badly damaged by early ranching practices. According to their critics, the BLM and Forest Service allowed too many cattle to graze with too few restrictions on the roughly 31,000 allotments these agencies administer. Critics cited a 1988 GAO finding that, on about 18 percent of the BLM's allotments and 21 percent of Forest Service allotments, the authorized grazing levels exceeded the carrying capacity of the land.[37] Moreover, critics claimed that, by undercharging ranchers for grazing, federal land managers encouraged overgrazing, a claim supported by numerous studies that found federal grazing fees were well below rates charged on comparable private lands.

Environmentalists also contended that a third aspect of grazing policy, range improvement—a euphemism for developing water supplies, fencing, seeding, and making other investments to enhance the forage supply—had serious ecological consequences. They pointed out that fences limited the movement of wildlife; water development reduced the supply of water for wildlife and depleted aquifers; predator control had extirpated species such as wolves, mountain lions, and bears from their historic ranges; and vegetation controls—such as herbicide spraying and plowing and seeding—had reduced plant species diversity. Environmentalists noted that ranchers generally condoned such operations while opposing riparian restoration projects—those most important to environmentalists—because the latter involved prohibiting access to stream banks.[38]

In an effort to undermine the cowboy symbolism deployed by ranchers, environmentalists alleged that low grazing fees, overstocking, and range improvements constituted a subsidy to wealthy hobby ranchers, not the rugged individuals that the word "rancher" connoted. While acknowledging that some of the approximately 23,000 western BLM and Forest Service permittees were small ranchers, environmentalists pointed out that the main beneficiaries were large ranchers with historical ties to the public lands. As evidence, they cited GAO studies saying that 3 percent of livestock operators in the West used 38 percent of the federal grazing land, while less than 10 percent of federal forage went to small-time ranchers (those with fewer than 100 head of cattle); only 15 percent of BLM permittees had herds of 500 or more animals, but they accounted for 58 percent of BLM's AUMs; and stockmen with more than 500 animals comprised only 12 percent of Forest Service permittees but accounted for 41 percent of the agency's AUMs.[39] To make their argument more concrete, environmentalists named the wealthy individuals, partnerships, and corporations that worked many of the largest allotments: Rock Springs Grazing Association of Wyoming, for example, controlled 100,000 acres; J. R. Simplot, said to be the wealthiest man in Idaho, ranched 964,000 acres; the Metropolitan Life Insurance Company had permits for 800,000 acres; and the Zenchiku Corporation of Japan controlled 40,000 acres. Union Oil and Getty Oil, as well as the Mormon Church, were also large permit holders.[40]

Although some environmentalists advocated eliminating grazing on BLM and Forest Service land, others had more modest goals. They demanded that public lands ranchers pay grazing fees equivalent to the market value of the forage, noting that existing fees covered only 37 percent of BLM and 30 percent of Forest Service program administration costs.[41] They also insisted that federal land managers reduce the number of cattle to a level the land could comfortably support. And they asked for more opportunities for public participation in rangeland management decisions.

Ranchers' Resistance to Reform

Public lands ranchers did not take environmentalists' assaults lying down. They responded that public and private fees were not comparable because it cost more to run cattle on public lands. Those who ranch on public lands, they argued, had to provide capital improvements such as fencing and water that property owners supplied on private lands. Moreover, they pointed out, the value of their lease or permit had long ago been capitalized into the cost of the associated ranch, and a cut in cattle numbers would therefore devalue their ranches. Permittees portrayed themselves as an important part of the livestock industry, even though public lands supply only about 2 percent of the forage consumed by beef cattle.[42] They further contended that because many of them were marginal, family operations, an increase in grazing fees could put them out of business and destabilize the communities in which they operated. Ranchers' most formidable rhetorical weapons were not reasoned arguments, however, but the iconographic symbols they wielded: cowboys, rugged individualism, and freedom from control by a distant and oppressive federal government.

In truth, though, permittees' influence derived less from their public arguments than from their historic ties to members of Congress who sat on the public lands authorizing committees and appropriations subcommittees. Legislators from the Rocky Mountain states in particular had secured positions on the House and Senate Interior (House Natural Resources and Senate Energy and Natural Resources as of 2011) committees, from which they fended off efforts at grazing policy reform. As a result, in spite of their small numbers and their minority status in the livestock industry as a whole, public lands ranchers retained substantial power. As journalist George Wuerthner marveled, "There are more members of the Wyoming Wildlife Federation than there are ranchers in the entire state of Wyoming, but it is ranchers, not conservationists, who set the agenda on public lands."[43]

Permittees' influence was enhanced by their association with the academic community in western land grant colleges and universities. As one commentator observed, "This community specializes in rangeland management and has, with few exceptions, been solidly allied with ranching interests, which, in turn, have the political power to determine higher education budgets and sometimes serve as regents of various schools."[44] The permittees' clout also derived in part from their ties to allies in banking and real estate. According to Charles Wilkinson,

> Ranches are usually valued for loan purposes based on AUMs, and the appraised value will drop if the AUMs drop. A decrease in AUMs thus will reduce a rancher's ability to raise capital and will weaken the security on existing loans. . . . As Charles Callison, longtime observer of range policy, has told me: "It's one thing when western congressmen hear from the ranchers. But they really leap into action when the bankers start getting on the telephone."[45]

Finally, many prominent stock raisers affected policy directly by taking an active role in Republican Party politics or actually holding elected or appointed office: former president Ronald Reagan's BLM director, Robert Burford, was a millionaire BLM rancher; Wyoming's former senators Clifford Hansen and Alan Simpson, as well as New Mexico Representative Joe Skeen, were cattlemen; former Nevada Senator Paul Laxalt was a sheep rancher; and Representatives Robert Smith of Oregon and Jim Kolbe of Arizona came from ranching families.

The Federal Land Policy Management Act

Although their track record was not encouraging, in the early 1970s environmentalists hoped to capitalize on the policy window opened by Earth Day and began to press Congress for grazing policy reform. The most obvious route by which to alter grazing policy was through the organic act that BLM had been prodding Congress to pass for a decade.[46] Between 1971 and 1974 the Senate approved several bills granting BLM statutory authority, but proposals foundered in the House, primarily on the issue of grazing fees. Finally, in the 94th Congress (1975–1976), the Senate approved a bare-bones authorization bill sponsored by Henry Jackson, D-Wash., and supported by the Ford administration. The House managed to settle on a bill as well, although its version clearly favored ranchers: it gutted BLM law enforcement authority and increased local control over BLM and Forest Service planning.[47] It also contained four controversial grazing provisions, all beneficial to ranchers. First, it established a statutory grazing fee formula based on beef prices and private forage costs, with a floor of $2.00 per AUM. Second, it improved ranchers' tenure by making ten-year permits the rule, rather than the exception. Third, it required the BLM to compensate ranchers for improvements they had made to the land if it canceled their permits. Fourth, it resurrected and prescribed the composition of local grazing advisory boards, which Congress had replaced with multiple-use advisory boards one year earlier.[48] The Interior Committee narrowly approved the bill, 20–16; and the full House—deferring to the committee—also approved it, 169–155.

Because the House and Senate versions were substantially different, it was unclear whether the conference committee could reconcile them. House negotiators offered to implement the rancher-friendly fee formula for two years but in the meantime conduct a study of the issue, a deal Senate conferees refused. House members then offered to drop the grazing fee formula altogether but freeze current fees, an option also unpalatable to Senate conferees. Finally, the Senate conferees made a counteroffer that the House conferees accepted: a one-year freeze on grazing fees, accompanied by a one-year study of the fee issue conducted jointly by the agriculture and interior departments.[49] Once the study was completed, Congress could take up the issue again. The conference bill passed the House on September 30, 1976, and the Senate on October 1, just hours before the 94th Congress adjourned. Grazing interests objected to the deletion of their preferred grazing fee formula and encouraged the president to pocket veto the bill, but President Gerald Ford signed the Federal Land Policy and Management Act (FLPMA) into law on October 21. Although disappointed about the grazing fee provision, overall, livestock interests were satisfied with the outcome.

The Carter Administration, 1977–1981

The following year, the expiration of the grazing fee freeze and the election of a proenvironmental president presented reformers with another opportunity, and a flurry of legislative activity ensued. In the autumn of 1977 President Jimmy Carter's agriculture and interior departments released their congressionally mandated study of grazing fees and proposed raising fees administratively to a uniform $1.89 per AUM for the 1978 season. The fees would rise thereafter no more than 12 percent per year until they reached a market value of $2.38 per AUM. Livestock groups called the proposal "unfair and unrealistic" and, worried about the Carter administration's propensity for reform, opted for their favorite tactic: deferral. They urged the administration to let Congress decide the grazing fee question. The administration responded by freezing fees at the 1977 level, while ranchers pressured Congress to adopt another one-year moratorium on increases.[50]

By the fall of 1978 Congress had produced a new grazing bill, the Public Rangelands Improvement Act (PRIA), which had two major provisions: range improvement funding, which met with near-universal support, and a statutory grazing fee formula, which was highly controversial. Environmentalists, as well as both the BLM and the Forest Service, supported a market value-based fee similar to the one that had technically been in place since 1969; livestock interests agitated for a fee with built-in profits based on forage costs and beef prices. After bitter debates in the House and Senate, both chambers passed the PRIA with the rancher-approved formula and a provision prohibiting annual fee increases or decreases of more than 25 percent of the previous year's fee. Its language reflected the authorizing committees' rationale: "To prevent economic disruption and harm to the western livestock industry, it is in the public interest to charge a fee for livestock grazing permits and leases on the public lands which is based on a formula reflecting annual changes in the cost of production."[51] President Carter signed the bill on October 25, 1978.

Although constrained by the PRIA, Carter's BLM nevertheless embarked on a program of intensive, conservation-oriented range management in which Director Frank Gregg encouraged agency staff to conduct range inventories and cut grazing permit allocations if they found forage supplies insufficient to support assigned levels. The reduction program was short-lived, however. Public lands ranchers sounded the alarm about what they regarded as the agency's heavy-handed tactics, and rancher ally Sen. James McClure, R-Idaho, succeeded in attaching an amendment to the 1980 interior appropriations bill that mandated a two-year phase-in for any stock reduction of more than 10 percent.[52]

McClure's amendment notwithstanding, the BLM initiative fueled ranchers' antagonism toward federal land managers, which instigated the Sagebrush Rebellion, a vocal but disjointed attempt to persuade Congress to transfer ownership of federal lands to western states. The event that officially touched off the rebellion, which had been brewing since Carter took office, was Nevada's passage in 1979 of a law declaring state control over all land within its borders. (Eighty-seven percent of the land in Nevada is federally owned.) Proponents framed the rebellion as a constitutional challenge, citing the "equal footing" doctrine, according to which newer states should be admitted to the Union under the same conditions as the original thirteen colonies. Although it had no national leadership of formal administrative structure, the rebellion appeared to be an authentic, populist uprising against the values of environmentalism.[53] The rebels did not accomplish the transfer (nor is it clear the majority even wanted to), but they did succeed in drawing attention to their plight.[54] Their cause was sufficiently visible that, during the 1980 presidential campaign, Republican candidate Ronald Reagan declared himself a sagebrush rebel.

The Reagan Administration, 1981–1989

In the early 1980s federal grazing fees, now indexed to ranchers' costs and expected returns, fell to $1.37 as a consequence of sluggish beef prices (see Table 7.1). In the meantime, reformers and livestock interests mustered their forces for another confrontation when the PRIA expired in 1985. Although the battle lines were drawn in Congress, the position of the White House had shifted considerably. In contrast to his predecessor, newly elected President Reagan was a strong supporter of public lands ranchers.[55] Upon taking office, he appointed livestock industry supporters to prominent positions in the interior department. Secretary James Watt was the former president of the Mountain States Legal

Table 7.1 Annual Nominal Grazing Fees for the Bureau of Land Management and the Forest Service, 1940–2015

Year	BLM Fee ($ per AUM)	Forest Service Fee ($ per AUM)
1940	$0.05[a]	$0.15
1945	0.05	0.25
1950	0.10	0.42
1955	0.15	0.37
1960	0.22	0.51
1965	0.30	0.46
1970	0.44	0.60
1975	1.00	1.11
1976	1.51	1.60
1977	1.51	1.60
1978	1.51	1.60
1979	1.89	1.93
1980	2.36	2.41
1981	2.31	2.31
1982	1.86	1.86
1983	1.40	1.40
1984	1.37	1.37
1985	1.35[b]	1.35[b]
1986	1.35	1.35
1987	1.35	1.35
1988	1.54	1.54
1989	1.86	1.86
1990	1.81	1.81

Year	BLM Fee ($ per AUM)	Forest Service Fee ($ per AUM)
1991	1.97	1.97
1992	1.92	1.92
1993	1.86	1.86
1994	1.98	1.98
1995	1.61	1.61
1996	1.35	1.35
1997	1.35	1.35
1998	1.35	1.35
1999	1.35	1.35
2000	1.35	1.35
2001	1.35	1.35
2002	1.43	1.43
2003	1.35	1.35
2004	1.43	1.43
2005	1.79	1.79
2006	1.56	1.56
2007	1.35	1.35
2008	1.35	1.35
2009	1.35	1.35
2010	1.35	1.35
2011	1.35	1.35
2012	1.35	1.35
2013	1.35	1.35
2014	1.35	1.35
2015	1.69	1.69
2016	2.11	2.11
2017	1.87	1.87
2018	1.41	1.41

Source: U.S. Forest Service and Bureau of Land Management.

a The BLM began charging fees in 1936, and from 1936 to 1946 the fee was 5 cents per AUM.

b The Public Rangeland Improvement Act of 1978 set a minimum fee of $1.35 per AUM.

Foundation, which litigates on behalf of ranchers and other resource extraction interests. BLM director Robert Burford was a Colorado rancher "who had jousted repeatedly with the BLM, most often over his own grazing violations," and believed there was a tremendous capacity for increasing beef production on America's rangeland.[56]

The administration drew on all its resources to alter the BLM's range management agenda, changing the mix of professionals (including laying off ecologists), the budget, and the structure of the agency. In addition, Burford used informal rulemaking processes to increase ranchers' security. The centerpiece of his approach was the Cooperative Management Agreement (CMA) program. In theory, the program provided exemplary ranchers with CMAs, which allowed them to manage their allotments virtually unimpeded. In 1985, however, a federal district court declared that the CMA approach illegally circumvented BLM's statutory obligation to care for overgrazed public rangeland. Although forbidden to transfer authority for the public range to ranchers formally, the interior department managed to accomplish the same result through neglect: under pressure from Reagan's political appointees, BLM range managers abandoned the stock reductions and range restoration projects begun under President Carter.

Rebuffed by the administration, environmentalists could only hope to make inroads into grazing policy when the grazing fee issue returned to the congressional agenda in 1985. By that time, environmental groups, fiscal conservatives in the administration, and some BLM and Forest Service officials had formed a loose coalition to press for higher grazing fees. Reformers launched an all-out campaign in the media, labeling public lands ranchers as "welfare cowboys" and deploring low grazing fees as a subsidy in an era of high deficits and fiscal austerity. Hoping to shame public officials and reduce the influence of livestock interests, journalists investigated stories about BLM or Forest Service officials who had been fired or transferred for trying to implement environmental reforms.[57]

Buttressing the reformers' case were the results of yet another grazing fee analysis. In March 1985 the BLM and Forest Service completed a four-year study in which twenty-two professional appraisers collected data for every county in the West that had rangeland. The researchers discovered that fees on private lands averaged $6.87 per AUM, nearly five times higher than fees for comparable public land, and that the fees charged by other federal agencies averaged $6.35 per AUM.[58] The study presented for congressional consideration five alternative fee formulas, all involving a fee hike.

Congress, however, failed to come up with grazing legislation that satisfied both environmentalists and ranchers, and when the PRIA formula expired at the end of 1985, authority to set the fee reverted to the Reagan administration. Although environmentalists directed their best efforts at influencing President Reagan, they never had a chance. On December 13, 1985, Senator Laxalt, a close friend of the president's, delivered a letter from a group of mainly western, Republican senators urging the president to freeze the grazing fee. Eventually, twenty-eight senators and forty representatives joined the lobbying effort.

In the face of this assault, in late January 1986 the Office of Management and Budget (OMB) pulled out of its alliance with environmentalists. OMB director James Miller wrote to President Reagan that he was backing off in recognition of the political sensitivity of the issue. Miller did, however, recommend freezing fees for no more than one year in order to pressure Congress to act.[59] To the dismay of environmentalists, Reagan ignored Miller's advice and issued Executive Order 12548 extending the PRIA fee formula indefinitely. The executive order effectively hamstrung BLM and Forest Service efforts to rehabilitate the range because the cost of administering the grazing program substantially exceeded revenues from grazing fees under the PRIA system, and the enormity of the federal deficit precluded any additional funding for either agency.

The George H. W. Bush Administration, 1989–1993

In 1988 advocates of higher grazing fees began to prepare a run at the 101st Congress, hopeful that changes in the composition of the House Interior Committee, which had been a major obstacle to reform, would improve their prospects. In the 1970s Phillip Burton, a reform-minded California Democrat, had taken over as the committee chair and begun recruiting environmentally oriented members. For a brief period, ranching allies were in the minority; by 1991 only nine of the committee's twenty-six Democrats hailed from the West, and three of those were from urban areas of Los Angeles, Oakland, and Salt Lake City.

In 1990 the House adopted (251–155) an amendment to the interior department's appropriations bill that would have increased the grazing fee sharply. But the provision was deleted in conference with the Senate, where proranching western senators narrowly prevailed on the issue in return for rescinding their opposition to oil exploration on the Outer Continental Shelf. The same year, Rep. Bruce Vento, D-Minn., tried to attach his BLM authorization bill, which contained a grazing fee increase, to the budget reconciliation bill in an effort to circumvent western senators. The bill did not promise enough deficit reduction to satisfy the Budget Committee, however, and was dropped.[60]

In 1991 opposition from westerners on the House Interior Committee forced Rep. George "Buddy" Darden, D-Ga., to abandon an amendment to a BLM reauthorization bill that would dramatically increase grazing fees. Darden and Rep. Mike Synar, D-Okla., then took their case to the Appropriations Committee, where they were more successful, and in late June the full House voted 232–192 to quadruple grazing fees, bringing them up to market rates over a four-year period. (New information from the GAO had bolstered the claims of fee-increase advocates: the GAO reported that the existing formula double-counted ranchers' expenses, so that when ranchers' expenses went up, the fee went down. As a result, the fee was 15 percent lower in 1991 than it had been in 1975, whereas private grazing land lease rates had risen 17 percent in the same period.[61]) Then, in July, the House voted overwhelmingly to add a grazing fee amendment to the BLM reauthorization bill despite Interior Committee opposition. These votes signaled changes in the proportion of proenvironmental western representatives in the chamber, which in turn reflected the shifting demographics of the West: as the region urbanized, environmentalists, recreation advocates, and tourism interests were gaining a political voice. But despite these changes in the House, a handful of western senators, who continued to speak primarily for the traditional economic interests, managed to deflect attempts to raise grazing fees, and the BLM reauthorization bill simply disappeared in the Senate without a hearing. The negotiation over the appropriations bill was more convoluted, but the result was the same.[62]

Tentative efforts to reform rangeland management during the Bush administration rekindled the embers of the Sagebrush Rebellion, and a new force emerged: the wise use movement, which officially declared itself at a 1989 meeting in Reno, Nevada. At that meeting, policy entrepreneurs Ron Arnold and Alan Gottlieb formulated "The Wise Use Agenda," a list of twenty-five goals that mainly involved enhancing access to federal lands by resource extraction and motorized recreation interests. Wise use leaders articulated an antiregulatory storyline, denigrating environmentalists as elites and eco-fascists out to destroy the property rights and liberty of westerners. The real environmentalists, they claimed, were "the farmers and ranchers who have been stewards of the land for generations, the miners and loggers and oil drillers who built our civilization by working in the environment every day, the property owners and technicians and professionals who provided all the material basis of our existence."[63] They argued that environmental regulations would destroy western economies, warning, "People will lose jobs, rural communities will become ghost towns, education for our children will suffer, and state and local governments will forfeit critical income for police, fire protection, and social services."[64] The wise use movement gained momentum in the early 1990s, driven in part by globalization and automation, both of which had eroded employment and payrolls in the natural resource industries of the West.

A related movement, the county supremacy movement, also emerged in the early 1990s. Like the Sagebrush Rebellion, the county supremacy movement—which took hold primarily among state and county officials—sought local control over federal lands. The first effort to institutionalize the goals of the movement was the Interim Land Use Policy Plan of Catron County, New Mexico. Adopted in 1991, this ordinance mandated that federal managers "shall comply with" the county's land-use plan or face prosecution. Subsequently, Nye County, Nevada, went even further, declaring that the State of Nevada, not the federal government, owned the public lands within its boundaries. The courts uniformly rejected the tenets of the county supremacy movement, but its proponents remained disgruntled—and susceptible to mobilization.

The Early Clinton Years, 1993–1994

With the election of Bill Clinton to the presidency and Al Gore to the vice presidency, environmentalists believed they finally had a genuine chance to reform federal grazing policy. President Clinton confirmed their expectations when, immediately upon taking office, he announced his plan to cut subsidies for grazing, as well as logging, mining, and water development on the public lands. Bruce Babbitt, Clinton's secretary of the interior, also adopted a markedly different tone from that of his predecessors. Babbitt got a standing ovation from a roomful of federal employees when he proclaimed, "I see us as the department of the environment. . . . We are about the perpetual American love affair with the land and the parks."[65] Clinton also delighted environmentalists with his appointment of Jim Baca as head of the BLM. Baca had been New Mexico's public land commissioner, a former board member of the Wilderness Society, and an outspoken critic of overgrazing.

Some members of Congress found Clinton's position refreshing. George Miller, D-Calif., head of the House Interior Committee, claimed that most westerners embraced the changes that Clinton was proposing. According to Miller, "Reagan and Bush were just holding back the future. They were the last gasp of an outdated philosophy."[66] But others were not so optimistic about prospects for reform, especially when ranchers banded together with other resource users to form the wise use movement, with the objective of promoting unfettered access to public lands. These interests continued to have fiercely protective and influential congressional sponsors, particularly in the Senate.

Although he favored reform, Clinton was wary of alienating proranching western senators. On March 16, 1993, capitalizing on the president's frail majority, seven western senators led by Democrat Max Baucus of Montana met with Clinton to discuss trading their support for his economic stimulus and deficit reduction program for his dropping reform of public land management. Two weeks later, the president backed off his initial proposal to raise fees for commercial uses of public resources as part of the budget but promised instead to pursue an increase administratively.

After a series of public hearings on the issue throughout the West, Secretary Babbitt proposed to add to the interior appropriations bill a provision that more than doubled grazing fees on federal lands over a three-year period and imposed tough environmental standards on ranchers. In response, Sen. Pete Domenici, R-N.M., and Sen. Harry Reid, D-Nev., proposed an amendment placing a one-year moratorium on Babbitt's ability to spend any money to implement his grazing policy. Because of the issue's low salience outside the West, the Senate deferred to Domenici and Reid, 59–40. The House instructed its negotiators on the bill to reject the Senate moratorium, however. In response, Reid—hoping to put the grazing policy issue to bed—worked out a compromise with Babbitt and House Democrats to resolve the differences between the two chambers by increasing grazing fees (to $3.45, far short of the $4.28 Babbitt wanted, well below most state lands' fees, and barely one-third of the average on private lands in the West) but imposing fewer land management requirements than the original proposal. Babbitt agreed, and the House and Senate conferees approved Reid's amendment.[67]

In late October the full House approved the bill containing Reid's grazing compromise, but the Senate was unable to muster the sixty votes necessary to head off a filibuster by Domenici, who viewed the bill as too proenvironment. Republicans achieved near-perfect unity in support of Domenici's filibuster, and five western and northern plains Democrats joined them. "This proposal threatens a rural way of life," said the embattled Domenici. "If we lose we'll go down with every rule in the Senate to protect our people."[68] After failing three times to invoke cloture (54–44 on the final try), Reid agreed to drop the grazing compromise, Domenici stopped stalling the bill, and on November 9 the Senate and House sent the revised version without grazing fee language to President Clinton.[69] Again, ranching interests had averted reform.

Following this series of highly publicized congressional debacles, Babbitt vowed to raise grazing fees and institute management reforms administratively. Faced with the prospect of a lawsuit and extremely hostile press coverage in the West, Babbitt adopted a conciliatory stance. In February 1994 he ousted his outspoken BLM director to placate western governors and senators who had complained about Baca's aggressive approach to rangeland management. Then, in April, after traveling extensively throughout the West to gain local buy-in and defuse a second Sagebrush Rebellion that was bubbling up in several western counties, Babbitt unveiled a proposal that retained some of the protectiveness of the original plan but made several concessions to ranchers. The proposal called for creating local "resource advisory councils" to develop range management plans. The councils, designed to operate by consensus, would consist of five permittees or other commodity interests, five environmentalists, and five representatives of other public land uses and state and local government. The proposal also allowed environmental groups to purchase grazing permits for conservation, a practice that was prohibited under the existing rules; made grazing permits good for ten years, compared with five years in the initial plan; rendered decisions by BLM field officers effective immediately; and allowed any member of the public—not just those directly affected—to participate in and appeal BLM decisions. Although the modified proposal doubled grazing fees over three years, Babbitt sought to increase its palatability by adding a two-tier fee structure that charged small ranchers less and offered a 30 percent discount to ranchers who improved the land.

Despite the grazing fee increase, environmentalists sharply criticized the administration for backpedaling. Babbitt defended his position, saying it reflected his view that "those closest to the land, those who live on the land, are in the best position to care for it."[70] But no one expected a system of local advisory councils dominated by ranchers to result in any serious scaling back of grazing privileges. Meanwhile, ranchers also rejected the compromise rules: after issuing his revised proposal, Babbitt held a second round of public hearings that were, once again, dominated by ranchers who denounced the proposal as an attack on rural westerners. Confirming environmentalists' worst fears, in December 1994 Babbitt, after encountering serious opposition by powerful Republican members of Congress, retracted the grazing fee increase altogether and delayed the effective date of many of the proposed environmental regulations to allow Congress to vote on them. Ironically, in 1995, under the existing formula, grazing fees dropped from $1.98/AUM to $1.61 and fell again to $1.35 in 1996 (see Table 7.1).

The Republican Congress Retaliates, 1995–1997

Even as environmentalists' hopes of reform were dashed, ranchers saw a chance to expand their privileges when the Republicans assumed control of Congress in 1995. The Republican takeover owed much of its success to voters in the West. There, Republican politicians had capitalized on the hostility emanating from the wise use and county supremacy movements. When the Clinton administration tried to phase out government subsidies for mining, logging, and grazing on federal land, Republicans labeled the effort a "war on the West"—a slogan that moved quickly to bumper stickers and became a rallying cry in the 1994 election campaigns.[71] Enhancing the clout of ranchers and other constituents who favored resource extraction on federal lands, the Republican leadership appointed western members sympathetic to the wise use movement to chairmanships of the major environmental committees.

In the early spring Senator Domenici introduced the Public Rangeland Management Act (S 852), a bill designed to preempt Secretary Babbitt's proposed rule changes. The bill raised grazing fees by a nominal amount, thereby heading off the substantial increase sought by Babbitt and the majority of members of Congress, and enhanced permittees' control over federal grazing allotments. Most important, the bill excluded from land management decisions anyone but ranchers and adjacent property owners by creating 150 advisory boards consisting solely of ranchers. In July the Senate Energy and Natural Resources (formerly Interior) Committee approved the bill over the objections of the BLM, whose acting director, Mike Dombeck, fumed, "This bill takes the public out of public lands. It returns land management to an era of single use at taxpayers' expense."[72] Domenici's proposal stalled, however, when it became clear it could not garner the sixty votes necessary to break an anticipated filibuster by Democrats and moderate Republicans.

In the House, the Resource Subcommittee on Public Lands approved a similar bill (HR 1713), which barred the interior secretary from setting national rangeland standards; gave ranchers proportional title to improvements, such as fencing, landscaping, and ponds; and lengthened the term of a grazing lease from ten years to fifteen years. Like Domenici's failed effort, however, HR 1713 never reached the floor, thanks to a parliamentary maneuver by New Mexico Democrat Bill Richardson. Sponsors of both bills then tried to insert a modest fee increase provision into the budget reconciliation bill, again in hopes of averting a much more substantial one, but that rider was dropped before the bill went to the president. In a final attempt at an end-run around Babbitt's rules, the Senate attached a provision to interior's appropriation bill to postpone the rules' implementation, but President Clinton vetoed the bill.

In the meantime, in August 1995, Babbitt finally issued his new grazing package, which did not contain the most controversial element, a grazing fee increase. The regulations did, however, set federal standards for all rangelands and allowed the federal government to claim title to all land improvements and water developments made by ranchers on public lands. They also established regional resource advisory councils, with guaranteed spots for environmentalists, to help the BLM devise grazing guidelines for each state and write comprehensive plans for preserving rangeland ecosystems. And they limited ranchers' rights to appeal BLM decisions to reduce the number of animals on an allotment. Not surprisingly, environmentalists applauded the regulations, and ranchers were enraged.

Ranchers' Senate allies were determined to thwart Babbitt's reforms, and in 1996 they managed to pass a bill (S 1459) similar to S 852 that aimed to replace Babbitt's initiative with a more rancher-friendly version. The House Resources (formerly Interior) and Senate Energy and Natural Resources committees approved a similar bill, but it met resistance in the full House from a coalition of environmentalists and fiscal conservatives and faced a certain presidential veto. The bill died at the end of the session. Frustrated westerners in the House also attempted to insert prograzing provisions into the omnibus parks bill by holding hostage funding for New Jersey's Sterling Forest, a priority for many easterners. In September western representatives abandoned that effort as well, recognizing, according to Republican James Hansen of Utah, that pressing forward would carry an unacceptable political price.[73]

Although their legislative efforts to thwart Babbitt stumbled, ranchers got a boost from the courts. Scheduled to go into effect in March 1996, the new rules had been held in abeyance until legal challenges by five livestock groups were resolved. In June Judge Clarence Brimmer of the U.S. District Court in Wyoming rejected several of the reforms on the grounds that they would "wreak havoc" on the ranching industry and exceed the BLM's legislative authority. Brimmer ruled that the provision to weaken ranchers' rights to renew federal grazing permits was illegal because the Taylor Grazing Act specified "grazing preference" to ensure that ranchers and their creditors had some certainty about their tenure. The judge also rejected the rules giving the government title to future range improvements and allowing conservationists to acquire grazing permits. He did uphold the agency's right to check on permittees' compliance with regulations and to suspend or cancel a permit if the lessee was convicted of violating environmental laws. Ranchers' judicial victory was short-lived, however. In September 1998 the Tenth Circuit Court of Appeals rejected major parts of the Brimmer ruling: it affirmed the BLM's authority to reduce the number of cattle allowed by permits and retain title of range improvements on public lands, but it concurred with Brimmer's rejection of conservation-use permits. In 2000 the Supreme Court unanimously upheld the appeals court's ruling.[74]

As Babbitt's rules wended their way through the courts, Congress continued to wrestle with the issue of grazing on federal lands. In October 1997 the House passed a "grazing reform" measure (HR 2493) that raised grazing fees by 15 percent—again in hopes of deflecting attempts to raise fees more substantially—lengthened ranchers' lease terms, and eased restrictions on ranching permits. The bill's passage was a victory for ranching advocates, as the House had been a "major burial ground for grazing bills" in the 104th Congress.[75] The Senate Energy and Natural Resources Committee marked up and reported HR 2493 without amendment on July 29, 1998, but the bill did not reach the Senate floor. At that point legislative efforts to make grazing policy less restrictive retreated to the back burner, while environmental reformers once again focused on the administrative arena.

Administrative Policymaking Redux

Finding their efforts to impose across-the-board reforms temporarily waylaid by the courts, the Forest Service and the BLM began targeting individual sites on which to implement ecological improvements. For example, BLM officials proposed in July 1997 to reduce grazing by one-third across 1.3 million acres in Owyhee County, Idaho, and to restrict off-road vehicles to marked trails. In addition, lawsuits filed by the Santa Fe-based Forest Guardians and the Tucson-based Southwest Center for Biological Diversity against both the Forest Service and the BLM prompted the agencies to review the environmental impacts of grazing on hundreds of allotments throughout the Southwest and reduce cattle numbers on many of them.[76] Although some ranchers adapted to land managers' conservation efforts, others were aghast and vowed to resist. Revealing the depth of some locals' antipathy toward federal regulators, Owyhee sheriff Gary Aman warned that federal agents risked being thrown in jail if they ventured into the county to enforce grazing reductions.[77] In another high-profile skirmish, Nevada ranchers reacted with fury to a 2001 BLM roundup of the cattle of two ranchers who refused to pay hundreds of thousands of dollars in back grazing fees and fines. Ranchers and their supporters employed a variety of tactics to thwart land managers, including intimidation and violence. Nor were incendiary tactics limited to ranchers: radical environmentalists began cutting fences and otherwise sabotaging ranching operations.[78]

Ranchers hoped that with the presidency of George W. Bush in 2001 they would regain some of their privileges, particularly after Bush expressed his solidarity with ranchers who resented the interference of land managers in their efforts to make a living and appointed rancher-friendly officials to the interior department.[79] To ranchers' delight, in 2003 the BLM began considering grazing policy changes that, reflecting the Bush administration's antiregulatory stance, "would provide more management flexibility and promote innovative partnerships."[80] The new approach, called the "Sustaining Working Landscapes" initiative, encouraged ranchers to make voluntary improvements to their grazing allotments. In addition, a set of rules proposed in late 2003 sought to limit the BLM's discretion and enhance ranchers' security by giving permit holders part ownership of any fences or wells they installed, allowing them five years to comply with a request to remove more than 10 percent of their cattle, and prohibiting the BLM from improving damaged rangeland in the absence of extensive monitoring data.

In mid-June 2005 the BLM released a draft of its new grazing regulations. The rancher-friendly rules retained the provision giving ranchers an ownership share in fences or other "improvements" to their allotments; required managers to consider the social, economic, and cultural effects of their decisions on communities; required the agency to carry out detailed monitoring before asserting that an allotment does not meet rangeland standards and then allow up to twenty-four months before demanding any change in damaging grazing practices; and phased in over five years any decreases (or increases) of more than 10 percent in the number of livestock allowed on an allotment. Public lands watchdog groups complained that the new rules made life easier for ranchers while making it more difficult for the public to have a say in federal range management. BLM spokesperson Tom Gorey responded, "We believe the changes are going to improve our working relationships with public land ranchers."[81]

But the administration's credibility was severely damaged after reports that in preparing its environmental impact statement (EIS), the BLM had ignored or excised criticisms of the rules from its own as well as EPA and Fish and Wildlife Service (FWS) scientists. According to a sixteen-page FWS report, "The proposed revisions would change fundamentally the way the BLM lands are managed, temporally, spatially and philosophically. These changes could have profound impacts on wildlife resources."[82] The BLM's internal report also warned, "The proposed action will have a slow, long-term adverse impact on wildlife and biological diversity in general."[83] By contrast, the final EIS said that the rule changes would, at worst, harm wildlife only in the short run and only in a few cases. In some cases, it asserted, wildlife would benefit from the change. The report also claimed that riparian areas would remain in the same condition or even improve slightly under the new rules—a finding that directly contradicted concerns expressed by both BLM and EPA experts in their reviews.

Unsurprisingly, when the final rules were published, in mid-July 2006, the Idaho-based Western Watersheds Project promptly sued, charging that the regulations violated NEPA, the Endangered Species Act, and FLPMA. U.S. District Court Judge B. Lynn Winmill agreed; he struck down the rules on the grounds that the BLM had limited public comment and ignored evidence that they could harm rangeland. The Ninth Circuit Court of Appeals upheld the district court ruling in September 2010. While the rules were in court, however, the Bush administration made another rancher-friendly policy change: in mid-August 2008, it granted eight new "categorical exclusions" from a NEPA review to speed up approvals for grazing, logging, oil and gas development, and other activities on federal land. Among the changes was a streamlining of the renewal process for grazing permits.[84]

OUTCOMES

Although ranching allies have managed to prevent major curtailments in public lands grazing policy, annual authorized AUM levels declined between 1970 and the late 2000s, from about 12.8 million to less than 9 million on BLM lands and from around 8.5 million to 6.5 million on Forest Service lands.[85] The number of AUMs authorized was considerably less than the number permitted because of resource protection needs, forage depletion caused by drought or fire, and economic and other factors.[86] It is difficult to translate these numbers into assessments of the ecological health of rangeland, however, because data on rangeland health are inconsistent and difficult to compare over time.[87] In the 1990s the agencies moved away from the "excellent/good/fair/poor" system of evaluating rangeland because they said it was too simplistic and controversial; instead, they began focusing on "ecosystem function" relative to management objectives. Based on the ecosystem function criterion, *Rangeland Reform '94: Draft Environmental Impact Statement* described the condition of vegetation on BLM lands in the following way: of the 86 percent of its land that had been assessed, 67 percent were static or had reached management objectives; 20 percent were moving toward management objectives; and 13 percent were moving away from management objectives.[88] Using the same classification scheme, a Forest Service assessment in 2000 found that by 1997, 49 percent of the allotments it had evaluated met its management objectives; 38 percent were moving toward those objectives; and 13 percent were failing.[89]

Although these reports portrayed BLM and Forest Service rangelands as generally stable or improving, both agencies acknowledged they lacked the funding and the personnel to evaluate their allotments systematically or to repair past damage. The BLM's *Rangeland Reform* '94 asserted, "There is still much progress to be made. Rangeland ecosystems are still not functioning properly in many areas of the West. Riparian areas are widely depleted and some upland areas produce far below their potential. Soils are becoming less fertile."[90] The report concluded that the all-important public land riparian areas "have continued to decline and are considered to be in their worst condition in history."[91] Moreover, the Forest Service's 2000 *RPA Assessment* pointed out that the agency lacked the necessary information, both historical and current, to evaluate broad ecological processes. In November 2004 the Forest Guardians released the results of its own study of more than 6,000 Forest Service records for 1999 to 2003. The group looked at how often the agency was completing the required monitoring and how many allotments were being overgrazed. It found that one-half to three-quarters of all Arizona and New Mexico allotments were out of compliance during that five-year period, either because they were not monitored or because monitoring turned up violations.[92] For their part, federal scientists have struggled to collect systematic information on how grazing has affected the rangeland ecosystems of the West over time, foiled by ranchers and their agency allies who resist such efforts.[93]

Frustrated with the pace and contentiousness of attempts to promote rangeland conservation and discouraged by the polarization between ranchers and environmentalists, in the 1990s some groups began experimenting with a variety of alternative approaches. The National Public Lands Grazing Campaign, a coalition of regional conservation groups, began promoting the Voluntary Grazing Permit Buyout Act, which would compensate public lands ranchers who agreed to relinquish their grazing permits at a price of $175 per AUM. (Under this program a rancher with a federal permit to graze 500 cows for five months—2,500 AUMs—would receive $437,500.) The federal government would then retire the permits and manage the land for its biodiversity values. If passed, the law would formalize an approach that environmental groups such as the Grand Canyon Trust, the National Wildlife Federation, and the Oregon Natural Desert Association have employed for years. In fact, by 2005 the Conservation Fund alone had purchased permits covering 2.5 million acres.[94]

Not all environmentalists were on board with the buyout concept, which they regarded as rewarding ranchers for abusing the land. Similarly, while some ranchers supported such buyouts, traditional ranching associations—such as the National Cattlemen's Beef Association, the Public Lands Council, and the American Farm Bureau Federation—vehemently opposed them, fearing they would lead to the eviction of ranchers from the public lands altogether. "We've never objected to any transfer of ownership of allotments," said Doc Lane of the Arizona Cattle Growers Association. "Now if for some reason Grand Canyon Trust decides they're not going to use the allotments as the law says they're supposed to, then we'd have a problem."[95] To appease these groups, some land managers have required environmental groups who have acquired permits to run cattle on the land, at least at first. Skepticism notwithstanding, in spring 2009 Congress passed the Omnibus Public Land Management Act, which included language enabling conservationists to pay ranchers to retire grazing leases for more than 2 million acres of public land in Oregon and Idaho.[96] However, Congress has not passed more comprehensive buyout legislation, although the idea remains popular among "free market environmentalists."[97]

Other environmentalists have endorsed a different approach that involves collaborating with ranchers to enhance ecological values on their land. One of the earliest and best known collaborative is the Malpai Borderlands Group. This began in the mid-1990s on an 800,000-acre territory in southern Arizona and New Mexico along the U.S.–Mexico border. The group is a nonprofit organization with nine members, eight of whom are ranchers, dedicated to restoring and maintaining "the natural processes that create and protect a healthy, unfragmented landscape to support a diverse, flourishing community of human, plant, and animal life in our Borderlands region."[98] A pivotal member of the group is a representative of the Nature Conservancy, which in 1990 bought the Gray Ranch in southern New Mexico as part of its Last Great Places campaign. The group has assembled scientists, as well as federal and state land managers, to help it stake out what it calls "the radical center." Among other land management techniques, the Malpai Borderlands Group uses controlled burns to rejuvenate the range. It also pioneered the concept of grass banking, a means by which ranchers can trade a promise to protect their land from suburban development in exchange for the right to graze their livestock on grass-rich land during dry years.

The notions of controlled burns, grass banking, rest-rotation grazing, and collaborative management spread throughout the West during the 2000s, and other organizations—including the Six-Six Group, the Northern Lights Institute, and the Santa Fe–based Quivira Coalition—sprang up to promote these ideas. In fall 2000 the Conservation Fund, the Northern New Mexico Stockmen's Association, the Forest Service, the Cooperative Extension Service of New Mexico State University, the Quivira Coalition, and the Malpai Borderlands Group sponsored a conference to introduce the grass bank concept to ranchers and community organizers from seven states. Some groups on both sides of the issue were suspicious of collaborative problem solving, however. Caren Cowan, executive director of the 2,000-member New Mexico Cattle Growers' Association, did not attend the conference, saying, "We don't oppose grass banks as a tool per se, but we don't want anyone to tell us how to run our business."[99] Four years later Cowan remained skeptical that nonranchers could understand the land or that collaboration could be applied broadly.[100] Similarly, hard-line environmental groups such as the Forest Guardians of Santa Fe, the Western Watersheds Project, and the Southwest Center for Biodiversity (now the Center for Biological Diversity) did not attend the conference, and they continued to oppose livestock grazing on public lands altogether.[101]

A third idea that caught on among environmental groups in the 2000s was paying ranchers to preserve the private lands to which federal grazing allotments are attached. Most ranchers are land rich and cash poor, and in order to pay inheritance taxes, their heirs may have to sell off some property, at which point ranching becomes infeasible. To address this problem, environmentalists began buying development rights, an approach that gives ranchers money while reducing the value of their ranches—and hence their taxes. Similarly, the use of conservation easements expanded dramatically in the 1990s and early 2000s: between 1985 and 2005 about 260,000 acres were protected in easements in California alone, with 85 percent set aside between 1995 and 2005.[102] Such measures, although important, captured only a small fraction of the land at risk. For example, the number of land trusts ballooned during the early 2000s, but as journalist Jon Christensen warned, "The tributaries of the land trust movement are being overwhelmed by the great tidal force of development in the West."[103]

A new threat emerged during the Bush administration's tenure that unified ranchers and environmentalists. The acceleration of oil and gas development across the West was denuding vast swaths of desert, polluting water supplies, and spreading noxious weeds.[104] At the same time, however, the potential listing of the sage grouse under the Endangered Species Act (ESA) threatened the viability of both ranching and oil and gas development. Once ubiquitous, the sage grouse—a mottled brown, black, and white bird about the size of a chicken—was declining across eleven western states as the landscape became more fragmented and degraded. Among the threats to the sage grouse, ranching was the most contentious and least well understood.[105] In the 1990s concern about the political and economic implications of listing sage grouse under the ESA had prompted the BLM to sponsor stakeholder working groups in hopes that such initiatives would generate conservation plans sufficient to render a listing unnecessary. The Western Watersheds Project, unimpressed with the progress made using the voluntary approach, filed suit in 2008, challenging eighteen separate resource-management plans devised by the BLM in collaboration with stakeholders and covering about 25 million acres across the West (*Western Watersheds Project v. Dirk Kempthorne, U.S. Department of the Interior*).

While that case was pending, in late June 2010, the Western Watersheds Project, the Center for Biological Diversity, and WildEarth Guardians filed suit against the FWS for delaying ESA protection for the sage grouse. In March 2010, the agency had determined that the grouse warranted protection under the act but that other species faced more imminent threats and therefore had a higher priority. In January 2013, under legal pressure, the FWS finally proposed listing the sage grouse as endangered and setting aside critical habitat for the bird. In November 2014, however, it officially listed the grouse only as threatened—a move that would not force restrictions on landowners or energy companies taking voluntary steps to help the bird.[106] However, President Obama's interior secretary Sally Jewell announced in 2015 a bi-state great sage grouse effort that brought together vested interests to preserve the sage grouse. Because of conservation efforts, it was possible to avoid range listing. Despite these efforts, President Trump has pledged to roll back collaborative efforts to open oil and gas development near sage grouse habitat.[107]

And, as noted previously, national attention focused once again on grazing in the West when President Trump pardoned Oregon ranchers who, in 2014, were in an armed standoff with federal officials. Cliven Bundy and his allies, heavily armed, assembled on the BLM land in Clark County, Nevada. They rounded up hundreds of cattle that Bundy had grazed there illegally for twenty years, racking up $1 million in fees and fines. Bundy defended his resistance by saying that he believed in the sovereign state of Nevada but did not recognize the United States as existing.[108] Although the BLM had the backing of the federal courts, it ultimately backed down rather than engage in a violent confrontation.

CONCLUSIONS

The more than 100-year history of federal grazing policymaking has been a tug-of-war between environmentalists and ranchers. The overall pattern that emerged over the first sixty-five years of grazing management was one in which western members of Congress were able to stave off most reforms and blunt the impacts of those that did pass. In the statutes that govern BLM and Forest Service grazing policy, they mandated a rancher-friendly approach to planning and management, empowering rancher-dominated advisory boards and keeping grazing fees low. When land managers got out of line, ranchers' congressional allies on agency oversight committees used intimidation to promote greater deference—holding hearings to embarrass agency officials; delaying administrative action by moratoria, studies, and reports;

and threatening to cut the agencies' budgets. In a more subtle exercise of control, congressional supporters of the livestock industry contacted agencies' Washington staff and urged them to discipline or even transfer aggressive range managers.

The repeated use of such tactics over time established the context in which environmentalists launched modern grazing policy reform efforts. By the early 1970s ranchers' congressional allies had achieved firm control over the public lands, receiving appropriations by subcommittees in both chambers. They had subdued BLM and Forest Service range managers with their punitive responses to efforts at environmentally protective range management, their unwillingness to raise grazing fees, and their minimal budgetary allocations. As a result, when environmentalists challenged the permissive grazing policy regime, they encountered a dispirited federal bureaucracy and a well-guarded congressional fortress. Unable to provoke widespread public outrage about the deterioration of the western range, and lacking focusing events that might attract public attention, environmentalists consistently found themselves foiled.

The tenacity with which western members of Congress continue to protect their constituents reflects both their convictions and the intense electoral pressure they face. Sen. Malcolm Wallop, R-Wyo., himself a rancher, articulated the rationale for western senators' ardent defense of grazing privileges when he said, "The public lands were reserved for the expansion of the economy. The variety of uses on public lands provides for economic stability."[109] Others were more circumspect, acknowledging the force of constituency pressure. For instance, pro-environment representative Morris "Mo" Udall, D-Ariz., abandoned his proposal to raise grazing fees in the early 1980s, saying, "I haven't seen the light, but I have felt the heat."[110] Apparently, Udall judged ranchers' electoral clout accurately; the National Cattlemen's Association lobbied heavily against Representatives Synar and Darden after they tried to raise grazing fees, and both lost their seats in 1994.

The political power of ranchers is not immutable, however, and signs suggest the possibility of change. Since the 1990s, demographic changes in the Rocky Mountain West have reduced the electoral consequences of opposing ranching interests for representatives in some districts. As the region urbanizes, constituents' demands for outdoor recreation and for ecologically healthy rangeland have intensified, although it is important to note that recreation can be as damaging as extractive use, particularly if it is motorized.[111] Moreover, ranching on public lands continues to be a money-losing proposition from the point of view of U.S. taxpayers: in 2005 the GAO reported that in fiscal year 2004 the BLM and Forest Service together spent about $132.5 million per year on grazing-related activities, while grazing fees generated only about $17.5 million.[112]

Nevertheless, unless environmentalists can muster sufficient political influence in the public lands states to counteract the clout of resources users, the institutions of Congress will provide numerous opportunities to stymie reform. For this reason, perhaps, some of the most promising recent developments in rangeland management are not the result of traditional congressional, or even administrative, reforms;

instead, they have grown out of the voluntary, collaborative ventures springing up in the Rocky Mountain West.[113] These initiatives may not be adequate, however, in the face of a rapidly changing climate that threatens to exacerbate the problems caused by cattle and sheep ranching.[114]

QUESTIONS TO CONSIDER

- What approaches should be adopted for grazing lands to reconcile protection for species and economic concerns for industry?
- How do you think federal land management agencies should deal with the issue of grazing on federal land, given the history of congressional intervention?
- Is overgrazing on federal lands a sufficiently important environmental problem to warrant environmentalists' attention? Why, or why not?
- What should environmentalists do if they want to bring about changes in grazing on the western range, and why?

NOTES

1. Judith Goldstein, "Ideas, Institutions, and American Trade Policy," *International Organization* 42 (Winter 1988): 179–217; Christopher M. Klyza, *Who Controls Public Lands? Mining, Forestry, and Grazing Policies*, 1870–1990 (Charlotte: University of North Carolina Press, 1996).
2. Theda Skocpol, *Protecting Soldiers and Mothers: The Political Origins of Social Policy in the United States* (Cambridge, Mass.: Harvard University Press, 1992), 58.
3. Francis Rourke points out that agencies work to cultivate client groups because the ability to command strong political support is an important source of agency power. Francis E. Rourke, *Bureaucracy, Politics, and Public Policy*, 3rd ed. (Boston: Little, Brown, 1976).
4. Richard F. Fenno Jr., *Congressmen in Committees* (Boston: Little Brown, 1973); Kenneth A. Shepsle, *The Giant Jigsaw Puzzle: Democratic Committee Assignments in the Modern House* (Chicago: University of Chicago Press, 1978).
5. Kenneth A. Shepsle and Barry R. Weingast, "The Institutional Foundations of Committee Power," *American Political Science Review* 81 (March 1987): 85–104; Randall L. Calvert, Matthew D. McCubbins, and Barry R. Weingast, "A Theory of Political Control and Agency Discretion," *American Journal of Political Science* 33 (August 1989): 588–611.
6. Herbert Kaufman, *The Administrative Behavior of Federal Bureau Chiefs* (Washington, D.C.: Brookings Institution, 1981).

7. Keith Hamm, "The Role of 'Subgovernments' in U.S. State Policy Making: An Exploratory Analysis," *Legislative Studies Quarterly* 11 (August 1986): 321–351; James Q. Wilson, *Bureaucracy* (New York: Basic Books, 1989). Wilson argues that the extent to which Congress can constrain a bureaucracy also depends on the type of task the agency performs and the level of support in its political environment.

8. Bryan D. Jones, *Reconceiving Decision-Making in Democratic Politics: Attention, Choice, and Public Policy* (Chicago: University of Chicago Press, 1994).

9. Quoted in Debra L. Donahue, *The Western Range Revisited: Removing Livestock to Conserve Native Biodiversity* (Norman: University of Oklahoma Press, 1999), 2.

10. The term *public domain* refers to land that the federal government had neither disposed of nor set aside in federally managed reserves.

11. Charles F. Wilkinson, *Crossing the Next Meridian: Land, Water, and the Future of the West* (Washington, D.C.: Island Press, 1992), 90.

12. Congress established the 160-acre limit with the Land Ordinance of 1785, which divided western land into 6-mile-square townships, subdivided into 1-mile squares containing four 160-acre plots. The goal was to create a system of small freeholders that would ultimately become a prosperous republican society. According to historian Richard White, "It was an ideal more suited to the East than to the West and more appropriate for the American past than the American future." Richard White, *"It's Your Misfortune and None of My Own": A New History of the American West* (Norman: University of Oklahoma Press, 1991), 142.

13. Wilkinson, *Crossing the Next Meridian.*

14. Ibid.

15. William D. Rowley, *U.S. Forest Service Grazing and Rangelands: A History* (College Station: Texas University Press, 1985).

16. William Voigt Jr., *Public Grazing Lands: Use and Misuse by Industry and Government* (New Brunswick, N.J.: Rutgers University Press, 1976).

17. Samuel T. Dana and Sally K. Fairfax, *Forest and Range Policy*, 2nd ed. (New York: McGraw-Hill, 1980), 137.

18. Rowley, *U.S. Forest Service Grazing.*

19. At the time, the interior department was responsible for disposing of the public domain but did not have any regulatory authority over unclaimed lands. Like any good bureau chief, Ickes wanted to retain control of his "turf" and therefore sought ways to avoid turning over public domain land to the Agriculture Department's Forest Service. Getting the support of ranchers, through rancher-friendly policies, was the key to retaining interior department control.

20. Phillip Foss, *Politics and Grass* (Seattle: University of Washington Press, 1960).

21. E. Louise Peffer, *The Closing of the Public Domain* (Stanford, Calif.: Stanford University Press, 1951), 264.

22. Easterners, disgruntled with the extent to which grazing fees subsidized ranchers (fee receipts were one-fifth of program expenditures), unwittingly collaborated with western,

anti-Grazing Service interests to slash the agency's budget. See William L. Graf, *Wilderness Preservation and the Sagebrush Rebellions* (Savage, Md.: Rowman & Littlefield, 1990).

23. Dana and Fairfax, *Forest and Range Policy*.

24. Ibid.

25. Bernard DeVoto, *The Easy Chair* (Boston: Houghton Mifflin, 1955), 254–255.

26. Quoted in Paul W. Gates and Robert W. Swenson, *History of Public Land Law Development* (Washington, D.C.: U.S. Government Printing Office, 1968), 629.

27. Dana and Fairfax, *Forest and Range Policy*, 186.

28. Voigt, *Public Grazing Lands*, 132.

29. Klyza, *Who Controls Public Lands?*

30. Reed F. Noss and Allen Y. Cooperrider, *Saving Nature's Legacy: Protecting and Restoring Biodiversity* (Washington, D.C.: Island Press, 1994).

31. Quoted in George Wuerthner, "How the West Was Eaten," *Wilderness* (Spring 1991): 28–37.

32. Ibid.

33. U.S. GAO, *Rangeland Management: Comparison of Rangeland Condition Reports*, GAO/RCED-91-191 (July 1991). One prominent dissenter, Thadis Box, pointed out that two issues confounded efforts to assess range conditions. First, he claimed that data on range conditions were old or nonexistent. Second, he argued that range conditions had to be evaluated with respect to some management objective. If the goal was to provide better forage for cattle, the trend for rangelands had, on average, been upward over a number of decades, and the range was in the best condition of the twentieth century. See Thadis Box, "Rangelands," in *Natural Resources for the 21st Century* (Washington, D.C.: American Forestry Association, 1990), 113–118. In 1994 a National Academy of Sciences panel concluded that because many reports on range conditions depended on the judgment of field personnel, rather than on systematic monitoring data, they could not determine whether livestock grazing had degraded western rangelands. See National Research Council, *Rangeland Health: New Methods to Classify, Inventory, and Monitor Rangelands* (Washington, D.C.: National Academy Press, 1994).

34. U.S. GAO, *Public Rangelands: Some Riparian Areas Restored but Widespread Improvement Will Be Slow*, GAD/RCED-88-105 (June 1988); EPA report cited in Wuerthner, "How the West Was Eaten."

35. Harold E. Dregne, "Desertification of Arid Lands," *Economic Geography* 3 (1977): 322–331.

36. Some of the disagreements among scientists over the impact of grazing on the arid western range arise out of differences in their orientation. For decades, range scientists were trained at land grant colleges whose programs were closely affiliated with, and often funded by, the ranching industry. Studies of rangeland ecology, to the extent they existed, were primarily concerned with forage quality and availability. By contrast, conservation biologists are primarily concerned with maximizing biodiversity. For conservation biologists' perspective, see Noss and

Cooperrider, *Saving Nature's Legacy;* Thomas L. Fleischner, "Ecological Costs of Livestock Grazing in Western North America," *Conservation Biology* 8 (September 1994): 629–644. For arguments over whether intensively managed grazing is actually beneficial for rangeland (the "herbivore optimization" theory), see E. L. Painter and A. Joy Belsky, "Application of Herbivore Optimization Theory to Rangelands of the Western United States," *Ecological Applications* 3 (1993): 2–9; M. I. Dyer et al., "Herbivory and Its Consequences," *Ecological Applications* 3 (1993): 10–16; S. J. McNaughton, "Grasses, Grazers, Science and Management," *Ecological Applications* 3 (1993): 17–20; Allan Savory, "Re-Creating the West . . . One Decision at a Time," in *Ranching West of the 100th Meridian,* ed. Richard L. Knight, Wendell C. Gilgert, and Ed Marston (Washington, D.C.: Island Press, 2002), 155–170.

37. U.S. GAO, *Rangeland Management: More Emphasis Needed on Declining and Overstocked Grazing Allotments*, GAO/RCED-88-80 (June 1988).

38. U.S. GAO, *Public Rangelands*.

39. U.S. GAO, *Rangeland Management: More Emphasis Needed*; U.S. GAO, *Rangeland Management: Profile of the Forest Service's Grazing Allotments and Permittees*, GAO/RCED-93-141FS (April 1993).

40. William Kittredge, "Home on the Range," *New Republic,* December 13, 1993, 13–16.

41. U.S. Congress, House Committee on Government Operations, *Federal Grazing Program: All Is Not Well on the Range* (Washington, D.C.: U.S. Government Printing Office, 1986).

42. Private lands in the East, which are far more productive per acre, support 81 percent of the livestock industry, and private lands in the West sustain the remaining 17 percent. The small western livestock industry is quite dependent on public grazing privileges; however, about one-third of western cattle graze on public land at least part of the year.

43. Wuerthner, "How the West Was Eaten," 36.

44. Philip L. Fradkin, "The Eating of the West," *Audubon,* January 1979, 120.

45. Wilkinson, *Crossing the Next Meridian*, 108.

46. An organic act articulates an agency's mission. Recall that the BLM was created as part of an executive reorganization, so it did not have an overarching, congressionally defined purpose.

47. Irving Senzel, "Genesis of a Law, Part 2," *American Forests,* February 1978, 32–39.

48. Klyza, *Who Controls Public Lands*?

49. Ibid.

50. Ibid.

51. PL 95-514; 92 Stat. 1803; 43 U.S.C. Sec. 1901 et seq.

52. Wilkinson, *Crossing the Next Meridian*.

53. R. McGreggor Cawley, *Federal Land, Western Anger: The Sagebrush Rebellion and Environmental Politics* (Lawrence: University Press of Kansas, 1993).

54. Robert H. Nelson, "Why the Sagebrush Revolt Burned Out," *Regulation*, May–June 1984, 27–43.

55. The Reagan administration's primary goal for the public lands was to make the nation's resources more accessible to those who wished to exploit them. (More generally, the administration aimed to reduce government intervention in the economy.) See Robert F. Durant, *The Administrative Presidency Revisited: Public Lands, the BLM, and the Reagan Revolution* (Albany: State University of New York Press, 1992).

56. Ibid., 59.

57. For example, *The New York Times*, *High Country News*, and *People* magazine all covered Forest Service District Ranger Don Oman's story. When Oman, a twenty-six-year veteran of the Forest Service, tried to regulate grazing in the Sawtooth Forest in southern Idaho, ranchers threatened to kill him if he was not transferred. Instead of accepting a transfer, Oman filed a whistle-blower's complaint with the inspector general's office of the Agriculture Department. See Timothy Egan, "Trouble on the Range as Cattlemen Try to Throw Off Forest Boss's Reins," *The New York Times*, August 19, 1990, 1.

58. Klyza, *Who Controls Public Lands?*

59. Cass Peterson, "OMB Urges Freezing Fees for Grazing Federal Land," *The Washington Post*, January 28, 1986, 4.

60. "BLM Reauthorization Died in Senate," 1991 *CQ Almanac* (Washington, D.C.: Congressional Quarterly, 1992), 216.

61. U.S. GAO, *Rangeland Management: Current Formula Keeps Grazing Fees Low*, GAO/RCED-91-185BR (June 1991).

62. Phillip A. Davis, "After Sound, Fury on Interior, Bill Signifies Nothing New," *Congressional Quarterly Weekly Report*, November 2, 1991, 3196.

63. Quoted in Frederick Buell, *Apocalypse as Way of Life* (New York: Routledge, 2003), 23.

64. Quoted in Kate O'Callaghan, "Whose Agenda for America," *Audubon*, September-October 1992, 84.

65. Quoted in Timothy Egan, "Sweeping Reversal of U.S. Land Policy Sought by Clinton," *The New York Times*, July 21, 1995, 1.

66. Ibid.

67. Catalina Camia, "Administration Aims to Increase Grazing Fees, Tighten Rules," *Congressional Quarterly Weekly Report*, August 14, 1993, 2223.

68. Quoted in Steve Hinchman, "Senate Dukes It Out with Babbitt," *High Country News*, November 15, 1993.

69. Catalina Camia, "The Filibuster Ends; Bill Clears; Babbitt Can Still Raise Fees," *Congressional Quarterly Weekly Report*, November 13, 1993, 3112–3113.

70. Ibid.

71. Timothy Egan, "Campaigns Focus on 2 Views of West," *The New York Times*, November 4, 1994.

72. Quoted in Timothy Egan, "Grazing Bill to Give Ranchers Vast Control of Public Lands," *The New York Times*, July 21, 1995, 1.

73. "Grazing Rules Bill Fizzles in House," 1996 *CQ Almanac* (Washington, D.C.: Congressional Quarterly, 1997), Sec. 4, 14–16.

74. Pamela Baldwin, *Federal Grazing Regulations:* Public Lands Council v. Babbitt (Washington, D.C.: Congressional Research Service, 2000).

75. "House Backs Grazing Fee Increase," 1997 *CQ Almanac* (Washington, D.C.: Congressional Quarterly, 1998), Sec. 3, 32.

76. The plaintiffs charged that federal grazing programs were jeopardizing endangered species such as the loach minnow and the willow flycatcher, a songbird.

77. Hal Bernton, "Grazing-Cutback Proposal Meets Trouble on the Range," *The Seattle Times*, July 6, 1997, B8.

78. James Brooke, "It's Cowboys vs. Radical Environmentalists in New Wild West," *The New York Times*, September 20, 1998, 31.

79. Michael Grunwald, "BLM Attacked for Inaction on Tortoise Land," *The Washington Post*, May 12, 2001, 3.

80. See http://www.blm.gov/nhp/news/releases/pages/2003/pr030325_grazing.htm.

81. Quoted in Joe Bauman, "Battle Brewing over BLM's New Grazing Rules," *Deseret Morning News*, June 23, 2005.

82. Quoted in Julie Cart, "Federal Officials Echoed Grazing-Rule Warnings," *Los Angeles Times*, June 16, 2005, 14.

83. Quoted in Tony Davis, "New Grazing Rules Ride on Doctored Science," *High Country News*, July 25, 2005.

84. B. Christine Hoekenaga, "Free Range," *High Country News*, July 16, 2008.

85. U.S. Forest Service, *2000 RPA Assessment of Forest and Range Lands* (Washington, D.C.: U.S. Department of Agriculture, 2001); U.S. Forest Service, "Grazing Statistical Summary FY2013," March 2014, available at http://www.fs.fed.us/rangelands/ftp/docs/GrazingStatisticalSummary2013.pdf; U.S. Department of the Interior, Bureau of Land Management, "Fact Sheet on the BLM's Management of Livestock Grazing," January 30, 2015, available at http://www.blm.gov/wo/st/en/prog/grazing.html.

86. Carol Hardy Vincent, "Grazing Fees: Overview and Issues," Congressional Research Service, June 19, 2012, RS21232.

87. For example, in 2011, the USGS attempted a range-wide survey of the impact of grazing on sagebrush ecosystems managed by the BLM. The agency found that "data collection methodologies varied across field offices and states, and we did not find any local-level monitoring data . . . that had been collected consistently enough over time or space for range-wide analyses." See Kari E. Veblen et al., "Range-Wide Assessment of Livestock Grazing Across the Sagebrush Biome," U.S. Geological Survey, Open-File Report 2011-12630. Available at http://pubs.usgs.gov/of/2011/1263/pdf/ofr20111263.pdf.

88. Donahue, *The Western Range Revisited.*

89. John E. Mitchell, *Rangeland Resource Trends in the United States*, Report No. RMRS-GTR-68 (Fort Collins, Colo.: U.S.D.A. Forest Service, Rocky Mountain Research Station, 2000).

90. Quoted in Donahue, *The Western Range Revisited*, 58–59.

91. Ibid., 59.

92. Tania Soussan, "Report: Federal Land Overgrazed," *Albuquerque Journal*, November 10, 2004, B3.

93. Felicity Barringer, "The Impact of Grazing? Don't Ask," *The New York Times* Green Blog, December 1, 2011.

94. April Reese, "The Big Buyout," *High Country News*, April 4, 2005.

95. Quoted in ibid.

96. Mark Salvo and Andy Kerr, "Ranchers Now Have a Way Out," *High Country News*, April 9, 2009.

97. Shawn Regan, "A Peaceable Solution for the Range War Over Grazing Rights," *The Wall Street Journal*, April 22, 2014.

98. Jake Page, "Ranchers Form a 'Radical Center' to Protect Wide-Open Spaces," *Smithsonian*, June 1997, 50–60.

99. Quoted in Sandra Blakeslee, "On Remote Mesa, Ranchers and Environmentalists Seek Middle Ground," *The New York Times*, December 26, 2000, F4.

100. Carrie Seidman, "Not His Father's Ranch," *Albuquerque Journal*, December 3, 2004, 1.

101. At the heart of the dispute regarding grazing is a difference over whether grazing can be managed in ways that make it ecologically beneficial. Some experts argue that rest-rotation grazing, in which cattle graze an area in short, intensive bursts and then are moved, favors native vegetation because it mimics the behavior of elk and bison that once inhabited the western range. Critics see little scientific support for this claim and point out that it may be relevant only in specific portions of the western range, such as the Great Plains, that evolved in the presence of ungulates.

102. Glen Martin, "Easy on the Land," *San Francisco Chronicle*, July 2, 2006.

103. Christensen observes that the amount of developed land in the thirteen western states rose from 20 million acres in 1970 to 42 million acres in 2000 and that developers prefer the same land that is best for ranching and wildlife: the midelevation, well-watered area. See Jon Christensen, "Who Will Take over the Ranch?" *High Country News*, March 29, 2004.

104. Gail Binkly, "Cowboys Fight Oil and Gas Drillers," *High Country News*, December 9, 2002. Thanks to a complex pattern of western landownership, people who owned the land on which they live often do not own the mineral rights beneath the surface. When the BLM sells leases for those mineral rights, energy companies erect small industrial camps to gain access to the troves of natural gas beneath people's homes. See Timothy Egan, "Drilling in West Pits Republican Policy Against Republican Base," *The New York Times*, June 22, 2005, 11.

105. According to journalist Hal Clifford, many scientists assert that the cow is well documented as the single greatest threat to sage grouse. See Hal Clifford, "Last Dance for the Sage Grouse?" *High Country News*, February 4, 2002; Hal Clifford, "Can Cows and Grouse Coexist on the Range?" *High Country News*, February 4, 2002.

106. Joshua Zaffos, "A Grouse Divided," *High Country News*, December 8, 2014.

107. Available at https://www.sltrib.com/news/environment/2018/05/03/trump-proposes-easing-oil-gas-leasing-restrictions-in-west/.

108. As Matt Ford points out, Bundy's position was ironic given that the Nevada constitution itself recognizes the superior position of the federal government. See Matt Ford, "The Irony of Cliven Bundy's Unconstitutional Stand," *The Atlantic*, April 14, 2014.

109. Quoted in Phillip A. Davis, "Cry for Preservation, Recreation Changing Public Land Policy," *Congressional Quarterly Weekly Report*, August 3, 1991, 2151.

110. Ibid.

111. The 2000 census showed that the West is now nearly as urban as the Northeast, with more than three-quarters of its residents living in cities. Political developments reflect these trends: as the 2000s wore on, even some Republicans began embracing environmental themes—a sharp break from the traditions of the wise use movement. See Blaine Harden, "In West, Conservatives Emphasize the 'Conserve,'" *The Washington Post*, December 2, 2006, 1.

112. U.S. Government Accountability Office, *Livestock Grazing: Federal Expenditures and Receipts Vary, Depending on the Agency and the Purpose of the Fee Charged*, GAO-05-869 (September 2005).

113. For testimonials about collaborative environmental problem solving in the West, see Phil Brick et al., *Across the Great Divide: Explorations in Collaborative Conservation and the American West* (Washington, D.C.: Island Press, 2001); Richard L., Knight, Wendell C. Gilgert, and Ed Marston, eds. *Ranching West of the 100th Meridian: Culture, Ecology, and Economics* (Washington, D.C.: Island Press, 2002).

114. Robert L. Beschta et al., "Adapting to Climate Change on Western Public Lands: Addressing the Ecological Effects of Domestic, Wild, and Feral Ungulates," *Environmental Management* 51 (2013): 474–491. The authors argue that removing or reducing livestock across large areas of public land would alleviate a widely recognized and long-term stressor and make those lands less susceptible to the effects of climate change. A group of critics disputes these conclusions, however, arguing that it is prohibitively difficult to generalize about the impact of grazing given the legacy impacts of homestead-era overgrazing and the potential impacts of climate change. See Tony Svejcar et al., "Western Land Managers Will Need All Available Tools for Adapting to Climate Change, Including Grazing: A Critique of Beschta et al.," *Environmental Management* 53 (2014): 1035–1038.

CHAPTER EIGHT

JOBS VERSUS THE ENVIRONMENT

Saving the Northern Spotted Owl

In the late 1980s the federal government became embroiled in one of the most notorious environmental controversies in the nation's history—a conflict that was ostensibly resolved in the early 1990s but whose impacts continued to reverberate in 2018. At issue was the government's obligation to protect the northern spotted owl, a creature that makes its home almost exclusively in the old-growth forests of the Pacific Northwest. But the debate transcended disagreement over the fate of a particular species; instead, it was yet another eruption of the long-standing confrontation between fundamentally different philosophies about the relationship between humans and nature. It pitted those determined to preserve the vestiges of a once-abundant forest ecosystem against those who feared losing their way of life and the region's historical economic base.

The spotted owl case vividly illustrates the evolving role of science and environmental values in decision making by the nation's natural resource management agencies. Each federal bureau with jurisdiction over the environment has a distinctive orientation that arises out of its founding principles and subsequent development. Congress created many of these agencies, including the U.S. Forest Service and the Bureau of Land Management (BLM), to pursue the dual objectives of conserving natural resources for exploitation by future generations and maintaining the stability of the industries and communities that depend on revenues from exploiting those resources. Such missions predisposed the agencies to treat resources as commodities and pay little attention to their aesthetic or ecological values. As the grazing policy case in Chapter 7 shows, pressure from members of Congress sympathetic to industries that extract resources reinforced this orientation, as did the fact that for many years agency personnel dealt almost exclusively with ranching, timber, and mining interests in formulating policy. In addition, the agencies relied on experts trained in forestry and range sciences, both of which were rooted in the utilitarian goal of maximizing resource yields. A variety of agency socialization practices—such as hiring from one profession, promoting from within, and conducting frequent lateral transfers—further increased the homogeneity of agency professionals.[1]

In the 1970s, however, the Forest Service and the BLM began incorporating environmental science and values into their decision making. The change occurred in part because in order to comply with new environmental and land-management laws, both agencies had to hire more environmental scientists; collect biological data on species, water quality, and other ecosystem amenities; and put more emphasis on considerations such as fish, wildlife, and watershed health. In addition, by the 1980s the agencies were employing a new generation of professionals who were more diverse demographically and had grown up in an era when environmentalism was part of mainstream American culture.[2] In the Forest Service, for example, even though many foresters continued to adhere to a timber philosophy, the infusion of environmentally oriented personnel, combined with directives to take factors other than resource extraction into account in decision making, contributed to a gradual shift in district rangers' and forest supervisors' attitudes and values and hence to changes in the agency's organizational culture.[3] Some agency employees, particularly scientists, began to make more ecologically risk-averse assumptions when generating information on which decisions were based.

Enhancing the influence of environmentally oriented employees on agency decision making was the increasing number, growing vigilance, and expanding clout of environmental groups. Just as the grazing policy case exemplifies the way members of Congress can constrain agencies on behalf of commodity interests, the spotted owl case shows how environmentalists can successfully challenge the dominance of extractive interests in administrative decision making. In hopes of raising the salience of threats to the spotted owl and old-growth forest, and thereby increasing pressure on the agencies to act protectively, environmentalists transformed scientific claims into a compelling story. Frustrated with the slow pace of BLM and Forest Service responses to the owl's plight, they resorted to a tactic that had become a staple of the environmental movement: litigation.

Judicial involvement can change the dynamics of bureaucratic decision making by raising the level of scrutiny to which an agency's calculus is exposed. During the 1970s the courts became more inclined to scrutinize agencies' regulatory decisions, justifying their behavior on the grounds that congressional delegation of vast authority made agencies more susceptible to "capture" by particular interests and that judicial intervention was necessary to ensure fairness.[4] Scholars and judges subsequently debated the merits of reviewing the substance or merely the procedures of agency decision making. But both approaches had the same goal: getting agencies to "elaborate the basis for their decisions and to make explicit their value choices in managing risks."[5] That requirement, in turn, elevated environmentally oriented scientists who could provide solutions to the challenges posed by new statutory requirements.

In addition to forcing agencies to justify their decisions more explicitly, lawsuits can raise an issue's public visibility, which in turn often prompts intervention by high-level political appointees. Such officials, if part of an administration sympathetic to commodity interests, may try to suppress an agency's efforts to incorporate environmental values. But when increased visibility exposes narrow subsystems, it generally benefits previously excluded groups wielding public-spirited arguments. Naturally, defenders of the status quo have not simply acquiesced in the face of environmentalists' challenges to their prerogatives. Forced into the limelight, resource extraction interests have retaliated with arguments about the reliance of regional economies on extractive jobs, the importance of low-cost resources for the nation's well-being, and the ostensible trade-off between economic growth and environmental protection. In response, environmentalists generate economic projections of their own. As with scientific predictions, the two sides make very different assumptions about the likely impact of regulations and therefore reach divergent conclusions.[6] As previous cases have made clear, neither side is persuaded by the other's evidence; rather, the purpose of both sides' efforts is to win over the uncommitted public and persuade elected officials to see the problem (and therefore its solution) the way they do.

BACKGROUND

In the eighteenth, nineteenth, and early twentieth centuries, timber companies, railroads, and homesteaders cleared most of the nation's primeval forest as they moved westward. But for many years the enormity and impenetrability of the nearly 20 million acres of the Pacific Northwest's old-growth forest daunted explorers. To facilitate westward expansion and economic development, in the late 1800s the federal government gave most of the West's most spectacular old-growth forest—the biggest trees on the flattest, most fertile land—to timber companies and railroads. (In all, states and the federal government gave railroads about 223 million acres. The rationale was that the railroads would enhance the value of the surrounding, government-owned land, and subsequent purchasers would pay more for it.) The government also transferred western forestland to the states and to Native Americans. At the same time, Congress began setting aside some acreage in forest reserves, in hopes of averting a repeat of the timber industry's cut-and-run practices in the Midwest. In 1905 Congress established the Forest Service within the Department of Agriculture to administer these newly created national forests.[7] Forest Service Chief Gifford Pinchot articulated the agency's founding principle: "The continued prosperity of the agricultural, lumbering, mining, and livestock interests is directly dependent upon a permanent and accessible supply of water, wood, and forage, as well as upon the present and future use of these resources under businesslike regulations, enforced with promptness, effectiveness, and common sense."[8]

Although the national forests were open to logging, timber harvesting in the Pacific Northwest did not begin in earnest until the lumber requirements of World War I pushed a railroad out to the farming village of Forks, Washington. Then, after World War II, demand for Pacific Northwest timber skyrocketed.

By the mid-1980s, when the spotted owl controversy was coming to a head, private companies had virtually denuded the region's privately owned old growth. (In the early 1980s the upheaval in New York's financial markets had prompted a spate of corporate takeovers, and timber companies with uncut assets became prime takeover targets for raiders, who then clear-cut their holdings to pay off debts.) State land management agencies, responding to state policies to manage their forests for maximum dollar benefits, logged most of the state-owned old growth during the 1980s as well.

As a result of these logging practices, when the spotted owl controversy erupted, nearly 90 percent of the remaining old-growth forest was on federal lands. Federally managed lands fall under a variety of designations and receive varying degrees of protection. The law prohibits logging in national parks and monuments, which are managed by the National Park Service. Congress also has designated roadless portions of the land managed by the Forest Service and the BLM as wilderness and therefore off-limits to timber companies. The protected old growth in both the parks and wilderness areas tends to be on rocky, high-altitude lands, however, so nearly all of the remaining low-elevation old growth was on Forest Service and BLM lands that were eligible for logging.[9] By the mid-1980s many observers were noting with alarm that the old-growth forests on Forest Service and BLM land were going the way of those on state and private property. Despite a host of laws passed in the 1960s and 1970s requiring those agencies to incorporate recreation, watershed, and wildlife concerns into their land management, logging remained the dominant use of federal forests in the region. The primary reason was that the timber industry had become inextricably bound up with the region's economy: in 1985 the industry accounted for almost 4 percent of the workforce in western Oregon and 20 percent of the area's total manufacturing-sector employment; in 1988 the Forest Service estimated that 44 percent of Oregon's economy and 28 percent of Washington's were directly or indirectly dependent on national forest timber.[10] In many small communities and some entire counties, the timber industry was central not only to the economy, but to the culture. In 1990 a journalist described the integral role of logging in Douglas County, Oregon, saying, "Oregon produces more lumber than any state, and Douglas County boasts that it is the timber capital of the world. . . . There one can tune in to KTBR, feel the roads tremble beneath logging trucks, and watch children use Lego sets to haul sticks out of imaginary forests."[11]

The enormous old-growth trees occupied a particular niche within the logging economy of the Pacific Northwest. Timber companies use huge, scissors-like machines called "feller bunchers" to log second-growth forests (which have been harvested and replanted and thus contain smaller trees), whereas harvesting old-growth stands is labor-intensive and relies on highly skilled cutters. Many of the region's small, independent mills use old-growth logs to make specialty products. And old-growth timber is highly profitable; it provides long stretches of clear-grained lumber and therefore sells for triple what second-growth timber is worth. In the 1980s, in about seventy towns in the two states, the local sawmill was the largest private taxpayer and employer.[12] In addition to providing jobs, logging on federal lands generated revenues for local governments. Under the Twenty-Five Percent Fund Act of 1908, the federal government returned 25 percent of revenues derived from timber sales on federal lands to the county in which the forest was located. These revenues were earmarked primarily for schools and roads. In the late 1980s ten Oregon counties earned between 25 percent and 66 percent of their total income from federal timber sales.[13]

Federal law requires the Forest Service and the BLM to harvest trees at a sustainable pace—that is, by the time the last tree of the virgin forest is cut, the first tree of the regrown forest should be big enough to harvest.[14] During the 1980s, however, under pressure from Reagan administration appointees, the agencies accelerated the rate at which timber companies were allowed to cut, and logging substantially exceeded new growth. Congress, hoping to preserve community stability, forced the Forest Service to cut even more than the agency itself deemed sustainable.[15] As the rate of cut increased during the 1980s, environmentalists became convinced that not only were federal land managers ignoring the principles of sustainable yield but that all the trees over 200 years old would be gone in less than thirty years.

THE CASE

Conflict over preserving the spotted owl raged during a fifteen-year period from the mid-1980s through the 1990s, but the issue actually arose a decade earlier when scientists first sounded the alarm. In 1968 a twenty-two-year-old Oregon State University student named Eric Forsman and his adviser, Howard Wight, began to study the biology and ecology of the owl, and they quickly became concerned about Forest Service harvesting practices. By the early 1970s Forsman was pestering anyone who might be able to help him protect spotted owl habitat: the Corvallis City Council, the Audubon Society, the Forest Service, and the BLM. Although federal land management agencies were gradually becoming more receptive to scientists' ecological concerns, their long-standing timber bias circumscribed their willingness to act protectively. Eventually, environmentalists—impatient with the pace of policy change—took the agencies to court to force their hands. The result was a national controversy that pitted protecting species against preserving regional economies.

Land Managers Try to Solve the Problem Quietly, Locally

Reacting to Forsman's inquiries and spurred by the imminent passage of a stringent federal endangered species protection law, the director of the Oregon State Game Commission established the Oregon Endangered Species Task Force. At its first meeting, on June 29, 1973, the task force formally acknowledged the importance of the old-growth forest for wildlife and expressed concern about its disappearance from Oregon. The group also recommended protecting a minimum of 300 acres per nest for northern spotted owls—an estimate based on Forsman's best guess, as no one knew much about either the nongame animals on Forest Service land or the extent of the old-growth forest.[16] But neither the Forest Service—which held about 68 percent of the owl's habitat in northern California, Oregon, and Washington (see Map 8.1)—nor the BLM was particularly interested in abiding by the task force's recommendations; agency managers feared the precedent that setting aside habitat for individual species might establish. So when the task force

Map 8.1 Spotted Owl Habitat

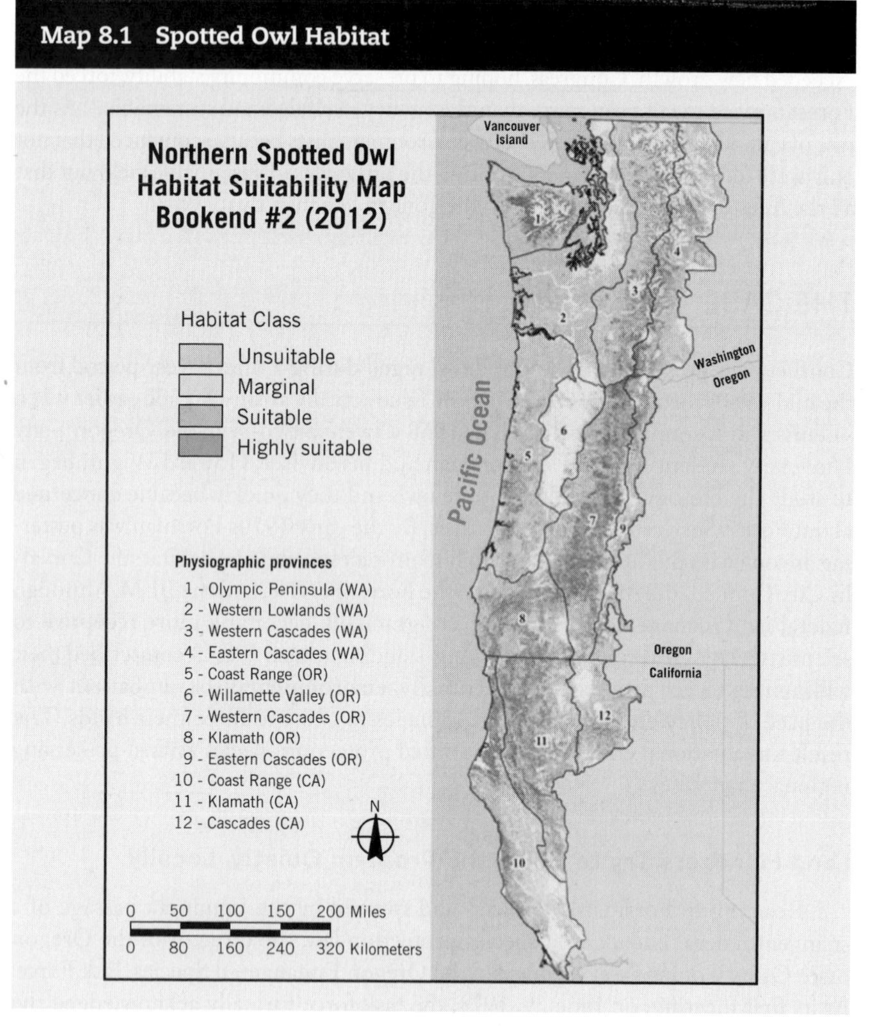

Source: U.S. Forest Service [cartographer]. (2012). Northern Spotted Owl Habitat Suitability Map Bookend #2 (2012). Retrieved from https://www.fs.fed.us/pnw/pubs/pnw_gtr929.pdf

sent plans around, the agencies responded that there were insufficient scientific data on which to base a management decision with such potentially devastating economic consequences.

A number of statutory changes in the mid-1970s complicated matters for the agencies, however. In December 1973 Congress passed the Endangered Species Act (ESA), and the Fish and Wildlife Service (FWS) included the northern spotted owl on a list of potentially endangered species. In addition, the Sikes Act, passed in 1974, required federal land managers to help protect state-listed sensitive species. Finally, both the 1976 National Forest Management Act (NFMA) and the Federal Land Policy and Management Act (FLPMA) required the agencies to develop comprehensive forest- and district-level management plans and formally designated wildlife as a major use of Forest Service and BLM lands. Thus, in October 1977, when the task force recommended that land managers protect 400 pairs of owls (290 on Forest Service land, 90 on BLM land, and 20 on state and private land) by setting aside 300 acres of contiguous old growth for each pair, both agencies reluctantly agreed. They hoped by doing so to avert a decision to list the owl as an endangered species, which would severely curtail their management options.

Local environmentalists were dissatisfied with the degree of owl protection in the agencies' spotted owl plans, however. Led by the Oregon Wilderness Coalition, they filed an administrative appeal with the Forest Service in February 1980 on the grounds that the agency had implemented its plan without preparing an environmental impact statement (EIS) as required by the National Environmental Policy Act (NEPA).[17] The regional forester, whose decision was supported by the chief forester, rejected the appeal on the grounds that the agency had not taken a major federal action (which would require an EIS) but simply an affirmative step to protect the owls until a formal management plan, as required under the NFMA, was complete. The agency declared its intent to conduct a "proper biological analysis" in preparing the official regional plan.

Shortly thereafter, in spring 1981, the BLM became embroiled in a controversy over its owl management plan for the Coos Bay district.[18] When environmentalists appealed the plan, the timber industry retaliated with a local media campaign portraying spotted owl protection as a threat to industry and the region's economy. Under the leadership of Reagan appointees, the BLM was disposed to side with timber, but local agency officials were acutely aware that environmentalists could escalate their demands by petitioning to list the spotted owl as an endangered species. Facing intense pressure from the Oregon Department of Fish and Wildlife, in 1983 the BLM agreed to a compromise in which it would manage the land to maintain habitat sufficient for ninety pairs of spotted owls and revisit the issue within five years.[19]

The Emerging Science of Owls and Old-Growth Forests

Even as Forest Service and BLM managers were wrangling with environmental and timber interests over their owl protection guidelines, agency scientists were requesting more acreage for the owl. In developing its Region 6 Guide, which would direct forest-level planning in the Pacific Northwest, the Forest Service had designated the spotted owl as the indicator species for the old-growth forest ecosystem; in accordance with NFMA regulations, government biologists had begun grappling with analyses of its viability.[20] Additional research had clarified the owl's habits. For example, radio telemetry data—gathered by fitting radio transmitters into tiny backpacks strapped on owls—suggested that it required a much more extensive habitat area than Forsman had initially suspected. By 1981 scientists were suggesting that the agencies expand their owl reserves from 300 acres to 1,000 acres.

In addition to studying the owl, government biologists were learning more about the ecological value of the old-growth forest and biological diversity in general. Forest Service ecologist Jerry Franklin was among the first to draw attention to the potential value of old growth. In 1981 he released the first comprehensive ecological study of the Pacific forest. His research defined a "classic" old-growth forest as one that is more than 250 years old, with enormous trees, big downed logs, and standing snags (dead trees). Franklin noted that some trees in these forests had survived 1,000 years or more and reached heights of 300 feet and diameters of 6 feet. He found that in late-succession forests there was also a vigorous understory, a mixed and layered canopy, and light-filled gaps, as well as an abundance of ferns, moss, lichens, and other epiphytic plants.[21]

Franklin discovered that old growth, once assumed to be biologically barren because it lacked big game, was actually teeming with life. He described a system rich with symbiotic connections. Lichen and moss are not parasites but metabolize nitrogen from the air and feed the trees that support them. Some tree voles eat truffles buried in the forest humus, excreting the undigested spores throughout the forest. The truffles in turn enable the roots of trees to extract nutrients from the soil. Woodpeckers rely on standing dead trees because the wood of young trees is too hard and gums their beaks with sap. Fallen dead wood houses a multitude of amphibians. And the wider spacing of trees in old-growth stands provides flying room for predators and sunlight for a second layer of young trees.[22] Finally, Franklin suggested that, in addition to storing carbon and thereby serving as a hedge against climate change, old growth plays an integral role in regulating water levels and quality, cleaning the air, enhancing the productivity of fisheries, and enriching and stabilizing the soil. While they recognized its ecological significance, however, Franklin and his colleagues were uncertain about how much old-growth forest was needed to sustain the intricate ecosystem of the Pacific Northwest; in fact, they were not even sure how much viable old growth existed.

The Forest Service Region 6 Guide

Notwithstanding these scientific advances, in 1984 the Forest Service produced a Final Region 6 Guide that, according to environmental policy scholar Steven Yaffee, clearly looked like it had not fully incorporated information generated in the previous four years, and . . . was walking a line between what was seen as biologically legitimate and what was seen as politically and economically correct. . . . The level of owl protection at the forest level remained fairly minimal and dependent on the benevolence of the individual forest managers.[23]

Environmental groups filed another administrative appeal in the fall of 1984, challenging the methodology of and management measures contained in the Region 6 Guide. The forestry chief again supported the regional forester's rejection of the appeal, but this time the deputy assistant secretary of agriculture overruled him and required the region to prepare a supplemental environmental impact statement (SEIS) on spotted owl management.

With its plans already behind schedule, the Forest Service conducted its assessment relatively quickly. After intense negotiations among factions within the agency, in July 1986 the service released its two-volume draft SEIS. Within months, the Forest Service had received 41,000 comments on the report, only 344 of which supported its owl recommendations.[24] Environmentalists, aware that they needed powerful scientific ammunition, not just rhetoric, to challenge the agency's preferred approach, had recruited eminent Stanford population biologist Paul Ehrlich and population geneticist Russell Lande of the University of Chicago to create alternative owl viability models. The Forest Service tried a modest revision of the plan, hoping to satisfy both environmentalists and timber interests, but the final SEIS—released in April 1988—did little to quell rumblings on either side.

Pressure Mounts to List the Owl as Endangered

Although many within the agencies were beginning to see the value of the old-growth forest and the spotted owl, to environmentalists and concerned scientists the rate of actual policy change seemed glacial. As scientific knowledge advanced, it became clear that federal managers' actions fell far short of what was needed to protect the owl and its habitat. In fact, forest managers—under pressure from the White House and members of Congress—more than doubled the federal timber harvest in the Pacific Northwest between 1982 and 1988 from 2.5 billion to more than 5 billion board feet.[25] Although mainstream national environmental groups, such as Audubon and the Sierra Club, were concerned about the rapid demise of the old-growth forests, they were reluctant to confront the region's biggest industry. These groups suspected there was little public support in the region for their cause, and they did not want to provoke a backlash against the ESA or the environmental movement more generally. Andy Stahl of the Sierra Club Legal Defense Fund (SCLDF) even made a special trip to Northern California to dissuade a pair of earnest environmentalists from petitioning the FWS to list the owl as endangered.[26]

In October 1986, however, GreenWorld, an obscure environmental organization based in Cambridge, Massachusetts, forced the issue by submitting a petition requesting that the FWS list the northern spotted owl. After finding further action might be warranted, the FWS Region 1 (Pacific Northwest) director assigned three biologists to prepare a status report. But less than a year after the process began, under pressure from the Reagan administration and against the advice of its own biologists, the Region 1 director signed a finding that, although declining in number, the spotted owl was not endangered. Mainstream environmentalists, who had joined the fray by submitting their own petition just months after GreenWorld's submission, were livid at what they charged was a political rather than a science-based decision. In May 1988 the SCLDF sued the interior department on the grounds that the FWS had ignored scientific evidence clearly showing the owl to be endangered in the Olympic Peninsula and Oregon Coast Range. Six months later, Judge Thomas Zilly of the U.S. District Court for the Western District of Washington found that the FWS had acted in an "arbitrary and capricious" manner by failing to demonstrate a rational connection between the evidence presented and its decision. The court gave the service until May 1, 1989, to provide additional evidence.

Affirming Judge Zilly's ruling and further arming environmentalists, in February 1989 a review of the listing process by the U.S. General Accounting Office (GAO, later the Government Accountability Office) found that "Fish and Wildlife Service management substantively changed the body of scientific evidence. . . . The revisions had the effect of changing the report from one that emphasized the dangers facing the owl to one that could more easily support denying the listing petition."[27] The GAO found that two of the three scientists had concluded that the owl was already endangered, but FWS management had ignored them. Administrators also had deleted a section warning that Forest Service logging would lead to the owl's eventual extinction and had excised a twenty-nine-page scientific appendix supporting this conclusion, replacing it with a new report prepared by a forest industry consultant. Finally, although the decision-making process was largely undocumented, the Region 1 director admitted that his decision was based in part on a belief that top FWS and interior department officials would not accept a decision to list the owl as endangered. According to the GAO, "These problems raise serious questions about whether FWS maintained its scientific objectivity during the spotted owl petition process."[28]

Interest Group Confrontations: Lawsuits and Public Relations Campaigns

While the FWS struggled with the listing question and BLM and Forest Service officials tried to navigate a middle course in planning, environmental activists pressed ahead on other fronts. They had already persuaded both Washington and Oregon to designate the owl as a state-listed endangered species, and they were gumming up the timber sales process by appealing hundreds of individual sales.[29] Now they began pursuing injunctions against logging on the federal lands inhabited by the owl, claiming that neither the Forest Service nor the

BLM had satisfied its obligations under the NEPA, the NFMA, the O&C Lands Act, the Migratory Bird Treaty Act, and other laws.[30] In late 1987 the SCLDF represented the Portland Audubon Society and other environmental groups in a lawsuit challenging the BLM's logging of spotted owl habitat in Oregon (*Portland Audubon Society v. Lujan*). In response Portland Judge Helen Frye issued a temporary injunction that halved timber harvesting on BLM lands in Oregon. In 1988 the Seattle chapter of the National Audubon Society, the Oregon Natural Resources Council, and more than a dozen other plaintiffs filed suit challenging the adequacy of the Forest Service's plans to safeguard the owl (*Seattle Audubon Society v. Robertson*). They convinced Federal District Judge William Dwyer to enjoin the Forest Service from conducting timber sales scheduled for 1989. These injunctions had dramatic effects: they temporarily halted more than 150 timber sales throughout the Pacific Northwest, slashing the amount of timber available for harvesting on federal lands in 1989 from 5.4 billion board feet to 2.4 billion board feet.[31]

Environmentalists recognized that, in the long run, they would need more than legal support for their position. Furthermore, they knew the battle could not simply be waged in Oregon and Washington, where timber was a mainstay of the economy. Ultimately, they would have to create a vocal, national constituency for the old-growth forest by persuading millions of Americans that, even if they did not live in the Pacific Northwest, they needed to protect the old-growth forest because it was a national treasure. So they initiated a multifaceted national public relations campaign: they coined the term *ancient forest* to describe the area; sponsored hikes, tours, and flyovers to capitalize on the shocking visual impacts of clear-cutting; wrote articles for national magazines from *The New Yorker* to *National Geographic;* and toured the country with a giant redwood in tow to dramatize the plight of the Northwest's trees.

While mainstream environmental groups pursued these conventional means, more radical groups engaged in guerrilla tactics. Earth First! members, for instance, camped out on plywood platforms in trees scheduled to be cut down. They sat on boxes of company dynamite to prevent blasting, spiked trees, set their feet in cement-filled ditches, chained themselves to timber equipment, and buried themselves in rocks to stop bulldozers from moving up logging roads. Their approach did not always engender sympathy for their cause, but it made mainstream environmentalists appear reasonable by comparison.

Although the timber industry would have preferred to resolve the conflict locally, where its influence was greatest, it did not react passively to environmentalists' efforts to nationalize the issue. Timber workers quickly organized themselves into coalitions, such as the 72,000-member Oregon Lands Coalition. Like the environmentalists, these groups recognized the power of rhetoric and display. On July 1, 1989, the First Annual American Loggers Solidarity Rally came to Forks, Washington, the self-proclaimed timber capital of the world. A reporter for *Audubon* magazine described the scene, as hundreds of logging trucks rolled into town honking their horns,

Yellow balloons and flags and legends. "No timber, no revenue, no schools, no jobs." Over a picture of a mechanical crane: "These birds need habitats, too." On the side of a truck: "Enough is enough!" In the hands of a child: "Don't take my daddy's job." On a sandwich board: "Our Ancient Trees are Terminally Ill."[32]

Speaker after speaker at the rally derided environmentalists as frivolous and selfish.

Political organization was less straightforward for the timber industry than for loggers, however, because the timber industry consists of at least two components with different economic needs and political agendas. Six large timber companies led by Weyerhauser had already logged the old growth on the more than 7 million acres they owned in the Northwest, and their future in the region lay in harvesting managed stands of smaller trees. Therefore, although some leading timber companies had a stake in ensuring access to the old-growth forests, the companies most affected were the small sawmills that were entirely dependent on federal timber for their log supply.

Ultimately, big timber—fearing their lands would be scrutinized next—worked aggressively behind the scenes to counteract environmentalists' pressure. They made substantial campaign contributions to candidates supporting their position and asked their lobbying organizations, the American Forest Resource Alliance and the National Forest Products Association, to assemble evidence supporting their argument that the government should not protect the owl unless and until scientists were certain the bird was endangered. (Forest industry consultants insisted that they were finding northern spotted owls on cutover lands, not just in old growth, but most scientists concluded those findings were relevant to only a handful of forests in Northern California that included some mature elements.[33]) To mobilize popular sentiment, industry groups submitted editorials and took out advertisements in local newspapers. Because environmentalists had nationalized the issue, logging supporters also worked to transform it into a concern of carpenters, builders, and consumers nationwide. They activated allies in industries that relied on cheap wood, such as the National Homebuilders Association, which ran a full-page ad in national newspapers blaming the spotted owl for "soaring lumber prices."

The Region's Elected Officials Get Involved

With the BLM and Forest Service paralyzed by court-ordered injunctions and interest groups battling for public sympathy, politicians from California, Oregon, and Washington began to seek legislative solutions to the impasse. Members of Congress wanted to appease timber interests, but they were loath to propose modifying existing environmental laws, recognizing that such actions would attract negative publicity and therefore garner little support from members outside the region. So in the summer of 1989, as the controversy was heating up, Sen. Mark Hatfield, R-Ore., and Neil Goldschmidt, Oregon's Democratic governor, convened a meeting of timber industry representatives, environmentalists, and federal officials. After an acrimonious debate, Hatfield offered a one-year compromise plan that protected some areas of the forest while allowing old-growth harvesting to continue in others. A major provision prohibited anyone from seeking injunctions to prevent logging. Hatfield and Sen. Brock Adams, D-Wash., discreetly introduced the "compromise" as a rider to the interior department appropriations bill, and President George H. W. Bush signed it into law in October 1989.[34]

The Hatfield–Adams amendment forced Judge Dwyer in Seattle to rescind his temporary injunction and allow timber sales to proceed. It also compelled Judge Frye in Portland to dismiss the pending case against the BLM. During the subsequent nine months, while the SCLDF appealed those rulings to the Ninth Circuit Court of Appeals, more than 600 timber sales went forward, leaving only 16 fiscal year 1990 sales that could be challenged. Congress had found a short-term solution while avoiding the generic problem of reconciling habitat protection with timber harvesting—a strategy legislators hoped would placate timber advocates without unduly arousing environmentalists.

The Thomas Committee Report

The Hatfield–Adams amendment was clearly inadequate as a long-term solution, however, especially because in June 1989 the FWS reversed itself and announced its intention to designate the northern spotted owl as a threatened species under the ESA, thereby compelling the federal government to devise a recovery plan to protect the bird and its habitat. Meanwhile, complying with a provision in the Hatfield–Adams amendment requiring the agencies to develop scientifically credible owl-protection plans, in October 1989 the secretaries of agriculture and interior named an interagency scientific committee to formulate a strategy to save the owl. To enhance the group's legitimacy, the department heads chose veteran Forest Service biologist Jack Ward Thomas as its chair, and thereafter the group was known as the Thomas Committee. Thomas selected for his core team the five foremost spotted owl experts in the world, including Eric Forsman and Barry Noon, a specialist in mathematical modeling of bird populations.

After six months of study, the Thomas Committee recommended creating a network of spotted owl habitat conservation areas on federal lands. The plan prohibited logging in these areas and allowed cutover lands within the reserves to return to old-growth status. It preserved 7.7 million acres of habitat, of which 3.1 million acres was timberland designated for harvest. The remainder was already in national parks or wilderness areas or otherwise too remote, steep, high, or scenic for logging.[35] Heavily influenced by the scientific study of island biogeography, which emphasizes the importance of conserving large, interconnected habitat patches, the plan constituted a radical departure from previous conservation practice. On the other hand, Thomas made it clear that in its deliberations the committee did not consider "how much old growth shall be preserved, where, and in what form."[36] In fact, 41 percent of the land within the habitat conservation areas consisted of recent clear-cuts, nonforested lands and deciduous areas, and private lands (most of which had been logged). The committee acknowledged that its plan constituted a minimum, not an optimal, strategy to prevent the owl's extinction; it would still result in a 40 percent to 50 percent reduction in the owl population over the next century. After a panel of scientists convened by the Bush administration conceded it could not challenge the report's merits, the Forest Service announced its intention to manage its land consistently with the report's recommendations.

The Argument Shifts to Economic Costs

With an authoritative scientific document like the Thomas Committee report on the table, the only recourse open to owl-protection opponents was to demonstrate that implementation costs would be astronomical. Timber industry groups hastened to project the potential impact of the Thomas Plan on timber harvest levels and employment in the region. On July 9, 1990, the American Forest Resource Alliance sponsored a gathering at the National Press Club in Washington, D.C., at which a panel of statisticians contended that over the next decade the plan would cost the region 102,757 jobs. By contrast, a group of experts from the Wilderness Society, the Forest Service, and the Scientific Panel on Late-Successional Forest Ecosystems—known as the Gang of Four and assembled by two congressional subcommittees in 1991—estimated the region would lose between 30,000 and 35,000 jobs as a result of technological change, federal forest plans, and Thomas Committee recommendations combined. The Bush administration quoted a Forest Service projection of 28,000 jobs lost for all three spotted owl states, and some members of Congress cited a figure of only 13,000.[37]

Two factors accounted for most of the differences among these job-loss estimates. First, each faction independently estimated the extent of the timber harvest reductions that forest protection measures would cause. Some analysts assumed that mid-1980s harvest levels would continue in the absence of environmental restrictions, while others recognized that such levels were unsustainable even in the absence of regulations. Second, groups used different estimates to translate changes in the harvest level into changes in employment (the "multiplier"). For example, most analysts agreed that for every billion board feet of timber harvested, an estimated 9,000 jobs were created in direct woods work, milling, and production. But there was far less consensus on how many indirect jobs, such as work in restaurants and stores, resulted. Some put the ratio at 1:1, while others put it at 4:1 or 5:1. As a result, the estimates of job impact for each billion-board-foot change ranged from 18,000 to 50,000.[38]

Moreover, the American Forest Resource Alliance concluded that ESA restrictions would result in significant timber-harvest reductions on private lands, eliminating more than 62,000 additional jobs. By contrast, the Forest Service, the Gang of Four, and the Wilderness Society predicted an increase in private timber harvest levels, as well as a reduction in log exports, in response to harvest declines on public land and rising timber prices. Finally, groups that opposed spotted owl protection predicted a smaller number of jobs lost to technological change in the timber industry (8,000) than did groups supporting owl protection (12,000). (It is noteworthy that *none* of these economic studies contemplated the possibility that protecting the owl and the old-growth forest could spur job creation—in the restoration, recreation, and retirement sectors, for example.)

Regardless of which projections they believed, many commentators foresaw economic and social catastrophe for the region if the government undertook preservation measures. A reporter for *Time* magazine speculated that

> [r]eal estate prices would tumble, and states and counties that depend on shares of the revenue from timber sales on federal land could see those funds plummet. Oregon would be the hardest hit, losing hundreds of millions of dollars in revenue, wages and salaries, say state officials. By decade's end the plan could cost the U.S. Treasury $229 million in lost timber money each year.[39]

A joint Forest Service–BLM study predicted the very fabric holding some communities together would unravel and that "in severe cases of community dysfunction, increased rates of domestic disputes, divorce, acts of violence, delinquency, vandalism, suicide, alcoholism and other social problems are to be expected."[40]

Congress Takes Up the Issue

By 1990 the spotted owl was front-page news across the country, and some ambitious members of Congress crafted legislative measures to protect the old-growth forest in hopes of capitalizing on national environmental concern. The House Interior Subcommittee on National Parks and Public Lands and the House Agriculture Subcommittee on Forests, Family Farms, and Energy each debated forest protection bills in 1990. Neither pleased environmentalists, who argued the timber levels in both bills were too high, or timber lobbyists, who argued they were too low. Unable to move a bill out of committee, it appeared likely that Congress would once again pass a one-year measure specifying timber levels and limiting judicial review.

But on September 19, 1990, plans to attach another rider to the appropriations bill were scrapped when the Ninth Circuit Court of Appeals in San Francisco sided with environmentalists and declared the judicial review provision in the Hatfield–Adams amendment unconstitutional. Judge Harry Pregerson wrote for a unanimous panel that the amendment "does not establish new law, but directs the court to reach a specific result and make certain factual findings under existing law in connection with two cases pending in federal court," which contravenes the principle of separation of powers.[41] It was the first time in 120 years that an act of Congress had been overturned on such grounds, and the unexpected ruling—coming only twelve days before the previous year's timber harvest bill was to expire—left Congress in a bind. Should it waive environmental laws for a year and risk provoking a public furor? Or should it approve a package that would conform to environmental laws but antagonize timber interests?

On September 21 the Bush administration unveiled a plan to break the impasse by doing the former. An administration study group, which included the secretaries of agriculture and interior, as well as representatives of the EPA and the OMB, proposed to exempt timber sales in the Pacific Northwest from the two major laws (NFMA and NEPA) governing Forest Service management. The study group also called for a 20 percent reduction in the 1991 timber harvest, recommending that Congress approve a timber sale program of 3.2 billion board feet in Forest Service Region 6 but permit no timber sales from the habitat conservation areas designated by the Thomas Committee. Finally, the group requested that Congress convene an endangered species committee, or God Squad, to conduct a full review of federal timber sales and land management plans.[42] In yet another sign of the parties' polarization, the plan inflamed both environmentalists and timber interests.

Congress had only ten days from receipt of the administration's proposal to pass a timber program for the next fiscal year. Rather than adopt the administration's approach, congressional leaders chose to rely on the advice of federal scientists and devise their own proposal to allow harvesting of about 3 billion board feet annually, one-third less than the amount offered in 1990. By the end of the session, however, the sponsors of that plan had failed to gain majority support, so yet another year passed with no resolution to the controversy. During the winter and spring of 1991, Congress considered additional alternatives, but opponents managed to tie them up in committees, and the courts continued to be the arbiters of the agencies' timber policy.

The Crisis Escalates

At this point, two events outside of Congress increased the urgency of the spotted owl issue. First, on April 26, 1991, the FWS announced plans to designate as critical habitat, and thus ban logging on, 11.6 million acres of forest in the Pacific Northwest to ensure the owl's survival.[43] Three million of those acres were on private land and included not only old growth but land where old growth might exist in the future to link up habitat areas. Furthermore, unlike the Thomas Committee's plan, the FWS's set-aside did not even include the nearly 4 million acres already reserved in parks and wilderness areas. The federal government had never proposed anything so sweeping in its entire history. Although the FWS later reduced its critical habitat figure to a little less than 7 million acres—under pressure from the timber industry and the BLM to exclude private, state, and Native American lands—the protected area remained considerable.

Second, the courts stopped both agencies' old-growth timber sales. In May 1991 Judge Dwyer issued an injunction blocking all new Forest Service timber sales in the old-growth forest until the agency could present an acceptable plan to protect spotted owl habitat. In October, still unable to engineer a long-term solution, Congress acceded to interior secretary Manuel Lujan's request to convene the God Squad, composed of seven cabinet-level officials, to review forty-four BLM timber sales in spotted owl habitat in Oregon. (In June, the FWS had issued a "jeopardy" opinion under the ESA on 44 of 175 timber sales proposed for 1991 on BLM land in Oregon.) On May 14, 1992, after a month of evidentiary hearings, the God Squad voted to allow logging on thirteen of the forty-four disputed tracts. The BLM was unable to implement the committee's decision, however, because Judge Frye issued an injunction against the agency until it submitted a credible plan to protect the owl. Compounding the problem for policymakers, in July Judge Dwyer in Seattle made his injunction against the Forest Service permanent. He ordered the agency to draft plans protecting not only the owl but a number of other species dependent on the ancient forest, as required by the NFMA. (The Forest Service promptly commissioned a new Scientific Analysis Team to conduct assessments of the viability of all vertebrate species associated with late-successional forests.)

To lift these injunctions, the Forest Service and the BLM had to revise their spotted owl protection plans to satisfy the criteria of both the NFMA and NEPA, or Congress had to pass new legislation to override existing land management laws. The Bush administration offered no salvation. In mid-May Secretary Lujan had released the FWS's official owl recovery plan, which—as mandated by the ESA—sought to revive the owl's population to levels high enough that it could be removed from the list of threatened species. Lujan estimated the revised plan, although more modest than the agency's original proposal, would cost up to 32,000 jobs—a level he called unacceptable. He simultaneously released his own "preservation plan," prepared by an ad hoc committee comprising five political appointees from the Departments of the Interior and Agriculture, which emphasized saving regional timber jobs: it proposed setting aside only 2.8 million acres of timberland and 2 million acres of parks and wilderness. Lujan's plan protected an area that could support only 1,300 pairs of owls, a level most biologists regarded as unsustainable.[44] The Cooper Ornithological Society denounced the Lujan plan as a formula for extinction.[45]

Neither the official FWS plan nor Lujan's alternative received much support in Congress. Environmentally oriented legislators argued that neither would protect the owl or the old-growth forest, while industry supporters complained that both plans would cost too many jobs. In the meantime, for the third year in a row, subcommittees of the House Interior and Agriculture committees drafted ancient forest protection bills. Although an October 1991 report by the expert Gang of Four affirmed the importance of habitat conservation and proffered fourteen land-management options, none of the nine forest-conservation bills considered actually emerged from their respective committees, and Congress failed yet again to pass a comprehensive forest management plan.

The Northwest Forest Plan

By 1992 the spotted owl controversy had become sufficiently visible to be a campaign issue in the presidential race, and the election of Bill Clinton buoyed environmentalists hoping to gain permanent protection for the old-growth forests.

In April 1993, as they had promised on the campaign trail, President Clinton and Vice President Al Gore held a televised summit in Portland, Oregon, with four cabinet members, as well as scientists, environmentalists, timber workers, and industry officials. The president ordered a scientific and technical advisory team—the Forest Ecosystem Management Team (FEMAT)—to come up with a plan that not only would satisfy the warring parties but also would serve as the EIS required by Judge Dwyer.

While FEMAT worked on a new plan, proponents of habitat conservation got an unexpected boost in early May: in an uncharacteristically pointed report, the Society of American Foresters said that cutting trees at their rate of regrowth would not protect the forests over time; instead, the society recommended an "ecosystem approach" to forestry that would base logging decisions on protection of wildlife, water quality, and overall ecological health. Frances Hunt, a forester for the National Wildlife Federation, said of the report, "If you read between the lines, what it is saying is what the profession was taught, and what it helped teach, has turned out to be wrong and we are going to have to make amends for past mistakes."[46]

On July 1 President Clinton unveiled his long-awaited Northwest Forest Plan. Of the variety of forest management options proposed by FEMAT, the one Clinton endorsed—Option 9—allowed annual timber harvests of 1.2 billion board feet from old-growth forests on 24.5 million acres of federal lands, down from a high of more than 5 billion board feet per year in 1987 and 1988, and substantially less than the approximately 3 billion board feet per year of the early 1980s. Option 9 also set up reserve areas for the owl in which logging was limited to some salvage of dead or dying trees and some thinning of new trees, but only if it posed no threat to the species. It established ten "adaptive management" areas of 78,000 acres to 380,000 acres each; in these areas managers could experiment with new forest management approaches, with the goal of better understanding the impact of different types of interventions. And it tried to protect entire watersheds in an attempt to head off controversies over endangered salmon and other fish species.

In addition to its owl protection measures, the Northwest Forest Plan provided $1.2 billion over five years to assist workers and families in Oregon, Washington, and Northern California. It also supported retraining or related logging activities, such as cleaning up logging roads and streams. The White House estimated that 6,000 jobs would be lost immediately under its plan but anticipated that employing displaced timber workers to repair streams and roads would create more than 15,000 new jobs over five years. Most elements of the plan could be implemented administratively, without congressional approval. Thus, the plan shifted the status quo: if Congress wanted to raise timber harvest levels from those designated in the Clinton plan, it would have to change existing environmental laws.

Not surprisingly, critics on both sides immediately lambasted the Northwest Forest Plan: environmentalists thought it was insufficiently protective and objected in particular to the provision allowing timber salvage in owl reserves; timber advocates believed it was overly restrictive. Then, in December 1993, about four dozen scientists released a report providing strong evidence that the spotted owl population was declining and the trend was accelerating. Of particular concern, the scientists said, was that the survival rate of adult females had declined 1 percent per year between 1985 and 1993, a time when capturing and banding of individual owls provided reliable data.[47] According to the report, new data suggested that "the Northwest's old-growth forest ecosystems may already be approaching the extinction threshold for the northern spotted owl" and cast serious doubt on whether the population could survive *any* additional habitat loss. Moreover, the report's authors argued that "political compromise and scientific uncertainty should not be used to justify overexploitation for short-term economic and political gain at the cost of future sustainability."[48]

Scientists' reservations and the timber industry outcry notwithstanding, in early March 1994 the Clinton administration affirmed a slightly amended Option 9 as the final blueprint for the Pacific Northwest and announced its intention to present the plan to Judge Dwyer. For the plan to pass muster, the judge had to find it scientifically sound, so the question before the court was whether the plan represented a scientific consensus or a politically expedient compromise. Judge Dwyer approved Option 9 in December, setting forth a seventy-page opinion that addressed every substantive objection raised by environmentalists and the timber industry and therefore left little room for legal challenge. At the same time, Dwyer made his reservations plain. "The question is not whether the court would write the same plan," he explained, "but whether the agencies have acted within the bounds of the law."[49] He also said that any more logging sales than the plan contemplated would probably violate environmental laws and admonished the government to monitor owl populations carefully in the future. Notwithstanding the judge's caution, the Northwest Forest Plan was finally clear of legal hurdles, and it took effect in early 1995.

The Timber Salvage Rider

Protection advocates' triumph was not only limited but short-lived because Republicans assumed control of Congress in 1995, giving renewed hope to the timber industry. Shortly into the new session Sen. Larry Craig, R-Idaho, introduced a bill to accelerate logging in areas where trees had been damaged by fire or insects but still retained some commercial value (a practice known as salvage logging). The bill eliminated citizens' rights of administrative and judicial appeal and suspended provisions of the ESA and NEPA.[50] It also promoted logging in roadless areas and opened up sections of the forest that had been closed because of spotted owl restrictions. Rep. Charles Taylor, R-N.C., a tree farmer and staunch property rights advocate, hastily assembled a similar bill in the House. Taylor's bill directed the Forest Service to triple its current salvage timber volume over a two-year period, requiring the sale of an unprecedented 6.2 billion board feet of "salvage" timber over two years—an amount approximately double the 1994 yield from the entire national forest system.[51] In tacit recognition that no such quantities of salvage existed, the bill authorized the Forest Service to sell not just dead or dying trees but any "associated" green trees.

The premise of these initiatives was that there was a forest health crisis that needed to be addressed. At hearings on the bill, however, many scientists disputed the existence of such an emergency and challenged the appropriateness of thinning and salvage logging as a remedy for forest ills in any case.[52] Professional organizations, such as the American Fisheries Society, the Society for Conservation Biology, the Wildlife Society, and the Ecological Society of America, protested the salvage program as it was rushed through Congress as a rider to the 1995 Emergency Supplemental Rescissions Act. But according to journalist Kathie Durbin, congressional Republicans used the specter of wildfires to alarm constituents who were largely ignorant of forest ecology, and even skeptical members were concerned about jeopardizing the seats of western Democrats by voting against the rider.[53]

After years of increasingly serious efforts to protect the spotted owl and old-growth forest, federal land managers were stunned by the potential impact of the timber salvage rider.[54] On March 11, 1995, Lydon Werner, head of the BLM timber sale program for western Oregon, wrote a memo to BLM State Director Elaine Zielinski claiming that the BLM "would suffer a severe setback in the implementation of the Northwest Forest Plan [if the rider passed]. We support the need to improve forest health and expedite the salvage of diseased, infested, or dead and dying timber; however we are opposed to this amendment. . . . We believe [the sales] should occur in compliance with existing laws and management plans."[55] Even agency officials who supported salvage logging were dubious about the scale of the program Congress envisioned. "Physically, there's no way we could get it done," said Walt Rogers of the Lowman Ranger District in Idaho.[56]

Responding to a deluge of mail from forest activists and a spate of editorials opposing the timber salvage provision, President Clinton vetoed the budget rescissions act on June 7, citing as one reason the antienvironmental riders. In a series of backroom negotiations orchestrated by Senator Hatfield, however, Clinton eventually reached an agreement with Congress on a budget-cutting bill that included the timber salvage rider. Despite the vehement opposition of top advisers, including Vice President Gore, Clinton signed the measure on July 27. In doing so, he assured environmental activists that his administration would adhere to environmentally sound practices when implementing the rider.

But the bill's congressional sponsors had no intention of allowing that to happen. The day Clinton signed the rescissions act, a group of three senators and three representatives sent a letter to agriculture secretary Dan Glickman and interior secretary Bruce Babbitt reminding them that the rider applied to all unawarded timber sales in western Oregon and western Washington, regardless of their status as endangered species habitat. When President Clinton sent a directive to his department heads to begin implementing the provision in an "environmentally sound manner," Senate authors of the rider went on the attack, berating administration representatives in a hearing and threatening to cut off the agencies' funding. In a series of rulings issued throughout the fall of 1995, U.S. District Judge Michael Hogan affirmed the congressional sponsors' interpretation of the law, dismissing challenges to timber sales released under the rider, and in April 1996 the Ninth Circuit Court of Appeals upheld Judge Hogan's ruling. Senior Clinton administration officials, including interior secretary Bruce Babbitt, admitted that "unprepared White House negotiators gave away the store when they agreed to a so-called timber salvage rider."[57]

Bowing to congressional pressure, the Forest Service began ramping up timber sales east of the Cascade Range, in the Rockies, the Great Lakes, and the Southeastern Coastal Plain. Initially, the response from the timber industry was anemic: the Forest Service did not get a single purchase bid in the first three months after the law was signed. But timber companies responded enthusiastically to the opening up of old-growth stands west of the Cascades. There, logging proceeded on several previously closed stands, over the objections of government and university scientists as well as environmental protestors. The Forest Service estimated that 600 million board feet of old-growth forest—more than triple the amount that had been cut in any of the preceding five years—would be logged under the rider.[58] Although they received relatively little attention in the national press, the sales initiated under the timber salvage rider stimulated tremendous concern in the Pacific Northwest, where environmental activists were up in arms about what they dubbed "logging without laws."[59] Events in the Pacific Northwest in February 1996 made the environmentalists' case even more credible, as flooding caused devastating mudslides in heavily logged parts of Idaho, Montana, Oregon, and Washington. The floods ripped out logging roads, triggered massive landslides, and dumped soil, debris, and giant conifers into streams. During a tour of the Pacific Northwest, after being greeted in Seattle by more than 1,000 demonstrators protesting the timber salvage rider, Clinton called the rider "a mistake" and advocated its repeal.[60]

On June 14, 1996, however, the Ninth Circuit Court of Appeals overturned a decision by Judge Hogan, who had ruled that the habitat of the marbled murrelet, another listed species, could be logged unless a nest was found in the targeted trees.[61] The appeals court ruled that the Forest Service and BLM were required to use scientific criteria in determining where marbled murrelets were nesting. The ruling saved several thousand acres of coastal old growth, as well as stopping four hotly contested sales in the Umpqua and Siskiyou national forests.[62] By early summer, there were signs that environmentally damaging timber sales were slowing down, although congressional advocates continued to push for more liberal fire salvage policies. The prognosis for the owl and the old-growth forest it inhabits was still tenuous but improving. Tom Tuchmann, the Clinton administration's choice to direct the implementation of its Northwest Forest Plan, insisted that the timber sales associated with the timber salvage rider would have a minimal ecological impact in the long run. "What we're talking about is less than 600 million board feet of the last of the old sales," he said. "It's less than one percent of all late-successional and old-growth habitat."[63]

In July 1996 agriculture secretary Glickman acknowledged that the Forest Service had suffered a severe loss of credibility over the timber salvage rider. He announced strict new guidelines that restricted both logging in roadless areas and cutting of healthy trees as part of its "forest health" treatments. The agency subsequently withdrew some of its largest and most controversial roadless area sales, actions that were expected to reduce 1996 timber sale levels by 12 percent in Oregon and Washington alone. In June 1997 Forest Service Chief Mike Dombeck announced that the agency was shifting its focus, saying, "We are in the midst of a profound change—a change of values and priorities. . . . Our challenge is to make watershed health, ecosystem health, the health of the land—whatever you wish to call it—our driving force."[64]

In January 1998, after the timber salvage rider had expired, Dombeck announced an eighteen-month moratorium on logging in roadless areas in the national forests, reiterating that the Forest Service was changing its emphasis from timber extraction to stewardship. Furious, congressional Republicans threatened to reduce the Forest Service to a "custodial role" if it did not manage the national forests primarily for logging. President Clinton was unperturbed, and on November 13, 2000—shortly before leaving office—his administration issued a final rule banning virtually all commercial logging from 58.5 million national forest acres.[65] The incoming administration of President George W. Bush refused to defend the Roadless Area Conservation Rule against an industry challenge in court, and in May 2005 formally repealed it and replaced it with a rule giving governors discretion over decisions concerning inventoried roadless areas in their states. But in August 2009 the Ninth Circuit Court of Appeals rejected the Bush administration's plan and reinstated the Clinton-era rule—except in Idaho and Alaska's Tongass National Forest, where separate rules were in effect as of 2003.[66]

The Bush Administration Relaxes Timber Harvest Restrictions

Although the agencies balked, congressional allies of the timber industry showed no signs of giving up on their efforts to open old-growth forests to logging; between 1996 and 2000, they introduced dozens of "fire salvage" and "forest health" bills. Scalded by the timber salvage rider experience, however, Democrats and moderate Republicans resisted, suspecting these bills were thinly veiled efforts to increase logging in the national forests. After being foiled for years, timber industry supporters were delighted when newly elected President Bush introduced his Healthy Forests Initiative and pledged to double the amount of logging on federal lands in the Pacific Northwest. As political scientists Jacqueline Vaughn and Hanna Cortner point out, the Bush administration was able to shift the emphasis of national forest management by effectively redefining both the problem and the solution. In the administration's version of the story, failure to harvest trees was the problem and logging the solution; administrative appeals were obstacles, rather than opportunities for public participation; environmentalists were a threat to public safety, not guardians of public resources; and restoration, not regulatory rollback, was the objective of the policy process.[67]

The administration also made use of its bureaucratic discretion to redirect forest policy. When President Bush took office, the Forest Service and BLM were harvesting less than half the 1.2 billion board feet projected by the Northwest Forest Plan. In early 2002 the administration announced its intent to raise logging levels in the region by relaxing two rules that were hampering timber sales: the plan's "survey and manage" rule, which required the agencies to safeguard about 350 species linked to old growth, including fungi, lichens, and insects, and its mandate that the agencies maintain extra wide buffers along streams and rivers to protect salmon. Combined, Forest Service Chief Dale Bosworth argued, these provisions would lead to lawsuits and "analysis paralysis." Scientists reacted furiously to the administration's proposals, and in April 2002, 200 prominent biologists and ecologists sent a letter to the White House saying that logging on national forests should be banned altogether. Three months later six scientists fired off a letter to the House Subcommittee on Forests and Forest Health denouncing Bosworth's claims that unwieldy procedures and questionable science were bogging down timber sales. Instead, the scientists wrote, the Forest Service had caused the delays by ignoring the best available science, which supported fire prevention by thinning small trees and reducing roads, and instead pursuing controversial postfire salvage sales.[68] Nevertheless, in March 2004 the administration officially terminated the survey and manage provision and adjusted river and stream protections, on the grounds that doing so would save money.

At the same time, the agencies began implementing the Healthy Forests Restoration Act, which had been passed by Congress and signed by the president in December 2003. Over scientists' objections, the new law allowed logging of big trees to finance thinning of smaller ones; it also limited judicial review of timber sales. Thanks to these provisions, in the five months after the law's passage the government won seventeen straight court cases favoring timber cutting over environmental challenges.[69] Then, in late December 2004, the Bush administration issued comprehensive new rules for managing the national forests. The new rules gave economic activity and the preservation of ecological health equal priority in management decisions. They also eliminated the requirement that managers prepare an environmental impact statement with each forest management plan or use numerical counts to ensure the maintenance of "viable populations" of fish and wildlife under the NFMA. Instead, said Bush administration officials, federal managers would focus on the forest's overall health—although it was unclear how they would measure this.

Even the administration's best efforts were not enough to reverse the trend in the Pacific Northwest toward protecting old growth, however. Environmentalists did not simply acquiesce to the Bush administration's policy changes but instead initiated a new publicity campaign. For example, over the summer and fall of 2003 two former BLM employees toured the country with a 500-pound, six-foot-wide slab of a 400-year-old tree in tow to protest the Healthy Forests Initiative. And environmental groups persisted in challenging timber sales in roadless areas and endangered species habitat. The courts continued to be sympathetic to their arguments. For example, in August 2004 the Ninth Circuit Court of Appeals ruled unanimously that, in formulating its biological opinions on Forest Service and BLM timber sales, the FWS had to apply a stricter standard to protect habitat for the spotted owl and other listed species than merely gauging the "impact on species' survival." The judges pointed out that the ESA was enacted "not merely to forestall the extinction of species . . . but to allow a species to recover to the point where it may be delisted."[70] And in August 2005 U.S. District Court Judge Marsha Pechman in Seattle rejected the Bush administration's elimination of the Northwest Forest Plan's survey and manage rules.

In any case, timber companies' response to proposed logging sales was pallid; for instance, by summer 2005 the largest and most controversial postfire timber salvage plan, following the 500,000-acre Biscuit Fire in Oregon's Siskiyou National Forest, had not come close to generating the 370 million board feet anticipated.[71] Frustrated, the Bush administration in 2007 pushed logging to a high not seen in years: in April, halfway through the budget year, the administration gave forests in Oregon and Washington an extra $24.7 million to boost logging levels, while continuing to cut funding for other activities, such as recreation programs.[72] At the same time, the BLM proposed a major increase in logging on a 2.4 million-acre parcel of forest land known as the Oregon and California Railroad Revested Lands (the O&C Lands) in western Oregon—provoking harsh criticism by scientists who reviewed the plan.

Even as the Bush administration sought to reverse the forest conservation initiatives advanced during the Clinton era, several innovations emerged that promised to defuse some of the conflict over logging in old-growth forests. Efforts by environmentalists and loggers to devise collaborative forest-management plans sprang up, as environmentalists began to adjust their position on logging—at least in places where timberland was threatened by suburban and exurban development. These initiatives followed the model of the Quincy Library Group, a forest management plan developed in the 1990s that sought to balance the demands of forest-protection and timber interests in Northern California's Sierra Nevada. In addition, a handful of timber sales in the Corvallis-based Siuslaw National Forest and in the Mount Hood National Forest employed stewardship contracts. Those agreements, which Congress approved in 2002, allowed loggers to thin trees as long as they simultaneously undertook environmental restoration by erasing deteriorated roads, replacing failed stream culverts, or making other improvements for fish, wildlife, and water quality.[73]

OUTCOMES

The combined impact of the spotted owl injunctions, the Northwest Forest Plan, the timber salvage rider, and the Healthy Forests Restoration Act on the ecological health of the Pacific Northwest forest ecosystem remains unclear. As judges began imposing injunctions in response to environmentalists' lawsuits, timber harvests on federal lands in Washington and Oregon dropped by more than 50 percent—from more than 5 billion board feet in 1987, 1988, and 1989, to just over 2 billion board feet in 1992 (see Figure 8.1 and Table 8.1). The annual timber harvest dwindled steadily from then on: by 2000 it had fallen to 409 million board feet, and between 2001 and 2004 it ranged from 277 million board feet to 500 million board feet. (As of 2010 the Forest Service and BLM had sold about 300 million board feet annually under the Northwest Forest Plan—less than one-third the level allowed.) A FWS survey of owl habitat lost to logging, completed in January 2002, showed that logging had removed only 0.7 percent of the 7.4 million acres set aside under the Northwest Forest Plan, well under the 2.5 percent the plan had projected.[74] And in 2005, after ten years under the plan, scientists found that 600,000 more acres of old-growth forests stood on the west sides of Oregon, Washington, and California compared with 1995. According to journalist Michael Milstein, "The findings reveal that much larger expanses of the region's prized older forests are growing back than are being cut down or burned in wildfires."[75]

Despite this dramatic reduction in logging of its habitat, the spotted owl population continued to decline at a rate faster than forecast in the Northwest Forest Plan. In 1998 the owl population was declining at a rate of 3.9 percent annually, not the 1 percent predicted in the plan.[76] A review of the spotted owl's status completed in 2004 reported that between 1998 and 2003, owl numbers declined only slightly in Oregon and California (about 2.8 and 2.2 percent per year, respectively) but fell so fast in Washington (7.5 percent per year) that the population as a whole fell by 4.1 percent.[77] The review, conducted by the Sustainable Ecosystems Institute of Portland, noted that, although the hazard posed by habitat loss had been reduced by federal protection, other threats loomed: the barred owl, an invader from the Midwest, was preying on spotted owls; West Nile virus could spread to spotted owls; and Sudden Oak Death could attack tanoak, a tree favored by spotted owls in southern Oregon and Northern California.[78] By 2007 scientists were acknowledging that the spotted owl probably would never fully recover.[79]

In hopes of reversing the spotted owl's decline, which was continuing at a rate of 3 percent annually in Oregon and 7 percent in Washington, in 2007 the Bush administration proposed selectively killing more than 500 barred owls, anticipating that it could lure spotted owls back into historic nesting areas.[80] At the same time, however, the FWS under Bush sought to protect logging with a proposal to cut by nearly 25 percent the amount of acreage designated as critical habitat for the owl—from 6.9 million to 5.3 million acres—as part of a legal settlement with the timber industry. In mid-August of that year an agency-ordered peer review charged that in drafting its spotted owl recovery plan, the FWS had ignored the best-available science to justify focusing on the barred owl while reducing protected habitat.[81] So it was not surprising when, in early September 2010, U.S. District Court Judge Emmet Sullivan threw out the Bush administration's plan and ordered the FWS to come up with a new one within nine months; Judge Sullivan also ordered the FWS to redo its critical habitat designation.

Table 8.1 Forest Service Timber Sales in the Pacific Northwest, 1981–2004 (in millions of board feet)

Year	Timber Offered	Timber Sold	Timber Harvested
1981	5,488	5,482	3,382
1982	4,857	4,642	2,264
1983	4,746	4,915	3,868
1984	4,926	4,962	4,539
1985	5,367	4,753	4,760
1986	5,271	5,060	4,965
1987	5,271	5,273	5,597
1988	5,056	4,919	5,408
1989	4,413	2,811	5,231
1990	5,048	3,997	3,879
1991	1,094	2,106	3,166
1992	684	741	2,140
1993	630	787	1,666
1994	436	434	1,127
1995	777	401	877
1996	908	940	776
1997	951	871	768
1998	790	652	662
1999	419	434	570
2000	255	242	409
2001	317	269	307
2002	335	306	277
2003	438	400	321
2004	506	491	500

Source: U.S. Forest Service.

Figure 8.1 Oregon's Timber Harvest and Adult Spotted Owl Trends

Oregon's timber harvest

The spotted owl's listing under the Endangered Species Act in 1990 sharply reduced logging on Oregon's federal forests, which cover nearly 60 percent of the state's forestland. Today, three-quarters of Oregon's timber comes from privately owned forests.

Harvest in billions of board feet

10,000
8,215
8,000
Total harvest
6,000
4,566
4,000
Federal lands
3,441
2,000
323
0
'87 '90 '95 '00 '05 '08

Adult spotted owl trends

For two decades, researchers have counted spotted owls in study areas scattered through the bird's U.S. range from Washington to Northern California. Below are counts, all declining, in six Oregon and Washington study areas that reported results through 2009:

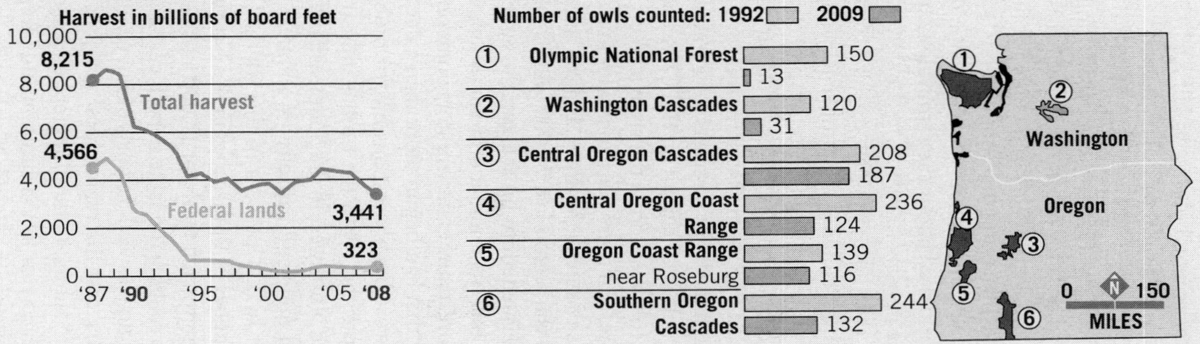

Number of owls counted:

	Study area	1992	2009
①	Olympic National Forest	150	13
②	Washington Cascades	120	31
③	Central Oregon Cascades	208	187
④	Central Oregon Coast Range	236	124
⑤	Oregon Coast Range near Roseburg	139	116
⑥	Southern Oregon Cascades	244	132

① ② ③ ④ ⑤ ⑥
Washington
Oregon
0 150
MILES

Source: U.S. Forest Service [cartographer]. (2012). Northern Spotted Owl Habitat Suitability Map Bookend #2 (2012). Retrieved from https://www.fs.fed.us/pnw/pubs/pnw_gtr929.pdf

Under President Obama the FWS dramatically expanded the amount of critical habitat for the spotted owl to 9.6 million acres across Washington, Oregon, and Northern California—although it allowed some logging within areas designated as critical habitat. More controversially, the administration's revised recovery plan included an experimental project to kill 3,603 barred owls in four study areas over a four-year period. If spotted owls returned to their nests after barred owls were returned, the FWS was likely to kill barred owls over a broader area. Although animal rights activists staunchly opposed killing barred owls, which routinely outcompete spotted owls for habitat, some ethicists argued that barred owls were able to expand into the Pacific Northwest only because humans had drastically modified the landscape, so managing their populations was simply an effort to repair human-caused damage. Eric Forsman, whose research drew national attention to the spotted owl, was conflicted about the removal of barred owls, in part because it would have to be done in perpetuity.[82]

Although the spotted owl's long-term prospects remain uncertain, the economic impacts of reducing the timber harvest are less murky. When the courts ordered restrictions on logging, Oregon mill owner Michael Burrill said, "They just created Appalachia in the Northwest."[83] Many commentators envisioned the timber communities of Oregon and Washington turning into ghost towns, and at first it seemed as though the bleak forecasts were coming true. In April 1990 a *U.S. News & World Report* article on the town of Mill City reported that those loggers who found jobs often had to travel across the state to keep them; their wives were taking jobs for the first time; and there were increases in teenage pregnancy, spouse abuse, the number of calls to suicide hotlines, and the number of runaway children. Between 1989 and 1996 the region lost 21,000 jobs in the forest products industry. Because the spotted owl injunctions hit at about the same time as a national economic recession, however, analysts found it difficult to separate the effects of the two.[84]

Furthermore, many commentators point out that two major structural changes in the industry that had begun more than a decade before the spotted owl controversy erupted were largely responsible for timber-related job losses in the region. First, throughout the 1980s the timber industry had been shifting its operations from the Pacific Northwest to the Southeast: the seven largest forest-products companies reduced their mill capacity by about 35 percent in the Pacific Northwest while raising it by 121 percent in the South.[85] Second, automation dramatically reduced the number of timber jobs in the Northwest: timber employment in Oregon and Washington fell by about 27,000 jobs between 1979 and 1989, even though the harvest was roughly the same in both years.[86] Other factors affected the region's timber employment as well. Exports of raw (unmilled) logs overseas, particularly to Japan and China, cost the region local mill jobs.[87] And during the 1980s and early 1990s urban and suburban development encroached on an average of 75,000 acres of timberland each year in Washington and Oregon.[88]

Other changes in the Pacific Northwest suggested that it was a region in transition away from a timber-based economy toward a more diversified one. Between

1988 and 1994 the total number of jobs in the Pacific Northwest grew by 940,000, and earnings rose 24 percent, according to a 1995 study endorsed by dozens of Northwest economists.[89] After a slowdown in the early 2000s, mirroring the national recession, by 2004 the region's economy was picking up again, and economists were forecasting a rosy future.

Oregon, the most timber-dependent of the spotted owl states, offers some instructive lessons about the complexity of economic projections based on the fortunes of a single industry. In October 1995, three years into a drastic curtailment of logging on federal land, Oregon posted its lowest unemployment rate in more than a generation—just above 5 percent. Although thousands of logging and forest-products jobs disappeared, the state gained nearly 20,000 jobs in high technology, with companies such as Sony opening up factories and Hewlett Packard expanding in the state. By early 1996, for the first time in history, high tech surpassed timber as the leading source of jobs in Oregon. Even some of the most timber-dependent counties in southern Oregon reported rising property values and a net increase in jobs—in part because of an increase in recreation and tourism-related employment. In fact, only two of the thirty-eight counties in the spotted owl region experienced a decline in total employment between 1990 and 1996.[90]

As the number of logging jobs fell, the average wage in Oregon rose. In 1988, the peak year for timber cutting, per capita personal income in Oregon was 92 percent of the national average, but in 1999 it was more than 94 percent of the national average.[91] Conditions continued to improve through the 1990s, and even places such as the remote Coos Bay, once the world's largest wood-products shipping port, cashed in on the high-tech boom.[92] The state made these gains without sacrificing its role as the nation's timber basket, producing 5 billion board feet of mostly second-growth lumber a year. Even though numerous small mills closed because they could no longer obtain the big trees, by the mid-1990s operations like Springfield Forest Products had retooled and were hiring.[93]

When the housing market plunged in the late 2000s, rural areas suffered badly: lumber mills closed, and unemployment in some counties approached 20 percent. According to journalist William Yardley, some residents tried to seize the opportunity to promote a more environmentally benign economy. Mills that once processed old-growth trees were producing energy from wood by-products; the federal stimulus package passed in early 2009 included funding that enabled unemployed loggers to work thinning federal forests. And some local officials were promoting mechanisms to credit forests for absorbing carbon dioxide, the main culprit in global warming (see Chapter 12).[94]

There were costs associated with the transition from a resource-based to a service-based economy, however. In isolated pockets, such as Burns, Oregon, and Hoquiam, Washington, poverty intensified. For example, journalist Jeff Barnard reports on hard times in Josephine County, Oregon, where the last remaining sawmill closed in May 2013. With the expiration of federal subsidies to replace declining revenue from logging, the county was nearing bankruptcy.[95] And while the service economy produced a cleaner industrial base, it also had a dark side: the gap between rich and poor widened. The wealthy bought enormous luxury homes as well as SUVs—which by the late 1990s made up more than half the new vehicles sold in the Northwest—trends that threatened to undermine the environmental gains associated with reduced natural resource extraction.[96] The same trends exacerbated the political polarization that accompanied the urban–rural divide.

However, in 2018, timber communities continue to hold out hope for a return of lumber jobs. During his presidential campaign, Donald Trump expressed his intent to revive the coal and timber industries. Trump lamented, "Timber jobs (in Oregon) have been cut in half since 1990," he said. "We are going to bring them up, folks, we are going to do it really right, we are going to bring them up, OK?"[97] By way of comparison, environmentalists have been wary since President Trump's election, and they remain unsure of how he intends to revamp the industry. Easing restrictions on environmental regulations, or replacing key Forest Service personnel, appear conceivable options.

CONCLUSIONS

The spotted owl challenge was a litmus test for federal land managers trying to adjust to environmentalism. With its spectacular old-growth forests, combined with a historic economic dependence on timber, the region was a powder keg waiting to be ignited, and the spotted owl provided the spark. For more than a decade, the Forest Service and BLM—the two agencies primarily responsible for managing the owl's habitat—wrestled quietly with protecting the bird while maintaining timber harvest levels. Although agency scientists accumulated compelling evidence suggesting the owl and its dwindling old-growth habitat were in trouble, the agencies' long-standing commitments to timber harvesting, as well as political pressure from the region's elected officials, made a dramatic departure from the status quo unlikely. Scientists and environmentally oriented managers found themselves isolated, as political appointees tried to steer policy in directions determined by political expediency rather than science.

Frustrated with the agencies' slow response to science indicating the ecological value of the old-growth forest, environmentalists filed lawsuits in hopes of changing the political dynamics of the issue. Court rulings in Washington and Oregon shifted the status quo radically, from extensive logging to virtually no logging of old-growth forests. They also elevated ecosystem protection relative to timber harvesting and raised the status of agency biologists, who had been overruled in the 1970s and early 1980s but now offered the only way through the impasse at which the agencies found themselves. Court-ordered injunctions also helped environmentalists raise the national salience of preserving old-growth forests, a phenomenon that was manifested in the candidates' attention to the issue during the 1992 presidential campaign, as well as in a burgeoning number of congressional proposals to protect the region's forests.

The national campaign that ensued revealed the political potency of "environment versus economy" rhetoric. Whenever environmental regulations are proposed, advocates of natural resource development publicize projections of massive job losses and dire economic repercussions, and the spotted owl controversy was no exception. Wielding studies conducted at the region's universities, as well as by government agencies and the timber industry, opponents of spotted owl protection measures crafted a powerful case about the human costs of such interventions. Environmentalists responded with their own studies suggesting that owl protection measures were not the primary culprit behind lumber price increases or job losses in the region's timber industry. Neither side took into account the broader changes in the economy of the Pacific Northwest or the industry's dynamic response to the new restrictions, however. Over time, the relationship between environment and economy has proven to be more complicated than either side portrayed it. According to Bill Morisette, mayor of Springfield, Oregon, "Owls versus jobs was just plain false. What we've got here is quality of life. And as long as we don't screw that up, we'll always be able to attract people and businesses."[98]

QUESTIONS TO CONSIDER

- Why do you think the northern spotted owl gained such national prominence?
- Have the Forest Service and the BLM done a good job of handling the demands that they take environmental, and not just extractive, considerations into account when managing the nation's land and natural resources? If not, how might they do better?
- What strategies should environmentalists focus on in trying to redirect the nation's land management agencies, and why?
- How can agencies like the Forest Service better manage missions with conflicting ideals?

NOTES

1. Reed F. Noss and Allen Y. Cooperrider, *Saving Nature's Legacy: Protecting and Restoring Biodiversity* (Washington, D.C.: Island Press, 1994); Herbert Kaufman, *The Forest Ranger: A Study in Administrative Behavior* (Baltimore: Johns Hopkins University Press, 1960); Ben Twight and Fremont Leyden, "Measuring Forest Service Bias," *Journal of Forestry* 97 (1989): 35–41.
2. Greg Brown and Charles C. Harris, "The Implications of Work Force Diversification in the U.S. Forest Service," *Administration & Society* 25 (May 1993): 85–113.

3. Ibid.; Paul A. Sabatier, John Loomis, and Catherine McCarthy, "Hierarchical Controls, Professional Norms, Local Constituencies, and Budget Maximization: An Analysis of U.S. Forest Service Planning Decisions," *American Journal of Political Science* 39 (February 1995): 204–242; Greg Brown and Chuck Harris, "Professional Foresters and the Land Ethic Revisited," *Journal of Forestry* 96 (January 1998): 4–12.

4. Agency capture refers to the situation in which agencies are responsive to a particular interest rather than considering the public interest in decision making (see Chapter 2).

5. David M. O'Brien, *What Process Is Due?* (New York: Russell Sage Foundation, 1987), 159.

6. As legal scholar David Driessen points out, however, both sides conduct their analyses within a "static efficiency" framework—that is, they do not consider the economy's dynamic response to changes in the regulatory climate. For example, predictions rarely capture the impacts of technological innovation or how communities and industries adapt to changes in external conditions. See David M. Driessen, *The Economic Dynamics of Environmental Law* (Cambridge, Mass.: The MIT Press, 2003).

7. Congress actually passed the Forest Service's Organic Act in 1897 but did not establish a management system until it transferred the national forests from the interior department to the agriculture department in 1905.

8. The Pinchot Letter, quoted in Charles F. Wilkinson, *Crossing the Next Meridian: Land, Water, and the Future of the American West* (Washington, D.C.: Island Press, 1992), 128.

9. Forest Service Region 6 (the Pacific Northwest) comprises twelve national forests: the Olympic, Mt. Baker-Snoqualmie, and Gifford Pinchot National Forests in western Washington; the Mt. Hood, Willamette, Umpqua, Rogue River, Siuslaw, and Siskiyou National Forests in western Oregon; and the Klamath, Six Rivers, and Shasta-Trinity National Forests in Northern California. The BLM oversees forestland in six management districts. In Oregon and California, the BLM manages land reclaimed by the federal government during the Great Depression from the O&C Railroad. These lands are exempt by statute (the O&C Lands Act of 1937) from much of the restrictive legislation governing the national forests and are heavily logged.

10. Jeffrey T. Olson, "Pacific Northwest Lumber and Wood Products: An Industry in Transition," in *National Forests Policies for the Future*, Vol. 4 (Washington, D.C.: The Wilderness Society, 1988); U.S. Department of Agriculture-Forest Service, *Final Supplement to the Environmental Impact Statement for an Amendment to the Pacific Northwest Regional Guide*, Vol. 1, *Spotted Owl Guidelines* (Portland: Pacific Northwest Regional Office, 1988).

11. Ted Gup, "Owl vs. Man," *Time*, June 25, 1990, 60.

12. Roger Parloff, "Litigation," *The American Lawyer*, January–February 1992, 82.

13. Ibid. In 2000, Congress passed the Secure Rural Schools and Community Self-Determination Act, which increased federal support to rural communities formerly dependent on logging while decoupling those payments from revenues derived from extractive activities on federal land. (Under the new law, payments came directly from the U.S. Treasury.) But that law expired in 2015, and the payment scheme reverted to the one established in 1908.

14. A sustainable-yield harvest is one that can be regenerated in perpetuity. Sustainable yield is not a static concept, however; as scientists learn more about the factors that affect forest health, they revise their ideas about what harvest levels are sustainable.

15. Research by the Portland *Oregonian* showed that in 1986 Congress ordered the Forest Service to sell 700 million more board feet than the agency proposed; in 1987 it ordered an extra billion board feet; in 1988 the increase was 300 million; and in 1989 it was 200 million. (A board foot is an amount of wood fiber equivalent to a one-inch thick, one-foot wide, one-foot long board. A billion board feet of lumber is enough wood for about 133,000 houses.) See William Dietrich, *The Final Forest: The Battle for the Last Great Trees of the Pacific Northwest* (New York: Simon & Schuster, 1992).

16. Steven L. Yaffee, *The Wisdom of the Spotted Owl: Policy Lessons for a New Century* (Washington, D.C.: Island Press, 1994).

17. An administrative appeal is the first stage in the formal process by which a citizen can force an agency to reexamine and explicitly justify a decision.

18. BLM timber management plans are revised every ten years. The plans due in the early 1980s had to conform to the Sikes Act, the Federal Land Policy and Management Act, and the Endangered Species Act. See Yaffee, *The Wisdom of the Spotted Owl.*

19. Ibid.

20. NFMA regulations require the Forest Service to identify indicator species, whose health reflects the condition of the entire forest ecosystem, and to determine the viability of those species' populations.

21. An epiphytic plant is one that grows on other plants. See Jerry F. Franklin et al., "Ecological Characteristics of Old-Growth Douglas-Fir Forests," General Technical Report PNW-118 (Portland, Ore.: Pacific Northwest Forest and Range Experiment Station, 1981).

22. Ibid. Scientists are uncertain about the extent to which various species depend exclusively on old-growth tracts. Research has yielded ambiguous information, mostly because wildlife biologists have compared young and old "natural" forests but have not systematically compared the robustness of species in natural and managed stands. Recent research suggests that for many species it is not the age of the stand but its structural components that are critical.

23. Steven L. Yaffee, "Lessons About Leadership from the History of the Spotted Owl Controversy," *Natural Resources Journal* 35 (Spring 1995): 392.

24. Yaffee, *The Wisdom of the Spotted Owl.*

25. Timothy J. Farnham and Paul Mohai, "National Forest Timber Management over the Past Decade: A Change in Emphasis for the Forest Service?" *Policy Studies Journal* 23 (June 1995): 268–280.

26. Dietrich, *The Final Forest.*

27. U.S. GAO, *Endangered Species: Spotted Owl Petition Evaluation Beset by Problems*, GAO/RCED-89-79 (February 1989).

28. Ibid.

29. David Seideman, "Terrorist in a White Collar," *Time*, June 25, 1990, 60.

30. Courts issue injunctions on timber sales if a plaintiff proves that "irreparable harm" will occur and if there is a "substantial likelihood" that the plaintiff will prevail at trial.

31. Michael D. Lemonick, "Showdown in the Treetops," *Time*, August 28, 1989, 58–59.

32. John G. Mitchell, "War in the Woods II," *Audubon*, January 1990, 95.

33. Tom Abate, "Which Bird Is the Better Indicator Species for Old-Growth Forest?" *Bioscience* 42 (January 1992): 8–9. For a political interpretation of these scientists' results, see James Owen Rice, "Where Many an Owl Is Spotted," *National Review*, March 2, 1992, 41–43; Gregg Easterbrook, "The Birds: Spotted Owls Controversy," *New Republic*, March 28, 1994, 22–29.

34. A rider is an amendment that is not germane to the law. Sponsors of such amendments use them to avoid public scrutiny.

35. Kathie Durbin, "From Owls to Eternity," *E Magazine*, March–April 1992, 30–37, 64–65.

36. Quoted in ibid.

37. Sylvia Wieland Nogaki, "Log Industry Fears Major Loss of Jobs—Others Dispute Study's Claims," *The Seattle Times*, July 10, 1990, D4; Bill Dietrich, "Experts Say Owl Plan Omits Other Species," *The Seattle Times*, December 18, 1991, D3; David Schaefer and Sylvia Wieland Nogaki, "Price to Save Owl: 28,000 Jobs?" *The Seattle Times*, May 4, 1990, 1.

38. Neil Samson, "Updating the Old-Growth Wars," *American Forests*, November–December 1990, 17–20.

39. Gup, "Owl vs. Man," 57.

40. Quoted in ibid., 58.

41. Quoted in Parloff, "Litigation," 82.

42. The ESA provides for the creation of a cabinet-level endangered species committee (informally known as the God Squad) to grant exemptions from the act's provisions for economic reasons. Congress can pass legislation to convene the God Squad only if an agency can prove to the interior secretary that it has exhausted all other alternatives. (Otherwise, the law requires the federal government to come up with a plan based on science.) In the seventeen years prior to the spotted owl controversy, Congress had convened the God Squad only twice.

43. The ESA requires the designation of critical habitat for any species listed as threatened or endangered.

44. Manuel Lujan Jr., "Bush Plan Protects Both Owl and Logging," *The New York Times*, September 28, 1992, 14.

45. Bruce G. Marcot and Jack Ward Thomas, "Of Spotted Owls, Old Growth, and New Policies: A History Since the Interagency Scientific Committee Report," U.S. Forest Service General Technical Report PNW-GTR-408 (September 1997).

46. Quoted in Scott Sonner, "In Switch, Foresters Push Ecosystem Policy," *The Boston Globe*, May 2, 1993, 2.

47. John H. Cushman Jr., "Owl Issue Tests Reliance on Consensus in Environmentalism," *The New York Times*, March 6, 1994, 28.

48. Quoted in ibid.

49. Quoted in John H. Cushman Jr., "Judge Approves Plan for Logging in Forests Where Rare Owls Live," *The New York Times*, December 22, 1994, 1.

50. More precisely, the bill required the Forest Service to consider the environmental impacts of the sale program but specified in advance that the sales satisfied the requirements of various environmental laws. See U.S. Congress, Senate, *Hearing Before the Subcommittee on Forests and Public Land Management of the Committee on Energy and Natural Resources*, 104th Congress, March 1, 1995 (Washington, D.C.: U.S. Government Printing Office, 1995).

51. Tom Kenworthy and Dan Morgan, "Panel Would Allow Massive Logging on Federal Land," *The Washington Post*, March 3, 1995, 1.

52. Most scientists concurred that many western forests were in trouble, particularly the dry, low- to medium-elevation forests once dominated by fire-dependent species like ponderosa pine and western larch. A history of fire suppression, combined with timber harvesting practices, had dramatically altered such forests. But although in some forests thinning and salvage logging may have been appropriate, scientists maintained that there was no "one size fits all" prescription. See Tom Kenworthy, "Forests' Benefit Hidden in Tree Debate," *The Washington Post*, April 18, 1995, 1.

53. Kathie Durbin, *Tree Huggers: Victory, Defeat, and Renewal in the Ancient Forest Campaign* (Seattle: Mountaineers, 1996).

54. The Forest Service had undertaken a comprehensive environmental study of the ten national forests in the area and, pending completion of that study, recommended interim protection for the remaining patches of old-growth trees. The agency's scientific panel wanted a comprehensive study of the effectiveness of thinning and salvage before undertaking such a program. See Kenworthy, "Forests' Benefit Hidden."

55. Quoted in Durbin, *Tree Huggers*, 255.

56. Quoted in Kenworthy, "Forests' Benefit Hidden."

57. Quoted in Susan Elderkin, "What a Difference a Year Makes," *High Country News*, September 2, 1996.

58. Timothy Egan, "Clinton Under Attack by Both Sides in a Renewed Logging Fight," *The New York Times*, March 1, 1996, 10.

59. Brad Knickerbocker, "The Summer of Discontent for Greens, Monks in West," *The Christian Science Monitor*, July 24, 1995, 3.

60. Egan, "Clinton Under Attack by Both Sides in Renewed Logging Fight."

61. The marbled murrelet, a seabird, also relies on old-growth forest and was listed as threatened in 1992.

62. Durbin, *Tree Huggers*.

63. Quoted in ibid., 263. Only about half of the total national forest harvest in the Pacific Northwest between 1995 and 1997 came from the western forests covered by the Northwest Forest Plan.

64. Remarks of Mike Dombeck, Outdoor Writers Association of America Public Lands Forum, June 25, 1997.

65. Between the draft and the final rule, the administration added roadless areas in the Tongass National Forest in Alaska to the ban. See Hal Bernton, "Forest Chief Asks Reduced Logging," *The Oregonian*, March 30, 1999, B1; Douglas Jehl, "Expanded Logging Ban Is Proposed for National Forests," *The New York Times*, November 14, 2000, 8. The final rule was published in the *Federal Register* on January 12, 2001.

66. In October 2011 the Tenth Circuit Court of Appeals also upheld the Roadless Rule, and the following year the Supreme Court declined to review the legality of the rule. (In 2013 the D.C. Circuit Court ended a challenge by the State of Alaska to the Roadless Rule nationwide, saying the statute of limitations had run out.) Meanwhile, in March 2011 Alaska challenged the application of the Roadless Rule on the Tongass, after the Alaska District Court vacated that exemption and reinstated the rule on the Tongass. In mid-December 2014 the Ninth Circuit reheard that case *en banc*, but as of early 2015 the court had not rendered a decision.

67. Jacqueline C. Vaughn and Hanna J. Cortner, *George W. Bush's Healthy Forests: Reframing Environmental Debates* (Boulder: University Press of Colorado, 2005).

68. Michael Milstein, "Scientists Chastise Forest Service Chief," *The Oregonian*, July 27, 2002, B1.

69. Matthew Daly, "Forest Cases Going Cutters' Way Under New Law," Associated Press, May 20, 2004.

70. Quoted in Joe Rojas-Burke, "Ruling on Spotted Owl May Hinder Logging," *The Oregonian*, August 8, 2004, D3.

71. Jeff Barnard, "Three Years After, Salvage from Biscuit Fire Limping Along," Associated Press, July 13, 2005.

72. Michael Milstein, "Cash Infusion Accelerates NW Logging," *The Oregonian*, August 9, 2007, 1.

73. Jim Kadera, "Timber Sale Has Potential for Peace," *The Oregonian*, December 30, 2004, 1.

74. Michael Milstein, "Federal Forests in Line to See More Logging," *The Seattle Times*, January 16, 2002, 1.

75. Michael Milstein, "Old-Growth Forests Gain Ground," *The Oregonian*, April 18, 2005, B1. About 17,300 acres of old growth was clear-cut during this period; about 287,000 acres of national forest were thinned in the region; and wildfires burned about 101,500 acres of old growth.

76. Associated Press, "Owl Disappears," July 10, 2000, ABCNews.com.

77. "Analysis Shows Northern Spotted Owl Still Declining," Associated Press, May 12, 2004.

78. Timber industry spokespeople seized on these threats to argue that harvest restrictions were not scientifically justified given the other threats to the owl's survival. See Hal Bernton, "Spotted Owl Faces Nonlogging Threats," *The Seattle Times*, June 23, 2004, B1.

79. Michael Milstein, "So Much for Saving the Spotted Owl," *The Oregonian*, July 29, 2007, 1.

80. Michael Milstein, "Owl vs. Owl Sets Stage for Intervention," *The Oregonian*, April 27, 2007, B1. Barred owls are larger, more aggressive, and more adaptable than spotted owls; they are less particular about their prey, so they require smaller territories, and they nest more often than spotted owls.

81. Dan Berman, "FWS's Spotted Owl Plan Based on Shaky Science—Review Panel," *Greenwire*, August 14, 2007.

82. Jeff Barnard, "Feds Plan to Shoot Barred Owls," *The Seattle Times*, September 23, 2013; Isabelle Groc, "Shooting Owls to Save Other Owls," *National Geographic*, July 17, 2014.

83. Quoted in Timothy Egan, "Oregon Foiling Forecasters, Thrives as It Protects Owl," *The New York Times*, October 11, 1994, 1.

84. Ed Niemi, Ed Whitelaw, and Andrew Johnston, *The Sky Did NOT Fall: The Pacific Northwest's Response to Logging Reductions* (Eugene, Ore.: ECONorthwest, 1999). Niemi and his coauthors estimate that about 9,300 workers in Washington and Oregon lost their jobs between 1990 and 1994 as a consequence of spotted owl restrictions. See also William R. Freudenburg, Lisa J. Wilson, and Daniel J. O'Leary, "Forty Years of Spotted Owls? A Longitudinal Analysis of Logging Industry Job Losses," *Sociological Perspectives* 41 (February/March 1998): 1–26. Freudenburg and his coauthors find no statistical evidence for a spotted-owl effect on timber employment in the Pacific Northwest.

85. "Log On," *The Economist*, November 5, 1992, 26.

86. Niemi et al., *The Sky Did NOT Fall.* Moreover, because industry had forced the timber workers' unions to take a pay cut during the 1980s, in 1989 timber workers received paychecks that were less than two-thirds of those they had received a decade earlier.

87. Recognizing the severe impact of raw log exports, the federal government banned such exports from federal and state lands in the West, but timber companies circumvented the ban by substituting logs from private land for export and cutting national forest timber for domestic sale. See Michael Satchell, "The Endangered Logger," *U.S. News & World Report*, June 25, 1990, 27–29.

88. Dietrich, *The Final Forest*.

89. Thomas M. Power, ed., "Economic Well-Being and Environmental Protection in the Pacific Northwest: A Consensus Report by Pacific Northwest Economists" (Economics Department, University of Montana, December 1995). Another study found 27 percent and 15 percent increases, respectively, in the region's total employment and per capita income. See Niemi et al., *The Sky Did NOT Fall*.

90. Niemi et al., *The Sky Did NOT Fall.* Sociologist William Freudenburg and his coauthors note that "the period since the listing of the spotted owl has . . . been one of soaring job growth in the Northwest." See Freudenburg et al., "Forty Years of Spotted Owls?"

91. Oregon.gov, "Per Capita Income, US: Oregon's Per Capita Personal Income Expressed as a Percentage of the US Per Capita Personal Income." Available at http://www.oregon.gov/10yearplan/Economy_Jobs/Pages/Personal_Income_US.aspx.

92. Sam Howe Verhovek, "Paul Bunyan Settling into His New Cubicle," *The New York Times*, August 21, 2000, 1.

93. Egan, "Oregon Foiling Forecasters."

94. William Yardley, "Loggers Try to Adapt to Greener Economy," *The New York Times*, March 28, 2009.

95. Rob MacWhorter, supervisor of the Rogue River-Siskiyou National Forest said the Forest Service there was focused on restoration and reducing the threat of wildfire. Whereas his predecessors spent their days filling timber quotas, he met regularly with community groups looking to diversify the local economy. See Jeff Barnard, "Ore. Timber Country Ponders Future With Fewer Logs," *The Seattle Times*, May 18, 2013.

96. Alan Thein Durning, "NW Environment Hurt by Raging Consumerism," *Seattle Post-Intelligencer*, July 13, 1999, 9.

97. Darling, D. (2016, November 14). "Timber Interests Welcome Trump Win; Environmentalists Worry." *The Washington Times*. Retrieved from https://www.washingtontimes.com/news/2016/nov/14/timber-interests-welcome-trump-win-environmentalis/.

98. Quoted in Egan, "Oregon Foiling Forecasters."

CHAPTER NINE

PLAYGROUND OR PARADISE?

Snowmobiles in Yellowstone National Park

In the winter of 1963 Yellowstone National Park staff allowed the first snowmobiles to enter the park. By the mid-1990s the National Park Service was reporting that wintertime air quality in some parts of the park was the worst in the nation and that the drone of snowmobiles was perpetually audible at Old Faithful, the park's premiere attraction. In fall 2000, after five years of studying the issue, the Clinton administration decided to phase in a ban on snowmobiles in Yellowstone and most other national parks on the grounds that allowing them was inconsistent with the Park Service mission as well as with laws and regulations governing park management. Just a day after his inauguration in 2001, however, President George W. Bush suspended the Clinton-era ban, and the following year he substituted a plan that limited the maximum number of snowmobiles entering the park but allowed an increase in their daily average number. Eventually, the courts threw out the Bush-era winter use plan, and the administration of President Barack Obama ushered in a new planning process.

The case of snowmobiles in Yellowstone exposes one of the schisms that have arisen among self-described environmentalists. Both sides claim to appreciate the amenities nature provides—wildlife, scenic beauty, and open space—but their values are actually quite different. For motorized recreation enthusiasts, the nation's public lands, including the parks, are playgrounds; nature is theirs to enjoy, and improved technology can resolve any problems posed by human use. By contrast, those who prefer passive recreation value nature for its own sake and believe managers should limit the form and amount of access to ensure the health of the parks' flora and fauna. This division poses a serious dilemma for environmentalists: on the one hand, they espouse the view that contact with nature fosters appreciation and concern; indeed, that is part of the rationale for establishing such preserves. On the other hand, they worry that overly intensive use by large numbers of people threatens the integrity and long-term survival of natural areas.

In many respects this debate resembles the grazing (Chapter 7) and spotted owl (Chapter 8) cases. As is true in nearly all instances of natural resource policymaking, efforts to protect ecological values—even in national parks—face enormous obstacles. Once a resource-dependent activity becomes entrenched, its beneficiaries believe they have rights to the resource. Furthermore, the businesses that provide services to resource users claim that the local economy depends on their prosperity. Elected officials feel compelled to support those with a direct economic stake, who—although often a minority numerically—tend to be organized and mobilized. Protection advocates hoping to overcome these forces rely heavily on the courts' willingness to interpret agencies' statutory mandates—which are often ambiguous—in protective ways and on scientific evidence that the activity they hope to curb or eliminate damages human health or the environment.

In addition to exemplifying the dynamics of natural resource politics, this case illustrates both the importance of agency rules as policy instruments and the extent to which presidents and their appointees can use their rulemaking power to influence the content of public policy. Many scholars emphasize the difficulties presidents face in trying to change policy and bureaucratic behavior. Richard Neustadt, an influential scholar of the presidency, asserts that in the fragmented U.S. government system a president's power lies in the ability to persuade, not command.[1] Political scientists Harold Seidman and Robert Gilmour emphasize the difficulty for the president's political appointees of steering agencies, which have "distinct and multidimensional personalities and deeply ingrained cultures and subcultures reflecting institutional history, ideology, values, symbols, folklore, professional biases, behavior patterns, heroes, and enemies."[2] But other scholarship suggests that presidents have more influence over administrative decision making than once believed. Working in the tradition of "new institutionalism," political scientist Terry Moe argues that, because they can make unilateral decisions, presidents actually have considerable power over the bureaucracy. According to Moe, "If [the president] wants to develop his own institution, review or revise agency decisions, coordinate agency actions, make changes in agency leadership, or otherwise impose his views on government he can simply proceed—and it is up to Congress (and the courts) to react."[3] Scholars have also noted that presidents have "important and practical legal powers, and the institutional setting of the presidency amplifies these powers by enabling presidents to make the first move in policy matters, if they choose to do so."[4] These scholars have described the various ways that presidents historically and with increasing frequency have used their "powers of unilateral action" to shape policy in substantial ways.[5]

The case of snowmobiles in Yellowstone also demonstrates how advocates on both sides of environmental policy disputes use the courts strategically to challenge administrative decisions. After environmentalists and other public interest groups achieved many of their early successes through lawsuits in the 1960s and 1970s, conservative organizations quickly mobilized and adopted the same tactic.[6] As a result, "a wide range of groups regularly resort to the judicial arena because they view the courts as just another political battlefield, which they must enter to fight for their goals."[7] Advocates on both sides cite procedural requirements to challenge agency decisions they disagree with. For example, opponents of environmentally protective regulations have begun using the National Environmental Policy Act

(NEPA), the darling of environmentalists, to delay or block administrative decisions. To the extent possible, advocates also engage in forum shopping. This simply means selecting the court in which they expect to fare best because they recognize that judges' attitudes are important predictors of how they will rule.[8] Precedent, statutory language, and how lawyers frame an issue all constrain judicial decision making. In addition, judges are sensitive to the political context; because they cannot enforce their own rulings, they need to make decisions that are likely to be carried out in order to maintain their legitimacy. But judges' decisions also reflect their values and ideology.[9]

BACKGROUND

Yellowstone National Park, established in 1872, is the nation's oldest park and among its most spectacular. But from its inception, the purpose of Yellowstone was unclear: was it to be a playground or a shrine for nature? More than forty years after founding Yellowstone, Congress created the National Park Service (NPS) to manage the nation's inchoate park system and, in the process, reinforced this tension by endowing the agency with an ambiguous mandate: to encourage visitation while at the same time protecting the parks' natural amenities in perpetuity. After World War II, as the number of people using the national parks surged, the condition of the parks began to decline precipitously.

The Origins of Yellowstone National Park

The Yellowstone region, which spans Wyoming, Montana, and Idaho, was occupied for thousands of years before Europeans "discovered" it. Although little is known about how the area's early human inhabitants modified the landscape, it is clear that they affected it in a variety of ways: they set fires to manipulate vegetation and animals, used the plentiful obsidian left over from volcanic activity to make arrowheads, and hunted large game animals. At the same time, as former Yellowstone park ranger and historian Paul Schullery notes, there is "overwhelming evidence that most of the tribes that used the Yellowstone area . . . saw it as a place of spiritual power, of communion with natural forces, a place that inspired reverence."[10]

Reports from some early white visitors to the region suggest that, like the native people, they were awed by its spectacular and peculiar geological features. The first white man known to have visited Yellowstone was John Coulton, a member of the Lewis and Clark Expedition, who traveled there in 1807 and 1808. Subsequently, trappers and a handful of explorers spent time in the area, followed by prospectors looking for gold. Although few of these visitors kept records of their experience, a handful of later explorers did. For example, in a retrospective account of the 1870 Washburn-Langford-Doane Expedition published in 1905, Nathaniel Langford described the region in the following way: "I do not know of any portion of our country where a national park can be established furnishing visitors more wonderful attractions than here. These wonders are so different from anything we have seen—they are so various, so extensive—that the feeling in my mind from the moment they began to appear until we left them has been one of intense surprise and incredulity."[11] According to Schullery, in other early depictions of the region, "The weirdness fascinates and attracts us, and then the beauty rises to awe and stun. All is brilliant light and ominous shadow, alternative dazzling prismatic displays with the 'dark, dismal, diabolical aspect' of each place."[12]

The official NPS myth is that the idea for a park at Yellowstone emerged during a Washburn expedition camping trip. According to a sign at the park's Madison Junction, the members of the expedition "were grouped around a campfire discussing their explorations when Cornelius Hedges suggested the region be set apart as a national park."[13] In March 1872, after a year in which the members of the Washburn expedition energetically promoted the idea, President Ulysses S. Grant signed the bill establishing Yellowstone National Park, the world's first national park. From the outset the park's purpose embodied two contradictory ideals. Congress decreed that 2.2 million acres be "reserved and withdrawn . . . dedicated and set apart as a public park or pleasuring ground for the benefit and enjoyment of the people" (see Map 9.1). The area was to be managed for "preservation from injury

Map 9.1 Yellowstone National Park

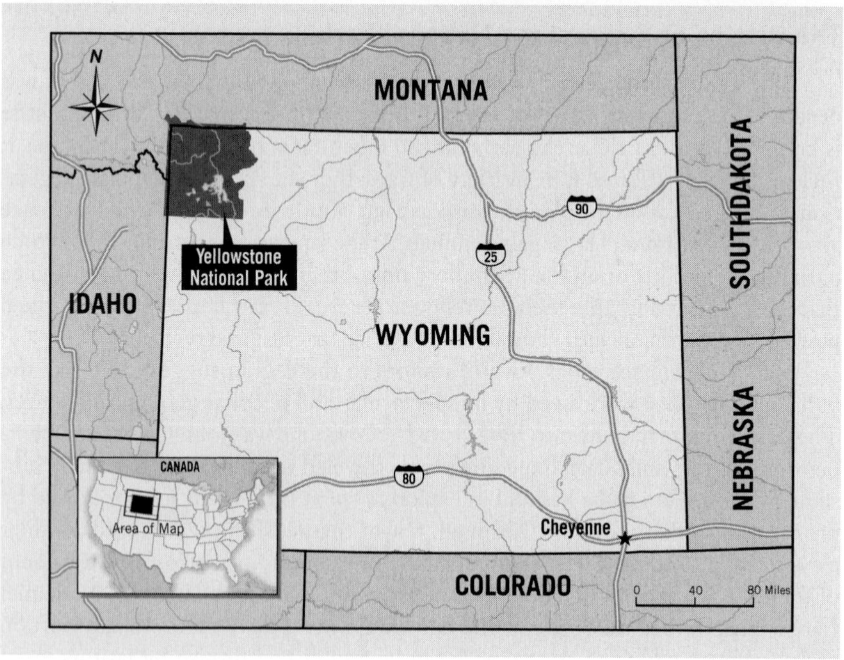

Source: National Park Service, U.S. Department of the Interior.

or spoliation" and retained in its "natural condition." At the same time, however, the act authorized the construction of "roads and bridle paths" in the park. And in 1906 Congress enlarged and extended the interior secretary's authority to allow leases for the transaction of business in Yellowstone "as the comfort and convenience of visitors may require" as well as to permit the construction of buildings in the park.[14]

The National Park Service Is Born

In 1910, nearly forty years after the creation of Yellowstone and after the designation of several other scenic areas, proponents began lobbying Congress to authorize a bureau to manage the entire park system. In 1912 President William Howard Taft urged Congress to create a bureau to oversee the parks, saying it was "essential to the proper management of those wonderful manifestations of nature, so startling and so beautiful that everyone recognizes the obligation of the government to preserve them for the edification and recreation of the people."[15] When Congress finally established the NPS in August 1916, the agency's Organic Act again contained dual, and potentially contradictory, aims: it required the NPS to manage the parks "to conserve the scenery and the natural and historic objects and the wildlife therein," while at the same time providing for "the enjoyment of the same in such manner and by such means as will leave them unimpaired for the enjoyment of future generations."[16]

According to Park Service historian Richard West Sellars, the park system's founders simply assumed that most natural areas would be preserved. Visitation was modest relative to the scale of the parks, and nature seemed resilient.[17] From the outset, however, NPS officials aggressively promoted park visitation because the agency's first director, Stephen Mather, recognized that tourists would form the base of public support that would allow the system to expand and thrive. As longtime Park Service observer Michael Frome writes, Mather was well aware that "bears and trees don't vote and wilderness preserved doesn't bring revenue to the federal treasury."[18] In hopes of raising the parks' profile and thereby establishing a secure place for the NPS in the federal bureaucracy, Mather encouraged the development of roads, hotels, concessions, and railroad access. Congress condoned the NPS's evolving management philosophy by sanctioning road building within and outside the parks and by declining to require (or fund) scientific research by the service.

Despite Mather's best efforts, the political constituency he envisioned never materialized. Instead, according to political scientists Jeanne Clarke and Daniel McCool, summer visitors to the parks remained "diverse, unorganized, and largely unaware of the political and funding problems facing the Park Service."[19] Although the burgeoning numbers of park visitors did not coalesce to lobby on behalf of the NPS, the parks' concessionaires, railroads, and automobile organizations quickly became vocal clients of the agency. The Organic Act allowed the NPS to "grant privileges, leases, and permits for the use of the land for the accommodation of visitors in the various parks, monuments, or other reservations."[20] The act placed minimal restrictions on concessionaires' twenty-year leases, and its provisions, taken together, "placed substantial qualifications on what Congress meant when it required the parks to be 'unimpaired.'"[21] The 1965 Concessions Policy Act modified concessionaires' mandate somewhat, allowing them to provide only those services deemed "necessary and appropriate." But it also cemented concessionaires' privileges by giving them a "possessory interest"—that is, all the attributes of ownership except legal title—to any capital improvements they erected on park land.

The consequences were predictable: it became virtually impossible to limit concessions in and around the parks, and the overall trend was toward allowing more access to more types of tourists. As Frome explains, over time, political necessity forced the parks to emphasize recreation "complete with urban malls, supermarkets, superhighways, airplanes and helicopters sightseeing overhead, snowmobiling, rangers who know they can get ahead by being policemen, and nature carefully kept in its place. Park service visitors expect comfort, convenience, and short-order wilderness served like fast food."[22] The rise in tourism took its toll, and from the 1950s through the 1980s observers decried the parks' deterioration. In 1953 Bernard DeVoto wrote a piece in *Harper's* titled "Let's Close the National Parks," in which he expressed dismay at the parks' inadequate staff and funding and their decrepit infrastructure. A fifteen-part series published by *The Christian Science Monitor* in spring and summer 1968 titled "Will Success Spoil the National Parks?" deplored overcrowding and fantasized about policies to reduce the intrusiveness of park visitation.[23] In 1980 the NPS issued a report that identified thousands of threats to the integrity of the park system, including air and water pollution, overuse, overcrowding, and unregulated private development at or near park boundaries. A 1980 U.S. General Accounting Office (GAO, later the Government Accountability Office) investigation looked at twelve of the most popular parks and found they were in extreme disrepair. A 1986 article in *U.S. News & World Report* argued that America's 209 million acres of forests, parks, and wildlife refuges were in the greatest peril since being saved from the robber barons during the conservation era at the turn of the twentieth century.[24] The same year a *Newsweek* article noted that the exponential growth in human ability to dominate nature presented the national parks with their greatest challenge ever.[25] A 1994 article in *National Geographic* ruminated on the problems facing the parks, from deteriorating air quality to invasive species to overcrowding and budget woes.[26] And another investigation by the GAO in 1994 reported on damage to the parks from activity beyond their borders.[27]

THE CASE

As with many of the activities that threatened the parks by the mid-1990s, the introduction of snowmobiles into Yellowstone National Park at first seemed innocuous. Within a decade, however, manufacturers were advertising bigger, faster machines; outfitters were aggressively promoting the sport; and affluent baby boomers were responding. By the mid-1990s the tiny gateway town of West Yellowstone, Montana (population 761), was billing itself as "the snowmobile capital of the world," and more than 1,000 snowmobiles were entering the park on peak weekend days. After failing to get a response to their requests, in 1997 environmentalists sued the NPS to force it to address the issue of increasing snowmobile use in Yellowstone. The lawsuit in turn triggered a series of administrative and judicial decisions that highlighted the intense differences between those who value the parks for their serenity and those who regard them as playgrounds.

Allowing Snowmobiles in Yellowstone

During its first 100 years, managing wildlife was the chief concern of staff at Yellowstone National Park, but they also had to regulate human visitation. In 1915 Yellowstone officially allowed automobiles into the park after several years of public debate in which local automobile organizations pressed resistant park managers for admittance. The number of visitors arriving in cars increased dramatically thereafter: in 1915 more than 80 percent of the park's 52,000 visitors arrived on trains, but by 1930 only about 10 percent of the park's 227,000 visitors came by rail. In 1940 the number of visitors rose to half a million, almost all of whom drove, and, with the exception of the war years, the number climbed steadily thereafter. One million visitors toured the park in 1949, and two million came in 1965.[28]

These visitors came in the summer, however, and the park's long winter months remained a time of snowy silence. All that changed in the early 1960s, when a few businesspeople from the town of West Yellowstone persuaded park officials to allow snowmobiles into the park as a way of creating some winter business. The number of snowmobilers remained modest through the 1960s, with only 10,000 riders entering the park during the winter of 1968–1969.[29] And, as journalist Todd Wilkinson points out, those intrepid early snowmobilers "encountered no open hotels or restaurants, no gas stations or warming huts, no grooming machines or fleets of vehicles used to shuttle skiers to the interior."[30]

In 1971, however, the NPS started grooming the park's snow-covered roads, and from then on the number of visitors began to increase: during the winter of 1973–1974, 30,000 snowmobilers entered the park; nearly 40,000 did so in the winter of 1986–1987.[31] Snowmobile outfitters were thrilled with the growing numbers. "People think it's great," said Joel Gough, an employee at the Flagg Ranch. "It's Yellowstone, it's the middle of winter, and they're on a snowmobile. They seem to thrive on it." But others were concerned. Park officials talked about the ecological stresses imposed by the soaring snowmobile population, and ranger Gerald Mernin fretted, "Sometimes it seems like you see more snowmobiles en route from West Yellowstone to Old Faithful than you do cars on a busy day in the summer."[32] In addition, the park was feeling the financial pinch of having to groom and patrol 180 miles of roads all winter.

In early spring 1987 officials began preparing a plan for managing winter operations that would address how much of the park should be accessible to snowmobiles. At the time, however, management assistant Judy Kuncl acknowledged that "politically, limiting snowmobiles or other visitation to the park is not a viable option."[33] Even so, the plan released in 1990 forecast that the park would have to consider curbing snowmobile access when the number of winter visitors reached an arbitrary threshold of 140,000, which it expected would be around 2000. Within just two years of making this projection, however, the number of winter visitors topped 143,000, and 60 percent of them used snowmobiles. Yellowstone had become "a premier snowmobiling destination."[34]

The Origins of the Snowmobile Ban

By the mid-1990s the effects of increased snowmobile use in Yellowstone were unmistakable. At Old Faithful, the park's most popular destination, the wait for gasoline at the service station reached forty-five minutes, and exhaust smell permeated the parking lots. In 1994 the park received 110 letters from winter visitors, almost all deploring the congestion and related air pollution and harassment of wildlife.[35] In early 1995, after employees complained of headaches and nausea during heavy snowmobile-traffic days, Yellowstone park officials set up monitors to gauge carbon monoxide levels in the West Yellowstone entrance booths. They also pressed Superintendent Bob Barbee to come up with a comprehensive monitoring program that would quantify the hazards facing employees and the park's biological resources.

In January 1995 the park began a two-year study to determine the effects of over-snow traffic. The following month park officials reported that during the President's Day weekend, air pollution levels at West Yellowstone violated federal air quality standards.[36] In early 1996 the NPS reported that fourteen months of measurement revealed that the nation's highest carbon monoxide (CO) levels occurred in Yellowstone on busy winter days.[37] In March researchers riding a snowmobile rigged with an air monitoring device had recorded CO levels of 36 parts per million (ppm) along the well-traveled fourteen-mile stretch between West Yellowstone and Madison Junction. The federal CO limit is 35 ppm, and the highest level recorded anywhere else in the nation in 1995 was 32 ppm in Southern California's Imperial County.[38]

Environmentalists and park officials expressed concern not just about emissions, but also about snowmobiles' impact on bison and other wildlife. In the winter of 1996–1997 Montana livestock officials killed 1,084 bison out of a herd of 3,500 when they wandered out of the park on roads groomed for snow machines.[39] (State livestock officials worried the bison carried brucellosis, a disease that can contaminate cattle herds and cause pregnant cows to miscarry.) According to research biologist Mary Meagher, grooming the roads removed the natural fence of snow depth and caused a population explosion among the bison. "We have screwed up the system royally," she told journalist Bryan Hodgson.[40]

Although its staff members were disgruntled, park officials feared the political repercussions of limiting snowmobile access to the park. On the other hand, their delays infuriated environmental activists, who had been warning the NPS for more than a year that they would sue if the agency did not conduct an environmental impact statement (EIS) for winter uses, and in May 1997 the Fund for Animals and the Biodiversity Legal Foundation filed a lawsuit against the agency. Having spent two years studying the impact of motorized recreation in winter, the groups alleged that the park's existing winter use plan violated the NEPA and the Endangered Species Act (ESA).

Four months later the NPS announced a settlement in which it agreed to start work on an EIS to consider a full range of winter uses for Yellowstone, Grand Teton National Park, and the John D. Rockefeller Memorial Parkway. The agency also agreed to consider closing a fourteen-mile stretch of road to see what effect it had on the bison—a decision that provoked fury among local economic interests. "I'm beyond upset," said Vikki Eggers of the Wyoming Chamber of Commerce, "I'm purple with rage." Sen. Conrad Burns, R-Mont., expressed his intent to weigh in on behalf of his development-oriented constituents, claiming that "[b]y caving in to a radical group . . . the Park Service has unnecessarily placed a lot of small businesses and the families that rely on them in jeopardy."[41]

Competing to Define the Problem in the Snowmobile Debate

The looming prospect of limits on snowmobile use in Yellowstone sparked a fierce debate that reflected the two sides' vastly different views about the appropriate use of the nation's parks. Environmentalists focused on snowmobiles' harmful impacts on air quality and wildlife and their inconsistency with the park's legal mandates. For them Yellowstone was a place where visitors should be able to wander through a landscape that had changed little since the 1800s and appreciate nature for its own sake. They regarded peace and quiet as central components of that experience. Scott Carsley, who guides skiers and hikers in Yellowstone, said, "There are just too many snowmobiles. They're loud. They stink. They ruin people's trips."[42] "There's no question the winter experience is diminished [by snowmobiles]," opined Jim Halfpenny, a Gardiner, Montana, scientist and guide.[43] Journalist Ben Long, who went to Yellowstone to try a snowmobile and see what the fuss was about, wrote,

> No matter how fast or far we went, we couldn't outrun the other snowmobilers. In three days of trying, we did not escape their racket. The engine noise muffled the earthy gurgles of the geysers and hot pots around Old Faithful. When we skied across wolf tracks along the Firehole River, we cocked our ears to listen for them. The only howling packs we heard were mechanical. We stopped to gaze at the haunting beauty of a trumpeter swan, swimming against the steaming current of the Madison River, as snowflakes showered down. But to do so we had to dodge a steady flow of rushing snowmobiles, like city pedestrians trying to cross against a traffic light.[44]

Long concluded that the biggest problem was "the carnival atmosphere that comes along with hundreds of zippy little machines. They quickly overshadow nature. The machine becomes the point of the visit. To snowmobile is to out-muscle nature, not enjoy it."[45]

Other critics disparaged the snowmobiling "culture." For example, Bob Schaap, a native of the area, said he lost his taste for snowmobilers as a result of his experience owning a motel in West Yellowstone, where they roared drunkenly through town after the bars closed. Some critics spoke of snowmobilers who ignored private property and wilderness area boundaries and were unconcerned with fragile winter ecology and traffic laws.[46] As the machines became faster and more powerful, snowmobilers were able to ride deeper into wilderness, and illegal routes multiplied, further tempting trespassers. Mike Finley, an outspoken former Yellowstone superintendent, said of snowmobilers, "They are not bad people, but they think they have a God-given right to engage in an activity that has adverse effects on human health and the environment. They are either in denial about the extent of their actions or they don't really care."[47]

By contrast, snowmobilers focused on Yellowstone's recreational value and asserted the right to experience it in whatever way they chose. Adena Cook of the Idaho-based off-road advocacy group, the Blue Ribbon Coalition, said, "It's a public place, just like Niagara Falls. It's a natural phenomenon, but natural doesn't equate to wild." She added, "If at certain times things are a little crowded, that means you're doing something right. I really resist people telling us something is for our own good, especially when it's the federal government."[48] Jeff Vigean, a Bostonian, said, "If we're banned from the park, they're taking away my right to bring my kids here someday, to show them Yellowstone in the wintertime." From this perspective, efforts to limit snowmobile access were exclusive and unfair. Clark Collins, executive director of the Blue Ribbon Coalition, said of environmentalists, "Theirs is a very selfish, biased and elitist viewpoint. Where do they get off deciding what types of access is appropriate?"[49] "Face it, there are a lot of folks who just don't have the time to go cross-country skiing to Old Faithful," said one Midwestern visitor. "It's our park too, and this is the only way we can see it in winter."[50]

Snowmobile boosters touted the freedom the machines allow. As one snowmobiler pointed out, "You can't see much of the park on cross-country skis. We saw coyotes, moose, buffalo, elk."[51] Entrepreneur Jerry Schmier suggested antisnowmobiling sentiment was nothing more than jealousy because snowmobilers could get to areas that cross-country skiers would never see. "In their minds, they'd like to get back into a quiet, pristine forest," he said. "They don't want to hear anything but the birds. But by the same token, many of these people never get more than a mile away from civilization. And the truth of the matter is they never will. With the snowmobiles you have freedom. You can go anywhere you want."[52] According to Brad Schmier of Yellowstone Adventures, "A snowmobile lets you be at one with the elements on your own little vehicle. The majority of the people who come here like that independence, being able to go at their own schedules—that will be a hard thing to take away from us Americans." He added, "Do I want to climb in a coach with 10 other people and somebody's whiny kid and sit there shoulder to shoulder with everybody and climb in and out when the driver decides we're going to stop? It doesn't sound like fun to me."[53]

Snowmobilers portrayed themselves as responsible, family-oriented people, pointing to industry surveys that said the average snowmobile rider was forty-one and married with children.[54] Jim Grogan, a real estate agent from Florida, described himself and his friends in the following way: "We're not yahoos; we're responsible people. We've got families and jobs and we're responsible. I'd call myself an environmentalist."[55] Many others depicted themselves as environmentalists as well, citing the trail work that snowmobile clubs perform. Ted Chlarson, a cattle trucker from Utah, said that environmentalists wanted to know how he could enjoy Yellowstone's beautiful scenery with all the noise from his snowmobile. He answered, "Well, I may not hear it, but I can see it. I can be riding on this noisy thing and I got my solitude."[56] To boost its image, the industry promoted safety standards, distributed rules of good conduct, and encouraged snowmobile clubs to get involved in local charities.

Snowmobile supporters also minimized the human health and environmental risks associated with the vehicles. Industry officials contended that snowmobile emissions occurred in remote areas and were dispersed by the wind. "There is no evidence to support the claim that snowmobile exhaust emissions cause damage to flora, fauna or the land or water," according to these officials. Roy Muth, who retired in late 1994 as president of the International Snowmobile Industry Association (ISIA), called the claims that a single snowmobile generates as much air pollution as 1,000 automobiles (a figure based on a California Air Resources Board study conducted in the 1980s) "irrational."[57] Snowmobile advocates also gleefully cited a study that showed cross-country skiers frightened elk and moose more than snowmobilers because they came upon them quietly and startled them, whereas the animals could hear a snowmobile a quarter of a mile away.[58]

While disparaging the claims of environmental impacts, snowmobile advocates emphasized the consequences of limiting the vehicles. Employing a slippery slope argument, some suggested that imposing a ban on snowmobiles in Yellowstone would lead inevitably to a prohibition of all forms of recreation on public lands. Ed Klim, president of the Snowmobile Manufacturers Association, warned, "This may be happening to snowmobiles today, but I tell you tomorrow it will be campers and the next day it will be sport utility vehicles."[59] But advocates' most potent threat was the prospect of local economic decline. To reinforce their claims, they cited the money snowmobilers poured into the regional economy, claiming, for example, that in the late 1980s snowmobilers in Canada and the United States spent more than $3.2 billion on their sport each year.[60] Proponents also cited the "happiness" factor, pointing out, "Snowmobile tourists spend a ton of money—between $200 and $300 a day for a family of four," said David McCray, manager of Two Top Snowmobile, Inc. "And most of them come back saying it was the best day of their trip, even the best day of their lives."[61]

Motorized Recreation Versus Protection

The controversy over snowmobiles in Yellowstone was part of a larger debate over the dramatic rise in motorized recreation on public lands. As legal scholar Robert Keiter observes, although once welcomed as an alternative to extractive industries, the new tourism-recreation economy posed environmental threats of its own:

> Rather than clearcuts and open pit mines, its legacy is suburban-like sprawl, new ranchettes, and mega-resorts that chop once pastoral landscapes into smaller and smaller fragments. As new homes and secondary roads spread across vacant agricultural lands, open space begins to disappear, winter wildlife habitat is lost, seasonal migration routes are disrupted, and erosion problems are exacerbated. . . . Unlike the site-specific impacts associated with a mine or timber sale, recreationists are ubiquitous; the mere presence of more people will generate more human waste, create more unauthorized travel routes, and disturb more wildlife. Motorized recreational users often compound those problems, particularly as they demand more roads and trails into undisturbed areas. The cumulative impact of myriad recreational users and related sprawling development places the very environmental qualities that lured them in the first place at risk.[62]

Conflicts among users became more intense in the 1980s and 1990s as people had more leisure time and money to pursue pleasure outdoors, and the development and sale of motorized recreational vehicles boomed. In 1972 an estimated 5 million Americans used ORVs; personal watercraft, such as Jet Skis, and all-terrain vehicles did not exist. By 2002 the Forest Service estimated 36 million people used these machines. Between 1992 and 2001 annual sales of all-terrain vehicles rose from 169,000 to 825,000. Snowmobile registration increased more than 60 percent during the same period.[63]

Federal agencies had the legal authority to restrict motorized recreation but rarely invoked it, choosing instead to ignore all but the most egregious abuses on their lands and watch passively as surrounding communities built their economic futures on such activities. As a result, by the 1990s motorized vehicles were pervasive throughout the nation's public lands, and relations between hikers and off-road vehicles (ORVs) were strained. Environmentalists pointed out that recreational machines had a range of negative impacts, from scarring the landscape to causing erosion and noise, air, and water pollution to disturbing wildlife and carving up habitat. They wanted public lands closed to motorized recreational use except for specifically designated trails. They argued that the national obsession with extreme sports increased the likelihood that some ORV users would break the rules if they had access to remote areas; they pointed to commercials for trucks that showed vehicles cutting across streams and fragile meadows or driving onto beaches. "The message is this is something you ought to be doing, and it doesn't matter what the damage to the land," said Jerry Greenberg of the Wilderness Society.[64]

By contrast, proponents of motorized recreation portrayed the conflict as a dispute between average families and a small number of privileged athletes. For example, Rep. Ron Marlenee, R-Mont., argued, "Mom and pop and their ice-cream-smeared kids have as much right to prime recreational opportunities as the tanned, muscled elitists who climb in the wilderness."[65] Proponents also portrayed their adversaries as extremists and said there was little prospect for cooperation as long as environmentalists wanted to turn large tracts of the West into federal wilderness areas.

The Evolution of the Snowmobile Ban

Throughout 1998, as the NPS collected data and met with local officials about Yellowstone's winter use plan, the controversy bubbled, and momentum built nationally for curbing snowmobile access to the park. An editorial in *USA Today* on February 19, 1998, suggested eliminating snowmobiles from Yellowstone altogether, saying, "Throughout the winter, there are days when the Yellowstone plateau looks and sounds more like Daytona than a national park. Thousands of snowmobiles cover its trails like swarms of two-stroke hornets, producing a chain-saw howl and leaving a pall of blue-white haze in the air." The editorial acknowledged that snowmobiles can be fun to ride but argued they should be banned from the park because "[t]he noxious fumes and obnoxious noise quickly can destroy the contemplative park experience sought by millions of others." The editorial added that although "it sometimes seems [Yellowstone] has been deeded over to narrow local interests," Yellowstone's mission is to preserve the natural beauty, not ensure local economic development.[66]

On January 21, 1999, the Bluewater Network—a coalition of sixty environmental groups—filed a petition with the NPS to ban snowmobiles from all U.S. national parks, saying, "They're killing our wildlife, ruining our air and water quality, poisoning the health of rangers exposed to snowmobiles' carbon monoxide exhaust, and destroying the solitude and peace cherished by other winter visitors."[67] Bluewater cited "adverse impacts to park wildlife, air and water quality, vegetation, park ecology, and park users." The network pointed to its analysis of government data, which showed that snowmobiles dumped about 50,000 gallons of raw gasoline into the snow pack in 1998.[68] The petition claimed that snowmobiles created more pollution around Old Faithful in one weekend than an entire year's worth of automobile traffic.

Manufacturers objected to such assertions, which they called "unscientific," but they had no data to refute them. Instead, they disparaged the litigants and tried to shift attention to the economic consequences of a snowmobile ban. Ed Klim described the lawsuit as "a rather extreme move by a small group that wants to limit access to the parks." And West Yellowstone businessman Brad Schmier said, "A ban would put me out of business. Snowmobiling is our community's only winter economy."[69]

In response to environmentalists' petition, in July the park released a draft EIS that contained seven alternatives, ranging from unrestricted snowmobiling to highly restricted use. None of the options eliminated snowmobiles or trail grooming. Nevertheless, at public meetings held around the West, the NPS raised the possibility of prohibiting snowmobiles from the park, and in mid-March 2000 Yellowstone officials told state and local officials that the agency was leaning toward a ban. They conceded that, although the agency had originally favored phasing in cleaner machines over ten years, and representatives of local governments had also proposed converting the fleet over time, neither approach seemed to resolve the park's legal obligations.

NPS officials pointed out that those obligations stemmed from several sources: the Yellowstone Act requires the NPS to preserve "from injury or spoliation" the "wonders" of the park and ensure "their retention in their natural condition"; the National Park Service Organic Act of 1916 directs superintendents to conserve scenery, national and historic objects, and wildlife and to "leave them unimpaired for future generations"; and the Clean Air Act requires national parks to preserve, protect, and enhance their air quality. In addition, two executive orders require the NPS to limit motorized recreation. No. 11644, signed by President Richard Nixon in 1972, mandated that each agency establish regulations designating specific zones of use for ORVs and that such chosen areas be located to "minimize harassment of wildlife and significant disruption of wildlife habitats." No. 11989, signed by President Jimmy Carter in 1977, strengthened the 1972 order, stating that if an agency head determines that the use of ORVs will cause "considerable adverse effects on the soil, vegetation, wildlife, wildlife habitat or cultural or historic resources of particular areas or trails of the public lands" the agency head shall "immediately close such areas or trails to off-road vehicles." Furthermore, NPS regulations prohibit the disturbance of any wildlife from their "natural state" and prohibit the use of snowmobiles "except where designated and when their use is consistent with the park's natural, cultural, scenic and aesthetic values, safety considerations, and park management objectives, and will not disturb wildlife or damage park resources."

As a ban began to appear imminent, newspapers around the country came out in support of it: an editorial in *The Oregonian* (Portland) on November 18, 1999, described "a slow-but-sure desecration of one of Nature's crown jewels. Just imagine some 60,000 snowmobiles belching blue smoke as they slice through white powder, bound over the hills and along streambanks, sounding like a reunion of jackhammer operators." (An outraged reader pointed out, correctly, that 60,000 snowmobiles are not all in the park at once.) As winter approached, editorials in the *Omaha World-Herald*, *The New York Times*, and the *San Francisco Chronicle* likewise lined up behind the proposed ban.

The NPS's inclination reflected not only the agency's legal constraints and public sentiment, but also the values of the interior department's political appointees. A flurry of similarly protective verdicts on mechanized recreation in the parks accompanied the announcement that the NPS was seriously considering a snowmobile ban. On March 15, 2000, at the agency's behest, Congress passed a bill banning tourist flights over Rocky Mountain National Park and requiring all national parks to complete air-tour management plans in cooperation with the Federal Aviation Administration. On March 21 the NPS issued a rule banning personal watercraft from all but 21 of its 379 parks and recreation areas. Previously, the watercraft had been allowed in 87 parks.[70] Two days later interior secretary Bruce Babbitt unveiled a plan to reduce auto congestion in Yosemite. The plan called for tearing out several parking lots inside the park, reducing the number of parking spaces from 1,600 to 550, and having visitors park in lots at the edges of the park and take shuttle buses into the valley. The following day, President Bill Clinton announced new restrictions on sightseeing flights over Grand Canyon National Park.

In line with this trend, in April the NPS issued a memo directing parks that currently allowed snowmobiling to review their regulations and within a year amend or replace them with regulations that complied with existing laws and regulations. The announcement specifically excluded Denali and eleven other parks in Alaska where snowmobiling is explicitly permitted by law; Voyageurs, where the law establishing the park permitted snowmobiling; and Yellowstone, Grand Teton, and the John D. Rockefeller Memorial Parkway, for which the NPS was in the midst of crafting a winter use plan. In the other parks that snowmobilers used, the agency planned to allow them only on small sections to gain access to adjacent lands. Announcing the ban, Donald Barry, assistant secretary of interior for fish and wildlife and parks, asserted, "Snowmobiles are noisy, antiquated machines that are no longer welcome in our national parks. The snowmobile industry has had many years to clean up their act, and they haven't."[71]

An approving editorial in *The New York Times* of April 29, 2000, said,

> What Mr. Barry's announcement underscores is the simple principle that America's national parks should lead the nation in adherence to environmental law. Recreation is not incompatible with strict adherence. It is enhanced. For years snowmobiling has been tolerated even though it violated the law, and its environmental impacts had scarcely been monitored until recently. But now that the impact of snowmobiling is clearly understood and the extent of its pollution clearly documented, overlooking the law has become intolerable.[72]

Editorials in *The Denver Post* and *The Seattle Times* also endorsed the decision. Nevertheless, in hopes of prompting Congress to pass a law preventing the NPS from phasing out snowmobiles in the parks, in late May legislative opponents of a ban convened hearings in the House and Senate. Defenders of the status quo disputed and downplayed the scientific evidence and highlighted the economic importance of snowmobiling to the region. By contrast, environmentalists emphasized the NPS's legal obligation to protect the park's resources, the scientific evidence that snowmobiles harmed those resources, and the public's overwhelming support for a snowmobile ban.

Debating the Science. In testimony before the Senate Energy Committee's Subcommittee on National Parks, Historic Preservation, and Recreation, Assistant Secretary Barry described the pollution caused by snowmobiles' two-stroke engines in the following way:

> First, up to one-third of the fuel delivered to the engine goes straight through and out the tailpipe without being burned. Second, lubricating oil is mixed directly into the fuel, and is expelled as part of the exhaust. Third, poor combustion results in high emissions of air pollutants as well as several toxic pollutants, such as benzene and aromatic hydrocarbons

> that the EPA classifies as known probable human carcinogens. When compared to other emissions estimates, a snowmobile using a conventional two-stroke engine, on a per-passenger mile basis, emits approximately 36 times more carbon monoxide and 98 times more hydrocarbons than an automobile.[73]

Barry added that winter meteorological conditions—particularly cold temperatures, stable atmospheric conditions, and light winds—exacerbated the accumulation of pollutants in the air. He cited the NPS study, "Air Quality Concerns Related to Snowmobile Usage," which found that "although there are 16 times more cars than snowmobiles in Yellowstone, snowmobiles generate between 68 and 90 percent of all hydrocarbons and 35 to 68 percent of all carbon monoxide released in [the park]."[74]

In addition to detailing air pollution impacts, Barry referred to noise studies showing that the "relentless whine" of snowmobiles was persistent and inescapable in the park. For example, in March 2000 two environmental groups, the Greater Yellowstone Coalition and the National Parks Conservation Association, released a study providing data that substantiated complaints about snowmobiles' noise. Using methods established by the NPS, volunteers had listened at a variety of often-used places in the Old Faithful area. Of the thirteen spots chosen, they documented eleven where the sound of snowmobiles was audible 70 percent of the time or more. At Old Faithful itself, listeners heard snowmobiles 100 percent of the time. They found that only one place, the most remote location in the sample, was free of snowmobile sound. (Vikki Eggers responded to the study by saying that the park's mandate was not to preserve quiet, pointing out that motorcycles in the summer are noisy too.[75]) Barry also warned of the possibility that unburned gasoline could reach park streams and lakes as snow melted. And he pointed out that snowmobilers regularly harass bison, noting that winter is already the most stressful time for Yellowstone's wildlife. He concluded that the new machines, although cleaner and quieter, were neither clean nor quiet and in any case did not eliminate the problem of wildlife harassment. In short, he said, snowmobiling "is not an essential, or the most appropriate, means for appreciating park resources in winter."[76]

Testifying before the Senate subcommittee, Mark Simonich, director of Montana's Department of Environmental Quality, disputed the NPS claim that air quality standards had been exceeded in the park. He added that there were short-term technological solutions—including the use of ethanol blend fuels and biodegradable lubrication oils—to the environmental problems posed by two-stroke-engine snowmobiles and claimed that cleaner, quieter four-stroke-engine vehicles would soon be available.[77] Although snowmobile groups did not contest the park's air pollution figures, they accused the NPS of rushing the study to comply with the lawsuit and fanning the flames of controversy by comparing Yellowstone's particulate concentrations with those of a Los Angeles suburb.

Local Economic Impacts. Snowmobilers and their supporters said they felt as though their concerns had been ignored, and they raised the specter of economic disaster if the ban were implemented. Kim Raap, trail-system officer with the Wyoming Department of Commerce, pointed out that snowmobiles contributed between $100 million and $150 million to the state's economy.[78] The Jackson Hole Chamber of Commerce said a ban would devastate the local economy, claiming that snowmobile outfitting in Jackson Hole brought in $1.3 million annually and that nearly 20 percent of West Yellowstone's total resort taxes were collected during the winter season.[79] Adena Cook of the Blue Ribbon Coalition claimed the ban would cost neighboring communities 1,000 jobs and $100 million a year in tourism revenues.[80] Testifying on July 13, 2000, before the House Small Business Committee's Tax, Finance, and Exports Subcommittee, Clyde Seely, owner of the Three Bear Lodge in West Yellowstone, warned, "Cuts will have to be made. The first cut would be employee insurance. The second cut would be employees."

But Kevin Collins of the National Parks Conservation Association pointed out that the mission of the Park Service was to preserve places in their original condition so they would be available for future generations, not provide economic benefits for nearby communities and businesses. Collins added that the economic impact of a snowmobile phaseout had been "greatly exaggerated," noting that an economist who assessed the rule's likely impact on West Yellowstone found it to be negligible—approximately $5 million—and temporary. And he concluded that in the long run the region's economy depended on the park's ecological health.[81]

To the chagrin of western lawmakers who were intent on demonstrating the adverse local impacts of a snowmobile ban, some residents of West Yellowstone went to Washington to make clear to Congress that locals were not monolithic on the issue. One resident carried a petition signed by 160 people (the town's entire population was less than 1,000) asking lawmakers to protect the park and help the town diversify. Craig Mathews, a West Yellowstone native who supported the ban, said people in town had stopped waving at or greeting him. He accused those people of being more concerned with making a profit than with the "air our kids breathe" and said that they were developing an "NRA [National Rifle Association] mentality, contending that a successful ban on snowmobiling in national parks will eventually lead to bans on walking, hiking, fishing, and camping on public lands."[82] Other local supporters of the ban said it was necessary to restore the pristine glory of Yellowstone, noting that, should it be lost, tourists would no longer visit the park. Still others said they were fed up with the constant roar of snowmobiles at all hours of the night on the city streets. (Earlier in the year a group of West Yellowstone High School students had petitioned city hall to pass an ordinance banning recreational snowmobile use in town between 11 p.m. and 5 a.m., but the town declined to take any action.)

A New Ideology and a New Rule

In October 2000 the NPS issued a final EIS on the winter use plan for Yellowstone and Grand Teton national parks. The plan selected the environmentally preferred alternative, which phased out snowmobile use altogether by 2003 to 2004 and restricted winter access to snowcoaches, snowshoes, and skis. Rep. Barbara Cubin,

R-Wyo., charged that the process was a "farce" whose "outcome was predestined by the Clinton–Gore administration and its extremist environmental cohorts" and vowed to use every regulatory and legislative means necessary to achieve a compromise.[83] Nevertheless, in November the NPS signed a record of decision, and in December the agency issued a proposed rule. The snowmobile industry immediately sued to block the new rule, saying it was not supported by the facts, and all three members of the Wyoming delegation denounced the ban. In response to the plea of Sen. Craig Thomas, R-Wyo., Congress approved a two-year delay in the phaseout, giving the newly elected president, George W. Bush, an opportunity to block it.

The Bush administration's ideology concerning public lands, which was diametrically opposed to that of the Clinton administration, manifested itself in a dramatic turnaround vis-à-vis snowmobiles. Just hours after he was sworn in, President Bush signed an executive order imposing a sixty-day moratorium on the ban in order to review it. That winter snowmobilers entered Yellowstone in record numbers: the number of sleds coming in through the west entrance jumped nearly 37 percent in January to 21,742.[84] (There was also a 32 percent increase in the number of people coming in by snowcoach, to 2,042.) On President's Day weekend, a total of 4,339 snowmobilers poured into the park—at a rate of one every ten seconds.[85] In mid-February Senator Thomas introduced legislation to rescind the snowmobile ban altogether.

In late April, as Earth Day neared, the Bush administration tried to shore up its environmental credentials by announcing that it would allow the snowmobile ban to stand. At the same time, however, the interior department was actively negotiating with snowmobile manufacturers, and in June 2001 the NPS settled the manufacturers' lawsuit by agreeing to conduct a supplemental environmental impact statement (SEIS) that considered data on new snowmobile technologies. In the meantime, the park would allow unlimited access in winter 2002 to 2003 but limit entries in 2003 to 2004 to 493 snowmobiles per day. Justifying its decision not to defend the Clinton-era rule, interior department spokesperson Mark Pfeifle said, "This administration feels strongly that greater local input, new information, scientific data and economic analysis and wider public involvement can only lead to better, more informed decisions."[86] Yellowstone Superintendent Mike Finley pointed out, however, that in formulating the settlement, administration officials had not consulted anyone working in the park.

In February 2002 the NPS released its draft SEIS. The new document did not furnish any evidence that contradicted the Clinton administration's decision; instead, it characterized the information submitted by the snowmobile industry about technology designed to cut noise and air pollution as "speculative and insufficient for analysis purposes."[87] It added that banning snowmobiles would reduce jobs and the local economy by less than 1 percent.[88] Nevertheless, the agency carefully avoided recommending proceeding with the ban, and in late June the Bush administration announced at a meeting with local state and county governments that it would allow snowmobiling in the park to continue but would restrict traffic volume and address noise and air pollution concerns. (In early July Democrats and eastern Republicans introduced legislation to restore the ban, but the move was pure politics; although the bill quickly garnered 100 cosponsors, the Republican-controlled House had no intention of letting the chamber vote on it.)

During the comment period, the NPS received more than 350,000 reactions to the SEIS, 80 percent of which supported a ban.[89] Nevertheless, on November 18, 2002, the NPS released a final rule delaying the implementation of a phaseout for another year. The agency also announced the administration's plan to allow 950 snowmobiles per day into the park, a 35 percent increase over the historical average of 840 per day—though well below the number that entered the park on peak weekend days. To soften the blow, the plan required that rented machines conform to best available technology standards and required that, beginning in the winter of 2003–2004, 80 percent of snowmobiles entering the park be accompanied by a guide. The administration justified its plan by saying there had been major improvements in snowmobile technology—in particular the introduction of the four-stroke engine, which, according to a park spokesperson, cut noise from 79 to 73 decibels and emissions by 90 percent.[90] The administration called its approach a "balanced" one that would allow the public to continue using the park in winter and the nearby tourism industry to survive.

A critical editorial in the Minneapolis *Star Tribune* noted that the plan contained no clear standards for determining its effectiveness. Instead, it set the number of snowmobiles and assumed that whatever noise and smoke that number of machines emitted would be fine. "A more sensible approach," the editor urged, "would start with some management goals and work backward to policy," making sure that snowmobile use was consistent with a national park's "natural, cultural, scenic and aesthetic values, safety considerations, park management objectives, and [would] not disturb wildlife or damage park resources." The editorial noted that the park's new vision was about starting points, not outcomes, and that the Bush administration had "[spun] the giveaway to snowmobile clubs, companies and local businesses as stewardship for balanced use of the park."[91]

In February 2003 the NPS issued a final SEIS containing five alternatives, including an "environmentally preferred" option identical to the original phaseout. But this time the NPS selected Alternative 4, which had not been included in the draft SEIS but contained the administration's new approach, as its preferred alternative. The NPS signed the 2003 record of decision affirming the administration's plan in March, issued a proposed rule to accompany it in April, and published the final rule in December.[92]

Legal Challenges

It appeared as though the Bush administration would be thwarted, however, when just hours before the advent of the 2003–2004 snowmobiling season, Judge Emmet Sullivan of the U.S. District Court of the District of Columbia overturned the Bush rule and reinstated the Clinton-era snowmobile phaseout. In December 2002—after the Bush administration released its proposed rule for snowmobiles in Yellowstone and Grand Teton—a coalition of environmental groups had filed suit in

Washington, D.C., claiming the new rule ignored requirements that national parks be preserved in a way that leaves them "unimpaired for the enjoyment of future generations." In his ruling on the case, Sullivan—a Reagan appointee—criticized the Bush administration's decision-making process, pointing to its reversal of the rule despite the fact that there had been no change in the scientific evidence supporting it. "The gap between the decision made in 2001 and the decision made in 2003 is stark," he said. "In 2001, the rule-making process culminated in a finding that snowmobiling so adversely impacted the wildlife and resources of the parks that all snowmobile use must be halted. A scant three years later, the rule-making process culminated in the conclusion that nearly 1,000 snowmobiles will be allowed to enter the park each day." Sullivan pointed out that the Park Service was "bound by a conservation mandate . . . that trumps all other considerations." He added that the NPS had not supplied a reasoned analysis for its changed position but relied exclusively on the prospect of improved technology. Yet the original EIS had considered and rejected the argument that newer, cleaner machines would alleviate the problems associated with snowmobiles in the park. Sullivan concluded that in 2003 the NPS rejected the environmentally preferred alternative in favor of an alternative whose primary beneficiaries were the "park visitors who ride snowmobiles in the parks and the businesses that serve them."[93]

Locals reacted with desperation to the last-minute reinstatement of the ban. Jerry Johnson, mayor of West Yellowstone (and a motel owner and snowmobile outfitter), said, "It's out of control here. People are frantic not to have their vacation ruined."[94] Describing the week after the ruling came down, David McCray, owner of Two Top snowmobile rentals, said, "I saw three grown men in tears. This is their livelihood. This is their identity. This is how they're going to send their kids to college. It's just so unbelievable that the judge would not take any of this into account." According to Marysue Costello, executive director of the West Yellowstone Chamber of Commerce, "The impact of [a ban] is going to be significant. The ripple effect of this, we're not just talking snowmobile people. It's the schools, the snowplows."[95] Since 1994, six new hotels had been built and at least 400 rooms added in anticipation of winter tourists.[96]

But snowmobile proponents' despair lifted in February 2004, when U.S. District Court Judge Clarence Brimmer, in Wyoming, responded to the lawsuit originally filed by the manufacturers' association and the Blue Ribbon Coalition challenging the snowmobile ban and reopened after Judge Sullivan issued his ruling. Brimmer issued a temporary restraining order requiring the NPS to suspend limits on snowmobiles in Yellowstone and Grand Teton. In overturning the Clinton-era limits, Brimmer ruled that the state of Wyoming and snowmobile touring companies would suffer irreparable harm that "far outweighed" any impact that would be suffered by the park's employees or wildlife. Brimmer credited a claim by Wyoming that the phaseout would cause millions of dollars in losses that could not be recovered or compensated, as well as its claim that, based on the revised (2003) rule, Wyoming outfitters had invested $1.2 million to convert their fleets, booked reservations, taken deposits, and contracted with employees for the 2003 to 2004 season.

In addition, the judge was swayed by the manufacturers' association argument that some businesses were incurring catastrophic losses as a result of the rule and that many subsidiary businesses were threatened as well. In deciding on the balance of harms versus benefits, Brimmer agreed with the plaintiffs that any competing harm to park staff and visitors would be much reduced by new snowmobile technology. He ordered the park to adopt temporary rules to allow four-stroke machines into the park for the remaining five weeks of the winter. Defending his decision, Brimmer said, "A single Eastern district court judge shouldn't have the unlimited power to impose the old 2001 rule on the public and the business community, any more than a single Western district court judge should have the power to opt for a different rule."[97]

In August 2004 the Park Service proposed letting up to 720 guided, four-stroke-engine snowmobiles enter the park each day through the winter of 2006–2007, while the agency devised a new winter use plan. In her defense of the agency's choice despite overwhelming public sentiment in favor of a ban, Yellowstone Superintendent Suzanne Lewis explained, "This is not a public opinion poll. This is not about majority votes."[98] The administration's "compromise" rule was bolstered by Judge Brimmer's decision in October 2004 to strike down the Clinton-era snowmobile ban. Brimmer said the rule had been imposed without allowing for meaningful public participation; he also asserted that the NPS had not adequately studied emissions, noise, and other impacts of an increased number of snowmobiles in the park. Turning Judge Sullivan's reasoning on its head, Brimmer argued that the government had ignored proper procedures and arrived at a "prejudged political decision to ban snowmobiles from all the national parks."[99]

Just days after the NPS's announcement on November 4 that it would go ahead and promulgate a three-year interim rule allowing 720 snowmobiles into Yellowstone (as well as 140 machines per day into Grand Teton and the John D. Rockefeller Memorial Parkway), advocates filed another series of lawsuits. On November 10 the Wyoming Lodging and Restaurant Association filed suit in federal court in Wyoming saying its members wanted *more* traffic in the park. The Greater Yellowstone Coalition filed suit in Washington, D.C., asking the NPS to monitor snowmobiles' impact and reduce their numbers if pollution thresholds were exceeded. The Fund for Animals and Bluewater Network filed a separate suit in Washington charging that the NPS had failed to address the impacts of trail grooming which, according to Bluewater's Sean Smith, changed the dynamics of the entire ecosystem.

In February 2005 interior secretary Gale Norton toured the park by snowmobile, making a clear statement about her sympathies. Concluding her trip, Norton said, "We, I think, have a better understanding of what the experience is here and why people are so excited about the opportunity to snowmobile here." Norton also disparaged snowcoaches, saying they offered a "much more ordinary kind of experience" that is "not as special as a snowmobile."[100] Norton's enthusiasm notwithstanding, the number of snowmobiles passing through the park's west entrance in the winter of 2004 to 2005 was down to 250 per day—about 70 percent less than two winters earlier and about one-third of the 720 allowed; 311 snowmobiles entered on the year's busiest day.[101] Although the decline was partly because of the meager snowfall, the main reason was that more passengers were electing to travel by snowcoach. Largely as a result of the decreased snowmobile traffic, the monitors at the West Yellowstone entrance had registered a 90 percent drop in carbon monoxide.[102] Nevertheless, an internal NPS report released in late October 2006 found that snowmobiles and snowcoaches were still audible around Old Faithful for two-thirds of the day, despite their smaller numbers.[103]

OUTCOMES

During the winters of 2005–2006 and 2006–2007, there was heavy snowfall, but the average number of snowmobiles entering the park daily was only 250 and 294, respectively. In March 2007 seven former NPS directors called on the interior secretary to disapprove Park Service plans to permanently allow up to 720 machines per day into Yellowstone. As a compromise, in November the Bush administration proposed adopting a limit of 540 snowmobiles per day. Dissatisfied, environmental groups promptly sued, noting that the Environmental Protection Agency (EPA) had said the draft plan would not go far enough to protect the park's human health, wildlife, air, or quiet spaces. Shortly thereafter, the state of Wyoming filed suit demanding an even higher limit.

On September 17, 2008, Judge Sullivan threw out the Bush administration's new approach in a harshly critical ruling. According to Sullivan, the plan clearly elevated recreational use over conservation of park resources and failed to explain why its "major adverse impacts" were necessary to fulfill the purposes of the park. Put simply, he said, "the [plan] provides 'no rational connection between the facts found and the choice made.'"[104] A few days later Judge Brimmer in Wyoming issued a separate decision. This time he deferred to the D.C. court's decision, albeit resentfully. He stated, however, that the Bush administration's interim rule (720 snowmobiles per day) should remain in force while the NPS revisited its winter use plan. In its next attempt at a permanent rule, issued in November 2008, the NPS proposed allowing an even smaller number of snowmobiles—318 per day—into the park. But in a dramatic twist, shortly after releasing that proposal, the agency reversed course and adopted Judge Brimmer's recommendation, reverting to a limit of 720 per day.

Environmentalists hoped that the administration of President Barack Obama would bring a new worldview to the management of public lands, and sure enough, in October 2009 the NPS announced it would allow only up to 318 snowmobiles and 78 snowcoaches per day for the next two winters while it worked on yet another longer-term plan. Although the interim plan was unlikely to reduce the average number of machines entering the park—which had ranged from 205 per day to 295 per day in the preceding five years—it promised to cut significantly the number entering on peak days. (The peak in 2006–2007 was 557, and in 2008–2009, it was 429.) The administration expected to complete its winter use plan in the winter of 2010–2011.

Even as the Obama administration contemplated its options for Yellowstone, interior secretary Ken Salazar faced other disputes over motorized recreation, which had raged across the West during the 2000s: as the decade wore on, conflicts between ORV recreationists and other public-land users had intensified, and the number of violent incidents had risen. With their enforcement budgets declining, Forest Service and Bureau of Land Management (BLM) rangers struggled to maintain order, often finding themselves outmanned or outgunned. Many rangers contended that motorized recreation was the biggest threat facing the public land; as one explained, "In many parts of the West, it's a Mad Max situation, with a quarter-million people on a weekend and one ranger to keep them from tearing the place up."[105]

In large measure, the conflict was the result of population growth at the borders of the public lands: by 2007 more than 28 million homes were less than thirty miles from federally owned lands, and the number of visitors to those lands had soared.[106] Crowding at the more accessible sites, combined with improved technology, spurred people to go even further into the backcountry seeking adventure—often on snowmobiles or all-terrain vehicles. In hopes of alleviating the confrontations that ensued, land managers requested that Congress grant them the authority to impose larger fines, confiscate vehicles, and suspend hunting and fishing licenses. But Meg Grossglass of the Off-Road Business Association (ORBA) insisted that "[i]nstead of outlawing the use [of ORVs] we need to manage the use. Off-roaders are environmentalists. We get on our bikes and ride because we want to be there in the outdoors." ORBA and the Blue Ribbon Coalition released a statement saying, "We believe that a collaborative effort should be employed to solve the problems associated with those that choose to recreate irresponsibly or are uneducated about the rules."[107]

In early November 2005 the Forest Service began a process of designating which national forest trails were suitable for ORV use, in an effort to more permanently quell the controversy over (and reduce the environmental damage associated with) ORVs on public land. While the Forest Service aimed to complete its process by 2009, the BLM gave itself between ten and fifteen years to designate appropriate trails. In the meantime, dozens of states proceeded with their own legislation to limit ORV use on federal and state lands.[108]

More ominously for ORV users, a court ruling in summer 2010 suggested that the judiciary would continue to play a critical role in the debate, particularly where national parks are concerned: U.S. District Judge Gladys Kessler quashed a pair of rules allowing personal watercraft at Pictured Rocks National Seashore in Michigan and at Gulf Islands National Seashore, which stretches from Florida to Mississippi. Judge Kessler said the agency's 2006 EIS was "profoundly flawed" because it ignored impacts on water and air quality, soundscapes, shoreline vegetation, and wildlife. Judge Kessler's ruling, like Judge Sullivan's two years earlier, raised an issue that was increasingly salient: the impact of noise from traffic, park buildings, and recreational equipment on park visitors and wildlife. In 2000 the Park Service had begun studying those impacts, and over the course of the decade a growing body of research showed that loud noises at parks have demonstrably harmful effects. They disrupt the peace and quiet that many park visitors say they are seeking; moreover, frequent noises prevent animals from warning each other about predators, disrupt breeding cycles, and may even discourage some bird species from singing during the day.[109] Such findings bolster environmentalists' contention that motorized recreation on public land imposes unacceptable costs, not only on people, but also on the resources those lands seek to conserve.

In fall 2013 the Obama administration issued a winter use plan for Yellowstone that relied heavily on technological improvements in snowmobiles and, miraculously, assuaged the concerns of both environmentalists and snowmobile enthusiasts. Under the new plan, fewer than fifty-one groups of snowmobiles with up to ten vehicles each were allowed into the park beginning in December 2014. As of December 2015 snowmobiles entering the park would have to pass stringent tests for noise and air pollution—tests that experts said few existing snowmobiles could pass. "This is the most reasonable, the most balanced plan that has ever been presented," said Clyde Seely, the West Yellowstone snowmobile operator.[110] Tim Stevens, Northern Rockies Regional Director for the National Parks Conservation Association, additionally asserted, "Absolutely, under this plan Yellowstone will be a cleaner and quieter place, and a place [where] park visitors can find the solitude that is unique to Yellowstone."[111]

CONCLUSIONS

The decision of whether to allow snowmobiles in Yellowstone and other national parks turns on the simple question posed by former NPS Director George Hartzog: what are the parks for? More concretely, a 1986 *Newsweek* magazine article asked, "Do the parks exist to conserve nature or to put it on display?"[112] Environmentalists contend that the NPS's highest duty is to protect the parks' unique and fragile natural resources for future generations. They tend to agree with the sentiment expressed by veteran national park observer Michael Frome, who writes, "National parks are sources of caring based on inner feeling, on emotional concern for wolf, bear, insect, tree, and plant, and hopes for the survival of all these species. National parks are schools of awareness, personal growth, and maturity, where the individual learns to appreciate the sanctity of life and to manifest distress and love for the natural world, including the human portion."[113] For protection advocates, as in the Arctic National Wildlife Refuge case (Chapter 6), a compromise is a loss because even a small number of motorized vehicles can dramatically change the experience for everyone else.

By contrast, snowmobile proponents contend that the parks belong to all Americans and that those who choose to enjoy them riding on a snowmobile have a right to do so. It is easier for advocates of motorized recreation to portray themselves as moderate (and environmentalists as extremists): by accepting limits on use, advocates can claim to have compromised. Commentary on the issue by Nicholas Kristoff, the iconoclastic *New York Times* columnist, reflects the appeal of this position. He explains that limiting snowmobile access is a "reasonable" solution.

He acknowledges that snowmobile manufacturers only developed cleaner, quieter machines when faced with eviction from the park but confirms that the new machines are, in fact, better. He concludes, "Some environmentalists have forgotten . . . that our aim should be not just to preserve nature for its own sake but to give Americans a chance to enjoy the outdoors."[114]

Michael Scott of the Greater Yellowstone Coalition describes the debate as a culture clash that pits "NASCAR vs. Patagonia." During the Clinton administration, the Patagonia crowd prevailed, as many park superintendents devised plans to curb overcrowding and limit the types and extent of park use. For example, the NPS announced plans to remove roads and parking lots in the Grand Canyon park and build a light-rail system to carry visitors to the South Rim. Yosemite devised a plan to limit the number of cars entering the park. Poll after poll showed public support for such changes: in a 1998 survey conducted by Colorado State University, 95 percent of 590 respondents said the NPS should limit the number of visitors if overuse was harming park resources; 92 percent said they would be willing to ride a shuttle bus to ease traffic congestion rather than drive; and 92 percent said they would be willing to make reservations to visit a popular park to reduce crowding.[115]

But opposition to such restrictions was intense and well organized, and advocates of unlimited access found sympathetic allies in the Bush administration, which moved quickly to reverse many of the Clinton-era reforms. For example, in February 2002 the Forest Service delayed closing Montana's Mount Jefferson to snowmobiles even though scientific studies had documented that the vehicles disturbed wolverine and other wildlife, and the public favored closure by two to one. After local politicians demanded that the Forest Service delay decision making in order to take snowmobilers' views into account and see if new information was available, Forest Supervisor Janette Kaiser acquiesced. In 2003 the BLM reversed a Clinton-era plan to put about 50,000 acres of California's Imperial Sand Dunes Recreation Area off-limits to ORVs, even though these vehicles clearly damage sensitive plants, and gatherings of their users have on occasion turned violent. The administration also made it easier for local and state authorities to claim rights-of-way through millions of acres of federal land, potentially opening vast areas to ORV use. And the NPS began drawing up plans to allow personal watercraft in several western lakes and eastern seashores, undoing a Clinton-era trend to ban their use.

Throughout the Clinton and Bush administrations, advocates who were excluded from or disadvantaged in the decision-making process turned to other venues, particularly the courts. Both sides chose their courts strategically: snowmobile advocates sought recourse in local courtrooms, where judges presumably were more sympathetic to local economic considerations, while environmentalists turned to D.C. judges, who historically have been more inclined to support their arguments. The judiciary has both advantages and disadvantages as a forum for resolving disputes over agency decision making. On the one hand, relative to other policymakers, judges are more insulated from public opinion and can—in theory at least—deliberate on the public interest without considering the short-term political ramifications. At the same time, they can take particular situations into account; they do not need to design a policy that is appropriate for every case. And, at least according to some observers, the adversarial structure of judicial decision making creates incentives for both sides to bring forward information, so decisions are likely to be well-informed.[116] On the other hand, judges are limited to reacting to issues in the ways that lawyers have defined them. Moreover, the legalization of issues allows judges to avoid hard questions and tackle the tractable ones—for example, by focusing on a procedural rather than a substantive aspect of the case. And judges are rarely experts in the areas they are making decisions about and so may fail to anticipate consequences of a ruling. The most vocal critics of "adversarial legalism" have focused not on these drawbacks, however, but on the win-lose aspect of judicial decision making, saying it promotes resistance, backlash, and endless appeals. Defenders of the courts reply that most court cases are settled, and that judges spend a great deal of time trying to get the two sides to negotiate with one another.

QUESTIONS TO CONSIDER

- What do you think is the appropriate role for motorized recreation on public lands and waters, and why?
- ORV users have argued strongly for a collaborative approach to resolving disputes over access to public lands. Why do you think they support a change in the process? Would a collaborative process have produced a "better" outcome in the case of snowmobiles in Yellowstone National Park than the adversarial one did? If so, why, and in what respects? If not, why not?
- Technological improvements, driven by the threat of regulation, contributed significantly to the resolution of the controversy over snowmobiles in Yellowstone. Can you describe other instances in which this has been (or could be) the case?
- The debate over snowmobiles in Yellowstone National Park highlights the contrasting goals in the National Park Service mission. Do you think our national parks should be preserved from human impact, or should they be cultivated for human enjoyment? Can a compromise exist between the two ideas?

NOTES

1. Richard E. Neustadt, *Presidential Power and the Modern Presidents* (New York: Free Press, 1990).
2. Harold Seidman and Robert Gilmour, *Politics, Position, and Power: From the Positive to the Regulatory State*, 4th ed. (New York: Oxford University Press, 1986), 166–167.

3. Terry Moe, "The Presidency and the Bureaucracy: The Presidential Advantage," in *The Presidency and the Political System*, ed. Michael Nelson (Washington, D.C.: CQ Press, 1995), 412–413.

4. Kenneth Mayer, *With the Stroke of a Pen: Executive Orders and Presidential Power* (Princeton, N.J.: Princeton University Press, 2001), 10.

5. William G. Howell, *Power Without Persuasion: The Politics of Direct Presidential Action* (Princeton, N.J.: Princeton University Press, 2003).

6. Karen O'Connor and Bryant Scott McFall, "Conservative Interest Group Litigation in the Reagan Era and Beyond," in *The Politics of Interests*, ed. Mark P. Petracca (Boulder, Colo.: Westview Press, 1992), 263–281; Steven M. Teles, *The Rise of the Conservative Legal Movement: The Battle for Control of the Law* (Princeton, N.J.: Princeton University Press, 2008).

7. Lee Epstein, *Conservatives in Court* (Knoxville: University of Tennessee Press, 1985), 148.

8. As legal scholar Lettie Wenner explains, the Northeast, Midwest, West Coast, and D.C. circuit courts have been favorably disposed toward environmental litigants since the 1970s; judges in the Southeast, Southwest, and Rocky Mountains have tended to favor development interests. "Interest groups," she notes, "are aware of the forums that are most favorably disposed toward their cause. Whenever they have a choice about where to raise an issue, they take their case to the court most likely to agree with them." See Lettie M. Wenner, "Environmental Policy in the Courts," in *Environmental Policy in the 1990s: Toward a New Agenda*, 2d ed., ed. Norman J. Vig and Michael E. Kraft (Washington, D.C.: CQ Press, 1994), 157.

9. Cass R. Sunstein, Lisa Michelle Ellman, and David Schkade, "Ideological Voting on Federal Courts of Appeals: A Preliminary Investigation" (John M. Olin Program in Law and Economics Working Paper No. 198, 2003). Available at http://chicagoun bound .uchicago.edu/cgi/viewcontent.cgi?article=1246&context=law_and_eco nomics.

10. Paul Schullery, *Searching for Yellowstone: Ecology and Wonder in the Last Wilderness* (Boston: Houghton Mifflin, 1997), 29.

11. Quoted in Paul Schullery and Lee Whittlesey, *Myth and History in the Creation of Yellowstone National Park* (Lincoln: University of Nebraska Press, 2003), 9.

12. Cornelius Hedges, quoted in Schullery, *Searching for Yellowstone*, 55.

13. Quoted in Schullery and Whittlesey, *Myth and History in the Creation of Yellowstone National Park*, 4. Note that historians have challenged the NPS myth and, in particular, the notion that the original promoters of the park idea were motivated by altruism rather than commercial ambitions.

14. George B. Hartzog Jr., *Battling for the National Parks* (Mt. Kisco, N.Y.: Moyer Bell Limited, 1988), 6.

15. Quoted in Michael Frome, *Regreening the National Parks* (Tucson: University of Arizona Press, 1992), 8.

16. Quoted in Hartzog, *Battling for the National Parks*, 6.

17. Richard West Sellars, "The Roots of National Park Management," *Journal of Forestry*, January 1992, 15–19.

18. Frome, *Regreening the National Parks*, 8.

19. Jeanne Nienaber Clarke and Daniel C. McCool, *Staking Out the Terrain: Power and Performance Among Natural Resource Agencies*, 2d ed. (Albany: SUNY Press, 1996), 7.

20. Quoted in Hartzog, *Battling for the National Parks*, 6.

21. Sellars, "The Roots of National Park Management."

22. Frome, *Regreening the National Parks*, 11.

23. Robert Cahn, "Will Success Spoil the National Parks," *The Christian Science Monitor*, May 1–August 7, 1968.

24. Ronald A. Taylor and Gordon Witkin, "Where It's Nature vs. Man vs. Machines," *U.S. News & World Report*, April 28, 1986, 68.

25. Jerry Adler, "Can We Save Our Parks?" *Newsweek*, July 28, 1986, 48–51.

26. John Mitchell, "Our National Legacy at Risk," *National Geographic*, October 1994, 18–29, 35–55.

27. U.S. GAO, *National Park Service: Activities Outside Park Borders Have Caused Damage to Resources and Will Likely Cause More*, GAO/RCED-94-59 (January 1994).

28. Schullery, *Searching for Yellowstone*.

29. Robert B. Keiter, *Keeping Faith with Nature: Ecosystems, Democracy, and America's Public Lands* (New Haven, Conn.: Yale University Press, 2003).

30. Todd Wilkinson, "Snowed Under," *National Parks*, January 1995, 32–37.

31. Thomas J. Knudson, "Yellowstone and Staff Are Strained as Winter Visitors Swell," *The New York Times*, March 24, 1987, 16.

32. Quoted in ibid.

33. Ibid.

34. Kevin McCullen, "Snowmobiles Kick Up Concern," *Rocky Mountain News*, March 7, 1994.

35. Lynne Bama, "Yellowstone Snowmobile Crowd May Hit Limit," *High Country News*, March 6, 1995.

36. James Brooke, "A Quiet, Clean, Solitary Winter in Yellowstone Park? Vroom! Cough! Think Again," *The New York Times*, February 18, 1996, Sec. 1, 16.

37. Dan Egan, "Yellowstone: Geysers, Grizzlies and the Country's Worst Smog," *High Country News*, April 1, 1996.

38. Ibid. Note, however, that the Montana Department of Environmental Quality disputed the method park officials used to measure air quality.

39. "Parks to Study Snowmobiles' Effect on Bison," *The New York Times*, September 28, 1997, 30.

40. Quoted in Bryan Hodgson, "Snowmobile Eruption: Vehicles Packing Yellowstone in Winter, Too," *Cleveland Plain Dealer*, February 27, 1995, 5E.

41. Quoted in Scott McMillion, "Snowmobiles Remain an Issue," *High Country News*, October 27, 1997.

42. Quoted in Ben Brown, "Snowmobile! Machines of Winter a Necessity for Some, a Nuisance to Others," *USA Today*, January 18, 1990, 8C.

43. Quoted in Kevin McCullen, "Winter Crowds Threaten Yellowstone," *Rocky Mountain News*, March 7, 1994, 6.

44. Ben Long, "Yellowstone's Last Stampede," *High Country News*, March 12, 2001.

45. Ibid.

46. Brown, "Snowmobile!"

47. Quoted in Blaine Harden, "Snowmobilers Favoring Access to Yellowstone Have Found an Ally in Bush," *The New York Times*, March 6, 2002, 16.

48. Quoted in Egan, "Yellowstone: Geysers, Grizzlies and the Country's Worst Smog."

49. Quoted in Bill McAllister, "Snowmobiles Can Stay; Park Service Says Yellowstone Decision Strikes Balance," *The Denver Post*, February 21, 2003, 2.

50. Quoted in Bryan Hodgson, "Environmentalists Decry Use of Snowmobiles in Yellowstone National Park," *The Tampa Tribune*, February 19, 1995, 5.

51. Quoted in Brooke, "A Quiet, Clean, Solitary Winter in Yellowstone Park?"

52. Quoted in Brown, "Snowmobile!"

53. Quoted in Katharine Q. Seelye, "Bush May Lift Park's Snowmobile Ban," *The New York Times*, June 24, 2001, 15.

54. Donald A. Manzullo, Chair, "Remarks Before the U.S. House of Representatives Small Business Committee, Subcommittee on Tax, Finance, and Exports," July 13, 2000.

55. Quoted in William Booth, "At Yellowstone, the Din of Snowmobiles and Debate," *The Washington Post*, February 6, 2003, 3.

56. Quoted in Harden, "Snowmobilers Favoring Access to Yellowstone Have Found an Ally in Bush."

57. Quoted in Joseph B. Verrengia, "Thin Snowpack Revs Up Arguments over Snowmobiles," *Rocky Mountain News*, January 19, 1995, 64.

58. Brooke, "A Quiet, Clean, Solitary Winter in Yellowstone Park?"; Tim Wade, Chair, Park County, Wyoming, Commission, "Testimony Before the U.S. Senate Energy Committee, Subcommittee on National Parks, Historic Preservation, and Recreation," May 25, 2000.

59. Quoted in Douglas Jehl, "National Parks Will Ban Recreational Snowmobiles," *The New York Times*, April 27, 2000, 12.

60. Brown, "Snowmobile!"

61. Quoted in Hodgson, "Environmentalists Decry Use of Snowmobiles in Yellowstone National Park."

62. Keiter, *Keeping Faith with Nature*, 262.

63. Tom Kenworthy, "Parkland Debate Keeps Trekking," *USA Today*, April 25, 2003, 3.

64. Quoted in Kim Cobb, "Invasion of Off-Road Vehicles," *Houston Chronicle*, July 23, 2000, 1.

65. Quoted in Taylor and Witkin, "Where It's Nature vs. Man vs. Machines."

66. "Snowmobiles Plague Yellowstone," *USA Today*, February 19, 1998, 14A.

67. Quoted in Reuters, "Snowmobile Ban Sought," *The Gazette* (Montreal, Quebec), January 22, 1999, 10.

68. Veronica Gould Stoddart, "Green Groups Fuming over Park Snowmobiles," *USA Today*, February 12, 1999, 4D.

69. Quoted in Stoddart, "Green Groups Fuming over Park Snowmobiles."

70. Shortly thereafter, Bluewater Network sued the NPS to exclude personal watercraft (PWC) from the twenty-one parks where they were permitted, and the Personal Watercraft Industry Association countersued to preserve access. On April 12, 2001, a federal judge agreed with Bluewater and required NPS to undertake park-specific EISs on PWC use. In April 2002 the NPS announced that five of the twenty-one remaining parks would ban PWC permanently; the other sixteen parks put temporary bans in place while they conducted environmental assessments and came up with regulations.

71. Quoted in Jehl, "National Parks Will Ban Recreational Snowmobiles."

72. "A Broad Ban on Snowmobiles," *The New York Times*, April 29, 2000, 12.

73. Donald J. Barry, assistant secretary for Fish and Wildlife and Parks, Department of the Interior, "Testimony Before the Senate Energy Committee, Subcommittee on National Parks, Historic Preservation, and Recreation," May 25, 2000.

74. Ibid.

75. Jim Hughes, "Snowmobiles Shatter the Sounds of Silence," *The Denver Post*, March 10, 2000, 1.

76. Barry, "Testimony Before the Senate Subcommittee on National Parks, Historic Preservation, and Recreation."

77. Mark Simonich, director, Montana Department of Environmental Quality, "Testimony Before the Senate Energy Committee, Subcommittee on National Parks, Historic Preservation, and Recreation," May 25, 2000.

78. Jim Hughes, "Snowmobile Ban Favored," *The Denver Post*, March 15, 2000, 1.

79. Rachel Odell, "Parks Rev Up to Ban Snowmobiles," *High Country News*, March 27, 2000.

80. Theo Stein, "Snowmobile Ban Set for 2003 at Two Parks," *The Denver Post*, November 23, 2000, B1.

81. Kevin Collins, legislative representative, National Parks Conservation Association, "Testimony Before the House Small Business Committee, Subcommittee on Tax, Finance, and Exports," July 13, 2000.

82. Kit Miniclier, "Town Roaring Mad over Snowmobiles' Future," *The Denver Post*, May 21, 2000, B6.

83. Quoted in Bill McAllister, "Snowmobiles to Get Boot in Yellowstone, Teton," *The Denver Post*, October 11, 2000, B1.

84. Kit Miniclier, "Noise vs. Nature: Record Number of Snowmobilers in Yellowstone," *The Denver Post*, February 18, 2001, B1.

85. Kit Miniclier, "Park Snowmobile Phase-Out Delayed; Yellowstone Anticipates Record Crowds," *The Denver Post*, February 6, 2002, B6; Todd Wilkinson, "Snowmobile Buzz Echoes in White House," *The Christian Science Monitor*, February 21, 2001, 3.

86. Quoted in Katharine Q. Seelye, "U.S. to Reassess Snowmobile Ban in a Park," *The New York Times*, June 30, 2001, 10.

87. Miniclier, "Park Snowmobile Phase-Out Delayed."

88. Katharine Q. Seelye, "Snowmobilers Gain Against Plan for Park Ban," *The New York Times*, February 20, 2002, 14.

89. Kit Miniclier, "Winter Ban in Parks Eased; Snowmobiles Still OK in Yellowstone, Teton," *The Denver Post*, June 26, 2002, 1.

90. Gary Gerhardt, "Proposal for Parks Unveiled," *Rocky Mountain News*, November 9, 2002, 14.

91. "Yellowstone: Status Quo on Snowmobiles," (Minneapolis) *Star Tribune*, November 14, 2002, 20.

92. In addition to relaxing the restrictions on snowmobiles in the park, the Bush administration undermined the EPA's ability to regulate pollution from snowmobiles. New standards released in April 2002 required a 30 percent reduction in emissions by 2006 and a 50 percent reduction by 2010. But snowmobile industry officials appealed to John Graham of the Office of Regulatory Affairs to eliminate the 50 percent limit. In response to this plea, Graham asked the EPA to conduct a full cost-benefit analysis of its snowmobile rule (the "nonroad" rule). See Arthur Allen, "Where the Snowmobiles Roam," *The Washington Post Magazine*, August 18, 2002, W15.

93. *Fund for Animals v. Norton*, 294 F. Supp. 2d 92-117, D.D.C. (2003).

94. Quoted in Jim Robbins, "New Snowmobile Rules Roil Yellowstone," *The New York Times*, December 22, 2003, 22.

95. Quoted in Steve Lipsher, "The Brink of Beauty, the Edge of Ruin," *The Denver Post*, January 11, 2004, 1.

96. Ibid.

97. *International Snowmobile Manufacturers Association v. Norton*, 304 F. Supp. 2d 1278, D. Wyo. (2004).

98. Quoted in Michael Janofsky, "U.S. Would Allow 720 Snowmobiles Daily at Yellowstone," *The New York Times*, August 20, 2004, 14.

99. Quoted in Felicity Barringer, "Judge's Ruling on Yellowstone Keeps It Open to Snowmobiles," *The New York Times*, October 16, 2004, 9.

100. Quoted in Felicity Barringer, "Secretary Tours Yellowstone on Snowmobile," *The New York Times*, February 17, 2005, 18.

101. Becky Bohrer, "Official: Snowmobile Average Falls Below Daily Cap," Associated Press, April 20, 2005.

102. Barringer, "Secretary Tours Yellowstone."

103. The report also found that snowcoaches had higher sound levels at higher speeds, but snowmobiles were more audible for more time. A subsequent study found that snowcoaches elicited greater wildlife responses than did snowmobiles, probably because they were bigger, but also because the NPS had not yet required newer, quieter snowcoaches. See Dan Berman, "Yellowstone Snowmobile Plan May Be Delayed," *Land Letter*, October 26, 2006; Scott Streater, "Yellowstone Snowmobile Debate Overlooks Major Pollution Source: Snowcoaches," *Land Letter*, August 13, 2009.

104. Noelle Straub, "Judge Tosses Yellowstone Snowmobile Plan, Berates Park Service," *Land Letter*, September 18, 2008.

105. Quoted in Timothy Egan, "Rangers Take On Urban Woes Spilling Into Wide Open Spaces," *The New York Times*, July 26, 2006, 1. Egan reports that between 1996 and 2006 the number of rangers with police power fell from more than 980 to 500; by 2006 there was one law enforcement agent for every 291,000 acres or for every 733,000 visitors.

106. Felicity Barringer and William Yardley, "A Surge in Off-Roading Stirs Dust and Debate in the West," *The New York Times*, December 30, 2007, 1.

107. Quoted in Dan Berman, "Retired Land Managers Call for Expanded ORV Regs, Fees," *Land Letter*, July 5, 2007.

108. Eric Bontrager, "States Move to Curb Off-Highway Vehicles as Federal Efforts Lag," *Greenwire*, December 24, 2008.

109. Scott Streater, "NPS Pursues Efforts to Protect 'Soundscapes,'" *Land Letter*, August 7, 2008.

110. Elizabeth Shogren, "15 Years of Wrangling Over Yellowstone Snowmobiles Ends," NPR, October 22, 2013.

111. Ibid.

112. Adler, "Can We Save Our Parks?"

113. Frome, *Regreening the National Parks*, 7.

114. Nicholas Kristoff, "Yellowstone in Winter a Snowmobiler's Paradise," *The New York Times*, December 26, 2002, 21.

115. John H. Cushman Jr., "Priorities in the National Parks," *New York Times*, July 26, 1998, Sec. 5, 11.

116. Others argue the opposite: that adversarial decision making creates incentives for each party to conceal information.

CHAPTER TEN

CRISIS AND RECOVERY IN THE NEW ENGLAND FISHERIES

In 1994 many New Englanders were startled to learn of a crisis in the cod, flounder, and haddock fisheries. The region's groundfish stocks were on the verge of collapse, and federal regulators intended to institute strict new rules to try and save them.[1] In the meantime, regulators were shutting down indefinitely Georges Bank—a 6,600-square-mile area more than 100 miles offshore that was once the most prolific fishing grounds in the world. The fishermen were in an uproar, and many charged that the federal management regime instituted in 1976, not overfishing, was the culprit. Regulators held firm, however, pointing out that government scientists had been warning of the groundfish's demise for well over a decade. Since the mid-1990s fishery rules have tightened substantially, and some groundfish species are rebounding. But a long-term recovery of the region's groundfish stocks depends on the political will to impose and maintain restrictions on the number of fish that can be caught.

Eliminating overfishing is essential not just in New England, but around the world. Of the fish stocks analyzed by the United Nations Food and Agriculture Organization, 57 percent are fully exploited, nearly 30 percent are overexploited, and only 12.7 percent are non-fully exploited.[2] About 90 percent of the world's large, predatory fish are already gone.[3] By eliminating top predators, fishing has destabilized and degraded marine ecosystems, rendering them less resilient in the face of pollution, climatic changes, and other assaults.[4] Moreover, vast quantities of nontarget species—including other fish, seabirds, marine mammals, and turtles—are caught and killed in fishing nets.[5]

Efforts to eliminate overfishing in New England illuminate the complexity of managing common pool resources. Garrett Hardin captured the special challenges of such management in a model he popularized in a 1968 *Science* magazine article titled "The Tragedy of the Commons." According to Hardin's model, when a resource is open to everyone, those who use it will inevitably overexploit it. "The logic of the commons remorselessly generates tragedy," he wrote.[6] The tragedy of the commons applies to fisheries in the following way: as more fishermen enter a fishery, each one eventually experiences declining yields for the same unit of effort.

As the profit margin shrinks, all rational fishermen redouble their efforts, recognizing that if they cut back for the good of the resource, someone else will catch the fish they leave behind. In doing what is individually rational, however, fishermen produce a result that is collectively disastrous.[7] From Hardin's perspective, the only way to head off the destruction of a common resource was for an authoritarian state to regulate its use or control access to it by converting it into private property. In the latter regime, owners of the resource theoretically have a stake in conserving it because they now have the sole rights to the "economic rents" it generates. One way to create property rights in a fishery is to distribute individual transferable quotas (ITQs) to individual fishermen or fishing enterprises. Under an ITQ system, the owner of a quota share has exclusive rights to a particular fishery, to take a certain proportion or amount of fish in the fishery, and—in most instances—to sell or lease that right to others.

The Australian lobster fishing regime is one example of a successful ITQ system. Beginning in the 1960s the Australian government set a limit on the total number of lobster traps and then assigned licenses to working fishermen. From then on, any newcomer had to buy a license from someone already working in the fishery. This is similar to how the New York City taxi medallion system works or Montana's liquor license system. By 2000 Australian lobster fishermen worked 187 days a year (compared to as many as 240 days per year for Rhode Island lobster catchers), tending sixty traps apiece (versus as many as 800 as is typical in Rhode Island). Proponents of the Australian system point out that it has not only rejuvenated the stocks and enhanced the lives of the fishermen, who make significantly more money than their American counterparts, but it has also eased tension between fishermen and scientists, who now work collaboratively to keep the fishery healthy.[8]

ITQs are currently the most widely discussed solutions for overcrowded fisheries. Implementing an ITQ system can have broad and controversial social consequences. For example, such systems can concentrate the benefits of a fishery in the hands of a privileged few. Political scientist Elinor Ostrom and others have questioned the assumptions that underlie the tragedy of the commons model, and hence the need for privatization, pointing out that people are not always as self-serving, individualistic, and short-sighted as the model posits.[9] Ostrom cites numerous examples of creative institutional arrangements devised by communities to solve the problems of managing common pool resources. For instance, the Maine lobster fishermen maintained a system of self-regulation for more than a century.[10] Although this and other examples suggest that the tragedy of the commons is not inevitable, they have limited applicability. As Ostrom explains, comanagement is more likely to succeed where common pool resources have been managed by a settled, homogeneous population under long-enduring institutions, so that the people believe they or their children will reap the benefits of conservation and are therefore more likely to adhere to agreements.[11]

The New England fisheries case illustrates the challenges posed by management schemes in which regulated interests play a major role in devising and implementing government policy. Federal law mandates that regional fishery councils, made up of members who are familiar with the resource, formulate the rules under which fisheries operate. In theory, such an approach enables regulators to take into account the perspectives of the regulated sector and therefore to make rules that are both fair and sensible; in addition, because each fishery has its own dynamics, local knowledge of fish behavior is essential to credible, effective management.[12] Furthermore, if they are *not* involved in decision making, regulated interests often perceive rules as arbitrary and coercive and resist complying with them. As is true of every policy—from provisions in the tax code to anti theft laws—regulators cannot simply institute and enforce a set of rules; they rely heavily on the voluntary cooperation of those who must abide by them.

That said, the struggle to manage groundfishing in collaboration with fishermen exemplifies some of the conflicts that arise over natural resource conservation: like other resource users, fishermen face strong incentives to support risk-tolerant interpretations of the available science, focus on short-term economic conditions rather than the long-term sustainability of the resource, and resist moving away from a degraded baseline. Institutional design can change the incentives faced by fishermen and therefore enhance cooperation between fishermen and fishery managers. But the political obstacles to instituting and implementing rules that genuinely transform incentives, and hence behavior, are enormous, and overcoming them requires a great deal of political fortitude.

This case increases our understanding of the pivotal role of litigation in bringing about major environmental policy shifts. As both the grazing policy and spotted owl cases (Chapters 7 and 8) reveal, once a single interest has become entrenched in a subsystem, it is difficult to dislodge. Attempts to alter policy from both outside and within an agency encounter resistance from its clientele, congressional overseers, and even its own personnel, who have become accustomed to a particular routine. A successful legal challenge can, however, shift the internal balance of power, giving those who previously were isolated—in this case, agency scientists—more authority. Moreover, by requiring an agency to move in a particular direction, the courts can narrow the range of options available to managers.

BACKGROUND

The history of New England is inextricably tied to the fish found off its shores. In the year 1500, explorer John Cabot described the Grand Banks area off Northeast Canada as "swarming with fish [that they] could be taken not only with a net but in baskets let down with a stone."[13] From Canada to New Jersey, the waters teemed with cod, supporting a relatively stable fishing industry for 450 years. Groundfish—not just cod but also yellowtail flounder, haddock, American plaice, and pollock—were the backbone of the region's fishing trade. In fact, the Sacred Cod, a commemorative wooden plaque, still hangs in the Massachusetts statehouse, symbolizing the state's first industry and the source of its early wealth. For many years, the ports of Gloucester and New Bedford were among the nation's most prosperous.

At the beginning of the twentieth century, the introduction of steam-powered trawlers that drag nets across the ocean floor prompted concern among scientists about bottom-dwelling animals and plants. Then, in 1930, the fleet engaged in sufficiently intensive fishing that it caused a crash in the Georges Bank haddock fishery. Still, fishing in New England remained a small-scale affair, and after fishermen shifted their effort northward, the haddock stocks recovered. The same pattern repeated itself for other species: because the overall harvests were modest, there was no persistent "recruitment overfishing," so depleted stocks could rebound when fishermen moved on to other species.[14]

In the mid-1950s, however, huge factory ships from Europe and the Soviet Union began to roam the North Atlantic just beyond U.S. coastal waters. Some of those boats exceeded 300 feet in length, brought in as much as 500 tons of fish in a single haul, and could process and deep-freeze 250 tons a day. They operated around the clock in all but the worst weather and stayed at sea for a year or more.[15] Called "factory-equipped freezer stern trawlers," or factory trawlers for short, the fleet was "a kind of roving industrial complex."[16] As fisheries expert William Warner describes the scene, the ships "paced out in long diagonal lines, plowing the best fishing grounds like disk harrows in a field."[17] Between 1960 and 1965 North American groundfish landings increased from 200,000 metric tons to 760,000 metric tons.[18]

For many years after the first factory trawler invasion, Canadian and American fishery officials clung to the hope that the fleets of both countries could withstand the assault. Some hoped that the International Commission for Northwest Atlantic Fisheries might reverse the trend of overfishing in the early 1970s when it instituted a management system allocating quotas by country, with the sum for each species equal to the total recommended catch. But by the time those quotas took effect, enormous damage had already been done: between 1963 and 1974, groundfish populations declined almost 70 percent, and by 1974 many species had fallen to the lowest levels ever recorded.[19] Haddock was at an all-time low throughout its Northwest Atlantic range; the species was so rare, in fact, that biologists feared the end of a commercially viable fishery, if not extinction. Even the bountiful cod was showing signs of decline, as were the redfish, yellowtail flounder, and many other prime market fish.

The turning point came in 1974 when 1,076 fishing vessels swarmed across the Atlantic to fish North American waters. Their total catch of 2.176 million metric tons was ten times the New England and triple the Canadian catch.[20] Although the total was huge, the catch per vessel was down, and the fish were running smaller than before, despite the fact that the foreign vessels were fishing longer hours with improved methods over a larger range for a greater part of the year. In fact, the foreign catch was better for five of the preceding six years—slowly declining from a peak of 2.4 million metric tons in 1968—with fleets of equal or lesser size.[21] American fishermen charged that the international commission was ineffectual and demanded that the United States assert control over the offshore fishery.[22]

In response to fishermen's complaints about the foreign fleets and scientists' concern about the precarious status of fish stocks, Congress passed the Magnuson

Fisheries Conservation and Management Act in 1976. The act had two contradictory goals: to rejuvenate the American fishing fleet and to restore and conserve fish stocks. It unilaterally asserted U.S. jurisdiction over fisheries within an exclusive economic zone that extended 200 miles from the coast and authorized the National Marine Fisheries Service (NMFS, pronounced "nymphs"), a line office of the U.S. Department of Commerce's National Oceanic and Atmospheric Administration (NOAA), to administer the nation's resources between 3 miles and 200 miles off the coast.[23]

The Magnuson Act also established eight regional councils to work with the five regional NMFS offices to develop management plans for the offshore fisheries. The act directed that regional councils comprise federal and state officials, as well as "individuals who, by reason of their occupation or other experience, scientific expertise, or training are knowledgeable regarding the conservation and management of the commercial and recreational harvest."[24] Although the act vests final authority for rulemaking with the secretary of commerce, Congress clearly intended the councils to play the dominant role. According to the legislative history of the act,

> The regional councils are, in concept, intended to be similar to a legislative branch of government. . . . The councils are afforded a reasonable measure of independence and authority and are designed to maintain a close relation with those at the most local level interested in and affected by fisheries management.[25]

THE CASE

The Magnuson Act allows the Department of Commerce to reject, accept, or partially approve a council's fishery management plan (FMP) but not to change it, an arrangement that from the beginning encouraged acceptance of lax plans. NMFS was reluctant to impose its own plan over council objections, as the rules would be nearly impossible to enforce without fishermen's compliance. Exacerbating NMFS's weak position was the tendency of members of Congress to intervene on behalf of fishing interests whenever the agency threatened to institute strict rules. Finally, although the region's fishing industry was not unified, it managed to dominate policy because it encountered little opposition. Aside from scientists from the Northeast Fisheries Science Center (NEFSC), few voices spoke out on behalf of fishery conservation and precautionary management. Most environmental groups were not paying attention to fisheries—they were far more interested in marine mammals and ocean pollution—nor were they alerting the public to the impending groundfish collapse.

Scientists' Assessments of New England's Groundfish

Like range management and forestry science, both of which historically were dominated by resource extraction rather than ecosystem health, fisheries science traditionally has been concerned with determining the "maximum sustainable yield" of a fishery. In practice, this emphasis has meant developing expedient indicators, primary among which are fish stock assessments. A stock assessment includes estimates of the abundance of a particular fish stock (in weight or number of fish) and the rate at which the fish are being removed as a result of harvesting and other causes (mortality). It also includes one or more reference estimates of the harvesting rate or abundance at which the stock can maintain itself in the long term. Finally, a stock assessment typically contains one- to five-year projections for the stock under different management scenarios.

Because they cannot actually count the fish, scientists use a variety of data to estimate the abundance of each stock and its population trends. They rely primarily on two sources: reports on commercial fish landings and NMFS resource surveys. The landing reports include the number of fish caught, their size, and the ratio of fish caught to the time spent fishing (catch per unit of effort). Landing data tell only part of the story, however, since many fish caught are discarded because they are too small to sell, exceed the catch limit, or belong to a species whose catch is prohibited. (These discards, known as bycatch, comprise a large portion of fishing mortality worldwide.[26]) Unlike fishing boats, which search for the largest aggregations of fish, NMFS's research trawlers conduct stratified random sample surveys of a wide range of locations. Such "fishery-independent" surveys are especially important for schooling species because fishermen can maintain high catch rates on them by selectively targeting areas where the fish congregate. The New England groundfish survey, which began in 1963, is the nation's oldest and most reliable and provides the longest time-series data.[27]

Although they acknowledged that their stock assessments were uncertain and their understanding of groundfish biology and life history incomplete when the Magnuson Act took effect at the end of 1976, NEFSC scientists nevertheless issued strong warnings about the need to cut back fishing of the region's groundfish. They advised the New England Fishery Management Council to set 1977 Gulf of Maine and Georges Bank commercial cod catches at approximately half the levels reached between 1970 and 1974. Scientists feared that fishing beyond the recommended level would lead to a precipitous decline. They also noted a pronounced decline in haddock since 1967 and found yellowtail flounder stocks "severely depressed." They recommended strict catch limits to enable these stocks to recover as well.[28]

Establishing a "Cooperative" Management Regime

The newly appointed council was reluctant to follow scientists' advice, however. In theory, the cooperative management regime mandated by the Magnuson Act would ensure that managers took both the best available science and the needs and expertise of fishermen into account when developing conservation measures. But the regional councils soon became a focal point for controversy because their members knew a lot more about catching and marketing fish than about marine biology or natural resource stewardship. The New England Council, in particular, was dominated by current or former commercial fishermen, who were understandably reluctant to impose stringent controls on their peers.[29] One former council member explained, "As members of the community, it's difficult for them to divorce themselves from the consequences of their actions."[30]

In response to scientists' warnings about decimated stocks, however, the council did put some restrictions in place. Modeling its approach on the regulatory regime of the International Commission for Northwest Atlantic Fisheries, the council imposed overall catch quotas, prohibited fishing in spawning areas, required large-holed mesh (to allow young fish to escape), enacted minimum fish-size requirements, and limited the amount of yellowtail flounder that could be caught per trip. Nevertheless, in the first year under the new regime, fishermen regularly exceeded the total catch quotas, so NMFS often had to close the fishery abruptly. As a result, "the trip limits were perceived as unfair; many fish were mislabeled and handled illegally; and the closures were extremely unpopular."[31] In addition, council meetings were unfocused and chaotic; council members vacillated and were reluctant to make difficult choices, which tarnished their image as decision makers.

Making matters worse for the council, many fishermen simply did not believe scientists' claims that the groundfish were in trouble. Fishermen had a host of complaints about stock assessments. They criticized the NEFSC for using ancient gear that did not detect or catch fish as effectively as modern gear. (NEFSC scientists explained that they had standardized the survey in a variety of ways, from using the same boats and gear to survey methods, to ensure a consistent time-series of comparable data.) Fishermen also regarded fishery science as inaccessible because it relied heavily on complex mathematical models that few but the most specialized scientists could understand. Finally, fishermen pointed out that fishery science was weakest in an area that was crucial for fishery management: understanding fish behavior, life history, and interaction with other species. This, they noted, was where their own local knowledge—which many scientists dismissed as anecdotal—was likely to be most valuable.

The underlying reason for the disjuncture was that unlike scientists, who were precautionary, the fishermen's perspective was cornucopian: they preferred to believe in the ocean's resilience and near-limitless bounty. And high groundfish landings in the early part of 1977 seemed to belie scientists' pessimism. According to one observer,

> Codfish landings throughout New England were nearly twice what they [had been] the previous spring (13.4 million pounds compared to 7.3 million pounds). . . . Piers in Gloucester and New Bedford were groaning under the load of spring landings and there was much concern that the processing houses for frozen fish were going to be grossly inadequate. Processors were running their facilities seven days a week in June and adding extra shifts, and still they could not keep up with the boats.[32]

The surge in landings was probably the belated fruit of the international commission's quota system, instituted in the mid-1970s; nevertheless, it widened the credibility gap between scientists and fishermen.

Although unified by their distrust of scientists and contempt for fishery managers, New England's fishermen were deeply split in other respects, which complicated fishery management even further. The fishermen were divided by gear type, vessel size, vessel ownership, fishing style, and port of origin. Large trawlers accused small boats of taking advantage of loopholes in the laws and lax enforcement of fishery regulations to fish out of season or use illegal nets. Small boat owners blamed trawlers with their advanced sonar tracking system for cleaning out the ocean. Gillnetters pointed out that, unlike trawlers, they did not catch juvenile fish and did not tear up the ocean bottom. Longliners said their gear did not affect the bottom either and was even more selective than gillnets. Both derided draggers (small trawlers) for pounding the bottom with their heavy doors. The otter trawl, other gear users pointed out, had the greatest bycatch problem because unwanted fish were damaged in the net, brought up too quickly, or not thrown back soon enough. Draggers, on the other hand, complained about ghost-net fishing, lost lines, and hooks "souring" the bottom. The Maine Yankees disparaged the New Bedford Portuguese and the Gloucester Sicilians, who in turn criticized each other.[33] Unable to reconcile the many factions, the council simply tried to devise rules that would be perceived as "democratic"—that is, affecting everyone equally.

Devising a Multispecies Groundfish Fishery Management Plan

During the first two years of council operations, differences between scientists and fishermen, uncertainty about the exact administrative procedures needed to ensure timely implementation of plans, apparent misunderstandings between the council and NMFS, the vagueness of the concepts on which the plans were supposed to be based, and the inexperience of many council members as fishery managers all lowered morale and undermined the fishery management process.[34] In hopes of moving away from the short-term, reactionary policymaking it had engaged in thus far, in summer 1978 the council set to work on a more comprehensive, long-term fishery management plan. Ironically, in 1982—nearly four years after it began—the council submitted for Commerce Department approval an Interim Groundfish FMP that substantially weakened controls on fishing. Encouraged by apparent improvement in the stocks and under heavy pressure from fish processors, the council abandoned trip limits and quotas; instead, the plan allowed open fishing and required only "age-at-entry" controls in the form of minimum fish sizes and minimum mesh sizes. In a nod to conservation, the plan retained some spawning area closures (March to May) and instituted voluntary catch reporting as a data collection device.

After sustained pressure from several (though not all) fishing groups, NMFS agreed to forgo the usual four- to six-month comment period and implement the interim plan immediately under emergency regulations, a procedure normally reserved for a resource, not an industry, in jeopardy. The decision came after the commerce secretary and federal fisheries officials met in Washington with fish dealers and processors who spoke of the "disaster" that had befallen their businesses in the past two years. The dealers described falling employment in their industry, while the processors claimed that the quotas had prevented the boats from bringing in enough fish to meet operating costs or consumer demand.

In the years following implementation of the interim plan, federal scientists continued to urge the New England panel to institute more restrictive measures. Early on, it became apparent that age-at-entry controls were insufficient to protect fish stocks. Many fishermen were not complying with minimum mesh size requirements, so juvenile fish were virtually unprotected. Making matters worse, in response to loan guarantee programs and generous depreciation and operating cost allowances added to the federal tax code, fishermen had begun investing in new boats and high-tech equipment.[35] Because entry into the fishery was unlimited, the number of otter trawlers doubled between 1976 and 1984, even as the size and efficiency of the boats increased substantially.[36] The result was that by 1980 the New England fleet was catching 100,000 metric tons of cod, haddock, and yellowtail flounder—double the 1976 level—off Georges Bank, the Gulf of Maine, and Cape Cod. According to scientists, the industry was taking 50 percent to 100 percent more than the already weakened groundfish stocks could sustain. But, even though NEFSC cruises started to show stocks dropping, catches (and fishermen's optimism) remained high until 1983, at which point they started to decline.[37]

Despite its apparent failure to curb fishing, the council proposed formalizing the interim plan's open fishing regime in 1985 as the Northeast Multispecies FMP. Initially, NMFS gave only conditional approval to the plan because agency officials, heeding their scientists' advice, were concerned that the lack of direct controls on fishing mortality would lead to overfishing. In 1987, however, after fishermen got the New England congressional delegation to weigh in heavily on the side of indirect controls, the agency approved the plan, requesting only modest adjustments.[38]

Not surprisingly, the new FMP did little to improve the prospects for New England's groundfish: stock assessments continued to show increases in mortality rates and corresponding decreases in stock sizes. This situation was not unique to New England, and NMFS officials were becoming increasingly frustrated with the regional councils' impotence. So in 1989, in response to the critical 1986 NOAA Fishery Management Study, NMFS scientists revised a crucial section of the Magnuson Act regulations: the Section 602 Guidelines for National Standards to assist in the development of FMPs. The original guidelines had not defined the term *overfishing*, which appears in the Magnuson Act only once: "Conservation and management measures shall prevent overfishing while achieving, on a continuing basis, the optimum yield from each fishery for the U.S. fishing industry."[39] By contrast, the 1989 guidelines mandated that each FMP define overfishing, a significant advance for fishery conservation.[40] The rationale for the change was clear: it was impossible to prevent overfishing if there was no standard against which to measure it, and therefore no way of saying for certain that it was occurring. The second important revision to the guidelines read, "If data indicate that an overfished condition exists, a program must be established for rebuilding the stock over a period of time specified by the Council and acceptable to the Secretary."[41]

The 602 Guidelines revisions had an enormous, albeit not immediate, impact on New England. At the time, the New England Groundfish FMP—like most others—did not define overfishing; in fact, the plan did not even specify an optimal yield.[42] The council had eliminated optimal yield figures because they were too controversial and instead had begun defining optimal yield as "the amount of fish actually harvested by U.S. fishermen in accordance with the measures listed below." In other words, any size catch was optimal, by definition. The FMP also contained biological targets for stocks covered by the plan: the total catch should not exceed 20 percent of the maximum spawning potential. Although these targets were intended to ensure sufficient reproductive potential for long-term replenishment of stocks, the plan's management measures were, in fact, inadequate to achieve them. Seeking to comply with changes in the 602 Guidelines, the New England Council proposed that its overfishing definitions for groundfish be the biological targets already contained in the FMP: 20 percent of the maximum spawning potential. At the same time, the council acknowledged that it was not meeting those targets—in effect, admitting that it was allowing overfishing to occur.

Under such circumstances, the revised guidelines required the council to develop a recovery plan for the overfished stocks, but because the guidelines did not specify a deadline, the council moved slowly. Among the many reasons for its tardiness, the most important was that representatives of the fishing industry remained dubious about the need for new fishing controls. Those council members who did perceive such a need anticipated strong resistance. The consequence was that "nineteen months after admitting that groundfish stocks were overfished, the council had not seriously begun to tackle the effort reduction that it had decided was necessary to end overfishing. Moreover, NMFS showed no signs of stepping in with a secretarial plan."[43]

The Conservation Law Foundation Lawsuit and Amendment 5

In early 1991 scientists reported that New England fishermen were catching less than half as many groundfish as they had a decade earlier. The Massachusetts Offshore Groundfish Task Force estimated that lost landings were costing the region $350 million annually and as many as 14,000 jobs.[44] According to the NEFSC, spawning stocks were now less than one-twentieth what they had been when the Magnuson Act was passed. Despite scientists' dire findings, council members remained reluctant to act, arguing that the series of halfway measures they had passed in the 1980s would, given time, enable stocks to rebound.

Frustrated with the council's inertia and hoping to force it to act, in June 1991 the Boston-based Conservation Law Foundation (CLF) filed a lawsuit against the secretary of commerce. CLF representatives had been attending council meetings for two years, but although they had challenged the council's planning measures, they had lacked a legal basis to hold the council accountable until the issuance of the new 602 Guidelines. The CLF complaint charged that NMFS had failed to prevent overfishing, in violation of its statutory mandate. In August 1991 the CLF and NMFS settled the case by signing a consent decree that compelled the council to develop a stock rebuilding plan. If it failed to do so by a specified deadline, the commerce secretary was required to devise a plan. In compliance with the settlement, the council began to develop Amendment 5 to the Northeast Multispecies FMP.[45]

The lawsuit had an important, if not immediately obvious, impact. Previously, fishing interests, despite their lack of cohesiveness, had managed to dominate fishery policy largely because neither environmentalists nor the general public were attentive to the issue of overfishing. By forcing the agency's hand, however, the lawsuit empowered conservation-oriented managers on the council and within NMFS to argue that they had no choice but to impose stringent regulations. NMFS was quick to use its new clout. On June 3, 1993, after receiving yet another report from its scientists that groundfish were severely depleted, NMFS issued an emergency order closing the eastern portion of Georges Bank to fishing for a month just two days after the area had opened for the season.[46]

Hoping to eliminate the need for more emergency closures, on June 30, 1993, the beleaguered New England Council approved Amendment 5. The amendment's ostensible goal was to reduce the groundfish catch by 50 percent in seven years. It included a moratorium on groundfishing permits and greatly limited the number of days fishermen could catch groundfish. Before June 30 they were allowed to be out year-round, with the exception of periodic closures; now the number of days allowed was to go down each year until by 1998 large boats could fish only 110 days each year. The amendment also limited fishermen to 2,500 pounds of haddock each trip and increased the mesh size of nets in most areas to 5.5-inch diamonds or 6-inch squares to allow young fish to escape. In addition, it required vessel owners and operators to possess valid fishing permits and to keep elaborate fishing logs detailing the species caught and bycatch.

As the council awaited Commerce Department approval of its proposal, the bad news on groundfish stocks continued to roll in. During a two-day meeting in early December, Richard Roe, NMFS's northeast regional director, sought to ban haddock fishing indefinitely off the East Coast in response to reports of drastic declines in landings—from 40,000 metric tons per year in the early 1970s to only 90 metric tons in 1993. Still, the council resisted; it condoned emergency measures to preserve the dwindling stocks but stopped short of endorsing a complete ban. Instead, by a ten-to-four vote, the council recommended that NMFS impose a 500- to 1,000-pound limit on haddock catches by commercial fishing boats and close the portion of Georges Bank where haddock spawn.

As the crisis deepened, however, the council found itself with less and less wiggle room. Responding to warnings from its scientists, at the beginning of Christmas week, Canada imposed sharp restrictions on haddock and other bottom feeders in nearly all of its Atlantic region. Following suit, on Thursday of that week NMFS ordered that New England boats be allowed no more than 500 pounds of haddock per trip and that haddock spawning grounds be closed in January, a month earlier than usual. Then, in late February 1994, federal regulators announced their intention to shut down even more valuable fishing grounds: in addition to Georges Bank, they planned to close a large swath of the Great South Channel and portions of Stellwagen Bank and Jeffreys Ledge (see Map 10.1).

Finally, on March 1 the provisions of Amendment 5 were scheduled to take effect. Fishermen found this series of apparently arbitrary regulations, culminating in the imposition of Amendment 5 rules, infuriating; they believed that government bureaucrats had gotten out of hand. They contended that the days-at-sea limits would impose exorbitant costs on fishermen (most fishing boat owners are independents who have high fixed costs—as much as $100,000 per month in loan

Map 10.1 New England's Fishing Ground

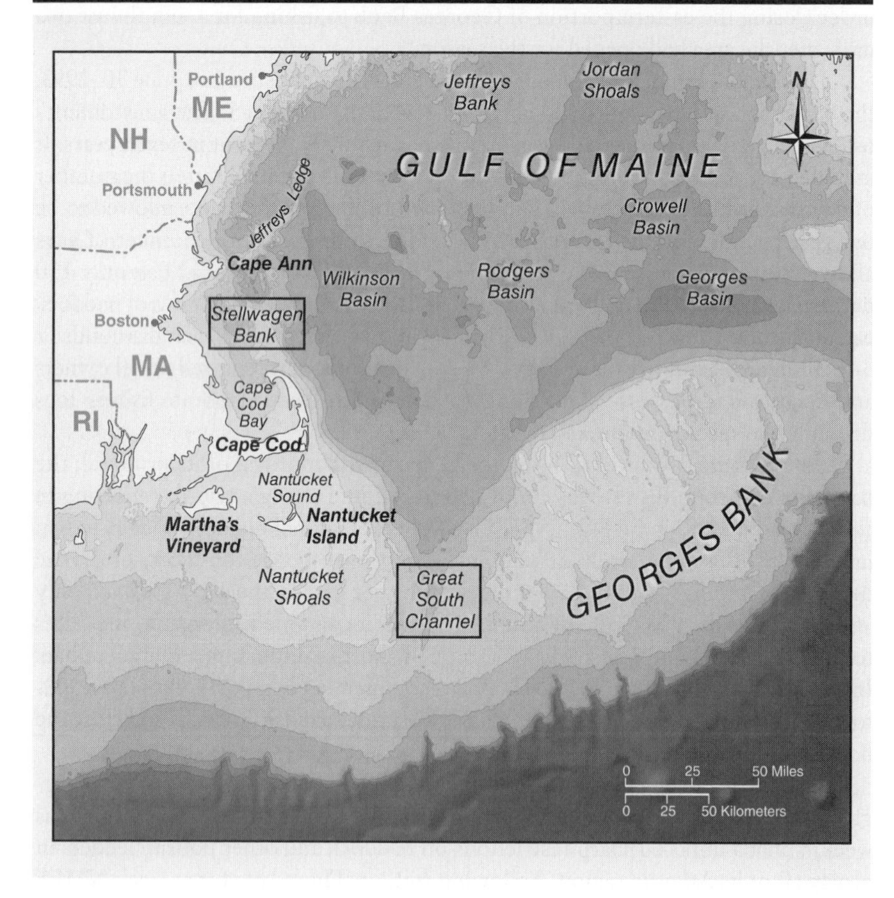

Source: NOAA.

payments—even if they never leave port). To attract public sympathy for their plight, an armada of more than 100 fishing vessels from Chatham, Gloucester, New Bedford, Provincetown, and elsewhere in the region jammed Boston Harbor.

Scientists Issue Another Warning—and Groundfish Stocks Crash

Adding to the pressure on regulators, Amendment 5 began to look obsolete even before it took effect. In August 1994—only months after the council adopted it—the NEFSC released more bad news. In an advisory report, scientists warned that reducing fishing by 50 percent over the next five to seven years, as Amendment 5 promised to do, probably would not be enough to save the groundfish. Of the yellowtail flounder, once the backbone of southern New England fishing ports, the report said, "The stock has collapsed! Fishing mortality on this stock should be reduced to levels approaching zero."[47] According to the report, of every 100 yellowtail flounder alive at the beginning of the year, only 8 survived the year. Under this pressure, the breeding population had declined to record lows.

The report was also skeptical about the potential for recovery of the Georges Bank cod. In 1993 cod mortality in the fishing ground hit a record high, while the number of mature adults dropped to a record low. Fishermen had hoped that the relatively large cod harvests of 1989 and 1990 meant that this fish was weathering the crisis, but NEFSC researchers believed those years were an aberration. Scientists noted that two years previously the Canadians had closed their cod fishery off the Newfoundland coast expecting a quick recovery, but the cod population instead had dropped another 25 percent in the first year of the closure. By 1994 the Canadians were estimating that a recovery was not likely until the late 1990s. And Georges Bank was in even worse shape than the Canadian fisheries because fishermen had been taking more than 70 percent of the cod swimming there each year. As a result, the Georges Bank catch had plummeted to its lowest level since the 1970s. Moreover, the cod's recovery was jeopardized because, as cod declined, predators, such as the spiny dogfish (a small shark), were becoming an ever larger proportion of the region's fish.

Alan Peterson of the Northeast Fisheries Science Center at Woods Hole estimated that it would take ten to twelve years before the Georges Bank cod stock would be healthy enough that regulators could afford to increase the harvest. Even a twelve-year comeback was no sure thing, he said, because the cod's fate was complicated by natural cycles such as below-average ocean temperatures or increased salinity, which could reduce the survival of young cod. Peterson urged the New England Council to go even further than Amendment 5 and shut down almost all fishing. It could then selectively reopen fisheries that were still healthy and where nets did not accidentally catch cod or other vanishing species.

On October 14, 1994, the council's groundfish committee began work on Amendment 7, fearing that if it did not act decisively, NMFS would institute its own plan. (Amendment 6 had simply made permanent the suite of emergency haddock measures enacted the previous year.) It had been less than six months since most of the Amendment 5 rules had taken effect, and fishermen complained bitterly that the council had not waited long enough to assess their impact. Nevertheless, nearly two weeks later the council approved an emergency measure to close indefinitely vast areas of the Gulf of Maine, including Georges Bank, as well as 4,500 square miles of fishing grounds in southern New England. To prevent trawlers displaced by the ban from overfishing elsewhere, the council also proposed a quota system limiting catches closer to shore. Then, on December 12, NMFS officially closed the entire 6,600-square-mile area of Georges Bank and announced its intention to lift the ban no earlier than March 1995.

Many fishermen were, predictably, incensed. Their complaints featured a familiar refrain: the scientists do not understand the condition of the fishery; they do not know where to look for the fish. "One scientist says one thing and another scientist says something else," scoffed Gloucester fisherman Jay Spurling. "They're not even out on the water; they don't see the things we see. . . . I think they're making a huge mistake."[48] Other fishermen lamented the end of the only way of life they had ever known. "I've been a fisherman for 25 years," said Vito Seniti, also of Gloucester. "My father was a fisherman, and his father, and his father before. What am I gonna do now, deliver pizzas?"[49] Fishermen's desperation notwithstanding, in January 1995 regulators extended the ninety-day closure indefinitely to give themselves time to come up with a real plan.

While the council wrangled over regulations, the formidable Massachusetts congressional delegation began seeking federal money to ameliorate the hardship, which was concentrated primarily around the ports of Boston, Gloucester, and New Bedford. Sen. John Kerry, D-Mass., and Rep. Gerry Studds, D-Mass., proposed legislation to create the New England Fisheries Reinvestment Program, which would disburse grants throughout the region. At the urging of New England's members of Congress, Commerce Secretary Ron Brown declared the Northeast fishery to be in an economic emergency and granted $30 million in aid to fishermen and their families.

Even more important, after a year of lobbying by Massachusetts legislators, in March 1995 the Commerce Department initiated a $2 million Fishing Capacity Reduction Program—or boat buyout—to compensate fishermen who retired their groundfish permits and fishing vessels and thereby reduced excess capacity in the fleet. Rather than embarking on a full-scale buyout, which could cost as much as $100 million, the department hoped to learn from the pilot program how to design an appropriate program. In August the department announced it would expand the buyout by $25 million but shrewdly made the money contingent on the New England Council showing progress on a groundfish stock rebuilding program, a potent incentive that NMFS hoped would entice fishermen to support new regulations.

Amendment 7 and Contrasting Definitions of the Problem

The buyout proposal came not a moment too soon. In mid-1995 the council received more alarming reports from the NEFSC: the measures taken under Amendment 5 and proposed since were insufficient to ensure stock recovery; in fact, things were getting worse. So the agency began to pressure the council to craft rules that would cut back fishing of groundfish by 80 percent rather than 50 percent. Fishermen's reaction ranged from disbelief to fury as meetings on the new Amendment 7 got under way in fall 1995. The fishermen had been blaming each other—different gear types, different ports—for the depletion of the fish, but the new proposals gave them a common foe: government regulators. As the council debated the terms of Amendment 7, the fishermen continued to plead that changes to Amendment 5 were premature and that it was too early to tell whether the new rules had been effective.

Underlying these debates were contrasting ways of defining the problem. Scientists and environmentalists emphasized the ecological risks and advocated a precautionary approach based on a risk-averse interpretation of the available science. By contrast, fishermen preferred a risk-tolerant interpretation of the science; they demanded proof that Amendment 5 was not working and that the stocks had, in fact, collapsed, and they argued that the real risk lay in imposing strict new rules that would cause certain economic pain and irreversible changes in the fishing industry. As Maggie Raymond, spokesperson for the Associated Fisheries of Maine, put it, if the plan takes effect "and then you realize you've gone too far, then it's too late. You've already put everybody out of business."[50]

Again the New England congressional delegation pressured the council on behalf of fishermen. Rep. James Longley Jr., R-Maine, wrote, "Make some modifications if you must, but do not destroy Maine's groundfishing industry solely to accomplish faster recovery rates."[51] Sen. Olympia Snowe, R-Maine, encouraged the council to resist pressure from NMFS to move more quickly. The Massachusetts delegation also weighed in. Democratic Senator Ted Kennedy, Democratic Representatives Barney Frank and Joseph Kennedy, and Republican Representative Peter Torkildsen urged NMFS to postpone further fishing restrictions until the socioeconomic impacts of the changes on fishing communities had been assessed.

The council was clearly in a bind. Empowered by the CLF lawsuit and backed by a coalition of environmental groups, NMFS Regional Director Andrew Rosenberg insisted that the council make conserving the fish its primary concern. NEFSC scientists were convinced that the groundfish decline was continuing unabated, despite measures instituted the previous year. Even a handful of fishermen doubted the wisdom of phased-in conservation rules and thought the fishermen needed to take responsibility for the health of the fishery and stop resisting protective measures. For example, John Williamson, a fisherman from Kennebunkport, Maine, pointed out that the council's year-and-a-half-long deliberation constituted a sufficient phase-in for the plan.[52] With NMFS threatening to withhold the money for the boat buyout unless it reduced fishing dramatically, and fishermen (backed by their congressional delegations) undermining its attempts to do so, the council spent meeting after meeting trying to arrive at an amendment that everyone could live with.

The council rejected the option of banning groundfishing altogether but then began to debate a mix of three alternatives: closing large fishing territories; reducing days at sea; and carving the fishery into inshore and offshore regions, each with its own quotas. Complicating the decision was the council's insistence that the rules affect all fishermen equally, regardless of gear type or boat size. Finally, in late January 1996 the council agreed on Amendment 7 by a vote of eleven to three and sent its proposal to Secretary Brown for review. Bowing to congressional pressure, the proposal gradually phased in days-at-sea limits for cod, haddock, and yellowtail flounder and instituted "rolling" closures of 9,000 square miles of fishing grounds.[53] Under the plan, a typical boat was allowed 139 days at sea in 1996 and 88 in 1997, a substantial reduction from the 250 to 300 days previously allowed.[54] The plan also limited the total allowable catch for cod to 2,770 metric tons in 1996, about one-third of the 1993 catch, and reduced it further in 1997.[55] Despite the accommodations made to their industry, many fishermen opposed Amendment 7, and congressional representatives notified Secretary Brown of their concerns about economic harm. Nevertheless, the Commerce Department approved the proposal, and it went into effect in July 1996.

Six months later, at the December 11 to 12 council meeting, the Multispecies Monitoring Committee (MMC) delivered its first assessment of Amendment 7.[56] The good news was that overfishing had been halted for all stocks except Gulf of Maine cod, and all were showing increases for the first time in years. The MMC remained cautious, however, noting that most stocks still needed to double or triple in size before they would reach minimum acceptable levels and that the measures contained in Amendment 7 would not be enough to accomplish this goal. It recommended an additional 62 percent cut in fishing for yellowtail and a 57 percent reduction in cod harvests on Georges Bank. Pressed by Rosenberg, the council voted to begin drawing up additional rules and to consider drastic proposals such as reducing the number of days at sea for cod fishing to as few as fourteen for the 1997 season.[57]

While the New England Council was wrestling with Amendment 7, proponents of a more protective legislative mandate for fisheries had been raising the salience of overfishing nationally, using events in New England as well as problems in the Gulf Coast and Pacific Northwest as cautionary tales. Environmentalists identified several failings of the Magnuson Act that they wanted Congress to address, but most important, they contended it was vague in its fishery conservation mandate. In particular, although the 1989 regulatory revisions required each FMP to contain an objective and measurable definition of overfishing and a recovery plan in the event that overfishing occurred, some critics wanted to see those guidelines delineated in the statute itself. They also wanted the act to specify the period of time within which councils were to address overfishing once it had been identified. Finally, they hoped to refine the concept of optimum yield to emphasize conservation, rather than economic and social considerations.

In 1996 Congress passed the Sustainable Fisheries Act (SFA), an amendment to the Magnuson Act that addressed that law's three biggest deficiencies: overfishing, bycatch, and habitat degradation. The SFA, more widely known as the

Magnuson–Stevens Act, mandated that each fishery management plan include an explicit definition of overfishing for the fishery, a rebuilding plan for overfished stocks, a timetable of less than ten years for reaching recovery, conservation and management measures to avoid bycatch, and a description of essential habitats and management measures to protect them. The law gave the councils two years to amend their existing plans and prepare new ones where necessary.

The Focus Shifts to Gulf of Maine Cod

Although NEFSC reports in late 1997 confirmed that several groundfish species had begun to recover, they also made it clear that New England faced a crisis in the Gulf of Maine cod fishery. This news, combined with the new Magnuson–Stevens Act requirements, severely restricted the council's options. In early January 1998, following yet another round of scientific reports documenting the cod's decline, the council announced that rules aimed at reducing the total cod catch by 63 percent would take effect in May 1998. Meanwhile, in April, Framework Adjustment 24 limited Gulf of Maine cod landings to 1,000 pounds per day and total days at sea to fourteen.[58] A month later, Framework Adjustment 25 reduced the cod landing limit to 700 pounds per day and instituted a four-step rolling closure of inshore fisheries. By June, when half of the total allowable catch of 1,783 metric tons was reached, NMFS reduced the cod landing limit to 400 pounds. And by late summer the council had no choice but to consider closing the Gulf of Maine fishery entirely.

In December 1998 the council heard yet another bleak scientific presentation by the MMC documenting serious overfishing of Gulf of Maine cod, which were at a record low despite strict fishing restrictions. Because the Magnuson–Stevens Act required that the cod population not drop below 7,500 metric tons, and scientists believed it was already down to around 8,300, the committee recommended an 80 percent catch reduction—to a total of 782 metric tons—in 1999.[59] "You want to get these [catches] as close to zero as possible," asserted MMC Chair Steven Correia.[60] Accomplishing such a goal without creating new problems would be no mean feat, however. Extremely low catch limits exacerbate the problem of bycatch: in New England, different species of groundfish swim together, so fishermen targeting one species typically catch others; if there are species with low limits in the haul, they must be thrown back. Furthermore, days-at-sea limits and ocean closures prompt fishermen to shift their efforts to alternative species or new fishing grounds. Partly as a result of pressure from displaced groundfishermen, shrimp, herring, and lobster fisheries were facing crises of their own.

In response to the MMC's recommendations, the council announced an emergency three-month closure, to begin in February 1999, of the cod fishery off the Massachusetts coast. The council's action, targeting the area where cod were concentrated, set off yet another round of protests. In January 1999, when the council took up proposals to address the Gulf of Maine cod problem more permanently, it faced an industry split by two major divides. First, Maine fishermen insisted that

Massachusetts and New Hampshire fishermen needed to share the burden of rescuing the cod; second, small boat owners, which are limited to inshore fisheries, demanded regulations that did not discriminate between them and large boats, which have more flexibility to pursue fish offshore. As the fishermen lined up behind competing approaches, their congressional representatives echoed their concerns, further highlighting the tension among states.

The council had three proposals before it, all aimed at meeting the goals for cod established by the MMC: (1) expand the existing regulatory regime of days at sea and daily catch limits, combined with rolling closures; (2) ban groundfishing all spring and summer in waters within forty miles of the coast between Cape Cod and south of Portland; and (3) divide the gulf into inshore and offshore fisheries and require every boat to limit itself to one or the other. After a long and rancorous debate, the council rejected all three options. According to journalist David Dobbs, by the final meeting, "no one trusted anyone. All the council members looked exhausted or scared or depressed or angry . . . and most of the audience appeared on the verge of rage or despair."[61] After the lunch break, security guards barely averted a violent confrontation between audience members and the council. At 1 a.m., the council finally came up with Framework Adjustment 27, which expanded the previous year's rolling closures and cut the daily catch limits to 100 pounds, while authorizing the administrator to cut trip limits to as low as 5 pounds if landings exceeded half the total allowable catch of 800 metric tons.[62]

Under pressure from Massachusetts and New Hampshire members and over the objections of members from Maine, the council opened inshore cod grounds off Portsmouth, New Hampshire, and Gloucester, Massachusetts. So it was hardly surprising when, less than a month after the measures went into effect, fishermen hit the 400-metric-ton trigger point, and NMFS reduced the daily catch limit on cod to thirty pounds. Fishermen were dumbfounded. "Fishermen all over the Gulf were catching cod no matter what they did to avoid them," Dobbs reports, describing one fisher who—despite his efforts to catch flounder without ensnaring cod—found that all he could do was catch the two in even proportions.[63] Fishermen found the waste horrific and complained bitterly to NMFS that its scientific assessments were inaccurate. The agency held firm, however, continuing to defend its view by saying that the cod populations had not increased overall but had contracted into their core areas, so that when those areas reopened, the fishermen were right on top of them.[64]

Amendment 9 and a Second Lawsuit

As it was wrestling with what to do about Gulf of Maine cod, the council—having acknowledged that Amendment 7 did not meet the goals of the Magnuson–Stevens Act—began working on a new, even more stringent amendment. In September 1998 the council submitted, and in November 1999 NMFS approved, Amendment 9, which established biomass and fishing mortality targets consistent with the Magnuson–Stevens Act mandate to rebuild overfished stocks within ten years. While awaiting NMFS approval, the council began work on a rebuilding plan consistent with the targets established in Amendment 9. But in April 2000, at the insistence of its regional office, NMFS implemented Framework Adjustment 33, which aimed to meet only the fishing mortality targets in Amendment 7, not the more ambitious ones contained in Amendment 9. A month later four environmental groups, including the CLF, sued the agency for failing to institute rules consistent with the new overfishing definitions in Amendment 9 and for failing to minimize bycatch, as required by the Magnuson–Stevens Act. Council officials conceded that Framework Adjustment 33 did not meet the standards set out by the Magnuson–Stevens Act but said that, anticipating a time-consuming public review process, the agency had wanted to get something in place by the deadline.[65]

In December 2001 U.S. District Court Judge Gladys Kessler issued her ruling. Citing "inaction and delay" by the council, she ordered NMFS to implement a rebuilding plan consistent with Amendment 9 so that groundfish stocks could rebound by the statutory deadline of 2009. Four months later, with the council still struggling to agree on its own measures, Judge Kessler handed down a draconian set of rules for the 2002 fishing season and required the council to put in place a new set of restrictions by August 2003. The judge acknowledged that "the livelihood . . . of many thousands of individuals, families, small businesses, and maritime communities [would] be affected" by her decision, but she noted that "the future of a precious resource—the once rich, vibrant and healthy and now severely depleted New England Northeast fishery—[was] at stake."[66] (Judge Kessler subsequently softened her rules somewhat in response to fishermen's objections.)

Ten days later more than 100 fishing boats crowded into Gloucester for a rally to protest the judicial decree. Fishermen and politicians denounced the new rules, saying, "To place further restrictions on fishermen at a time when the stocks are rebounding makes no sense."[67] Federal scientists and environmentalists responded that new data suggested fish stocks could rebound to a greater extent than once believed, and managers should impose sufficiently protective measures to allow fish populations to thrive. According to Eric Bilsky of Oceana, an environmental protection group, "[It] doesn't mean when fish show up things are OK. It just shows we are out of the critical care unit and into the intensive care unit."[68]

Disgruntled fishermen seized another opportunity to disparage the science on which the rules were based when in September 2002 they discovered an improperly rigged survey net on the *Albatross*, one of the NEFSC's two research trawlers. A detailed report issued the following month concluded that, although two years of equipment problems had affected the nets' performance, it did not cause scientists to underestimate groundfish populations. Those findings did not appease disaffected fishermen, who continued to refer to the incident as "Trawlgate." The NEFSC's Steven Murawski pointed out that even if the *Albatross* fish surveys were off by 100 percent, the overall picture would not change much. "We are so far from the goal post," he said, "that, even inserting a huge error in our computer models, the status does not change much."[69] But many fishermen remained skeptical and asserted that cod, haddock, and other groundfish were more plentiful than scientific reports indicated. In early November, after yet another demonstration by angry fishermen, the council asked Judge Kessler for a delay on new fishing rules to review the stock assessments. (Environmentalists supported the delay, hoping it would allow time to improve the public understanding of the science behind the rules.) In December the judge granted an eight-month delay but retained the 2009 stock rebuilding deadline.

In 2003, after another round of emotional and contentious public meetings, the council released Amendment 13, a modified version of a proposal devised by an industry group, the Northeast Seafood Coalition, and the least stringent alternative of the five the council had considered. Amendment 13 cut days at sea by 25 percent, to an average of fifty-two days per year, but it contained several provisions that worried environmentalists. First, it created a new category of fishing days ("B" days) that would allow fishermen to go after rebounding and therefore relatively plentiful species, such as haddock. Second, it allowed fishermen to lease fishing days from one another—a provision that environmentalists feared could undermine days-at-sea reductions. Oceana, the CLF, and the Natural Resources Defense Council immediately filed suit saying that Amendment 13 would not end overfishing of Georges Bank cod or four other significantly depleted groundfish stocks. But in March 2004 U.S. District Judge Ellen Segal Huvelle upheld most of the new rules, and they took effect in May.

Adopting a Sector-Based Approach

The years 2005 and 2006 featured a battle over reauthorizing the Magnuson–Stevens Act. In 2004 the U.S. Commission on Ocean Policy, appointed by President George W. Bush, had recommended broadening representation on regional councils and strengthening their scientific committees. A commission formed by the Pew Charitable Trust had reached similar conclusions. Both commissions urged the adoption of an ecosystem-based approach to fishery management, in which the multiple stresses in a system would be considered together. The New England congressional delegation, while supportive of these protective measures, strongly urged Congress to relax Magnuson–Stevens's ten-year stock rebuilding time frame. Maine's Senator Snowe insisted that the deadline had "required scientists to set speculative biomass targets that lead to drastic yet often biologically unnecessary reductions in fishing."[70] After months of wrangling, in late 2006 Congress passed a Magnuson–Stevens reauthorization that required regional councils to devise plans to end overharvesting of a stock within two years of its being declared overfished. Over the objections of the New England delegation, the law required that most stocks be rebuilt by 2014. (Subsequent legislative efforts to relax the rebuilding time frame led by Rep. Barney Frank also failed.[71])

While Congress was debating a new statutory framework, managers in New England struggled to end overfishing. Framework Adjustment 42, instituted in November 2006, imposed even smaller catch limits and further reductions in days at sea. Fishermen were increasingly frustrated because haddock was rebounding, but stringent controls on fishing—designed to curb fishing of more precarious stocks—meant that many haddock went uncaught. Yet, with rebuilding deadlines for vulnerable stocks looming, the council in spring 2007 began contemplating even more draconian reductions in fishing effort for spring 2009. Meanwhile, grim reports on the condition of those stocks continued to roll in: in late August 2008 the NEFSC reported that thirteen of nineteen Northeast groundfish stocks were still being overfished and remained at levels scientists deemed too low.[72] The 2008 assessment did contain some hopeful news: apparently haddock were bouncing back from their dangerously low levels, and Gulf of Maine cod, though still being fished too aggressively, had also rebounded.[73] Although some fishermen once again disparaged scientists' assessments, a report by NOAA's inspector general affirmed that the agency was basing its groundfish management on the best available science.[74]

After years of failed attempts to control fishing effort, many observers were delighted when, in June 2009, the New England council voted fourteen to one to try a new sector-based approach to fishery management. This was a radical departure from the input controls it had been using for more than twenty years. Under the new system, embodied in Amendment 16, the NEFSC sets an overall catch limit, and community-based cooperatives receive shares of the catch in the form of fishing quotas that they allocate among their members based on their fishing histories.[75] Participation in the cooperatives is voluntary, and fishermen pay a small percentage of the earnings after fuel costs to the association for legal fees and monitoring. Within each sector, members have flexibility as to when and where they fish; in addition, they are exempt from many gear restrictions. Proponents of the new approach hope it will create an incentive for fishermen to conserve because each sector's annual quota grows as fish stocks recover; in other words, fishermen have a stake in ensuring that fish stocks increase. The sector-based system should also eliminate the universally despised bycatch, since fishermen must land everything they catch. Because the fishermen in a sector are jointly and severally liable for any catch in excess of the sector's annual quota, the sector-based system creates an incentive to be self-policing. New England began experimenting with sectors in 2004, when NMFS allowed a group of Cape Cod hook-and-line fishermen to form a sector (under Amendment 13). A group of Cape Cod gillnetters received permission to do the same in 2006.

The Obama administration strongly endorsed the sector-based approach, widely known as catch shares, and between fiscal years 2009 and 2010 NOAA committed $47.2 million to the New England groundfish industry to aid the transition to sectors. Despite the cash infusion, as spring 2010 approached, resistance to the sector-based system became more vocal. Thousands of fishermen rallied in Washington, D.C., in hopes of persuading Congress to write more flexibility into the Magnuson–Stevens Act rather than rushing to embrace catch shares. Fishermen in New England were particularly concerned about "choke" species—pollock, Cape Cod yellowtail flounder, and Gulf of Maine cod—with very low annual quotas.

In New England, where groundfish school together, those quotas were likely to limit fishermen's ability to take healthier stocks because the act requires fishing to stop once any stock's limit is reached. Many fishermen were also concerned about consolidation in the industry; in legislative hearings, fishermen and their allies repeatedly invoked the small, independent fisherman and a way of life that was threatened. They complained that the new system would require a much bigger federal bureaucracy, along with greater expenses for monitoring, and more paperwork and potential enforcement actions against fishermen. NMFS held firm, however, and on May 1, 2010, Amendment 16 took effect. Within a week, five commercial fishing groups along with two of Massachusetts's largest fishing ports—Gloucester and New Bedford—had sued NMFS over the new management scheme, claiming that many boats had been unable to work for lack of software to manage their catch and that the total allowable catch had been distributed unfairly.[76] In 2012, the First Circuit Court of Appeals upheld a district court decision affirming the legality of the sector-based program established by Amendment 16.

Although President Obama argued for safeguards to replenish stocks, the Trump administration strongly supports expansion of seafood exports and recreational fishing. Under the Trump administration, we could see the largest overhaul of the Magnuson–Stevens Act. In July 2018, the U.S. House of Representatives voted (222–193) to remove annual catch limits and roll back requirements for overfished stocks. This legislation is awaiting review in the U.S. Senate.[77]

OUTCOMES

Many fishermen continue to blame government regulation, not overfishing, for the dire situation in the fisheries. In fall 2010, fisherman Brian Loftes released a film titled *Truth—Fishing Crisis or Government Mismanagement* that sought to document how regulations have harmed the fishing industry. In response, the Environmental Defense Fund released an assessment of the sector-based system. According to NOAA, landings were down 16 percent, while revenues were up 112 percent; sector fishermen were avoiding weaker stocks and targeting more robust ones.[78] Even some fishermen had compliments for the new system. Jim Odlin of Maine, owner of five boats, said the time pressure that the days-at-sea system produced was gone, while Chris Brown of Rhode Island contended that fishermen were more open to sharing information because their livelihoods now depended on each other.[79] In another positive review, researcher Michael Conathan noted that, in the first year of the new system, the portion of the groundfishery operating under sector-based management did not exceed its annual catch limit on any species.[80]

The shift to sector-based management put even more pressure on scientists, however. As researcher Jonathan Labaree points out, not only are the absolute numbers of each stock important but their abundance relative to one another is critical. Even if assessments for most stocks are accurate, if the assessment for a constraining stock is too low, industry will have difficulty harvesting other stocks. The Gulf of Maine cod fishery is a perfect example of how errors, even if quickly corrected, undermine faith in the science that underpins the regulatory regime. In 2008, the NEFSC released a relatively optimistic assessment of Gulf of Maine cod, anticipating the stock would be rebuilt by the target date of 2014, which was not the case.

In 2017, however, a revised assessment found that the cod were in poor shape, and fishing effort on that stock would have to be cut back dramatically.

Moreover, as fishermen feared, the industry has continued to consolidate under sector-based management. That consolidation was well under way, however, long before NMFS instituted the new system: according to NMFS, the number of groundfish boats in New England fell from 1,024 in 2001 to 477 in 2009.[81]

New England fishery managers hoped that cooperative research initiatives undertaken by NMFS and the council would enhance conservation efforts as well, by improving understanding of the fish stocks and the effect of fishery management regimes and by building mutual respect among fishermen and scientists. In the Northeast three main cooperative research programs got under way in the early 2000s: bottom-up planning among scientists and fishermen that aims to solve individual fishery management problems; the Research Partners program administered by the NMFS Northeast regional office; and the New England Consortium Cooperative Research, comprising the University of New Hampshire, the University of Maine, the Massachusetts Institute of Technology, and the Woods Hole Oceanographic Institution. Although NEFSC scientists emphasized that such efforts could not substitute for long-term, standardized resource surveys conducted by research vessels, they hoped that they would make "valuable and unique contributions to the science underlying fishery management."[82] One tangible achievement of cooperative research was the "eliminator trawl" introduced in 2008. This improved net allows fishermen to keep fishing once they have reached their quota on cod because it captures haddock, which swim upward when caught, while allowing cod, which swim downward, to escape.[83] With the introduction of sector-based management in 2010, NMFS began directing most of the region's collaborative research funding toward projects that would help the industry adjust to sectors. One such project is GEARNET, a network of fishermen and scientists that allocates money to test new nets, catch sensors, and mammal deterrents.[84]

CONCLUSIONS

Since 1976 New England fishery managers have struggled to institute effective schemes to save the groundfish while simultaneously preserving the character of the New England fishery. For two decades, cooperative management—as mandated by the Magnuson Act—translated into trying to devise conservation policies that did not antagonize fishermen. For the most part, the resulting policy consisted of reactionary measures prompted by crises; until recently, those measures did little more than slow the demise of the fishermen at the expense of the fish. The eventual shift to a more restrictive fishing regime was not the result of a recognition by the council or fishermen that scientists were right; in fact, as late as the early 1990s, exploitation rates of cod and yellowtail flounder were 55 percent and 65 percent of biomass, respectively, even though scientists recommended exploitation levels of 13 percent and 22 percent.[85] Rather, the CLF lawsuit empowered NMFS officials to force the council's hand. Twenty years later, although some fishermen had joined in cooperative research projects sponsored by NMFS, others continued to disparage the science produced by the NEFSC, probably because the results of scientists' stock assessments directly threatened their livelihoods.

Although litigation prompted a shift in regulatory practices, many observers believe that the only way to save New England's groundfish in the long run is to create private property rights in the fishery. This belief stems from their adherence to the assumptions about behavior embodied in the tragedy of the commons model. Certainly the actions of New England fishermen since the early 1970s are consistent with that model: individual fishermen reacted to whatever measures fishery managers put in place by expanding their effort in other ways. They pursued short-term economic gains at the expense of the long-term health of the fishery, and few seemed prepared to sacrifice their catch without assurances that others would do the same. Furthermore, the fishermen of New England appeared to be less-than-ideal candidates for voluntary comanagement because historically they did not trust each other, did not communicate regularly, and had little experience forming binding agreements with one another.

Workable solutions in New England are therefore likely to depend heavily on controlling access to the fishery rather than on voluntary cooperative schemes. But more than anything, New England fishermen value their independence and their egalitarian view that "anyone should be able to go out and fish," so they adamantly oppose limiting entry into the fishery.[86] They also worry that small fishermen fare poorly under quota-based approaches. Consolidation of the industry is not inevitable under an ITQ or catch-share system, but NOAA's own data show that quota-based fisheries have generally led to industry concentration: after twenty years the surf clam and ocean quahog fisheries in the mid-Atlantic have seen their fleets shrink by 74 percent and 40 percent, respectively. In Alaska the number of halibut boats fell by 70 percent and sablefish boats by two-thirds after those fisheries converted to catch shares in 1995. Only the red snapper fishery in the Gulf of Mexico, which converted in 2007, has experienced relatively minor increases in concentration.[87]

Officials have taken several steps to help small-scale fishermen in hopes of preventing excessive industry concentration. In 2009 federal and state officials set up a commercial fishing permit bank to support fishermen from rural ports along the Maine coast. The bank will buy and hold groundfish permits, divide up the rights, and lease them to eligible fishermen—in theory enabling them to stay in business until the fish stocks recover. (In March 2010 NOAA granted $5 million to help set up such permit banks.) In September 2010 the New England council voted to establish accumulation limits to prevent excessive concentration in the industry. Another innovation that began in New England is the community-supported fisheries (CSF) concept, in which customers buy shares of the catch and take deliveries throughout the fishing season. Port Clyde Fresh Catch, which originated the approach, has been tremendously successful, and others around the region (and the country) have picked up on the concept.

Although innovation and collaboration efforts are noteworthy, a 2018 article published in *Science* draws larger concerns. More specifically, climate scientist Scott

Doney and his research team suggest that due to rising global temperatures, the world's fisheries could be 20 percent less productive by 2300. This could have a dramatic impact on the North Atlantic with 60 percent less production.[88]

QUESTIONS TO CONSIDER

- One explanation for why federal managers delayed acting to conserve the fishery, despite repeated warnings from scientists, is that there was a time lag between the receipt of scientific advice and action. Does that seem like an adequate explanation? Why, or why not?
- What kind of regulation is most appropriate for New England to prevent the recurrence of overfishing? Does the political will exist to undertake the conservation measures you recommend?
- To what extent should those whose livelihoods depend on fish have a say in decisions about those regulations? What form should their participation take, how should their views be incorporated, and why?
- Climate change could lead to significant consequences for management of fisheries. Would an Australian model help with the adaptation or mitigate concerns? Why, or why not?

NOTES

1. Groundfish, or demersals, dwell at or near the bottom of the sea. The New England groundfish include members of the cod family (cod, haddock, hake, pollock), flounder, dogfish sharks, and skates.
2. United Nations Food and Agriculture Organization, "World Review of Fisheries and Aquaculture," Part 1 in *The State of World Fisheries and Aquaculture (SOFIA)*, 2016. Available at http://www.fao.org/3/a-i5555e.pdf.
3. Ransom A. Myers and Boris Worm, "Rapid Worldwide Depletion of Predatory Fish Communities," *Nature* 423 (May 15, 2003): 280–283.
4. Jeremy B. C. Jackson et al., "Historical Overfishing and the Recent Collapse of Coastal Ecosystems," *Science* 293 (July 27, 2001): 629–637; U.S. Commission on Ocean Policy, *An Ocean Blueprint for the 21st Century* (2004). Available at http://govinfo.library.unt.edu/oceancommission/documents/full_color_rpt/000_ocean_full_report.pdf.
5. Jennie M. Harrington, Ransom A. Myers, and Andrew A. Rosenberg, "Wasted Fishery Resources: Discarded By-Catch in the USA," *Fish and Fisheries* 6 (2005): 350–361; R. W. D. Davies, S. J. Cripps, A. Nickson, and G. Porter, "Defining and Estimating Global Marine Fisheries Bycatch," *Marine Policy* 33, 4 (2009): 661–672. Available at http://www.sciencedirect.com/science/journal/0308597X.

6. Garrett Hardin, "The Tragedy of the Commons," *Science* (December 13, 1968): 1243–1248.

7. For a simulation of how this process works in a fishery, see the Fishbanks game, originally developed by Dennis Meadows and refined by William A. Prothero. Available at http://earthednet.org/Support/materials/FishBanks/fishbanks1.htm.

8. John Tierney, "A Tale of Two Fisheries," *The New York Times Magazine*, August 27, 2000, 38–43.

9. Elinor Ostrom, *Governing the Commons: The Evolution of Institutions for Collective Action* (New York: Cambridge University Press, 1990); James R. McGoodwin, *Crisis in the World's Fisheries* (Stanford, Calif.: Stanford University Press, 1990).

10. James M. Acheson, "Where Have All the Exploiters Gone?" in *Common Property Resources: Ecology and Community-Based Sustainable Development*, ed. Fikret Berkes (London: Bellhaven Press, 1989), 199–217.

11. Ostrom, *Governing the Commons*.

12. For a discussion of the importance of local ecological knowledge to credible, effective natural resource management, see Fikret Berkes, *Sacred Ecology: Traditional Ecological Knowledge and Resource Management* (Philadelphia: Taylor & Francis, 1999).

13. Quoted in Carl Safina, "Where Have All the Fishes Gone?" *Issues in Science and Technology* (Spring 1994): 38.

14. Recruitment overfishing occurs when fishermen catch too many spawning adults and thereby affect the population's ability to reproduce. See Steven A. Murawski et al., "New England Groundfish," in National Marine Fisheries Service, *Our Living Oceans: Report on the Status of U.S. Living Marine Resources*, 1999, NOAA Technical Memo NMFS-F/SPO-41 (December 1999).

15. William Warner, *Distant Water: The Fate of the North Atlantic Fishermen* (Boston: Little, Brown, 1983).

16. Rodman D. Griffin, "Marine Mammals vs. Fish," *CQ Researcher*, August 28, 1992, 744.

17. Warner, *Distant Water*, viii.

18. Murawski et al., "New England Groundfish." A metric ton is equal to 1,000 kilograms, or approximately 2,200 pounds.

19. National Marine Fisheries Service, *Our Living Oceans: Report on the Status of U.S. Living Marine Resources*, 1993, NOAA Technical Memo NMFS-F/SPO-15 (December 1993).

20. Warner, *Distant Water*.

21. Ibid.

22. Margaret E. Dewar, *Industry in Trouble: The Federal Government and the New England Fisheries* (Philadelphia: Temple University Press, 1983).

23. The coastal fisheries (zero to three miles off the coast) are managed by states and by interstate compacts.

24. PL 94-265, 90 Stat. 331; 16 U.S.C. Secs. 1801–1882.

25. Quoted in H. John Heinz III Center for Science, Economics, and the Environment, *Fishing Grounds: Defining a New Era for American Fisheries Management* (Washington, D.C.: Island Press, 2000), 87.

26. In 1996 the Food and Agriculture Organization of the United Nations estimated that 27 million metric tons of marine fish were killed as bycatch worldwide each year. In 2004 that bycatch estimate dropped to about 7.3 million tons, in part due to adoption of measures by fishermen to avoid bycatch, but also as a result of better use of the catch and improved reporting. See "New Data Show Sizeable Drop in Number of Fish Wasted," press release, United Nations Food and Agriculture Organization, September 14, 2004. More recently, scholars have suggested that bycatch represents more than 40 percent of global marine catches. See Davies et al., "Defining and Estimating Global Marine Fisheries Bycatch."

27. National Research Council, *Review of Northeast Fishery Stock Assessments* (Washington, D.C.: National Academy Press, 1998).

28. David E. Pierce, "Development and Evolution of Fishery Management Plans for Cod, Haddock, and Yellowtail Flounder" (Boston: Massachusetts Division of Marine Fisheries, 1982).

29. The New England Council has eighteen voting members: the regional administrator of NMFS; the principal state officials with marine fishery responsibility from Connecticut, Maine, Massachusetts, New Hampshire, and Rhode Island; and twelve members nominated by the governor and appointed by the secretary of commerce. The council also has four nonvoting members: one each from the U.S. Coast Guard, the Fish and Wildlife Service, the Department of State, and the Atlantic States Marine Fisheries Commission.

30. Quoted in Heinz Center, *Fishing Grounds*, 88.

31. Eleanor M. Dorsey, "The 602 Guidelines on Overfishing: A Perspective from New England," in *Conserving America's Fisheries*, ed. R. H. Stroud (Savannah, Ga.: National Coalition for Marine Conservation, 1994), 181–188.

32. Quoted in Pierce, "Development and Evolution of Fishery Management Plans," 13.

33. Madeleine Hall-Arber, "'They' Are the Problem: Assessing Fisheries Management in New England," *Nor'Easter* (Fall–Winter 1993): 16–21.

34. Sonja V. Fordham, *New England Groundfish: From Glory to Grief* (Washington, D.C.: Center for Marine Conservation, 1996).

35. The standard investment tax credit was available to anyone, but it was especially appealing to capital-intensive industries like fishing. The fishery-specific Capital Construction Fund program allowed fishing boat owners to set aside and invest pretax dollars for later use in upgrading or buying fishing boats. The Fishery Vessel Obligation Guarantee program provided government-guaranteed boat-building loans at lower interest rates and longer payback periods than traditional five-year loans. These and other incentives—such as fuel tax relief, gear replacement funds, and market expansion programs—attracted fishermen into the industry and encouraged existing boat owners to expand and upgrade their boats. See David Dobbs, *The Great Gulf* (Washington, D.C.: Island Press, 2000); Heinz Center, *Fishing Grounds*.

36. Murawski et al., "New England Groundfish."

37. Dobbs, *The Great Gulf*.

38. Dorsey, "The 602 Guidelines."

39. PL 94-265, 90 Stat. 331.

40. Dorsey, "The 602 Guidelines."

41. Ibid.

42. In theory, optimal yield is the level of catch that will allow the stocks to sustain themselves, modified by any relevant economic, social, or ecological factor.

43. Dorsey, "The 602 Guidelines," 185.

44. Massachusetts Offshore Groundfish Task Force, *New England Groundfish in Crisis—Again*, Publication No. 16, 551-42-200-1-91-CR (December 1990).

45. Previous amendments had involved nongroundfish species.

46. In prior years NMFS had closed Area 2 of Georges Bank between February 1 and May 31 to protect spawning groundfish. The emergency order extended the closure and enlarged the covered area. See John Laidler, "U.S. Emergency Order Shuts Part of Georges Bank to Fishing," *The Boston Globe*, June 4, 1993, 31.

47. Northeast Fisheries Science Center, Report of the 18th Northeast Regional Stock Assessment Workshop: Stock Assessment Review Committee Consensus Summary of Assessments, 1994, NEFSC Ref. Doc. 94-22.

48. Quoted in Sam Walker, "Georges Bank Closes, Ending an Era," *The Christian Science Monitor*, December 12, 1994, 1.

49. Ibid.

50. Quoted in Edie Lau, "Panel Backs Cutting Days for Fishing," *Portland Press Herald*, October 26, 1995, 1B.

51. Quoted in Linc Bedrosian, "Portland: Amendment 7 Is Sheer Lunacy," *National Fisherman*, December 1995, 17.

52. Edie Lau, "Council to Phase in Restraints," *Portland Press Herald*, December 15, 1995, 1.

53. Rolling closures are area closures that are instituted one after the other, as stocks migrate.

54. Andrew Garber, "Lawsuit Put on Fast Track," *Portland Press Herald*, August 9, 1996, 2B.

55. The actual 1996 cod catch was 6,957 metric tons, however. See Andrew Garber, "Report Calls for Deeper Cuts in Catches to Save Groundfish," *Portland Press Herald*, December 7, 1996, 1.

56. The monitoring committee includes scientists, managers, and a fisherman.

57. Garber, "Report Calls for Deeper Cuts"; "New England Fisheries News" (Boston: Conservation Law Foundation, December 1996).

58. A framework adjustment can be implemented more quickly than a full-blown plan amendment.

59. John Richardson, "Panel to Vote on Drastic Cod Limits," *Portland Press Herald*, December 10, 1998, 4B.

60. Quoted in John Richardson, "Fishery Closure Also Lifts Hopes," *Portland Press Herald*, December 12, 1998, 1B.

61. Dobbs, *The Great Gulf*, 153.

62. "Closure and Trip Limits Are Part of Framework 27," *NMFS Northeast Region News*, April 30, 1999.

63. Dobbs, *The Great Gulf*, 174.

64. Ibid.

65. Associated Press, "Hard Line on Fishing Is Pushed in Lawsuit," *Portland Press Herald*, May 23, 2000, 2B.

66. Quoted in Susan Young, "Judge Sets Number of Fishing Days," *Bangor Daily News*, April 27, 2002, 1.

67. Quoted in Beth Daley, "New U.S. Fishing Rules Protested," *The Boston Globe*, May 6, 2002, B1.

68. Quoted in ibid.

69. Quoted in David Arnold, "U.S. Study Says Fish Numbers Accurate," *The Boston Globe*, October 26, 2002, B1.

70. "Snowe Pledges to Continue Fighting for Maine's Fishermen," press release, U.S. Senate Office of Olympia J. Snowe of Maine, March 7, 2005.

71. The Magnuson–Stevens Act allows the ten-year rebuilding period to be lengthened under one of three conditions: a fish has a long reproductive cycle, environmental conditions make the rebuilding period infeasible, or an international agreement dictates a longer timetable. But Representative Frank wanted to add more exceptions, including one to minimize the economic impacts on fishing communities.

72. Carolyn Y. Johnson, "Report Says Fish Stocks Still Too Low," *The Boston Globe*, August 30, 2008.

73. Ibid.

74. Todd J. Zinser, Memorandum to Dr. James W. Balsiger on the Northeast Fisheries Science Center, February 26, 2009. Available at http://www.nefsc.noaa.gov/program_review/background2014/TOR1RagoNOAA%20memo%20Feb%2026.pdf.

75. Nine species of groundfish are included in sector management. A small portion of the catch history is reserved for fishermen who want to remain in the tightly regulated common pool. Most fishermen have opted to participate in sectors; however, in 2010, 98 percent of the catch history ended up in sectors; that percentage increased to 99 percent in 2011. See Michael Conathan, "The Future of America's First Fishery: Improving Management of the New England Groundfishery," Center for American Progress, May 2012. Available at https://cdn.americanprogress.org/wp-content/uploads/issues/2012/05/pdf/ne_groundfishery.pdf.

76. The plaintiffs pointed out that the Magnuson Act specifies that, in order to establish a catch-share (or "limited access privilege") program in either New England or the Gulf of Mexico, a council must conduct a referendum of permit holders. But the New England council circumvented this requirement by allocating quotas to sectors, rather than to individual fishermen.

77. Chris D'Angelo and Alexander C. Kaufman, "House Republicans Vote to Gut Lauded Law that Saved America's Fisheries," July 2018. Available at https://www.huffingtonpost.com/entry/magnuson-stevens-fishing-don-young_us_5b4628d9e4b07aea754696d7.

78. "NMFS, NOAA and NE Fishery Management Council Work Together to Address Initial Hurdles of Groundfish Sectors," States News Service, October 1, 2010.

79. May Lindsay, "As Species Rebound, Skippers Bemoan 'Underfishing,'" Associated Press, October 25, 2010.

80. Michael Conathan, "The Future of America's First Fishery: Improving Management of the New England Groundfishery," Center for American Progress, May 2012. Available at https://cdn.americanprogress.org/wp-content/uploads/issues/2012/05/pdf/ne_groundfishery.pdf.

81. Beth Daley, "Change in Fishing Rules Altering Storied Industry," *The Boston Globe*, January 27, 2011.

82. Michael Sissenwine, "On Fisheries Cooperative Research," Testimony Before the Committee on Resources Subcommittee on Fisheries Conservation, Wildlife and Oceans, U.S. House of Representatives, December 11, 2001.

83. Peter Lord, "A Better Fishnet—It Nabs Haddock but Lets Cod Escape," *Providence Journal*, April 12, 2008.

84. Labaree, "Sector Management in New England's Groundfish Fishery." Andrew Rosenberg points out that cooperative research is most productive when the question is posed well, scientists could not otherwise get the required sampling intensity, or the objective is to test a gear type. Cooperative research is less likely to be useful when the goal is to provide support for or undermine a proposed regulation. Andrew Rosenberg, personal communication, October 21, 2013.

85. National Research Council, *Sustaining Marine Fisheries* (Washington, D.C.: National Academy Press, 1999).

86. Hall-Arber, "'They' Are the Problem."

87. Richard Gaines, "Resistance on the Waterfront," *Gloucester Daily Times*, June 29, 2010.

88. Fariss Samarrai, "Global Fisheries to Decline," 2018. Available at https://news.virginia.edu/content/study-global-fisheries-decline-20-percent-average-2300

CHAPTER ELEVEN

THE DEEPWATER HORIZON DISASTER

The High Cost of Offshore Oil

Around 10 p.m. on the night of April 20, 2010, there was chaos aboard the Deepwater Horizon, an oil exploration rig in the Gulf of Mexico, about 50 miles off the coast of southeast Louisiana. Just seconds earlier, a series of massive explosions hurled crew members across rooms and buried them in debris. Those who were able to move staggered or crawled toward the lifeboats, hoping to escape the inferno that was consuming the rig. Upon reaching a nearby rescue ship, the survivors learned that eleven of their colleagues had perished, fatally injured by the blasts or caught up in the raging fire that ensued. But the disaster was only just beginning to unfold; it soon became clear that crude oil was pouring from BP's Macondo well into the gulf. By the time the well was finally capped, 86 days later, the Deepwater Horizon oil spill was the largest in U.S. history by far: 4.9 million barrels had gushed from the stricken well into the gulf, causing untold damage to hundreds of miles of shoreline marsh, thousands of square miles of marine habitat, and the myriad creatures that depend on the gulf ecosystem for their survival.[1]

Some scholars argue that disasters like this one are a fact of modern life. In his book *Normal Accidents*, organizational theorist Charles Perrow famously argued that, as our technologies become more sophisticated and industrial operations more complex, catastrophic accidents are inevitable. Two factors make accidents in such high-risk systems "normal." The first is interactive complexity—that is, the numerous ways that failures interact. The second is that system elements are tightly coupled, meaning the system features time-dependent processes that must be completed in sequence, and there is little slack in terms of resources required for a successful outcome. Interactive complexity and tight coupling can propagate failure. In such systems, technological fixes that aim to mitigate risk-taking behavior, including safety devices, can actually make matters worse because they allow operators to run the system faster, in less hospitable conditions.[2]

In any organization, the inherent risks posed by complex production systems are exacerbated by the lack of an effective safety culture, also known as a "culture of reliability."[3] In her analysis of the decision to launch the space shuttle *Challenger*, sociologist Diane Vaughan emphasizes the role of organizational culture in causing the worst disaster in the history of space flight. (The *Challenger* exploded just over a minute into its tenth flight, killing all seven crew members aboard.) Over time, she argues, NASA's culture fostered complacency about risk; those operating within the system came to believe that because they had succeeded so frequently in the past, failure was impossible. Thus, even when the agency's own tests showed that a critical joint was flawed, engineers simply accepted the risk rather than trying to improve its design.[4]

Since the 1980s accident, prevention research has emphasized the importance of building an effective safety culture. To maximize the chances that an organization can recognize and respond to signs of potential emerging hazards, an effective safety culture should involve the following: a commitment among senior management to safety, shared care and concern about hazards and their impact on people, realistic and flexible norms for dealing with risk, and continual reflection on and improvement of practice. The final item on the list, organizational learning, is the most difficult to achieve. Environmental psychologist Nick Pidgeon argues that to counter rigidities in thinking, engineers must practice "safety imagination." That is, they must go beyond the record of the past and consider seemingly inconceivable worst-case scenarios and unexpected consequences of their decisions.[5] Such practices make sense given that—as veteran journalists John Broder, Matthew Wald, and Tom Zeller point out—"no one can predict what might upend all the computer models, emergency planning and backup systems designed to eliminate those narrow theoretical probabilities [of catastrophic failure] or mitigate their effects."[6]

In the oil and gas industry, there are well-established process-safety principles, yet many companies continue to engage in risky behavior.[7] Publicly held companies must always balance their short-term responsibility to deliver value to shareholders with concerns about health, safety, and environmental protection, all of which contribute to the company's long-term value but often impose costs in the short run. Each decision made in a drilling operation involves judgment about the likely risk of damages versus the cost—in time and money—of ensuring safety. In the absence of strict oversight, companies may well decide that the immediate financial benefits of cutting corners justify the risk. So, in addition to highlighting the risk-prone decision making of the companies involved, this case also lays bare the lack of regulatory oversight provided by the main government entity responsible for offshore oil and gas development: the Minerals Management Service (MMS) in the Department of the Interior.

The Deepwater Horizon disaster lays bare the perils of giving agency regulators multiple and conflicting mandates and then failing to respond to reports that the agency is unduly friendly with the industry it is supposed to regulate—a phenomenon known as "agency capture." When a natural resource agency is charged with both enabling resource development and ensuring that development is safe, the former mandate is almost certain to dominate (see, for example, Chapter 7). Private interests pay close attention to such agencies and have far greater resources with which to participate in agency decision making and lobby congressional overseers than those advocating for safety and environmental protection. One justification for adopting a conciliatory approach—a partnership—with industry is that, if regulators are not on good terms with the managers and CEOs of regulated companies, those companies will not furnish full information about their practices. A second is that industry has greater expertise than regulators and so can devise more effective safety and environmental protection practices. Although in theory these arguments makes sense, in practice a more prescriptive approach repeatedly has yielded better results in terms of compliance (see Chapter 4). As legal scholar Thomas O. McGarity points out, however, the antiregulatory ideas that gained currency in the 1980s and 1990s have exacerbated the government's difficulties enacting and enforcing strict controls on extractive industries, even as threats posed by those industries' activities grow.[8]

With industry failing to regulate itself and regulators failing to oversee industry, the accountability mechanism that remains is liability, as determined by the courts. Ostensibly, unlimited liability creates uncertainty about the costs associated with safety and environmental damage. That uncertainty, in turn, promotes risk-averse behavior among private-sector decision makers. Over the last several decades, however, the federal government and many states have adopted liability caps—either for specific industries or activities, or blanket caps—on the grounds that unlimited liability discourages industry, sending it to places that are less hospitable. But some scholars point out that such caps actually *encourage* risk-taking behavior.[9]

Finally, the Deepwater Horizon case illuminates the political dynamics associated with focusing events, which can open policy windows that may or may not present opportunities for major policy change. The *Exxon Valdez* oil spill of 1989—in which a tanker ran aground in Prince William Sound, Alaska, spilling 257,000 barrels of oil—prompted major policy changes: a legislative requirement that oil tankers have double hulls and that both government and industry develop spill-response plans and an executive order placing a ten-year moratorium on most offshore oil development. In the weeks after the Deepwater Horizon sank, environmentalists hoped that the ensuing oil spill would create similar momentum for a policy to limit offshore oil drilling and curb the nation's reliance on fossil fuels more generally. However, there was no lasting groundswell of outrage over the Deepwater Horizon spill and therefore no counterweight to the intransigent opposition among Republicans, who refused to consider additional regulations or more spending to prevent future disasters. Another impediment to change was the fact that the gulf states are among the poorest in the United States, and their economies rely heavily on the oil and gas industry. Their dependence on resource extraction has left them vulnerable to the boom-and-bust cycles associated with such activities.

Nevertheless, industry political representatives have consistently resisted restrictive regulations on oil and gas development, fearing that such rules would push the industry elsewhere. The net result is that environmentalists ultimately failed to capitalize on the Deepwater Horizon disaster to bring about major policy change. Instead, oil residue has negatively affected the biodiversity of the Gulf of Mexico. The future of the Bureau of Safety and Environmental Enforcement (BSEE), an agency within the Department of the Interior responsible for implementing safety regulations, is in doubt due to the Trump administration's stance to open more offshore drilling opportunities.

BACKGROUND

The Deepwater Horizon debacle occurred in water nearly a mile deep. In 2010 water more than 500 feet deep was the latest frontier for oil exploration, made viable and profitable by a combination of rising global oil prices and technological breakthroughs.[10] At $147 per barrel in July 2008, just prior to the worldwide recession, oil prices reflected surging demand. In an effort to meet that demand, and with extraction from conventional reservoirs beginning to decline, oil and gas companies increasingly sought unconventional plays, including tight rock and deep water. Just as advances in hydraulic fracturing and horizontal drilling enabled exploitation of reserves in tight rock (see Chapter 14), technical innovations made drilling in deep water possible: enhanced computer processing power allowed geologists to discern formations miles below the surface of the ocean; three-dimensional seismic imaging permitted them to see through thick layers of salt that previously obstructed visibility; and superstrong alloys made it possible for drill bits to work in the hot, high-pressure fields of the deep ocean.[11]

The Gulf of Mexico is an especially attractive place for deep water drilling for several reasons. Perhaps most important is the abundance of oil there.[12] Also appealing is the fact that the region already has the infrastructure for transporting and refining oil. Moreover, the United States provides generous subsidies for oil exploration and development, while capping liability for spills at $75 million.[13] Although representatives of other coastal states have vigorously defended their offshore waters from oil and gas development, Congress has repeatedly exempted the gulf from moratoriums and other protective measures. As a result of all these factors, the gulf is home to 99 percent of the nation's offshore oil production. In the 2000s, as production declined in the gulf's shallow waters, large companies began moving rigs into deeper water, where they quickly discovered major reservoirs of oil and gas. Between 1992 and 2008, the number of rigs drilling in the gulf's deep water jumped from three to thirty-six.[14] By 2010, the gulf was producing more than 1.6 million barrels of crude per day—nearly one-third of U.S. production, with about 80 percent of that coming from deep water.[15]

The most active deep-water driller in the gulf was BP, which by the early 2000s was the world's third largest oil company and the fourth largest company overall, behind Walmart, Royal Dutch Shell, and Exxon Mobil. Once a struggling and hidebound company, in the 1990s BP underwent a transformation after acquiring two U.S. oil companies: Amoco in 1998 and ARCO in 2000. These mergers dramatically increased BP's financial clout but also its complexity. In an effort to streamline operations, at the direction of then-CEO John Browne, the company engaged in waves of cost cutting, slashing the number of employees worldwide and relying more heavily on contractors.[16] Browne also sought to create a culture of entrepreneurship within the company by creating autonomous "business units" and devolving decision-making responsibility to the managers of those units. Managers contracted with BP to meet performance targets, and their compensation—as well as that of their employees—was tied to their performance.[17] This structure ensured that BP's emphasis on saving money percolated down from the top to the lowest levels of operations, often forcing managers to choose between profits and safety and environmental protection. The consequences of those decisions became apparent in the mid-2000s.

On March 23, 2005, an explosion and fire at BP's refinery in Texas City, Texas, killed 15 workers and injured more than 170. The refinery, which was built in 1934 and covered two square miles, had long been plagued by overdue maintenance and poor safety practices. The blast occurred as workers restarted a raffinate splitter tower, a facility within the isomerization unit that had been undergoing routine maintenance for four weeks.[18] In the hours leading up to the blast, workers continued to add heavy liquid hydrocarbons known as raffinate to the splitter tower despite having already exceeded the maximum level. They were unaware of the high fluid levels because several warning devices failed, while others were ignored. Even under optimal conditions, a restart is extraordinarily dangerous because everything is unstable; no one who is not directly involved in the operation should be nearby. But in this case, there was a trailer that housed a team of subcontractors just 120 feet away (rather than the 350 feet required by BP's own policies). The workers inside the trailer—who had no idea that the tower was being restarted—were oblivious to the mounting danger, as overheated hydrocarbons spilled out of the tower and into an antiquated "blowdown drum." When the hot fuel overflowed the drum, a gas cloud formed and was ignited by the idling engine of a nearby pickup truck.[19]

After the disaster, a panel headed by former Secretary of State James Baker III found that inspections at BP's refineries were long overdue. Near disasters had never been investigated, and known equipment problems had been left unaddressed for more than a decade. Tests of alarms and emergency shutdown systems were either not done or were conducted improperly. Rather than emphasizing process safety, BP focused on preventing personal accidents, while lacking a mechanism for hearing workers' more serious concerns. Carolyn Merritt, head of the Chemical Safety and Hazard Investigation Board (CSB), which also investigated the accident, concluded, "The combination of cost-cutting, production pressures and failure to invest caused a progressive deterioration of safety at the refinery."[20] One of the most fatal decisions made in the name of cost consciousness, according to the CSB, was BP's decision not to replace the blowdown drum, a 1950s-era safety valve. Had BP installed a flare system, as it had considered doing years earlier, the explosion would not have happened, or it would have been far less severe.[21]

While it was economizing on maintenance, BP poured money into exploration—particularly in the Gulf of Mexico. Yet on July 11, 2005, BP's state-of-the-art

Thunder Horse drilling rig nearly sank in the gulf after Hurricane Dennis blew through the region. Investigators determined that the main cause of this near-disaster was a one-way valve that had been installed backward, allowing seawater into the platform's enormous legs. As the rig destabilized, water rushed in, soaking the plastic seals that surrounded cables and electrical wiring. Those seals, too, had been improperly installed and failed in large numbers. BP had declined to conduct any of the standard inspections and processes that could have mitigated each of these problems.[22] Adding to the irony, just five months earlier, interior secretary Gale Norton had visited the rig and declared that it was "created to protect the blue waters that it stands in, no matter how great the storm."[23]

Then, in March 2006, another disaster: a section of BP's North Slope pipeline ruptured, spilling 267,000 gallons of crude onto the Alaskan tundra. The pipeline failure exposed the company's longstanding practice of neglecting to seek out and repair corrosion along the hundreds of miles of pipeline network. Having expected the North Slope field to last only until 1987, BP was reluctant to upgrade its equipment there, even as the oil kept flowing well beyond that date. And with cost-cutting a priority, pipeline maintenance and inspections were practically the only flexible budget items. Workers who complained about the neglected infrastructure and unsafe working conditions were harassed or suddenly found themselves unable to get work on the North Slope.[24]

Amazingly, all of these preventable accidents occurred despite the fact that BP had been warned repeatedly about its shoddy practices. In 1999, the company's Alaska division pleaded guilty to illegal dumping at Endicott Island in Alaska. To avoid having the division debarred—that is, banned from obtaining federal contracts and new U.S. drilling leases—BP agreed to a five-year probationary plan with the EPA. As part of that agreement, BP pledged to modify its safety culture, to establish a "revised corporate attitude."[25] Yet throughout the probationary period and beyond, the same problems persisted and were well documented. After the Texas City and Prudhoe Bay accidents, Tony Hayward, an engineer and long-time BP executive, replaced John Browne as CEO. Hayward again pledged to improve the company's safety record and to make process safety a central aspect of BP's corporate culture. The Deepwater Horizon accident suggests, however, that Hayward's efforts were (to put it mildly) unsuccessful.[26]

THE CASE

A culture of safety is particularly important in deepwater oil development, which involves an extraordinarily complex and delicate set of processes, most of which are not directly visible. When an oil well is drilled, it is built like a telescope of successively smaller pipes called conductor casing that run from the seafloor to the bottom of the well. The first section of the Macondo well, which runs from the seafloor to a depth of several thousand feet, is 36 inches wide; the final stretch, from 12,168 feet to 13,360 feet, is 7 inches across. At each transition from one casing width to the next, cement is pumped down through the bottom of the wellbore and up the sides to seal the space between the sides of the newly dug well and the casing, known as the annulus, as well as between each incrementally narrower section of pipe. Throughout the process, the crew must balance the hydrostatic pressure inside and outside the well. Doing this in deep water poses special difficulties because of the enormous pressure and high temperatures encountered at great depths. Drilling in the Gulf of Mexico can be particularly challenging because the gulf is notoriously "overpressured"; rapid layering of mud in deep, low-oxygen water has led to high rates of natural gas formation. Water gets trapped in the pores of rocks buried by successive layers of sediment, and imprisoned water increases the pressures in the rock formation.[27] Even among gulf wells, the Macondo well was considered tricky, and workers aboard the Deepwater Horizon often referred to it as the "well from hell." Nevertheless, when it came time to seal the well temporarily, for later production, BP and its subcontractors made a series of high-risk decisions that, in retrospect, seemed to make disaster inevitable.

Drilling and Cementing the Macondo Well

In March 2008, BP paid the federal government $34 million at auction for the right to drill in the nine-square-mile parcel known as the Mississippi Canyon Block 252. In early October 2009, a rig called the Marianas began drilling within that parcel in the prospect BP had named Macondo.[28] But after a late-season hurricane damaged the Marianas, forcing it into port for repairs, BP sent an alternative rig, the Deepwater Horizon, to take over the job. Owned and operated by Transocean, one of the world's largest deepwater service providers, the Deepwater Horizon was a massive, 33,000-ton drilling rig, 396 feet long by 256 feet wide, with a derrick 242 feet high, set atop two giant pontoons. On board were six 9,777-horsepower engines, along with two 15-foot boom cranes, four 7,500-psi mud pumps, and a host of other machinery. Two dynamic-positioning officers maintained the rig's position over the well by using global-positioning readings to control eight powerful thrusters. Transocean charged $525,000 a day for use of the Deepwater Horizon, but the full cost of operating the rig—including mud engineers, cement engineers, pilots for the remotely operated submersibles, caterers, and others—was more than $1 million a day.[29]

In February 2010, the Deepwater Horizon restarted the drilling process by placing a blowout preventer atop the wellhead. The main purpose of the blowout preventer is to avert uncontrolled releases of oil and gas; it also enables numerous tests and other drilling-related activities. This particular blowout preventer was a 300-ton, multistory steel stack with a series of valves that branched off a main pipe stem. The topmost valves, the donut-shaped annular preventers, could cinch tight around the drill pipe and close off the space between it and the well's casing. Below the annular preventers was the lower marine riser package, equipped with an emergency disconnect system that allowed the rig to separate from the blowout preventer in the event of an emergency. Lower still was the blind shear ram, whose blades are designed to sever the drill pipe and seal the well—the well-control mechanism of last resort. And below the blind shear ram were three variable bore rams, whose pincers would close around the drill pipe without shearing it, instead cutting off the annular space (for use at higher pressures than the annular preventers). The blowout preventer was controlled from the rig through two cables connected to two redundant control "pods"—blue and yellow. There were two control panels to operate the blowout preventer, one on the rig floor and one on the bridge.[30]

With the blowout preventer latched down, the crew ran a 21-inch pipe, known as the riser, up from the wellhead through the blowout preventer and nearly 5,000 feet of water, through the middle of the drill floor to the top of the derrick. Inside the riser ran the drill pipe with a drill bit at the end. The Deepwater Horizon was now ready to resume where the Marianas had left off.

Almost immediately, however, the crew began to encounter difficulties. Most important of these was that the well routinely lost drilling mud, which migrated into the porous sandstone rocks that surrounded the wellbore—a phenomenon known as "lost returns." Drilling mud, a combination of viscous synthetic fluids, polymers, oil, and chemicals specially mixed for each well, is used to cool the drill bit, lubricate the well, and carry to the surface bits of rock the drill bit has removed from the bottom of the well. But its most important function is to maintain hydrostatic pressure in the well; its weight counterbalances the upward pressure of the hydrocarbons below.

At the Macondo, the disappearance of drilling mud often set off violent "kicks" of oil and gas, sending crew members scrambling to control the well. On April 9, at 18,139 feet below sea level, the lost returns were so severe that the crew had to stop drilling and seal fractures in the formation around the wellbore. Despite the lost returns and other challenges, the crew was elated when they struck oil and gas (the "pay zone," which BP believed contained between 50 million and 100 million barrels of crude) between 18,051 and 18,223 feet.[31] Still, the engineers concluded that they would have to stop drilling at 18,360 feet rather than 20,200 feet as originally planned because of "well integrity and safety" concerns.[32]

By April 18 workers had finished drilling and were preparing to seal the well so it could be left unattended until a production rig arrived. At this point, however, they were well aware that the job was six weeks behind schedule and some $58 million over budget.[33] The first step in sealing the well was to cement the bottom. The efficacy of that cement job depended on several critical decisions. First, in selecting a well design, BP opted for a "long string" production casing, which entailed running a single tube through the center of the wellbore from the seafloor down to 13,360 feet underground. The alternative BP rejected was the "tie-back" method, which many experts prefer because it creates multiple barriers against hydrocarbons flowing into the well. By contrast, the long-string method relies exclusively on one barrier: the cement at the bottom of the well. In addition to being more risky than the tie-back design, the long string is significantly quicker and cheaper to install.[34]

Halliburton, one of BP's subcontractors, analyzed BP's well design and determined that the long string would not pose a problem, as long as BP used twenty-one centralizers to position the drill string evenly in the center of the hole. BP had only six centralizers on board the Deepwater Horizon, however. After running numerous models to project how the cement job would work, Jesse Gagliano, a Halliburton engineer, persuaded several BP engineers that more centralizers were essential; after communicating with Gagliano, Gregg Walz, the drilling operations manager, ordered fifteen additional centralizers. But John Guide, BP's Houston-based well team leader, (erroneously) argued that the additional centralizers lacked the right fittings to hold them in place; he also pointed out that it would take ten more hours to install them. Ultimately, the cement job proceeded with only six centralizers over the objections of Gagliano, who warned in a report attached to an April 18 e-mail that the well was likely to have a "SEVERE gas flow problem."[35]

After hanging the long-string casing and installing the centralizers, the next step was to cement the bottom of the well, the trickiest part of the well-completion process.[36] Here, ostensibly out of concern about fracturing the fragile formation surrounding the well, BP made a series of risky decisions that left little margin for error. First, in collaboration with Halliburton, BP opted for a nitrogen-foamed cement slurry that would be strong enough to seal the well but would form foamy bubbles that would make it less likely to leak into the porous walls of the formation and would be light enough to minimize the pressure on those walls. Second, rather than circulating 2,760 barrels of mud through the entire well in a bottoms-up circulation that takes six to twelve hours and is common industry practice prior to cementing, BP opted to circulate only 350 barrels in just thirty minutes. Finally, on the evening of April 19, workers sent sixty barrels of the specially formulated cement down the well, circulating it as they had the mud: pumping it down the middle and out the bottom, then up into the annulus between the casing and the wellbore. The crew finished pumping the cement at 12:40 a.m. on April 20. After observing that the amount of mud coming out of the well was the same as the amount of cement sent in, they were satisfied that this part of the job was complete.

Typically, the next-to-final step before sealing a well is a cement bond log, a test that determines the effectiveness of the bond between the cement, the formation, and the casing. After settling on a long-string well design, BP had ordered a bond log to ensure that the solitary cement barrier in the well would hold. The bond log test can take up to twelve hours, however, and would have cost an additional $128,000. Based on their observations of full mud returns, the BP engineers decided that it was not necessary, even though there is broad agreement among experts that full returns provide incomplete information about the success of a cement job. They sent the crew that was standing by to perform the test back to shore by helicopter late in the morning of April 20.[37]

Although it declined to perform the bond log, BP did conduct a battery of pressure tests to ensure the well's seals and casings were holding. The first, a relatively simple positive-pressure test that began around noon, was successful.

The negative-pressure test that began around 5 p.m. yielded disconcerting results, however. This test involved replacing the heavy drilling mud that had been circulating through the well with much lighter seawater. Next, the crew would shut the well to see if pressure built up inside it. If it did, that would mean oil and gas were seeping into the well—indicating an integrity problem with the well. As instructed, the workers displaced 3,300 feet of heavy mud from the wellbore with lighter seawater preceded by a "spacer" made up of leftover fluids that had been used to seal the well during the April 9 lost-returns event; next, they closed one of the blowout preventer's annular preventers to hold the mud in place outside the well (in the riser); then, having inserted it into the wellbore, they opened the drill pipe to release the pent-up pressure in the well to zero. The crew could not get the pressure in the drill pipe to bleed down below 266 psi, however, and as soon as the drillers sealed off the drill pipe, the pressure jumped to 1,262 psi. Thinking perhaps there was a leak between the riser and the wellbore, workers further tightened the annular preventer, bled the pressure down to zero, and then shut in the drill pipe. Twice more they measured the pressure, with even starker results: the pressure jumped to at least 773 psi on the first try and to 1,400 psi on the second.[38]

There was sharp disagreement among the team members about how to interpret this series of negative-pressure tests but an apparent determination to proceed. So, at the direction of Don Vidrine, one of BP's well-site leaders, the crew attempted an additional test that involved measuring the pressure not on the well itself, but on the kill line, one of several three-inch pipes that run from the drill rig down to the bottom of the blowout-preventer as a means to access the wellbore even when the blowout preventer is closed. Theoretically, at least, the pressure on the kill line should match the pressure on the well itself, as measured by the pressure on the drill pipe. The crew bled the pressure in the kill line down to zero and left the top of the line open for thirty minutes. They observed no increase in pressure, although the pressure in the drill pipe remained at 1,400 psi. Explaining away this pressure differential, possibly as the result of a "bladder effect," BP's well-site leaders decided that the well had integrity.[39] At 8:02 p.m. the crew began displacing the mud and spacer that remained in the riser with lighter seawater in preparation for running a final cement plug 3,300 feet below the wellhead and then installing a "lockdown sleeve" that locks the long-string casing to the wellhead.[40] Around 9 p.m. signs of rising pressure in the well began to manifest themselves, but the crew missed them—possibly because they were preoccupied with the process of testing the drilling mud and spacer and then dumping it overboard.[41]

The Disaster

The first unmistakable sign of trouble at the Macondo well came between 9:40 and 9:43 p.m., when mud began spewing out of the drill pipe. This appears to have been the first point at which the crew realized a gas kick had occurred. They reacted by trying to close the annular preventer around the drill pipe (at 9:41), but that did not stop the flow of gas, which was already above the blowout preventer;

next they tried to activate the variable bore rams (at 9:46), but those devices did not respond. Around 9:45, there was a loud hiss as the expanding bubble of gas reached the top of the drill pipe, and alarms began lighting up the control panel in the engine room.[42] In the office of Mike Williams, an electronics technician, the computer monitor exploded, after which the light bulbs began to pop one after another. On the bridge, the two dynamic-positioning officers, Andrea Fleytas and Yancy Keplinger, saw alarm lights showing magenta, the color code for the highest level of gas. At this point officials at the bridge could have sounded the general alarm; they also could have activated the emergency shutdown system, which would have stopped the ventilation fans and inhibited the flow of gas, turned off electrical equipment, and even shut down the engines. Paralyzed by indecision and uncertainty, they did none of these.[43]

Meanwhile, as the gas cloud on deck expanded, crew members heard the rig's engines revving faster and faster, emitting a high-pitched whine. Within seconds, there was a thunderous explosion that obliterated Engine 3 along with whole sections of the rig. Seconds later there was another deafening blast, as Engine 6 caught fire and exploded. As journalists David Barstow, David Rohde, and Stephanie Saul describe it, "Crew members were cut down by shrapnel, hurled across rooms and buried under smoking wreckage. Some were swallowed by fireballs that raced through the oil rig's shattered interior. Dazed and battered survivors, half-naked and dripping in highly combustible gas, crawled inch by inch in pitch darkness, willing themselves to the lifeboat deck."[44] As they made their way toward the lifeboats, workers saw an alarming sight: the derrick was enveloped by flames that leapt 300 feet into the air.

Finally, two minutes after the first gas alarms sounded and a full nine minutes after mud began shooting out of the riser, Andrea Fleytas hit the general alarm (at 9:47) and sent out a Mayday call to other vessels in the area (at 9:53). Amidst the chaos, Chris Pleasant, a subsea engineer, made his way to the bridge with the intention of activating the blowout preventer and disconnecting the rig from the well. An alarm on the control panel indicated that the pressure had fallen in the blowout preventer's hydraulics, however, disabling the blind shear ram. The disconnect button did not work either; nor did the "dead man's" switch that is supposed to seal the well automatically in the event that the blind shear ram cannot be triggered manually.

Although the crew had practiced the emergency evacuation process numerous times, the scene at the lifeboats was chaotic as well. Some men jumped from the rig into the dark, oily ocean 100 feet below, fearing that if they waited for the lifeboats to deploy they would die aboard the rig. One after the other, the two lifeboats closed their doors, dropped into the water, and started making their way toward the *Damon B. Bankston*, a nearby supply ship. The few crew members who remained on board got into a life raft. But they had difficulty figuring out how to lower and then free the raft, which was connected to the rig by a taut line. Finally, Captain Curt Kuchta—who had jumped into the water after the life raft departed—swam to a rescue boat, got a knife, and cut the raft free. It, too, headed for the *Bankston*.

Less than an hour after the explosion, the Deepwater Horizon crew members aboard the *Bankston* lined up for a head count. Eleven men were missing and never found. Of the 115 who escaped the flames, seventeen were injured; three were critically hurt and were evacuated by helicopter. The rest huddled on deck watching the Deepwater Horizon go up in flames. At 8:13 a.m. on April 21 the *Bankston* finally received permission to leave the scene, making several stops on the way to shore—including one to collect investigators from the Coast Guard and the interior department who interviewed survivors and asked them for complete written statements. Finally, at 1 a.m. on April 22 the *Bankston* delivered the Deepwater Horizon crew to Port Fourchon, Louisiana. There, each member of the crew was given forms to fill out along with a plastic cup and directed to the portable toilets, following the Coast Guard protocol for drug tests after a maritime accident. Survivors then boarded company buses for the two-hour drive to New Orleans, where they were reunited with their families.[45]

Responding to the Spill

As the crew headed to shore, a search-and-rescue effort was under way. On April 21 Coast Guard Rear Admiral Mary Landry took over as federal on-scene coordinator, a position she retained until June 1. Landry's main concern was a possible spill of the fuel oil on board the rig. On May 1, however, after it became apparent that oil was leaking continuously from the well, the Coast Guard designated the disaster a "Spill of National Significance" and established the National Incident Command to coordinate activities aimed at stanching the leak and cleaning up the widening oil slick. At the helm was Admiral Thad W. Allen, who became the public face of the response. Although officially the Coast Guard was in charge, day-to-day decisions were made jointly by the Coast Guard and their BP counterparts. BP played a prominent role for two reasons. First, under the Oil Pollution Act of 1990, BP—as the owner of the well—was responsible for most of the cleanup costs associated with the spill, as well as for damages to the environment and the local economy.[46] Second, the government simply lacked the expertise and equipment to manage the complex effort.

Despite an apparently clear hierarchy, the response was "bedeviled by a lack of preparation, organization, urgency, and clear lines of authority among federal, state, and local officials, as well as BP."[47] According to journalist Henry Fountain, documents from the early days after the explosion—which detail efforts to seal the well that were hampered by a lack of information, logistical problems, unexplained delays, and other obstacles—give a clear sense of the "improvisational, never-before-experienced nature of the work."[48] Subsequent investigations confirmed that the process of trying to plug the Macondo well was "far more stressful, hair-raising and acrimonious" than the public realized.[49] There were close calls and, as failures accumulated, a growing number of disputes that delayed the eventual killing of the well. As a result, the damages were worse than they might otherwise have been.

Trying (and Failing) to Stanch the Leak. Within days of the initial explosion, there was a sheen of emulsified crude about 42 miles by 80 miles, located less than 40 miles offshore.[50] The oil was coming from three leaks: one at the end of the riser where it broke off when the Deepwater Horizon sank into the sea on the morning of April 22, another where the riser was kinked at a 90-degree angle about five feet above the blowout preventer, and a third 460 feet away on a section of the riser that lay on the seafloor. At first BP downplayed the amount of oil spewing from the well, putting it at about 1,000 barrels per day. In late April the company grudgingly acknowledged the figure was probably closer to 5,000 barrels per day, the estimate offered by the National Oceanographic and Atmospheric Administration (NOAA). Ian MacDonald, an oceanographer at Florida State University, was suspicious of that figure as well; his analysis suggested the oil discharge was as much as 30,000 barrels per day.[51] As early as May 4, a senior BP executive conceded in a closed-door briefing for members of Congress that the Macondo well could spill as much as 60,000 barrels per day of oil, more than ten times the then-current estimate of the flow.[52] In late May the recently formed National Incident Command's Flow Rate Technical Group, a collection of government and independent researchers, acknowledged that far more oil than federal officials originally estimated was probably pouring into the gulf on a daily basis—as much 12,000 to 19,000 barrels.[53] Accurately determining the flow rate was critical because fines levied under the Clean Water Act are assessed by the barrel. If BP was found to be grossly negligent, it could face fines of up to $4,300 per barrel, rather than the minimum $1,200 per barrel.

BP's first efforts to stanch the leaks, initiated immediately after the explosion, involved having remote-control robotic submersibles operating 5,000 feet below the ocean's surface trying to activate the rig's damaged blowout preventer. The remotely operated vehicles (ROVs) tried to inject hydraulic fluid directly into the blind shear ram, but the blowout preventer's hydraulic system leaked so pressure couldn't be maintained in the ram's shearing blades. Even after the ROVs managed to repair the leak, attempts to activate the blind shear ram were unsuccessful. As oil and gas flowed through the blowout preventer, corroding its parts, it became increasingly unlikely the equipment would ever function properly. On May 5 BP finally gave up on the blowout preventer. (The previous day the ROVs did manage to seal one of the three leaks by clamping a specially designed valve over the end of the drill pipe lying on the seafloor, but that did not significantly reduce the flow of oil.)

With the blowout preventer hopelessly crippled, BP's Plan B was to construct a relief well that would intercept the original well. From the relief well, BP could push mud and concrete into the gushing cavity to plug it permanently. BP began drilling a relief well on May 2 and then, at the direction of interior secretary Ken Salazar, started work on a backup well two weeks later. Relief wells are routine even in deep water; they start out vertical but are then drilled at an angle to intersect the existing well. It would be relatively straightforward to find the bottom casing of the original well—a seven-inch diameter steel pipe roughly 18,000 feet below sea level—because engineers surveyed the existing well constantly while drilling it.[54] But drilling the relief wells would take two to three months, and in the meantime, BP needed a short-term alternative.

One option was to design a dome that could be submerged over the two remaining leaks from the riser. The dome—a 98-ton steel box four stories tall—would corral the oil and route it up to vessels, where it would be collected and taken to shore. This method had been used elsewhere but never at such depths. Nevertheless, late in the evening of May 6, workers began lowering a specially designed dome. They soon encountered setbacks, however: they discovered that the dome's opening was becoming clogged with methane hydrates, crystal structures that form when gas and water mix in very low temperatures and under high pressure. The hydrates were forming a slush that clogged the opening through which oil was to be funneled to the surface. In addition, because they are lighter than water, the hydrates threatened to increase the buoyancy of the dome and lift it out of place. Engineers had anticipated a problem with hydrates and planned to solve it by circulating warm water around the pipe that would connect the dome to the ship on the surface. But they underestimated how quickly large concentrations of hydrates would accumulate, effectively closing off the dome before the pipe was even attached. So the dome sat on the seafloor while engineers pondered their next move.[55]

The next option was a "junk shot," which involved reconfiguring the broken blowout preventer and injecting heavy material—including golf balls, pieces of rubber tire, and knots of rope—into it, then pumping drilling mud down the well at a rate fast enough to force the oil back into the reservoir (a "top kill"). If this procedure was successful, engineers could cement the well and stop the leak. The trick was to devise an approach that was most likely to work and least likely to make things worse. To that end, BP mobilized what it called a dream team of engineers from the oil industry and government agencies to flesh out the specifics of the plan. Meanwhile, technicians managed to put in place a stopgap measure: a mile-long pipe that captured about 1,000 barrels of leaking oil and diverted it to a drill ship on the surface. Gradually, BP increased the amount of oil it was siphoning through this pipe to more than 2,000 barrels per day.[56]

Finally, after days of delays and testing, in the early afternoon on Wednesday, May 26, BP began pumping as much as 50,000 barrels of heavy drilling fluids into the kill and choke lines above the well. After pumping for about nine hours, however, BP suspended the pumping throughout the day on Thursday while it scrutinized the initial results, which showed too much drilling fluid escaping along with the oil. Engineers began pumping again on Thursday evening, but despite injecting different weights of mud and sizes of debris down the well, they struggled to stem the flow of oil. On May 29 BP officials acknowledged that the junk shot/top kill had failed; the pressure of the escaping oil and gas was simply too powerful to overcome.[57]

Cleaning Up the Mess. While BP sought to plug the leak, the National Unified Command was leading an effort to clean up the oil and prevent it from reaching fragile coastlines. Planes dropped a chemical dispersant while ROVs applied it directly to the leaks in an effort to break down the oil into tiny droplets that would sink to the bottom or be diluted by the ocean currents. Coast Guard ships ignited surface fires to burn off some of the oil, while skimmers sought to collect the rest. High winds and choppy waters hampered burning and skimming efforts; on the other hand, they broke up the oil and impeded the coastward movement of the slick.[58] Nevertheless, by Sunday, May 2, oil was creeping into Louisiana's coastal wetlands. There it threatened to smother the spartina grasses that knit the ecosystem together. (If damaged only above the ground, spartina grass grows back quickly. But if the roots die, the plant dies and the ground underneath it sinks into the sea.) By the last week in May the expanding slick had hit 70 miles of Louisiana shoreline and was seeping into shellfish-rich estuaries near the mouth of the Mississippi River. *Washington Post* journalists Juliet Eilperin and David Fahrenthold reported that, although by May 26 more than 3 million feet of boom had been placed around sensitive coastlines, there was ongoing evidence that oil was getting past those barriers.[59]

As the spill dragged on, scientists began sounding the alarm about the potential ecological damage. The gulf is one of only two nurseries for the bluefin tuna, more than 90 percent of which return to their birthplace to spawn. Because they have to surface regularly to breathe, sea turtles and dolphins were in danger of becoming coated in oil or inhaling toxic fumes. Notably, the Gulf of Mexico is the sole breeding ground for the highly endangered Kemp's ridley sea turtle. The timing of the spill amplified its impacts. Spring is when the pelicans, river otters, shrimp, alligators, and other species native to the area reproduce. It is also the season when huge numbers of birds converge on the gulf's coastline marshes and sandy barrier islands. Oil damages birds by coating their feathers, destroying the natural chemistry that keeps them buoyant, warm, and able to fly; when birds try to clean themselves, they can swallow the oil and be poisoned.[60]

Scientists also raised concerns about the impact of releasing chemical dispersants in deep water, whose ecology is poorly understood. The dispersant used was Nalco's Corexit 9500A, which had never been deployed at depth.[61] Researchers wondered whether the dispersant would kill fish larvae and threaten filter feeders, like those for whale sharks. They suspected it might also harm oysters and mussels, which cannot move to escape it, as well as the tiny organisms that sustain larger marine creatures. Researchers speculated that damage to coral reefs, which take centuries to develop in the gulf's oxygen-poor depths, would occur. We now know that dispersants affected the health of cleanup workers (e.g., cough, wheezing, eye irritation) and ocean life (e.g., corals, sea turtles, birds).[62]

The Political Fallout. Like the ecological fallout, the political impacts of the spill were not immediately apparent. From the outset, the Obama administration struggled to show it was in charge, even as it relied heavily on BP to both stop the spill and orchestrate its cleanup. In an effort to demonstrate leadership, President Obama held regular conference calls with BP executives; he also instituted a temporary moratorium on new permits for gulf drilling in more than 500 feet of water and called a halt to the kind of environmental waiver that was granted to the Deepwater Horizon rig.[63] On May 22 Obama issued an executive order establishing a bipartisan national commission to investigate the spill to come up with a plan to revamp federal regulation of offshore drilling. For his part, Interior Secretary Salazar told CNN's *State of the Union*, "Our job basically is to keep the boot on the neck of British Petroleum."[64] However, at the same time, he met with oil and gas industry executives to appeal for ideas and help. On May 27 Secretary Salazar imposed a six-month moratorium on permits for new oil and gas exploration in water more than 500 feet deep, so that the cause of the Deepwater Horizon accident could be determined and stricter safety and environmental safeguards put in place.[65]

Environmentalists wanted the administration to do more, hoping to capitalize on the spill to redefine the problem—the disaster was an inevitable consequence of our dependence on fossil fuels—and thereby shift public opinion on climate and energy issues. "This is potentially a watershed environmental disaster," said Wesley P. Warren, director of programs at the National Resources Defense Council. "This is one gigantic wakeup call on the need to move beyond oil as an energy source."[66] While many environmentalists rushed to the gulf to help with the cleanup, others held news conferences, filmed TV spots, and organized protest rallies with the aim of persuading lawmakers to block new offshore drilling and pass legislation to curb U.S. greenhouse gas emissions. Activists held "Hands Across the Sand" events at gulf beaches and spelled out "Freedom From Oil" with American flags on the mall in Washington, D.C. As David Hirsch, managing director of Friends of the Earth, explained, "It's very difficult in our society to cut through the din and get people to listen and pay attention. Unfortunately, these are the times when it happens. These are the moments when you can be heard."[67]

Finally, on Friday May 21 President Obama began treating the spill as a political opportunity: he announced plans to impose stricter fuel-efficiency and emissions standards on cars and, for the first time, on medium- and heavy-duty trucks. He said the oil gushing from the crippled BP well highlighted the need to move away from fossil fuels toward a clean energy future. "We know that our dependence on foreign oil endangers our security and our economy," he said in a Rose Garden address. "And the disaster in the gulf only underscores that even as we pursue domestic production to reduce our reliance on imported oil, our long-term security depends on the development of alternative sources of fuel and new transportation technologies."[68] On May 26, as BP was beginning the top-kill operation, Obama focused on the hazards of fossil-fuel dependence in a speech in California. "With the increased risks and increased costs, it gives you a sense of where we're going," he said. "We're not going to be able to sustain this kind of fossil fuel use."[69]

Despite this more aggressive rhetoric, the Obama administration's comprehensive energy and climate change plan appeared doomed. A month before the disaster, the president had announced that his administration would expand offshore drilling, part of an effort to assemble a coalition large enough to pass climate change legislation. Just eighteen days before the spill, Obama had told an audience at a South Carolina town hall meeting, "It turns out, by the way, that oil rigs today generally don't cause spills. They are technologically very advanced."[70] By May, however, whatever fragile coalition his concessions had enabled was crumbling, as members of Congress—particularly Democrats Bill Nelson (Fla.) and Robert Menendez (N.J.)—issued calls for him to abandon his plans for expanded offshore drilling. Also discouraging, in July public opinion polls showed that only half the public was worried about climate change, down from 63 percent in 2008—almost certainly a result of the bad economy and the country's toxic politics.[71]

Perhaps the final nail in the coffin was that, although they voiced desperation after tar balls and oil slicks turned up on local beaches, state and local officials from the gulf region directed most of their ire not at BP but at the federal government, which they argued was moving too slowly in helping them hold back the oil. Moreover, gulf state politicians chafed against the Obama administration's moratorium, which they said would drive drilling rigs overseas and destroy thousands of jobs. Such loyalty to the oil and gas industry was not surprising. The industry is a major contributor to the campaigns of politicians from oil-producing states; in the three years leading up to the spill, lawmakers from Texas, Louisiana, Mississippi, and Alabama received an average of $100,000 from oil and gas companies and their employees, according to the Center for Responsive Politics.[72] In addition, the oil and gas industry has focused on hiring as its lobbyists former lawmakers from oil-producing states. (Fifteen of the eighteen former members of Congress who now lobby for oil and gas companies are from Texas, Louisiana, Mississippi, Oklahoma, or Kansas.) Dozens more previously worked as aides to lawmakers from those states. During a June hearing, Democratic Senator Mary Landrieu of Louisiana warned Secretary Salazar that a prolonged halt to deepwater drilling "could potentially wreak economic havoc on this region that exceeds the havoc wreaked" by the spill itself.[73] Even those who worked in the seafood industry, whose livelihoods were threatened by the prolonged spill, were unwilling to criticize the oil and gas industry.[74] The two industries have long coexisted, and residents recognize that oil is essential to their way of life and to the region's prosperity.

Finally, a Solution That Works. After the failure of the top kill in late May, BP began exploring another approach: a custom-fit containment cap known as the "top hat." The new cap was outfitted with pipes for injecting methanol, a kind of antifreeze, to avoid the formation of methane hydrates, which had doomed the previous containment cap.[75] On June 3, after a few setbacks, ROVs guided the cap into place. Journalists Clifford Krauss and Henry Fountain describe the scene:

> As the cap hit the oil and gas streaming with great force from the top of the well, it suddenly disappeared, hidden from the video cameras by clouds of hydrocarbons spewing everywhere. . . . The operation

was briefly flying blind. But a few tense moments later the cap was successfully centered on the wellhead, 5,000 feet below the surface, then lowered half a foot to make a seal. The crowd, including Energy Secretary Steven Chu, gave a cheer.[76]

Engineers began closing the vents atop the cap, working slowly to ensure the mounting pressure would not force it off the well. During its first twenty-four hours of operation, the cap recovered about 6,000 barrels of oil; the goal was to push that figure to 15,000 barrels per day, the capacity of the recovery ship. On June 16 BP moved an additional processing ship into position to help it capture the oil it was bringing to the surface through the choke line on the blowout preventer. BP officials expected that together the two containment systems would be able to collect 28,000 barrels per day, some of which was being transferred to a tanker, while the remainder was being flared. By the end of the month, BP intended to add a third system for capturing oil: a free-standing riser on the seabed floor. Combined, the three systems could collect about 50,000 barrels daily. This figure was critical because by mid-June, the Flow Rate Technical Group had raised its estimate of how much oil was actually spilling from the well to 60,000 barrels per day.

Although BP's efforts to capture the oil were finally bearing fruit, the public remained unimpressed. A CNN/Opinion Research poll released on June 18 showed that 48 percent of respondents believed the situation was getting worse, and only 14 percent thought it was improving. Half the respondents believed the gulf would never fully recover, and 53 percent favored criminal charges against BP employees and executives. A poll by *The New York Times*/CBS News published on June 22 revealed that the public had little faith in BP's ability to handle the cleanup in the public interest; by a two-to-one margin respondents trusted the federal government more than BP in this regard.[77] In mid-June, at the president's behest, BP sought to resuscitate its reputation by establishing a $20 billion escrow account from which the company could compensate individuals and businesses for spill-related losses. Kenneth R. Feinberg, the lawyer and mediator who ran the fund for the victims of the 9/11 attacks, agreed to administer the fund. (The fund did not preclude individuals or states from pressing claims in court, and it was separate from BP's liabilities for damages to the environment.)

On July 10, after extensive analysis and consultation among government and industry experts, BP began an operation to install an even tighter cap (the "capping stack") on the well, while also establishing a second free-standing riser pipe. After a minor setback in which the choke line sprung a leak and had to be repaired, on July 15 BP succeeded in capping the well completely, stopping the oil after eighty-six days. Engineers then initiated a forty-eight-hour integrity test to make sure the cap would hold. By July 31 engineers were preparing to pour heavy drilling mud into and then cement the Macondo well, part of a two-pronged strategy to kill the well once from above (the "static kill") and once from below (the "bottom kill," effected via the relief well). On August 3, the 107th day of the disaster,

BP announced that the static kill had been a success. Final sealing of the well was delayed until after Labor Day, however, so BP could replace the blowout preventer. (The new blowout preventer would be better able to handle any pressure changes that might occur when the relief well intercepted the stricken well and the bottom kill was deployed.[78]) In mid-September the first relief well intercepted the Macondo well, at which point BP pumped in cement to permanently seal the well. Finally, on September 19 the federal government declared the Macondo well officially dead. A pressure test confirmed that the cement poured into the well had formed an effective and final seal to prevent oil and gas from escaping the reservoir.

Investigating the Accident, Pointing Fingers

Throughout the summer, as the response team struggled to cap the Macondo well, the companies involved with drilling the well were quick to put forth their own definition of the problem—proffering causal stories that allocated blame elsewhere. They made their cases on TV news shows, in press conferences and interviews, and in congressional hearings. For example, on May 5 BP's Tony Hayward said on ABC's *Good Morning America*, "This wasn't our accident. This was a drilling rig operated by another company. It was their people, their systems, their processes. We are responsible not for the accident, but we are responsible for the oil and for dealing with it and for cleaning the situation up."[79] In a pair of Senate hearings on May 11 executives from BP, Transocean, and Halliburton agreed on one thing: someone else was to blame for the spill. BP America's president, Lamar McKay, blamed the failure of Transocean's blowout preventer and raised questions about whether Transocean disregarded "anomalous pressure readings" in the hours before the explosion. Transocean CEO Steven Newman said the blowout preventers clearly were *not* the root cause of the accident; instead, he blamed the decision made by BP to replace drilling mud with seawater and cited possible flaws in the cementing job done by Halliburton. And Tim Probert of Halliburton said his workers had simply followed BP's instructions, while the Transocean crew had started replacing drilling mud with seawater before the well was properly sealed.[80]

Its efforts to implicate other companies notwithstanding, BP clearly had an organizational culture that emphasized personal over process safety, largely as a way of cutting costs. The multiple investigations undertaken in the aftermath of the disaster made clear that BP was under severe pressure to close the Macondo well quickly and that the growing cost of the operation weighed heavily on both rig workers on the Deepwater Horizon and on-shore engineers and managers.[81] At numerous critical decision points, BP officials opted for the less costly or time-consuming option without calculating the accumulating risk of doing so. For example, in choosing a long-string casing for the Macondo well, BP selected the cheaper of the two options, saving $7 million to $10 million. Similarly, the decision to use six centralizers rather than twenty-one, as Halliburton recommended, saved both time and money. Subsequent investigations suggest these may not have been the fateful decisions, however.[82] More critical were the cementing process itself, which involved several shortcuts, and the decision not to perform a costly and time-consuming cement bond log, which would have corrected the crew's misinterpretation of the negative-pressure tests.

Also problematic was the decision to use unknown material as a spacer during the negative-pressure tests. BP had 424 barrels of a thick, heavy compound left over from its efforts to seal the areas in the wellbore where mud had been leaking weeks earlier. The drilling fluid specialists on the rig proposed using this leftover material as the spacer needed to complete the cementing process, even though such material had never been used or thoroughly tested for this purpose. By incorporating the heavy material into its spacer, BP could avoid the costs of disposing of it onshore, where it would be treated as hazardous waste. (Under the Resource Conservation and Recovery Act, if fluid is part of the drilling process, the crew is allowed to dump it at sea.) Some observers suspect the spacer may have clogged the kill line, accounting for the confusing results of the final negative-pressure test. It is also likely that the crew were distracted during the final process of circulating seawater through the well, missing signs of increasing pressure in the critical period before the blowout because they were focusing on dumping an unusually large amount of drilling fluid and spacer overboard.[83]

Transocean was culpable as well. Journalist Ian Urbina reports that a confidential survey of workers on the Deepwater Horizon in the weeks before the rig exploded showed that many of them were concerned about safety practices and feared reprisals if they reported mistakes or other problems. The survey, commissioned by Transocean, found that workers "often saw unsafe behaviors on the rig" and worried about equipment reliability because drilling seemed to take precedence over planned maintenance.[84] Workers' concerns appear to have been justified: a confidential audit conducted by BP seven months before the explosion revealed that Transocean had left 390 repairs undone, including many that were "high priority."[85] Another fact indicted Transocean as well: crew members on the Deepwater Horizon clearly were not trained for a worst-case scenario and were frozen by the sheer complexity of the rig's defenses; as a result, at the height of the disaster communications fell apart, warning signals were missed, and crew members in critical areas failed to coordinate a response.[86]

Halliburton, too, was responsible, particularly because it knew weeks before the explosion that the foamed cement compound it planned to use to seal the bottom of the well was unstable but it went ahead with the job anyway. Between February and mid-April, Halliburton conducted three sets of laboratory tests, all of which indicated that the custom-formulated cement mixture did not meet industry standards. On March 8 Halliburton submitted at least one of those tests to BP, which also failed to act on it. Another test, carried out about a week before the blowout, again found the mixture unstable, but those findings were never forwarded to BP. The final cement test was not even completed before Halliburton began the cement job on the Macondo. The presidential commission investigating the disaster obtained samples from Halliburton of the same cement recipe used on the failed well and sent the slurry to a laboratory owned by Chevron for independent testing. Chevron conducted nine separate stability tests intended to reproduce conditions at the BP well, and the cement failed them all.[87]

Beyond the companies involved, the MMS bore its share of the blame for both the disaster and the chaotic response. Since its creation in 1982, the 1,700-person MMS had struggled with its dual mandate to enable oil and gas drilling and collect royalties (an average of $13 billion per year), while policing the industry to make sure it fulfilled its responsibilities in terms of safety and environmental protection. Over time, the agency came to regard itself as a collaborator with, rather than an overseer of, the oil industry. It routinely adopted industry-generated standards as regulations; referred to the companies it was supposed to regulate as "clients," "customers," and "partners"; and abandoned in the face of industry opposition proposals that might have improved safety but would have increased costs.[88] Stark evidence of the MMS-industry "partnership" surfaced in late May 2010, when the interior department's inspector general issued a report faulting the MMS for corruption and complacency with respect to offshore drilling between 2005 and 2007.[89] The report found that, among other problems, federal regulators allowed industry officials to fill in their own inspection reports in pencil; regulators would then trace over them in pen before submitting them to the MMS. Regulators also accepted meals, tickets to sporting events, and gifts from at least one oil company while they were overseeing the industry. The result, according to testimony by witnesses in hearings as well as government and BP documents and interviews with experts, was that the agency routinely granted exceptions to rules and allowed risks to accumulate.[90]

One way the MMS enabled the industry's rapid expansion was simply to pay lip service to environmental hazards. Although numerous environmental laws seek to minimize the risks of offshore oil and gas development, the Outer Continental Shelf Lands Act (OCSLA) gives the interior secretary tremendous discretion with respect to how much weight to give environmental concerns. In September 2009 NOAA accused the MMS of a pattern of understating the likelihood and potential consequences of a major spill in the gulf. Consistent with that allegation, current and former MMS scientists contend they were routinely overruled when they raised concerns about the safety and environmental impact of particular drilling proposals. "You simply are not allowed to conclude that drilling will have an impact," said one veteran agency scientist. "If you find the risks of a spill are high or you conclude that a certain species will be affected, your report gets disappeared in a desk drawer and they find another scientist to redo it or they rewrite it for you."[91] Even if the agency wanted to highlight environmental concerns, parts of OCSLA made it difficult to do so. For example, the interior secretary must approve a lessee's exploration plan within thirty days of submission; moreover, the act expressly singles out the Gulf of Mexico for less rigorous environmental oversight under the National Environmental Policy Act (NEPA). There were other structural impediments as well. The MMS was understaffed for the job: it had fifty-five inspectors to oversee about 3,000 offshore facilities in the gulf; by contrast, on the West Coast five inspectors covered just twenty-three ships. Many inspectors were not properly trained and had to rely on company representatives to explain what they were looking at; salaries at MMS were far too low to attract engineers with the expertise and experience needed to oversee the increasingly complex deepwater activities in the gulf.[92]

In practice, both the industry and its regulators had become overconfident about the industry's capabilities and complacent about the dangers posed by deepwater drilling. Prior to the Deepwater Horizon disaster, industry spokespeople repeatedly pointed out that there had been no major accidents in the gulf in three decades. According to Amy Myers Jaffe, an energy expert at Rice University, in the fifteen years preceding the Deepwater Horizon disaster, there was not a single spill of more than 1,000 barrels among 4,000 active platforms off the shores of the United States.[93] On the other hand, between 2001 and 2010 the Gulf of Mexico workforce of 35,000 suffered 1,550 injuries, 60 deaths, and 948 fires and explosions.[94] The further companies pushed into deep water, the greater the unknowns—and the more significant the risks.

Yet BP's exploration plan for the Mississippi Canyon Block 252 lease, submitted to federal regulators in February 2009, minimized the risk of a spill, saying, "[I]t is unlikely that an accidental spill release would occur from the proposed activities." It acknowledged that a spill could "cause impacts to wetlands and beaches" but pointed out that because of the considerable distance to shore and the industry's response capabilities, "no significant adverse impacts are expected." The plan further asserted that the company "has the capability to respond to the appropriate worst-case spill scenario," which it defined as a "volume uncontrolled blowout" of 162,000 barrels per day.[95] Yet in an interview published on June 3, BP's Tony Hayward acknowledged that BP was not prepared for a blowout of this magnitude. "What is undoubtedly true," he said, "is that we did not have the tools you would want in your toolkit."[96] Admiral Thad Allen defended the company's plan, saying, "It's hard to write a plan for a catastrophic event that has no precedent, which is what this was."[97]

In addition to accepting BP's exploration plan without requiring any demonstration that it would work, the MMS granted BP's Macondo lease a "categorical exclusion" from NEPA, so the company did not have to conduct a costly and time-consuming environmental impact assessment of the project. The MMS did this despite the fact that it had never conducted a serious analysis of a major oil spill in the central and western gulf.[98] Where BP did consider environmental impacts, in the Oil Spill Response Plan for the well, it clearly did not take the exercise seriously. For example, the plan identified three worst-case scenarios but then used identical language to describe the potential shoreline impacts for each. Half of the plan's five-page resource identification appendix simply copied text from NOAA's website, with no effort to determine whether that information was applicable to the gulf. In fact, the plan mentioned sea lions, sea otters, and walruses, none of which frequent the gulf.[99]

Perhaps the best illustration of the complacency of both industry and regulators concerns the Deepwater Horizon's last line of defense: the blowout preventer, which failed to activate during the emergency.[100] After several studies indicated problems with blowout preventers, in 2003 the MMS adopted a regulation requiring companies to submit test data proving their blind shear rams could work on the specific drill pipe used on a well and under the pressure they would encounter.[101] In practice, however, agency officials did not demand proof of such testing before issuing drilling permits. As the industry moved into deeper waters, it pressed regulators to reduce requirements for testing blowout preventers, and the MMS complied, halving the mandatory frequency of tests. And although its own research documented more than 100 failures during testing of blowout preventers, the MMS never instituted a requirement that blowout preventers be equipped with two blind shear rams.[102] On the Deepwater Horizon, this laxity was catastrophic. After the accident investigators discovered that the rig's blowout preventer was crippled by weak batteries, a defective solenoid valve, and leaking hydraulic lines; in addition, they learned that Mike Williams, the rig's chief -electronics technician, had reported the blowout preventer was damaged during a routine test just before the explosion.[103] Both BP and Transocean officials were aware that federal regulations require that if a blowout preventer does not function properly, the rig must suspend drilling operations until it is fixed.

OUTCOMES

In early January 2011 the presidential panel named to study the Deepwater Horizon disaster issued its 380-page report, which concluded that the blowout "was an avoidable accident caused by a series of failures and blunders by the companies involved in drilling the well and the government regulators assigned to police them."[104] The report also found that, "[a]bsent major crises, and given the remarkable financial returns available from deepwater reserves, the business culture succumbed to a false sense of security." Most troubling, it argued that "the root causes [of the disaster] are systemic and, absent significant reform in both industry practices and government policies, might well recur."[105] The panel recommended that Congress enact sweeping new rules and approve major spending increases for oversight of offshore drilling to prevent, contain, and remediate major spills.[106] The panel also urged the creation of an industry-financed safety board and an independent monitoring office within the Department of Interior whose director would serve a set term and would not have to answer to the secretary, a political appointee. The changes that actually ensued were far more modest, however.

Policy and Political Consequences

Having disbanded the MMS over the summer, on October 1, 2010, interior secretary Salazar announced the creation of three new entities to govern offshore oil and gas development: one responsible for issuing leases and promulgating safety and environmental regulations, the Bureau of Ocean Energy Management (BOEM); another to ensure compliance with the regulatory regime, the Bureau of Safety and Environmental Enforcement; and a third, the Office of Natural Resources Revenue, to collect royalties. In relatively short order, the interior secretary issued new requirements for offshore energy development in U.S. waters. First, companies must develop and maintain Safety and Environmental Management Systems (SEMS) to protect worker safety and prevent accidents. Second, there are specific new drilling safety rules for cementing, blowout preventers, and other critical elements of wells. For instance, tie-back casing is now mandatory, and operators must get approval for their cementing plans and can no longer displace mud with seawater without regulators' approval. Third, an independent third-party certifier must verify compliance with these rules. For instance, independent inspectors must verify that the blind shear rams on a blowout preventer are capable of cutting any drill pipe under maximum anticipated pressure. Fourth, rig operators must generate worst-case scenarios and plan for blowouts. They must, for example, have quick access to a containment system like the one that ultimately succeeded on the Macondo well. And finally, both categorical exemptions from the requirement to assess environmental impacts and area-wide oil and gas leasing have been eliminated. The focus is now on targeted leasing of smaller tracts to allow for more accurate environmental impact assessments.

Although these changes were promising, skeptics argue that no matter how the division of responsibility is structured, the results depend on how the offices are staffed and the extent to which their organizational cultures—and particularly their relationships with industry—are transformed.[107] A host of forces beyond its dual mandate combined to push the MMS into the arms of industry. For decades, industry and its allies in Congress and several administrations fended off efforts to bolster the MMS's capacity to regulate safety and provide effective oversight. Tight budgets left the MMS increasingly unable to compete with industry for the best engineers; the rapid pace of technological change exacerbated that gap. If the new agencies are similarly hamstrung, they will struggle to produce better results. Certainly members of Congress have demonstrated little will to provide the BOEM with additional regulatory clout. Between April 20 and the end of July 2010, Congress held more than fifty hearings and considered more than eighty bills related to the Deepwater Horizon spill.[108] But any proposal to enact restrictive new legislation on offshore drilling or to raise the liability cap of $75 million withered in the face of concerted opposition from Republicans, who rejected both new regulations and additional spending, and from Democrats who hailed from oil-producing states.

Perhaps most tragic, from environmentalists' point of view, was that the spill failed to result in major changes to the nation's overall approach to fossil fuels. Americans generally support domestic development of oil and gas resources—although Democrats and people living on the coasts tend to be less likely than Republicans and those living in the South and Southwest to be supportive. Any reservations people had about offshore drilling after the Deepwater Horizon disaster soon faded: a Gallup poll taken immediately after the spill showed that 50 percent of Americans supported offshore drilling, while 46 percent opposed it. By March of 2011, 60 percent supported, while 37 percent opposed.[109] The sluggish economy, combined with huge Republican gains in the midterm elections, did not help matters. In the face of legislative gridlock, Obama was left to exercise his administrative discretion to discourage old, dirty power sources and vehicles and encourage clean, new technologies (see Chapter 12). However, public opinion has shifted. In 2018, the Pew Research Center conducted a poll asking respondents if the oil and gas drilling should occur in U.S. waters. The results suggest, "More Americans now oppose (51%) than favor (42%) allowing more offshore oil and gas drilling in U.S. waters."[110] These data are in contrast to the Trump administration's position to open more offshore drilling operations.

The Ecological Impacts

After the Macondo well was finally plugged, NOAA concluded that despite uncertainties there was increasing evidence that the gulf had escaped some of the worst possible outcomes. In early August 2010, federal scientists released their final estimates of the Macondo well's flow rate: about 62,000 barrels per day at first, falling to 53,000 barrels per day as the reservoir was depleted. Based on these figures, they concluded that 4.9 million barrels of oil had gushed from the well—with about 800,000 barrels captured by containment efforts.[111] Although sightings of tar balls and emulsified oil continued, the oil slick appeared to be dissolving faster than anyone expected, likely because of the warm water and natural capacity of the gulf, which is teeming with oil-eating bacteria.[112] NOAA estimated that about 40 percent of the oil probably evaporated once it reached the surface. Another 5 percent were burned at the surface, while 3 percent were skimmed and 8 percent were broken into tiny droplets by dispersants.[113] As of August 31, according to NOAA, 35 miles of shoreline were heavily oiled, 71 miles were moderately oiled, and 115 miles were lightly oiled. Of the oil that did reach the shore, most remained at the fringes of the marsh.

Some scientists were skeptical of NOAA's sanguine conclusions and expected to spend years trying to understand the full consequences of the spill. As of November 1, 2010, wildlife responders had collected 8,183 birds, 1,144 sea turtles, and 109 marine mammals affected by the spill.[114] Calculating the actual death toll was complicated, however, by the fact that many carcasses were carried out to sea, eaten by predators, or hidden on secluded islands. Efforts to discern the impacts of the spill also were hampered by the fact that the vast majority of animals found showed no visible signs of oil contamination. Much of the evidence in the case of sea turtles pointed to shrimping or other commercial fishing activities. But other suspects included oil fumes, oiled food, the dispersants used, or even disease. Oil, whether inhaled or ingested, can cause brain lesions, pneumonia, kidney damage, stress, and death. Oil can also have indirect effects, like disorienting animals that are then killed by other means. It can also reduce the amount of food available for animals, who are then driven closer to shore where they can more easily be snagged on fishermen's hooks.[115]

Despite methodological impediments, by 2014 scientists had begun to document the impact of the spill on fish larvae, which are exquisitely sensitive to oil and were probably killed by the millions during the prolonged spill. A NOAA study published on March 24, 2014, found the oil caused "serious defects" in the embryos of several species of fish, including tuna and amberjack.[116] Dolphins appeared to be adversely affected as well. In 2012 researchers reported that dolphins off the coasts of Louisiana, Mississippi, and Alabama were experiencing high rates of strandings. They concluded that several factors—a cold winter, an influx of cold freshwater, and the gulf oil spill—likely led to the surge in dolphin deaths.[117] In 2013 dolphins were found dead at more than three times the normal rates, and scientists were continuing to investigate the role of oil in this phenomenon.[118] After the spill an unusually high number of turtles died too: roughly 500 per year in the three years after the spill in the area affected by it. And a study released in early August 2014 documented damage to deepwater corals more than twelve miles from the spill site, suggesting a more widespread impact than previously suspected.[119]

As journalists Joel Achenbach and David Brown point out, ecosystems can survive and eventually recover from very large oil spills. In most cases the volatile compounds evaporate, are broken down by the sun, or dissolve in water. Microbes consume the simpler, "straight-chain" hydrocarbons, and the warmer it is, the more they eat. But the ecological effects of an oil spill can also extend far into the future. Damage to reproduction rates of sea turtles, for example, may take years to play out. Although it is rare to get funding to monitor the long-term impacts of oil spills, scientists have investigated the consequences of the 189,000-gallon spill off Cape Cod that occurred in September 1969. Fiddler crabs in the oiled marsh were sluggish and reproduced poorly immediately after the spill. Astonishingly, many of those problems remained thirty-five years later, the result of oil that still formed a visible layer four inches below the surface. Similarly, twenty-one years after the *Exxon Valdez* spill, populations of sea otters in the Prince William Sound have not recovered. Two intensely studied pods of killer whales in the sound suffered heavy losses from that accident and have struggled since; one of the two pods has no more reproducing females and is doomed to extinction. By 2003, nearly fifteen years after the spill, there were still 21,000 gallons of oil in Prince William Sound. It could be found simply by scraping three to six inches below the surface of the beach.[120]

Scientists continue to assess the damage caused by the Deepwater Horizon spill as part of the Natural Resource Damage Assessment (NRDA) required under the Oil Pollution Act of 1990. Consistent with that law, the interior department and other federal and state "trustees" are assessing the injuries to the gulf's natural resources and developing alternatives for restoration. BP and other responsible parties must provide the trustees with the funds to implement the final restoration plan. As of 2016, the trustees had undertaken (and BP had financed) more than fifty early restoration projects at a total cost of around $700 million.[121] BP will pay $8.8 billion for restoration efforts to be implemented over the next fifteen years.

On the legislative front, in 2012 Congress enacted legislation to allocate 80 percent of Clean Water Act penalties to restoration of the gulf region—the Resources and Ecosystems Sustainability, Tourist Opportunity, and Revived

Economies of the Gulf States (RESTORE) Act. As the act's name implies, however, this legislation gives the gulf states discretion to spend money on activities unrelated to ecological restoration.

Impacts on Deepwater Drilling in the Gulf—and BP

The dire predictions about the economic impact of the deepwater drilling moratorium—thousands of jobs lost and billions of dollars in foregone revenue—had not materialized. Oil-related unemployment claims were in the hundreds, not thousands. Oil production from the gulf was down, but supplies from the region were projected to increase in the coming years. Only two of thirty-three deepwater rigs operating in the gulf before the explosion actually left the region. Instead, rig owners used the hiatus to service and upgrade their equipment, keeping shipyards and service companies busy. Drilling firms retained most of their workers, knowing that if they released them it would be hard to restart operations when the moratorium was lifted. Oil companies shifted their operations to onshore wells.[122] Sure enough, with gas prices high (averaging $3.76 per gallon), by March 2012 exploration in the gulf had rebounded: there were forty rigs drilling in the gulf's deep water, compared with twenty-five a year earlier. The administration granted sixty-one drilling permits for wells in more than 500 feet of water in the twelve months ending on February 27, 2012, only six fewer than were permitted in the same period in 2009 and 2010, before the BP explosion.[123]

As for BP, it too was resuming activities in the Gulf of Mexico. On October 21, 2011, the federal government approved BP's first oil-drilling permit for the gulf since the Deepwater Horizon sank. In mid-July 2011, seeking to assure federal regulators that it had learned the lessons of the Deepwater Horizon disaster, BP promised to adopt strict new voluntary standards for any future drilling projects in the gulf. The company said it would require its wells to use more robust subsea blowout preventers with two sets of blind shear rams and would adopt new measures for testing the integrity of the cement used to seal wellbores and keep hydrocarbons from flowing to the surface.[124] Despite these pledges, in late November 2012 the EPA temporarily debarred BP from obtaining new federal contracts, including new leases for the Gulf of Mexico, citing the company's "lack of business integrity." BP promptly sued the agency, however, and in mid-March 2014 an agreement between BP and the EPA lifted the ban.[125] An independent auditor approved by the EPA will conduct an annual review and report on BP's compliance with new ethics, corporate governance, and safety standards.

With the threat of debarment removed, liability remained the last weapon available to critics of BP (and, by extension, the rest of the industry) in their efforts to curtail its risky practices. As of early 2015, BP estimated its total payout for the spill was likely to be around $42 billion—although at least 80 percent of that qualified as a tax deduction, according to the U.S. Public Interest Group.[126] In February 2011 charges from hundreds of civil cases were consolidated in a federal courtroom in New Orleans. Defendants included BP, Transocean, Halliburton, and Cameron International (the manufacturer of the blowout preventer). By early March BP and the lawyers for the plaintiffs had agreed to settle their case with individuals, businesses, and local governments, although not with the federal government or the states. The agreement entailed replacing the $20 billion escrow fund (known as the Gulf Coast Claims Facility) with a new fund to be administered by the court. BP expected the fund to distribute about $7.8 billion—although that figure rose quickly as the claims process unfolded.[127] (This settlement received final approval from the court in December 2012.)

In November 2012, BP agreed to pay $4.5 billion in criminal fines and other penalties and to plead guilty to fourteen criminal charges related to the rig explosion.[128] This settlement did not resolve the company's civil liability under the Clean Water Act, however. On September 4, 2014, at the end of the first phase of the trial to address those charges, Federal Judge Carl J. Barbier bluntly rejected BP's arguments that it was not chiefly responsible for the Deepwater Horizon spill, finding instead that BP was the primary culprit and had acted with "conscious disregard of known risk."[129] By finding that BP was grossly negligent in its actions leading up to the spill (although not in dealing with the accident), the judge opened the possibility of $18 billion in new Clean Water Act penalties for BP. During the second phase of the trial, Barbier determined that 3.19 million barrels of oil had escaped the Macondo well—less than the government had claimed because BP had captured nearly 1 million barrels—bringing BP's maximum liability down to $13.7 billion.

CONCLUSIONS

Journalists Joel Achenbach and David Hilzenrath conclude that the Deepwater Horizon spill "was not an accident in the sense of a single unlucky or freak event, but rather an engineered catastrophe—one that followed naturally from decisions of BP managers and other oil company workers on the now-sunken rig."[130] One burning question is whether BP will learn from its numerous mistakes during the first decade of the twenty-first century and transform its safety culture. As many observers have pointed out, after the *Exxon Valdez* spill in 1989, Exxon made a dramatic turnaround. According to Lustgarten, Exxon learned "to manage its risk and safety operations with a militaristic control and total conformity among its 104,000 global employees, setting an international example of good management."[131] Notwithstanding the relatively strong reputation of ExxonMobil, most investigators have concluded that the problems that arose with the Deepwater Horizon are systemic. This claim is supported by incidents that have occurred since 2010, including the 2012 beaching of Shell's state-of-the-art Kulluk rig in Alaska—an accident that exemplified the kind of risk-taking behavior driven by hubris and a desire to impress investors that has long characterized BP's operations.[132]

In light of the Obama administration's liberalization of policies toward drilling in the Arctic and off the Atlantic coast, ensuring that oil and gas companies face stiff regulatory controls is essential. For that to be the case, the nation's regulatory agencies must avert the kind of capture that afflicted the MMS for decades. One fundamental remedy proposed by legal scholar Sidney Shapiro is to rebuild the civil service, restoring its integrity and the sense of professionalism that has been corroded by decades of antigovernment rhetoric and inadequate budgets.[133] Likewise, removing caps on liability can help ensure that in their risk-benefit calculus, companies consider the potentially enormous cost of the environmental, human health, and economic damages they may inflict. The willingness of Judge Barbier to impose hefty Clean Water Act penalties may go a long way toward shaping the future behavior of BP and other oil and gas companies.

That said, the complexity of oil and gas development will only increase as companies pursue harder-to-access reservoirs. As Charles Perrow and other scholars make clear, the more complex the enterprise, the more likely the catastrophic failure, even if organizations pursue reliability with a vengeance. Environmental activists have much work ahead of them to persuade the public that relying on fossil fuels is folly. They need to help gird the public for higher prices for gasoline (and other carbon-based fuels) by transforming the nation's cities and suburbs so that transportation is no longer dependent on fossil fuels. This means far greater support for compact development, rail and bus rapid-transit systems, roads designed for walking and bicycling, and infrastructure that enables the use of electric vehicles. These transitions seem unlikely to come to fruition under the Trump administration. President Trump's approach is to roll back the Obama administration's policies and lift bans on offshore drilling in nearly all U.S. coastal areas. Safety is also a concern, because if there is an increase in drilling and a decrease in regulatory oversight, the next disaster is imminent.

QUESTIONS TO CONSIDER

- The lack of a prolonged shift in public opinion following the Deepwater Horizon spill suggests that Americans value access to cheap gasoline above the environmental consequences of obtaining it. How, in your view, can activists bring about a change in public attitudes toward this fuel?
- How exactly does "regulatory capture" arise, and what do you think are the best ways to combat it (and why)?
- In 2013, at the behest of Congress, the National Research Council assessed the ecological damages caused by the Deepwater Horizon spill using an "ecosystem services" approach, which focuses on the goods and services that natural resources supply to people. How might an ecosystem services approach strengthen the ongoing Natural Resource Damage Assessment process, and how might it weaken that process?

NOTES

1. A barrel is 42 gallons, so 4.9 million barrels translates to 206 million gallons.
2. Charles Perrow, *Normal Accidents: Living With High-Risk Technologies* (Princeton, NJ: Princeton University Press, 1999).
3. High-reliability organizations are those that perpetually seek to improve their reliability despite inherent risks. See Karl E. Weick, Kathleen M. Sutcliffe, and David Obstfeld, "Organizing for High Reliability: Processes of Collective Mindfulness," in *Research in Organizational Behavior*, ed. Robert S. Sutton and Barry M. Staw (Greenwich, CT: JAI Press, 1999), 81–123.
4. Diane Vaughan, *The Challenger Launch Decision: Risky Technology, Culture, and Deviance at NASA* (Chicago: Chicago University Press, 1996).
5. Nick Pidgeon, "Complex Organizational Failures: Culture, High Reliability, and the Lessons From Fukushima," *The Bridge* 42,3 (Fall 2012): 17–22.
6. John M. Broder, Matthew L. Wald, and Tom Zeller Jr., "At U.S. Nuclear Sites, Preparing for the Unlikely," *The New York Times,* March 28, 2011.
7. Process-safety management involves establishing procedures to prevent the unwanted release of hazardous chemicals in liquid or gas form; it ostensibly takes human error into account. See Nancy G. Leveson, "Risk Management in the Oil and Gas Industry," Testimony before the United States Senate Committee on Energy and Natural Resources, May 17, 2011.
8. Thomas O. McGarity, *Freedom to Harm: The Lasting Legacy of the Laissez Faire Revival* (New Haven: Yale University Press, 2013). I make a similar point in *Open for Business: Conservatives' Opposition to Environmental Regulation* (Cambridge, Mass.: The MIT Press, 2012).
9. Michael Greenstone, "Liability and Financial Responsibility for Oil Spills Under the Oil Pollution Act of 1990 and Related Statutes," Testimony before the House Committee on Transportation and Infrastructure, June 9, 2010; Andrew F. Popper, "Capping Incentives, Capping Innovation, Courting Disaster: The Gulf Oil Spill and Arbitrary Limits on Civil Liability," *DePaul Law Review* 60 (2011): 975–1005.
10. The industry considers less than 500 feet shallow water, 500 to 5,000 feet deep water, and more than 5,000 feet ultra-deep water.
11. Clifford Krauss, "The Energy Picture Redrawn," *The New York Times,* October 26, 2011.
12. The hydrocarbons in the Gulf of Mexico were deposited over millions of years by the Mississippi River and the ancient rivers that preceded it. The gulf also features temperatures, pressures, and geography that are hospitable to the formation of abundant oil reservoirs.
13. An examination of the U.S. tax code reveals that oil production is among the most heavily subsidized businesses, with tax breaks available at practically every stage of the exploration and extraction process. For example, according to the Congressional

Budget Office, capital investments like oil field leases and drilling equipment are taxed at an effective rate of 9 percent, significantly lower than the overall rate of 25 percent for businesses in general and lower than for most industries. See David Kocieniewski, "As Oil Industry Fights a Tax, It Reaps Subsidies," *The New York Times*, July 4, 2010. Periodically the government has subsidized development of the gulf in another way: by making such large swaths of land available that the average lease price has fallen dramatically. The Deepwater Royalty Relief Act of 1995, which suspends royalty payments on a portion of new production from deepwater operations, has helped sustain the industry as it has moved into deeper water. See National Commission on the BP Deepwater Horizon Oil Spill and Offshore Drilling, "Deep Water: The Gulf Oil Disaster and the Future of Offshore Drilling," Report to the President, January 2011. Available at http://www.gpo.gov/fdsys/pkg/GPO-OILCOMMISSION/pdf/GPO-OILCOMMISSION.pdf. In 2002, under George W. Bush, the interior department adopted a rule that allowed oil and gas companies to apply for additional royalty relief; two years later another rule reduced royalty payments for companies drilling in deep water. See Juliet Eilperin and Scott Higham, "Seeking Answers in MMS's Flawed Culture," *The New York Times*, August 25, 2010.

14. Christina Ingersoll, Richard M. Locke, and Cate Reavis, "BP and the Deepwater Horizon Disaster of 2010," MIT Sloan Management Case 10-110, revised April 3, 2012. On file with author.
15. Bob Cavnar, *Disaster on the Horizon: High Stakes, High Risks, and the Story Behind the Deepwater Well Blowout* (White River Junction, VT: Chelsea Green, 2010); Eilperin and Higham, "Seeking Answers in MMS's Flawed Culture"; Clifford Krauss, "Accidents Don't Slow Gulf of Mexico Drilling," *The New York Times*, April 23, 2010.
16. Abrahm Lustgarten, *Run to Failure: BP and the Making of the Deepwater Horizon Disaster* (New York: W.W. Norton & Company, 2012); Loren Steffy, *Drowning in Oil: BP and the Reckless Pursuit of Profit* (New York: McGraw Hill, 2011).
17. Ingersoll, Locke, and Reavis, "BP and the Deepwater Horizon Disaster"; Lustgarten, *Run to Failure;* Steffy, *Drowning in Oil.*
18. At the isomerization unit, hydrocarbons are upgraded into higher octane fuel blends. The raffinate splitter tower is a 164-foot-tall metal tower in which hydrocarbons are boiled to separate out the lighter molecules that constitute high-octane gasoline.
19. Ryan Knutson, "Blast at BP Texas Refinery in '05 Foreshadowed Gulf Disaster," Pro Publica, July 20, 2010. Available at http://www.propublica.org/article/blast-at-bp-texas-refinery-in-05-foreshadowed-gulf-disaster; Lustgarten, *Run to Failure;* Steffy, *Drowning in Oil.*
20. Quoted in Steffy, *Drowning in Oil*, 199.
21. Knutson, "Blast at BP Texas Refinery"; Lustgarten, *Run to Failure;* Steffy, *Drowning in Oil.*
22. Lustgarten, *Run to Failure*; Steffy, *Drowning in Oil.*
23. Quoted in Eilperin and Higham, "Seeking Answers in MMS's Flawed Culture."
24. Lustgarten, *Run to Failure;* Sarah Lyall, "In BP's Record, a History of Boldness and Blunders," *The New York Times*, July 13, 2010; Steffy, *Drowning in Oil.*

25. Quoted in Lustgarten, *Run to Failure*, 65.

26. In fact, despite fines and penalties BP exhibited a pattern of recurrent violations. OSHA imposed a record $21 million fine on BP for the Texas City disaster; subsequently, BP paid another $50 million to the Department of Justice for environmental violations stemming from the incident and promised to make the improvements OSHA required. Yet in 2009 OSHA inspectors identified more than 700 ongoing safety violations at Texas City and proposed a record fine of $87.4 million. (BP also paid $2 billion in claims on more than 1,000 lawsuits filed after the Texas City explosion.) Similarly, in early May 2011 federal officials announced that BP would pay $25 million in civil fines to settle charges arising from the two spills from its pipeline network in Alaska in 2006 and from a willful failure to comply with a government order to properly maintain the pipelines to prevent corrosion. The company had already paid more than $20 million in criminal fines and restitution and had been ordered by the Department of Transportation to perform extensive safety repairs to its system of 1,600 miles of pipeline. But BP failed to comply with that order, and the government sued the company again in 2009. See John M. Broder, "BP Is Fined $25 Million for '06 Spills at Pipelines," *The New York Times*, May 4, 2011; Cavnar, *Disaster on the Horizon;* Lyall, "In BP's Record"; Jad Mouawad, "Fast-Growing BP Also Has a Mounting List of Spills and Safety Lapses," *The New York Times*, May 9, 2010.

27. Joel Achenbach, *A Hole at the Bottom of the Sea: The Race to Kill the BP Oil Gusher* (New York: Simon & Schuster, 2011).

28. The name Macondo is from Gabriel Garcia Marquez's novel *One Hundred Years of Solitude*. According to ecologist Carl Safina, "Macondo is an accursed place, a metaphor for the fate awaiting those too arrogant to heed its warning signs." See Carl Safina, *A Sea in Flames: The Deepwater Horizon Oil Blowout* (New York: Crown Publishers, 2011), 18.

29. Achenbach, *A Hole at the Bottom of the Sea*.

30. Lustgarten, *Run to Failure*; Safina, *A Sea in Flames*.

31. Joel Achenbach, "What Went Wrong at the Bottom of the Sea?" *The Washington Post*, May 9, 2010; Safina, *A Sea in Flames*.

32. Lustgarten, *Run to Failure;* National Commission, "Deep Water."

33. The original plan for the well budgeted 51 days to completion, at a cost of $96 million. See National Commission, "Deep Water."

34. BP CEO Tony Hayward testified that BP chose the long-string design because it is more reliable over time. See Russell Gold and Tom McGinty, "BP Relied on Cheaper Wells," *The Wall Street Journal*, June 19, 2010. According to journalist Abrahm Lustgarten, the tie-back method would have cost BP between $7 million and $10 million more and added 37 hours to the job. See Lustgarten, *Run to Failure*.

35. Quoted in Steffy, *Drowning in Oil*, 168.

36. According to an MMS report, nearly half of all blowouts in the gulf trace back to faulty cementing. See Lustgarten, *Run to Failure*.

37. Lustgarten, *Run to Failure*; Steffy, *Drowning in Oil.*

38. National Commission, "Deep Water"; Lustgarten, *Run to Failure.*

39. Apparently veteran toolpusher Jason Anderson posited as an explanation for the anomalous readings the "bladder effect," in which the weight of the mud sitting above the isolated well causes pressure on the drill pipe and gives a false reading. See Lustgarten, *Run to Failure.*

40. National Commission, "Deep Water."

41. Safina, *A Sea in Flames.*

42. According to Lustgarten, the rig had a network of sensors designed to detect a combustible cloud of gas and sound alarms before it reached the engines and the control room. Those alarms should have triggered a series of valves to close, preventing the gas from burning up in the engines. But the sensors did not respond to the rising gas levels, and the valves did not shut. See Lustgarten, *Run to Failure.*

43. Achenbach, *A Hole at the Bottom of the Sea;* David Barstow, David Rohde, and Stephanie Saul, "Deepwater Horizon's Final Hours," *The New York Times,* December 26, 2010. According to Barstow, Rohde, and Saul, the technicians in the engine control room also could have activated the emergency shutdown when their control panel began to indicate the presence of gas; all they had to do was lift a plastic cover and hit a button. But the decision to do so was not straightforward. The crew did not know whether they were facing anything more than a routine kick, and there was a risk in overreacting: with no engines the Deepwater Horizon would drift from its position over the well, possibly damaging the drilling equipment and forcing costly delays. In the absence of communication with the bridge or the drill shack, the men in the engine control room did nothing.

44. Barstow, Rohde, and Saul, "Deepwater Horizon's Final Hours."

45. Achenbach, *A Hole at the Bottom of the Sea*; Antonia Juhasz, *Black Tide: The Devastating Impact of the Gulf Oil Spill* (New York: John Wiley & Sons, Inc., 2011).

46. Anadarko Petroleum and Mitsui Oil Exploration together owned 35 percent of the lease and were responsible for that share of the cleanup expenses.

47. Campbell Robertson, "Efforts to Repel Gulf Spill Are Described as Chaotic," *The New York Times,* June 15, 2010.

48. Henry Fountain, "Notes from Wake of Blowout Outline Obstacles and Frustration," *The New York Times,* June 22, 2010.

49. Clifford Krauss, Henry Fountain, and John M. Broder, "Behind Scenes of Oil Spill, Acrimony and Stress," *The New York Times,* August 27, 2010.

50. Steven Mufson, "Oil Spill in Gulf of Mexico Creates Dilemma in Nature and Politics," *The Washington Post,* April 27, 2010.

51. Steffy, *Drowning in Oil.* There are several ways to gauge the flow rate of an oil spill, and each method yields different results.

52. John M. Broder, Campbell Robertson, and Clifford Krauss, "Amount of Spill Could Escalate, Company Admits," *The New York Times*, May 5, 2010.

53. Tom Zeller Jr., "Federal Officials Say They Vastly Underestimated Rate of Oil Flow into Gulf," *The New York Times*, May 28, 2010.

54. Henry Fountain, "Best Bet to Fix Oil Leak in Gulf? Drill and Drill," *The New York Times*, June 4, 2010.

55. Campbell Robertson, "New Setback in Containing Gulf Oil Spill," *The New York Times*, May 9, 2010.

56. John M. Broder, "Warnings That Spilled Oil Might Spread More Widely," *The New York Times*, May 18, 2010; Shaila Dewan, "In First Success, a Tube Captures Some Leaking Oil," *The New York Times*, May 17, 2010.

57. Leslie Kaufman and Clifford Krauss, "BP Says Its Latest Effort to Stop Gulf Leak Failed," *The New York Times*, May 30, 2010.

58. Joel Achenbach and Anne E. Kornblut, "Forecast for the Gulf Coast Is Grim," *The Washington Post*, May 3, 2010.

59. Juliet Eilperin and David A. Fahrenthold, "Black Death Has Just Begun to Take Its Toll on Animals," *The Washington Post*, May 27, 2010. In late May the Army Corps of Engineers approved six permits to build about 50 of the 100 miles of sand berms proposed by Louisiana officials to prevent the spill from reaching the state's coastline. In early June the White House ordered BP to pay for the project, which was expected to cost an estimated $360 million. See Marc Kaufman, "U.S. Orders BP to Pay for More Barrier Islands," *The Washington Post*, June 3, 2010. (Most of these berms were never built. For details, see National Commission, "Deep Water.")

60. David A. Fahrenthold and Juliet Eilperin, "Toxic Slick Seeps Toward Treasured Gulf Coastline," *The Washington Post*, May 1, 2010.

61. In late May EPA administrator Lisa Jackson insisted that BP switch to a less toxic product, but the company resisted, arguing that no superior alternative was available. See Elisabeth Rosenthal, "In Standoff With Environmental Officials, BP Stays With an Oil Spill Dispersant," *The New York Times*, May 25, 2010.

62. National Institutes of Health. "Gulf Oil Spill Dispersants Associated with Health Symptoms in Cleanup Workers." Available at https://www.nih.gov/news-events/news-releases/gulf-spill-oil-dispersants-associated-health-symptoms-cleanup-workers.

63. Journalist Ian Urbina reported that as of late May, even after Obama announced the temporary moratorium and halt to environmental waivers, at least seven new permits for various types of drilling and five environmental waivers had been granted. interior department officials explained that the moratorium was meant only to halt permits for drilling new wells, not to stop new work on existing projects. But environmentalists were troubled by the waivers, which were supposed to be limited to projects that posed little or no environmental risk. They pointed out that at least six of the drilling projects that received waivers in the preceding four weeks were for waters deeper—and therefore more difficult and dangerous—than where the Deepwater Horizon was operating. See Ian Urbina, "Despite Obama's Moratorium, Drilling Projects Move Ahead," *The New York Times*, May 24, 2010.

64. Joel Achenbach and Anne E. Kornblut, "Forecast for the Gulf Coast Is Grim," *The Washington Post*, May 3, 2010.

65. On June 22, Federal Judge Martin L. C. Feldman issued a preliminary injunction against the enforcement of the administration's moratorium on the grounds that it violated the Administrative Procedures Act. Feldman, an appointee of Ronald Reagan, was clearly persuaded by the economic arguments made by a coalition of businesses that provided services and equipment to offshore drilling platforms and by the State of Louisiana, which filed a brief supporting the lawsuit on the grounds that the moratorium would cause irrevocable harm to its economy. The Fifth Circuit Court of Appeals denied the federal government's request to stay the district court's ruling. So Secretary Salazar modified the order slightly, and it remained in place until he lifted it on October 12.

66. Quoted in Jad Mouawad, "The Spill vs. a Need to Drill," *The New York Times*, May 2, 2010.

67. Quoted in Juliet Eilperin, "Oil Disaster Could Produce a Sea Change," *The Washington Post*, May 6, 2010.

68. Quoted in John M. Broder, "Obama Sketches Energy Plan in Oil," *The New York Times*, May 22, 2010.

69. Quoted in Joel Achenbach, "Effort to Plug Well 'Proceeding As We Planned,'" *The Washington Post*, May 27, 2010.

70. Quoted in Karen Tumulty and Juliet Eilperin, "Obama's Test: Contain Oil Spill and the Fallout," *The Washington Post*, May 15, 2010.

71. David A. Fahrenthold and Juliet Eilperin, "Climate Debate Unmoved by Spill," *The Washington Post*, July 12, 2010.

72. Dan Eggen, "Oil's Sway in Gulf States May Temper Response to Spill," *The Washington Post*, May 16, 2010.

73. Quoted in Dan Eggen and Kimberly Kindy, "Most in Oil, Gas Lobbies Worked for Government," *The Washington Post*, July 22, 2010.

74. As the spill dragged on, federal and state governments closed large swaths of fishing grounds. By June 2, at the peak of the closures, fishing was prohibited across 88,522 square miles, one-third of the gulf zone. See National Commission, "Deep Water."

75. Henry Fountain, "Best Bet to Fix Oil Leak in Gulf?"

76. Clifford Krauss and Henry Fountain, "BP Funneling Some of Leak to the Surface," *The New York Times*, June 5, 2010.

77. John M. Broder and Marjorie Connelly, "Even on Gulf Coast, Energy and Economy Surpass Spill Concerns, Poll Finds," *The New York Times*, June 22, 2010.

78. The main concern was that there were now about 1,000 barrels of oil trapped in the annulus between seals at the top of the well and the cement at the bottom as a result of the static kill. If more mud were pumped in through the relief well, the pressure in the original well could rise and the top seals or cement might be damaged, allowing oil and gas to travel up into the damaged blowout preventer and, potentially, into the gulf. See Henry Fountain and John M. Broder, "Maneuver Will Delay Final 'Kill' of Gulf Well," *The New York Times*, August 20, 2010.

79. Paul Farhi, "Title Wave," *The Washington Post*, May 6, 2010. Whereas environmentalists and government officials generally called the disaster the "BP oil spill," BP consistently referred to it as the "Gulf of Mexico response" and avoided mentioning the word "spill" altogether.

80. Steven Mufson and David A. Fahrenthold, "Oil Executives Pass the Blame for Spill," *The Washington Post*, May 12, 2010.

81. There were at least eight official probes into the spill: the Marine Board of Investigation, led by the Coast Guard and the MMS; the Department of Interior Outer Continental Shelf Safety Board, run by interior department officials; the National Academy of Engineering; the House Energy and Commerce Committee; the House Oversight and Government Reform Committee; the House Natural Resources Committee; the interior department's review of how the MMS conducted its procedures under NEPA, the Marine Mammal Protection Act, and the Endangered Species Act; and an independent, seven-member commission established by the White House. BP also conducted its own investigation, as did the American Petroleum Institute and other industry groups.

82. For example, the recommended twenty-one centralizers were meant to keep the bottom 900 feet of casing centered in the well. If they had had all twenty-one, workers would have put fifteen centralizers above the 175-foot span containing oil and gas, four in that zone, and two below it. As it was, BP placed the six centralizers so as to "straddle and bisect" the oil and gas zone: two above the zone, two in it, and two below. And where the cement failed was, in fact, in the pay zone, where the centralizers were. See Safina, *A Sea in Flames.*

83. Cavnar, *Disaster on the Horizon*; Safina, *A Sea in Flames.*

84. Ian Urbina, "Workers on Doomed Rig Voiced Concern on Safety," *The New York Times*, July 22, 2010.

85. Robbie Brown, "Siren on Oil Rig Was Kept Silent, Technician Says," *The New York Times*, July 24, 2010.

86. Barstow, Rohde, and Saul, "Deepwater Horizon's Final Hours."

87. John M. Broder, "Companies Knew of Cement Flaws Before Rig Blast," *The New York Times*, October 29, 2010.

88. Eilperin and Higham, "Seeking Answers in MMS's Flawed Culture."

89. Ian Urbina, "Inspector General's Inquiry Faults Actions of Federal Drilling Regulators," *The New York Times*, May 25, 2010. In fact, this report was the result of the second inspector general investigation into agency corruption. The first, which focused on activities in the agency's Lakewood, Colorado, office, resulted in several criminal cases involving federal contracting rules and conflict-of-interest laws. See Eilperin and Higham, "Seeking Answers in MMS's Flawed Culture."

90. Ian Urbina, "At Issue in Gulf: Who Was in Charge?" *The New York Times*, June 6, 2010.

91. Ian Urbina, "U.S. Said to Allow Drilling Without Needed Permits," *The New York Times*, May 14, 2010.

92. National Commission, "Deep Water"; Steffy, *Drowning in Oil.*

93. Campbell Robertson and Clifford Krauss, "Robots Working 5,000 Feet Underwater to Stop Flow of Oil in Gulf of Mexico," *The New York Times*, April 27, 2010.

94. National Commission, "Deep Water."

95. Quoted in Steven Mufson and Michael D. Shear, "Pressure Grows for Action by BP," *The Washington Post*, May 1, 2010.

96. Quoted in Joel Achenbach, "BP Installs Dome to Contain Oil Geyser," *The Washington Post*, June 4, 2010.

97. Quoted in Mufson and Shear, "Pressure Grows for Action."

98. Juliet Eilperin, "U.S. Exempted BP Rigs From Impact Analysis," *The Washington Post*, May 5, 2010.

99. National Commission, "Deep Water."

100. A 2011 study by a Norwegian company, Det Norske Veritas, found that a sudden rush of gas and oil forced the drill pipe upward so that a thick connecting portion became stuck near the top of the blowout preventer. With the pipe stuck in the annular preventer, additional forces from the blowout caused the pipe to buckle below it. That, in turn, stopped the blind shear ram from cutting the pipe. In June 2014 the Chemical Safety Board released its report, which reached a slightly different conclusion—that the blind shear ram probably activated as intended on the night of the accident, but rather than sealing the well it punctured the drill pipe, which had buckled under the tremendous pressure of the oil and gas from the initial blowout. See Clifford Krauss and Henry Fountain, "Report on Oil Spill Pinpoints Failure of Blowout Preventer," *The New York Times*, March 24, 2011; Clifford Krauss, "Fixes After BP Spill Not Enough, Board Says," *The New York Times*, June 6, 2014.

101. Subsequent studies yielded similar results. For instance, a 2009 Norwegian study found eleven cases where crews on deepwater rigs lost control of their wells and then activated blowout preventers. The wells were brought under control in only six of those cases, leading researchers to conclude that blowout preventers used by deepwater rigs had a failure rate of 45 percent. See David Barstow, Laura Dodd, James Glanz, Stephanie Saul, and Ian Urbina, "Between Blast and Spill, One Last, Flawed Hope," *The New York Times*, June 21, 2010. Other studies have demonstrated that shear rams can be disabled by the failure of a single, small part; that is why rigs built after 2000 generally have two blind shear rams.

102. Barstow et al., "Between Blast and Spill"; Cavnar, *Disaster on the Horizon;* Eric Lipton and John M. Broder, "Regulators' Warnings Weren't Acted On," *The New York Times*, May 8, 2010.

103. Antonia Juhasz, *Black Tide: The Devastating Impact of the Gulf Oil Spill* (New York: John Wiley & Sons, Inc., 2011).

104. John M. Broder, "Panel Points to Errors in Gulf Spill," *The New York Times*, January 6, 2011.

105. National Commission, "Deep Water," ix, 122.

106. John M. Broder, "Tougher Rules Urged for Offshore Drilling," *The New York Times*, January 12, 2011.

107. Tom Zeller Jr., "Mineral Agency's Split Follows Nations' Lead," *The New York Times*, May 12, 2010. Journalist Tom Zeller observes that the Nuclear Regulatory Commission

(NRC), which Congress created in the 1970s to separate the government's role as a regulator from its role as a promoter of nuclear energy, was an inherent conflict that bedeviled its predecessor, the Atomic Energy Commission. Yet forty years in, the NRC remains closely tied to industry, reluctant to deny relicensing applications and slow to address serious problems at nuclear power plants. See Tom Zeller Jr., "Nuclear Agency Is Criticized As Too Close to Industry," *The New York Times*, May 7, 2011.

108. Julia Werdigier and Jad Mouawad, "Road to New Confidence at BP Runs Through U.S.," *The New York Times*, July 27, 2010.

109. John M. Broder, "Americans Support Offshore Drilling, but Washington Wavers," *The New York Times*, June 17, 2011. A poll released by CBS News in late May 2010 found that whereas 64 percent favored increased drilling in 2008, just 45 percent did in this poll. Those who said the risk was too great increased from 28 percent in 2008 to 46 percent. See Peter Baker, "Obama Extends Moratorium; Agency Chief Resigns," *The New York Times*, May 28, 2010.

110. Bradley Jones, "More Americans Oppose than Favor Increased Offshore Drilling." Pew Research Center. Available at http://www.pewresearch.org/fact-tank/2018/01/30/more-americans-oppose-than-favor-increased-offshore-drilling/.

111. Campbell Robertson and Clifford Krauss, "Gulf Spill Is the Largest of Its Kind, Scientists Say," *The New York Times*, August 3, 2010.

112. Oil-eating bacteria are native to the gulf because there are myriad natural oil seeps. Justin Gillis and Campbell Robertson, "On the Surface, Oil Spill in Gulf Is Vanishing Fast," *The New York Times*, July 28, 2010.

113. Justin Gillis, "U.S. Report Says Oil That Remains Is Scant New Risk," *The New York Times*, August 4, 2010.

114. National Commission, "Deep Water."

115. Shaila Dewan, "Sifting a Range of Suspects as Gulf Wildlife Dies," *The New York Times*, July 15, 2010.

116. John P. Incardona et al., "Deepwater Horizon Crude Oil Impacts the Developing Hearts of Large Predatory Pelagic Fish," *PNAS*, March 24 (2014): E1510-E1515.

117. Leslie Kaufman, "Piecing the Puzzle Together on Dolphin Deaths," *The New York Times Blogs* (Green), July 20, 2012.

118. Douglas B. Inkley, "Four Years into the Gulf Oil Disaster: Still Waiting for Restoration," National Wildlife Federation, April 8, 2014. Available at http://www.nwf.org/News-and-Magazines/Media-Center/Reports/Archive/2014/04-07-14-Gulf-Report-2014.aspx.

119. Charles R. Fisher, "Footprint of Deepwater Horizon Blowout Impact to Deep-Water Coral Communities," *PNAS* 111,32 (August 2014): 11744–11749.

120. Joel Achenbach and David Brown, "The Spill's Long Reach," *The Washington Post*, June 6, 2010.

121. Adam Vann and Robert Meltz, "The 2010 Deepwater Horizon Spill: Natural Resource Damage Assessment Under the Oil Pollution Act," Congressional Research Service report R41972, July 24, 2013; National Oceanographic and Atmospheric

Administration, "Phase III of Early Restoration," Gulf Spill Restoration, n.d. Available at http://www.gulfspillrestoration.noaa.gov/restoration/early-restoration/phase-iii/.

122. John M. Broder and Clifford Krauss, "Dire Predictions for Drilling Ban Are Not Yet Seen," *The New York Times*, August 25, 2010.

123. Clifford Krauss and John M. Broder, "Oil Drilling in the Gulf Rebounds as Prices Promote Exploration," *The New York Times*, March 5, 2012.

124. John M. Broder, "BP to Adopt Voluntary Safety Standards in the Gulf of Mexico," *The New York Times Blogs* (Green), July 15, 2011. A newly formed consortium of oil companies known as the Marine Well Containment Company was working on a system that, by the end of 2011, was expected to be capable of operating in up to 10,000 feet of water and containing a spill of 100,000 barrels per day. See John M. Broder and Clifford Krauss, "Oil Drilling to Resume in the Gulf's Deep Waters," *The New York Times*, March 1, 2011.

125. Clifford Krauss, "U.S. Agrees to Allow BP Back into Gulf Waters to Seek Oil," *The New York Times*, March 14, 2014.

126. A tax loophole allows companies to deduct court-ordered punitive damages as an ordinary business expense, with the result that taxpayers subsidize corporate misconduct. In this case, the result of taking advantage of this loophole would be a roughly $245 million reduction in BP's tax bill, the U.S. Public Interest Group calculated. See Patricia Cohen, "When Company Is Fined, Taxpayers Often Share the Bill," *The New York Times*, February 3, 2015.

127. BP soon regretted agreeing to the settlement, particularly a clause that allowed businesses to claim damage without having to prove it was caused by the spill; instead, a business would simply have to show that its income dropped and rose again in a specific pattern during 2010. BP argued that unscrupulous lawyers and business owners were taking advantage of this provision and asked the court to reject the administrator's interpretation of the settlement. In early March 2014, however, the U.S. Court of Appeals for the Fifth Circuit ruled that BP would have to abide by its agreement.

128. Clifford Krauss and John Schwartz, "BP Will Plead Guilty and Pay Over $4 Billion," *The New York Times*, November 16, 2012.

129. Campbell Robertson and Clifford Krauss, "BP May Be Fined Up to $18 Billion for Spill in Gulf," *The New York Times*, September 5, 2014.

130. Joel Achenbach and David Hilzenrath, "From Series of Missteps to Calamity in the Gulf," *The Washington Post*, July 25, 2010.

131. Lustgarten, *Run to Failure*, 27.

132. McKenzie Funk, "The Wreck of the Kulluk," *The New York Times Magazine*, December 30, 2014.

133. Sidney A. Shapiro, "The Complexity of Regulatory Capture: Diagnosis, Causality, and Remediation," *Roger Williams University Law Review* 17 (Winter 2012): 221–256.

PART THREE

NEW ISSUES, NEW POLITICS

CHAPTER TWELVE

CLIMATE CHANGE

The Crisis of Our Time

In June 1988, James Hansen, a scientist for the National Aeronautics and Space Agency (NASA) testified before Congress that he was "ninety-nine percent confiden[t]" that "the greenhouse effect has been detected, and it is changing our climate now."[1] Shortly thereafter the United Nations (UN) established the Intergovernmental Panel on Climate Change (IPCC), comprising 2,000 leading experts from around the world to assess the causes, extent, and likely impacts of climate change. Since 1990 the IPCC—the most distinguished international group of scientists ever assembled to address a policy question—has reported with certainty that human-made emissions of greenhouse gases are causing rapid and potentially damaging changes in the global climate. For decades the United States has demonstrated little political will to address the problem. Although the Obama administration pledged support for the Paris Climate Agreement and administrative policies (e.g., Clean Power Plan), the Trump administration has made significant strides to roll back such endeavors. Even in the face of scientific consensus, politics has dominated the conversation in the United States curbing implementation of climate policy. However, the Paris Climate Agreement remains the centerpiece for international climate policy to date. December 2018 established international rules for tracking and reporting countries' emissions and climate policies. Patricia Espinosa, the UN Climate Chief, eloquently stated, "This is a roadmap for the international community to decisively address climate change."[2]

The climate change case vividly confirms that political factors shape the relationship between science and policy. Like the world's oceans, the global climate is a commons, so collective action problems hamper efforts to formulate policies to manage it—even in the face of scientific consensus. Because both responsibility for and the impacts of climate change are diffuse, individual nations have an incentive to free ride on the improvements made by others. The obstacles to collective action are likely to be even more formidable in the international arena than they are within or across regions of the United States because no international institution can enforce binding decisions on sovereign nations. Instead, nations must cooperate voluntarily, a prospect that some political scientists find improbable.

Traditionally, scholars who study international relations have portrayed nations as unitary actors with a single goal: survival. According to this "realist" perspective, countries do not cooperate with one another unless it is in their self-interest to do so. Furthermore, a nation's interest is self-evident: each wants to maintain its security and power relative to other countries. Therefore, "the potential for international cooperation is limited, and international laws and institutions are likely to be fragile and impermanent."[3] Extending this view, neorealists contend that international cooperation may occur if a single state with a preponderance of power (a "hegemon") is willing to use its resources to transform international relations.[4] By contrast, liberal theorists (and neoliberal institutionalists) argue that nations are interdependent and that their common interests lead them to work together. A third school of thought builds on the notion of interdependence and emphasizes the concept of international regimes, which consist of "principles, norms, rules, and decision making procedures around which participants' expectations converge in a given issue area."[5] International environmental scholars adopting the latter perspective have turned their attention to how and why such regimes develop and persist. In particular, they argue that a nation's self-interest—and therefore its willingness to participate in an international process—is not a given, but must be discovered.

From this perspective, forging international agreements involves what political scientist Robert Putnam has called a two-level game, in which policymakers try simultaneously to "maximize their own ability to satisfy domestic pressures while minimizing the adverse consequences of foreign developments."[6] According to the two-level game logic, the way the problem of and solutions to climate change are defined domestically is a primary determinant of the U.S. position on international agreements to address it. As the preceding cases in this book have made clear, environmentalists have had great success defining problems in ways that enable them to challenge policies favoring development interests. Enhancing their credibility have been highly reputable knowledge brokers—experts who translate scientific explanations into political stories—as well as the backing of authoritative scientific assessments. But those who oppose environmental regulation have not been passive; they have responded by forming interest groups and funding experts and studies of their own.

In response to a formidable campaign to defuse support for climate change regulations by a coalition of political conservatives and fossil fuel companies and trade associations, environmentalists have adopted some new tactics of their own. One approach has been to press for state-level policies to regulate greenhouse gas emissions and institute other climate-related policies. At one time such unilateral efforts by individual states to address an environmental problem—particularly a transboundary problem such as climate change—would have been unthinkable. In the 1960s few states had adopted strong measures to address air and water pollution (see Chapter 2), and many observers theorized that, left to their own devices, states would engage in a race to the bottom, competing to attract dirty industries and the economic benefits they bring. Environmentalists were also loath to fight on fifty fronts rather than one and so preferred federal policymaking to action at the state level. Since the late 1970s, however, states have greatly expanded their capacity to make and implement policy, and a growing chorus of scholars insists that states ought to play a larger role in solving the next generation of environmental problems.[7] Also important has been the growing role of cities, towns, and counties in combating climate change. With support from grassroots groups, local officials across the United States have made reducing their carbon emissions and becoming more environmentally sustainable a policy focus (see Chapter 15).

In addition to promoting city- and state-level action, environmentalists have tried to persuade business leaders that mandatory federal greenhouse gas emissions limits are in their interest. Some groups have adopted confrontational tactics. Greenpeace and the Rainforest Action Network have used consumer boycotts and negative publicity campaigns to tarnish the reputations of big companies who oppose policies to curb global warming. Environmentalists and religious groups have joined forces to submit shareholder resolutions demanding that corporations reveal their global warming liability. But other groups, such as the Environmental Defense Fund (EDF) and the World Wildlife Fund (WWF), have adopted a more conciliatory approach, working collaboratively to promote the environmentally friendly image of businesses that agree to reduce their emissions voluntarily.[8]

BACKGROUND

In 1827 French scientist Jean-Baptiste Fourier found that atmospheric gases trap thermal radiation in a fashion he likened to the role of glass in a greenhouse. The process begins when the Earth absorbs radiation from the Sun in the form of visible light; that energy is then redistributed by the atmosphere and the ocean and reradiated to space at a longer (infrared) wavelength. Most of that thermal radiation in turn is absorbed by greenhouse gases in the atmosphere, particularly water vapor, carbon dioxide (CO_2), methane, chlorofluorocarbons (CFCs), and ozone. The absorbed energy is then reradiated both downward and upward. The result is that the Earth's surface loses less heat to space than it would in the absence of greenhouse gases and therefore stays warmer than it would otherwise.[9]

A series of discoveries in the mid- to late nineteenth century laid the groundwork for subsequent investigations into the human impact on the climate. In 1860, a British scientist, John Tyndall, measured the absorption of infrared radiation by CO_2 and water vapor. In 1896, Swedish scientist Svante Arrhenius estimated that doubling CO_2 concentrations would raise the average global temperature by five to six degrees Celsius. Around the same time, American geologist T. C. Chamberlin warned independently that the fossil fuel combustion that accompanied industrialization could lead to an out-of-control greenhouse effect. In 1924 American physicist Alfred Lotka speculated that, based on 1920 coal use, industrial activity would double atmospheric CO_2 in 500 years. In 1938 British meteorologist G. D.

Callendar calculated the actual warming due to CO_2 from burning fossil fuels using data gathered from 200 weather stations around the world. Eleven years later Callendar speculatively linked the estimated 10 percent increase in atmospheric CO_2 between 1850 and 1940 with the observed warming of Europe and North America that began in the 1880s. Callendar's report was met with skepticism; the prevailing scientific view during the first half of the twentieth century was that climate remains constant, experiencing only short-term fluctuations.[10]

In the late 1950s, however, scientists revisited the possibility that greenhouse gases might accumulate in the atmosphere and eventually cause a runaway greenhouse effect. In 1957 Roger Revelle and Hans Suess of the Scripps Institute of Oceanography published a paper to that effect after they discovered that the oceans had not absorbed as much CO_2 as previously assumed. Revelle and Suess coined an expression that subsequently became a catchphrase of climate change policy advocates; they claimed that human beings were carrying out a unique, "large scale geophysical experiment."[11] Prompted by concerns about this experiment, in 1957 Revelle's graduate student, Charles David Keeling, instituted routine measurements of CO_2 at the observatory in Mauna Loa, Hawaii. By the early 1960s Keeling's instruments were detecting steady increases in CO_2 concentrations (see Figure 12.1). In 1963 the Conservation Foundation issued a report titled *Implications of the Rising Carbon Dioxide Content of the Atmosphere*, one of the first to speculate on the possible consequences of this trend. Shortly thereafter, a group of White House science advisers led by Revelle concluded that a projected 25 percent increase in atmospheric CO_2 concentrations could cause marked changes in the Earth's climate, with possibly deleterious consequences for humans.[12]

During the 1970s scientists debated whether changes in CO_2 concentrations were likely to produce global warming or the cooling effect of sulfur particles from coal combustion would dominate to produce global cooling. But scientific opinion quickly converged on the warming hypothesis. In the United States the National Academy of Sciences (NAS) launched a series of efforts to assess the scientific understanding of CO_2 and climate, all of which warned about the potentially severe impacts of changes in the global climate. A 1979 NAS report advised that a "wait-and-see policy may mean waiting until it is too late" to avoid significant climatic changes.[13] Also in 1979 the World Meteorological Organization, which in 1974 had begun to examine the evidence, convened an international conference on the topic in Geneva, Switzerland, and launched the World Climate Programme.[14] The final statement of the First World Climate Conference introduced the importance of factors besides greenhouse gas emissions and adopted a cautious tone: "We can say with some confidence that the burning of fossil fuels, deforestation, and changes of land use have increased the amount of carbon dioxide in the atmosphere . . . and it appears plausible that [CO_2 increases] can contribute to a gradual warming of the lower atmosphere, especially at high latitudes."[15]

By the early 1980s scientists were becoming more outspoken about their concerns. Delegates to a 1980 international climate conference in Villach, Austria, issued alarms about global warming, and in 1983 the U.S. Environmental

Figure 12.1 Atmospheric CO_2 at Mauna Loa Observatory

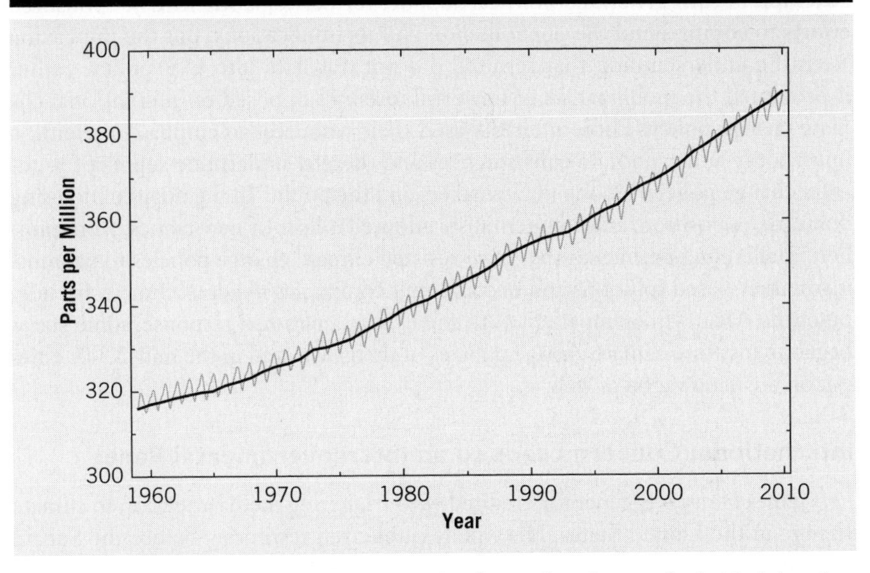

Sources: Scripps Institution of Oceanography; National Oceanic and Atmospheric Administration.

Note: The jagged line represents monthly mean atmospheric CO_2, which rises and falls seasonally. The smooth line represents the seasonally corrected data.

Protection Agency (EPA) released a report suggesting that global temperature increases could strain environmental, economic, and political systems.[16] The same year, the NAS confirmed that a doubling of CO_2 levels would eventually warm the Earth by 3 to 8 degrees Fahrenheit. In 1985 scientists from twenty-nine countries again met in Villach, where they agreed that human activity was causing increases in atmospheric concentrations of greenhouse gases and estimated that a doubling of atmospheric concentrations of CO_2 could lead to an increase in the global mean surface temperature of 1.5 to 4.5 degrees Celsius (or 3 to 8 degrees Fahrenheit). Because such a temperature rise would be unprecedented in the period since the beginning of Holocene, they encouraged policymakers to begin considering responses.[17] More scientific consensus that humans were altering the global climate emerged from two subsequent climate workshops in 1987, one in Villach and the other in Bellagio, Italy.

THE CASE

Despite their unusually urgent tone, the periodic scientific bulletins of the 1970s and 1980s generated little social or political response in the United States. Instead, a number of severe weather episodes, combined with the activism of scientific knowledge brokers, briefly focused public attention on the issue of climate change. That attention in turn generated political support for participating in UN-sponsored efforts to comprehend the phenomenon and its implications. But the improving scientific understanding that resulted did not translate into U.S. policy; rather it prompted the mobilization of powerful interests opposed to international climate change policy. Those interests used their resources to emphasize scientific uncertainty and economic consequences and thereby undermine support for climate change policies among policymakers and the public. In the hopes of defusing domestic opposition, environmentalists adopted a host of new tactics, from publicly challenging businesses that have resisted climate change policies to promoting market-based solutions and encouraging companies to adopt climate-friendly practices. Meanwhile, impatient with the lack of a national response, some states began to institute climate change policies of their own, and in the mid-2000s cities got on the bandwagon as well.

International Concern Leads to an Intergovernmental Panel

James Hansen is generally credited with triggering media attention to climate change in the United States. His widely publicized testimony before the Senate Committee on Energy and Natural Resources asserted that human-induced global warming was imminent and that the phenomenon was sufficiently well understood that policymakers should act to address it. His appearance in June 1988 also coincided with one of the hottest and driest summers on record in North America. Hurricanes and other freak meteorological events around the world further enhanced the public's receptivity to the idea that the climate was changing. Although scientists were reluctant to attribute the severe weather to global warming, it provided an obvious hook for the media.

Two weeks after Hansen made his statement, government officials, environmentalists, and industry representatives from forty-eight countries assembled at a conference in Toronto to discuss the global security implications of climate change. Adopting an alarming version of Revelle and Suess's phrase, the Toronto Conference Statement concluded that "humanity is conducting an unintended, uncontrolled, globally pervasive experiment whose ultimate consequence could be second only to global nuclear war."[18] The statement recommended that governments begin negotiating a global convention as a framework to protect the atmosphere.[19] It further recommended that governments agree to reduce global emissions of CO_2 to 20 percent below 1988 levels by 2005 and create a "world atmosphere fund" from taxes on fossil fuel consumption in industrial countries.[20] Shortly after the Toronto conference, several world leaders—including Britain's prime minister, Margaret Thatcher, a former skeptic—made statements about the need for a government response to climate change.

With the scientific consensus strengthening, the UN Environment Programme and the World Meteorological Organization jointly sponsored the creation of the Intergovernmental Panel on Climate Change to provide policymakers with a scientific foundation for international negotiations. The IPCC met for the first time in November 1988, elected Swedish scientist Bert Bolin as its chair, and formulated a threefold mandate to review the existing scientific literature on climate change and report on the scientific consensus (Working Group I), assess the environmental and socioeconomic impacts of climate change (Working Group II), and formulate response strategies (Working Group III).

The Bush Administration Demurs

Reflecting the heightened international attention, in 1989 the governing council of the UN Environmental Programme and the UN General Assembly adopted resolutions calling on national governments to prepare a framework convention on climate change, as well as protocols spelling out concrete commitments based on scientific knowledge and taking into account the needs of developing countries. In July 1989 the statement of the Group of Seven major industrial democracies' annual summit called for "the early conclusion of an international convention to protect and conserve the global climate."[21] Rather than lead the charge, however, the George H. W. Bush administration refused to propose or support climate policies, emphasizing instead the scientific uncertainties surrounding the issue.

To justify his position, President Bush relied heavily on a paper issued by the conservative George C. Marshall Institute, titled "Scientific Perspectives on the Greenhouse Problem," that downplayed the strength of the scientific consensus on climate change and concluded that it would be premature to impose policies to reduce greenhouse gas emissions. The president continued to invoke the theme of scientific uncertainty in his April 1990 opening speech to a seventeen-nation White House Conference on Science and Economics Research Related to Global Climate Change, in which he said that before acting, "what we need are facts."[22] In lieu of policy proposals, Bush called for further scientific investigation and a 60 percent increase in spending on research.

Hoping to pressure the administration to act, 49 Nobel Prize winners and 700 members of the NAS issued a public appeal saying, "[T]here is broad agreement within the scientific community that amplification of the Earth's natural greenhouse effect by the buildup of various gases introduced by human activity has the potential to produce dramatic changes in climate. . . . Only by taking action now can we insure that future generations will not be put at risk."[23] The scientists' petition did not have much impact on President Bush, however. Nor was he moved to act by the May 1990 presentation by the IPCC of its interim findings.

The IPCC's Working Group II tried to forecast the consequences of climate change, acknowledging that its predictions were highly uncertain and based on a number of simplifying assumptions. Along with an increase in severe weather events and a higher incidence of infectious diseases, among the most serious potential consequences of a rapid increase in global temperatures projected by the panel was a rise in global sea level of between four and twelve inches in the next fifty years. Even at the low end of such a rise, coastal areas everywhere would be more vulnerable to flooding, and some regions would be partially submerged.

The IPCC's tentative tone notwithstanding, many European leaders responded with alacrity to its projections; between May and December 1990, fourteen of the Organization for Economic Cooperation and Development's twenty-four member countries initiated policies to stabilize or reduce emissions of greenhouse gases.[24] By contrast, the United States continued to equivocate. In August the American delegation clashed with other nations attending a meeting in Sweden to finalize the policymakers' summary of the IPCC report because the Americans insisted on amendments emphasizing scientific uncertainty and refused to establish any timetables or targets for the stabilization of greenhouse gas emissions.[25] Their primary objective was to ensure that any actions taken would not curtail economic growth. Despite U.S. resistance, however, the Second World Climate Conference in Geneva held in October/November reported that a clear scientific consensus had emerged on the extent of global warming expected during the twenty-first century. It added that if the increase in greenhouse gas concentrations was not stemmed, the predicted climate change would place stress on natural and social systems unprecedented in the past 10,000 years.

The January 1992 release of the Supplementary Report to the IPCC Scientific Assessment precipitated another round of publicity for the scientific consensus on climate change. Between 1990 and 1992 scientists had begun to incorporate the cooling effects of stratospheric ozone depletion and aerosols (airborne sulfur dioxide particles) into their models, resulting in a much greater consistency between those models and observed temperatures. The supplement reaffirmed the conclusions of the 1990 report and added several new findings, among which was that the anomalously high global temperatures of the 1980s had continued into 1990 and 1991, the warmest years on record. (In 1991, Mount Pinatubo erupted in the Philippines, temporarily interrupting the increase in surface temperatures.)

In hopes of making progress on the policy front, the UN established the Intergovernmental Negotiating Committee to generate a convention in advance of the United Nations Conference on Environment and Development (the Earth Summit) scheduled for June 1992 in Rio de Janeiro, Brazil. The committee met at five two-week sessions between February 1991 and May 1992, during which the negotiators agreed on a process modeled after the one that produced the much acclaimed 1987 Montreal Protocol on Substances that Deplete the Ozone Layer: they would first establish a framework convention on the basic issues and then, at a later date, negotiate a protocol specifying the more concrete obligations of each country. During this period, however, President Bush made it clear that he would boycott the Earth Summit if negotiators produced a climate convention containing specific timetables or goals.

U.S. recalcitrance notwithstanding, at the UN conference in Rio, 154 governments signed the Framework Convention on Climate Change (FCCC), the primary goal of which was the "stabilization of greenhouse gas concentrations in the atmosphere at a level that would prevent dangerous anthropogenic [human] interference with the climate system."[26] The FCCC divided signatories into Annex I (developed) and Annex II (transitional and developing) nations, in recognition that the former were responsible for the bulk of the world's greenhouse gases, and created an obligation for Annex I nations to reduce their emissions to 1990 levels by 2000. The FCCC specified that policies devised under the convention should achieve equity through "common but differentiated responsibilities" and pay special attention to disproportionately burdened developing nations, such as small island states. Most notably, the FCCC stated that policies should be consistent with the "precautionary principle," which entails acting prudently in the absence of scientific certainty. The Bush administration signed the agreement reluctantly, with the understanding that it intended to encourage, but not mandate, emissions reductions. In October 1992 the U.S. Senate ratified the FCCC.

The Clinton Administration: Hopes for U.S. Leadership Bloom and Fade

With the inauguration of President Bill Clinton, advocates of climate change policies hoped the United States would take a more proactive role in negotiations. Bolstering these expectations, Clinton announced his support for the FCCC and for CO_2 emissions reductions. After failing to get a national fossil fuel use (Btu) tax through Congress, however, the president quickly retreated: the 1993 White House Climate Change Action Plan included about fifty voluntary federal programs aimed at promoting energy conservation but did not address greenhouse gas emissions directly. Still, the administration projected that its plan would reduce the country's emissions by about 109 million tons per year by 2000—enough to return them to their 1990 levels of 1.58 billion tons.

By the time government representatives gathered for the first Conference of the Parties (COP 1) in Berlin, Germany, in spring 1995, however, it was clear the Clinton administration's plan was failing to stem the tide of U.S. CO_2 emissions, which by then were almost 5 percent *above* 1990 levels. In fact, only Germany and the United Kingdom were on target to reduce their emissions to 1990 levels by 2000. Two factors in addition to the general failure of voluntary emissions controls lent urgency to the talks. First, in 1994 IPCC chair Bert Bolin had suggested that even if all Annex I governments met their commitments, it would not be sufficient to achieve the FCCC objective of preventing dangerous human interference with the climate system.[27] Second, in the spring and fall of 1994 negotiators had met five times to lay the groundwork for the upcoming COP, and each time the group was split by divisions between developed and developing nations.

Negotiators at COP 1 were therefore faced with two questions: whether Annex I countries should adopt binding emissions reductions and whether emissions reductions obligations should be extended to Annex II countries. The parties ultimately agreed on the Berlin Mandate, which specified that any legal instrument that resulted from formal negotiations scheduled to take place in Kyoto, Japan, would impose emissions reductions only on Annex I countries; the 134 developing nations, including China, India, and Mexico, would be exempt. They did not, however, resolve the issue of binding emissions reductions. The United States pressed for a mechanism that would allow industrialized countries to earn credit if they financed emissions reductions in other countries where, presumably, they could be made more cheaply. The parties ultimately agreed to devise such a mechanism; they also committed themselves to a schedule for adopting a protocol at COP 3, to be held in Kyoto in December 1997.

In August 1995, the Ad Hoc Group on the Berlin Mandate began convening to establish the emissions targets for industrialized nations in advance of the 1997 Kyoto meeting. In December, in the midst of these meetings, the IPCC produced its second assessment report. Since 1990 the IPCC had dramatically improved its climate models and was willing to make projections with greater confidence; it concluded that "the balance of the evidence suggests that there is a discernible human influence on global climate."[28] At the same time, the IPCC downgraded its estimate of the likely magnitude of global warming, projecting that the mean global temperature would increase between 1 and 3.5 degrees Celsius (2 to 6 degrees Fahrenheit) by 2100, with most scientists believing it would be in the lower half of that range. Nevertheless, the IPCC forecast that global mean sea level would rise between six inches and three feet and that substantial changes in spatial and temporal precipitation patterns would occur. Finally, the IPCC concluded that a 60 percent to 80 percent reduction in CO_2 emissions would be necessary just to stabilize atmospheric concentrations of greenhouse gases.[29]

Defining the Climate Change Problem in the United States

In the United States the 1995 IPCC report set the stage for a monumental battle to define the climate change problem—and thereby determine the U.S. position—in advance of the impending Kyoto meeting. Scientists had gotten the issue onto the public agenda, and environmentalists eagerly adopted a simplified version of the scientific story. Their most powerful weapon was the vocal support of prominent scientists and highly visible scientific consensus panels. But the story itself was not particularly compelling: the villains were ordinary Americans, with their wasteful lifestyles; the victims, at least in the short run, were small island nations; and the effects appeared to be at least a generation away—hardly an imminent crisis. Furthermore, powerful opponents, led by the oil and coal industries, retaliated with a well-financed lobbying campaign that featured a four-pronged attack: they argued that activist scientists and extremist environmentalists were using the highly uncertain results of climate change models to frighten people; that a warmer Earth would not be so bad, particularly for the United States; that imposing policies to avert climate change would cripple the American economy; and that imposing emissions limits on industrialized nations without holding developing countries to similar targets would unfairly disadvantage the

United States. Both sides launched all-out public relations campaigns, well aware that whoever succeeded in authoritatively defining the problem was likely to dictate the solution.

The national media initially furthered environmentalists' cause by publicizing Hansen's 1988 congressional testimony. More important, the media linked scientists' climate change predictions to the heat, droughts, and freak weather events of 1988, thereby generating public alarm. But by the early 1990s coverage of climate change had begun to feature conflict among dueling scientists, a reflection of the effort by those supporting the status quo to reframe the debate.

Environmentalists Capitalize on Scientists' Warnings. Among the earliest and most powerful proponents of policies to address climate change were members of the scientific community who were studying the global climate. Throughout the 1980s scientific consensus reports indicating that global warming was a real and serious problem were accompanied by recommendations that the international community formulate policies before the effects became irreversible. Proponents immediately portrayed such action as an "insurance policy against the potentially devastating and irreversible impacts of global warming."[30]

Environmentalists enthusiastically seized on these scientific warnings to advance their overarching goal of limiting human impact on the natural environment. As early as 1984, Environmental Defense's senior scientist Michael Oppenheimer wrote an op-ed piece for *The New York Times* featuring the evocative language that environmentalists are so adept at employing:

> With unusual unanimity, scientists testified at a recent Senate hearing that using the atmosphere as a garbage dump is about to catch up with us on a global scale. . . . Carbon dioxide emissions from fossil fuels combustion and other "greenhouse" gases are throwing a blanket over the Earth. . . . The sea level will rise as land ice melts and the oceans expand. Beaches will erode while wetlands will largely disappear. . . . Imagine life in a sweltering, smoggy New York without Long Island's beaches and you have glimpsed the world left to future generations.[31]

For environmentalists, climate change conveniently linked a host of concerns, from deforestation to air pollution, and implicated industrial nations' demand for growth and luxury.

Opponents Challenge the Scientific Consensus. In the early 1990s, alarmed by the prospect of economy-wide controls on fossil fuel use, a group of utility and coal companies created the Information Council on the Environment to promote arguments critical of climate change theory. The council ultimately disbanded when environmentalists exposed some of its unsavory tactics to the media, but it had already begun to accomplish its goal of creating public confusion about climate change. More enduring was the Global Climate Coalition (GCC), which spun off from the National Association of Manufacturers in 1989. The GCC boasted fifty-four industry and trade association members, primarily from the coal, oil, and automobile industries, and spent heavily on its campaign to redefine climate change.[32] A host of conservative think tanks published books, papers, and op-ed pieces critical of climate change regulation during the 1990s as well; among the most prominent were the Washington, D.C.-based Cato Institute, Competitive Enterprise Institute, and Heritage Foundation.

The skeptics' primary argument was that scientists had only a rudimentary understanding of the feedbacks in the climate system. For example, they noted that early climate models contained extremely crude estimates of the effects of the ocean, yet ocean circulation is coupled with atmospheric circulation in a complex and critical way. Furthermore, skeptics highlighted the potential impact of aerosols (small particles) and clouds on climate change. They argued that the magnitude of observed warming to date was modest compared to the large, natural variability of the system. And they pointed out that scientists' predictions were based on theories and general circulation models (GCMs) that are difficult to confirm.[33]

In addition to emphasizing scientific uncertainty, some skeptics accused environmentalists of using science to achieve political ends—scaremongering to promote radical solutions to a problem for whose existence there was little evidence. They charged that environmentalists' messages were invariably apocalyptic, that the media published such stories to sell newspapers and television time, and that the resulting publicity abetted environmental groups' fundraising efforts.[34] According to the skeptics, warnings about climate change were part of a larger, coordinated effort to "establish international controls over industrial processes and business operations."[35]

Climate change skeptics also went after the IPCC and its members directly. In 1994 Frederick Seitz, director of the conservative Marshall Institute and former president of the National Academy of Sciences, attacked the authors of the IPCC scientific summaries, contending that they had distorted the views of participating scientists and created the impression of a consensus where none existed. (Defenders of the summary responded that the process of writing it was cautious and consensual, that participants agreed to the summaries at plenary meetings, and that none had expressed subsequent dissatisfaction with the final product.) In May 1996 opponents launched personal attacks on two eminent scientists, Benjamin Santer, a climate modeler at the Lawrence Livermore Laboratory, and Tom Wigley, a senior scientist at the National Center for Atmospheric Research. Seitz accused Santer of deleting references to scientific uncertainty from the 1995 IPCC report. *The Wall Street Journal* and *The New York Times* published this allegation, although neither paper confirmed its veracity with any of the participants in the process. (In fact, not one IPCC scientist confirmed the charges against Santer, and forty-two signed a letter to *The Wall Street Journal* in his defense.[36])

Finally, some skeptics advanced the notion that climate change would be beneficial. Economist Thomas Gale Moore, the predominant exponent of this theme, pointed out that service industries could prosper equally well in a warm climate (with air conditioning) as in a cold one (with central heating). In fact, Moore argued (incorrectly, as it turned out), higher temperatures combined with more CO_2 in the atmosphere would enhance plant and crop growth, thereby providing more food for the burgeoning global population.[37] Western Fuels, a coal industry lobbying organization, spent $250,000 on a video titled *The Greening of Planet Earth*, which argued that global warming would improve the lot of the human race and the United States in particular. Western Fuels also founded an organization called the Greening Earth Society to promote this perspective.

Opponents Shift the Focus to Costs. In addition to portraying climate science as highly uncertain and biased, opponents of climate change policies highlighted the costs of adopting policies to limit greenhouse gas emissions. Such policies, they argued, would be exorbitant and lead to "worldwide recession, rising unemployment, civil disturbances, and increased tension between nations as accusations of cheating and violations of international treaties inflamed passions."[38] Taking precautionary action, wrote one journalist in *Forbes* magazine, could "spell the end of the American dream for us and the world."[39] The GCC's $13 million advertising campaign in advance of the Kyoto meeting in 1997 warned television viewers that

> strict reductions in greenhouse gases would have catastrophic economic consequences, endangering the lifestyle of every American. Gasoline would shoot up by fifty cents or more a gallon; heating and electricity bills would soar, while higher energy costs would raise the price of almost everything Americans buy. The livelihood of thousands of coal miners, autoworkers, and others employed in energy-related fields was on the line.[40]

Those hoping to prevent the imposition of greenhouse gas emissions limits seized on a 1990 study by the Bush administration's Council of Economic Advisors that estimated the cost of cutting carbon emissions by 20 percent by the year 2100 at between $800 billion and $3.6 trillion. The report concluded that until there was a solid scientific understanding of climate change, "there is no justification for imposing major costs on the economy in order to slow the growth of greenhouse gas emissions."[41] Opponents claimed that even a no-regrets policy, in which nations adopt such practices as conserving energy and increasing reliance on energy-efficient vehicles and public transit, would be nothing more than a "first, expensive, and ineffectual step down the road to programs that will cripple one of the most vital foundations of modern civilization—our energy supplies."[42]

The environmental Alliance to Save Energy responded with an analysis showing that U.S. carbon emissions could be cut by 25 percent by 2005 and by 70 percent by 2030 at a net savings of $2.3 trillion over forty years.[43] The 1991 NAS report, *Policy Implications of Global Warming*, concurred, arguing that the United States could reduce its greenhouse gas emissions between 10 percent and 40 percent of 1990 levels at low cost, or even net savings, if the proper policies were implemented. And a 1992 study by William Cline of the Institute for International Economics suggested that "social benefit-cost ratios are favorable for an aggressive program of international abatement."[44] Cline pointed out that opponents of climate change policies failed to take into account the possibility of cost-effective energy efficiency measures and technological innovation.

But the critics of climate change policies dismissed efforts to rebut their arguments and were adept at disseminating their competing perspective. At the Rio Earth Summit in 1992, the executive director of the GCC maintained that some of the proposals under consideration could cost the United States $95 billion and 550,000 jobs.[45] The Coalition for Vehicle Choice, financed by the U.S. auto industry and related groups, spent years trying to convince small business, labor, and local civic groups throughout the United States that the treaty would be "bad for America." In October 1997, immediately prior to the Kyoto meeting, the group ran an ad to that effect in *The Washington Post* featuring the endorsement of 1,300 groups.[46]

Opponents Raise the Equity Issue. To the delight of climate change policy opponents, by the early 1990s coverage of political debates and competing economic forecasts had begun to displace coverage of science.[47] To cement their advantage, opponents took a third tack: they began attacking the approach embodied in the Kyoto Protocol, framing it as inequitable—making the protocol, and not climate change, the problem. They insisted it would be unfair for developing countries to escape commitments to greenhouse gas emissions reductions because in the future they were likely to be the major emitters, while the industrialized nations' share of emissions would decline. They pointed out that several large developing countries—including China, India, Mexico, and South Korea—were already producing 44 percent of global fossil fuel emissions and were likely to surpass the emissions levels of the developed countries between 2020 and 2030. In addition, said critics, developing countries were responsible for much of the deforestation and other land-use practices that had eliminated carbon sinks.[48]

The developing nations and many environmentalists responded that the United States, with only 4 percent of the world's population, generated 25 percent of the world's greenhouse gas emissions and that industrialized nations were responsible for 70 percent of the human-made greenhouse gases currently in the atmosphere. Moreover, they noted that developing nations were likely to suffer the most serious consequences of climate change but were the least well-positioned, financially or technologically, to mitigate, adapt to, or recover from those impacts. Dr. Mark Mwandosya of Tanzania, chair of the developing country caucus at Kyoto, pointed out, "Very many of us are struggling to attain a decent standard of living for our peoples, and yet we are constantly told that we must share in the effort to reduce emissions so that industrialized countries can continue to enjoy the benefits of their wasteful life style."[49]

The Kyoto Protocol

As the Kyoto meeting drew near, the battle to shape the U.S. position intensified. In hopes of bolstering U.S. leadership, in June 1997 more than 2,500 American scientists endorsed the Scientists' Statement on Global Climatic Disruption. The statement claimed that

> further accumulation of greenhouse gases commits the earth irreversibly to further global climatic change and consequent ecological, economic, and social disruption. The risks associated with such changes justify preventive action through reductions in emissions of greenhouse gases. It is time for the United States, as the largest emitter of greenhouse gases, to . . . demonstrate leadership in a global effort.[50]

After receiving the statement, President Clinton told a special session of the UN General Assembly that "the science [of climate change] is clear and compelling," and he promised to bring to the Kyoto conference "a strong American commitment to realistic and binding limits that will significantly reduce our emissions of greenhouse gases."[51] The following month, Clinton launched an effort to convince the public that climate change was real by holding a conference of experts and convening a series of well-publicized regional panels.

A couple of domestic factors reduced the administration's international credibility, however. In 1996 alone U.S. emissions of greenhouse gases grew 3.4 percent, and by 1997 those emissions were about 7.4 percent greater than they had been in 1990.[52] At this rate of growth, U.S. greenhouse gas emissions promised to be an embarrassing 13 percent above 1990 levels by 2000.[53] Further hampering the administration's ability to negotiate was the unanimous (95–0) passage by the U.S. Senate, on June 12, 1997, of a nonbinding resolution (the Byrd–Hagel amendment) that it would not give its advice and consent to any agreement that did not require developing countries to reduce their emissions or that would result in "serious harm to the economy of the United States."[54] Concerned about building domestic support for the treaty, Clinton began emphasizing the importance of cooperation by China and other developing nations. At the seventh Ad Hoc Group on the Berlin Mandate meeting in Bonn, Germany, in October, U.S. negotiators pressed the other parties to commit both developing and industrialized nations to emissions reductions.

By the fourth day of the Kyoto meeting, a *New York Times* editorial declared that a "near miracle" would be required to salvage an agreement. The thorniest issue remained the degree to which developing countries would have to control their emissions. Led by the Chinese delegation, the developing countries adamantly resisted U.S. pressure. So, in a last-ditch effort to facilitate international agreement without provoking a domestic backlash, President Clinton dispatched Vice President Al Gore to give the American delegates more flexibility. On Monday, December 8, Gore told members of the conference that the president would allow the U.S. delegation to offer emissions reductions beyond those originally proposed (1990 levels between 2008 and 2012) in return for opening the door in Kyoto to language requiring emissions reductions by developing countries. COP chair Raul Estrada proposed a compromise that would allow developing countries to reduce emissions voluntarily and give Annex I nations the option of accepting differentiated emissions reduction commitments for 2008 to 2012. In the negotiations that followed, the Chinese led a bloc of developing nations (the G77) that vigorously opposed the compromise. The resulting protocol, which emerged just as the meeting was closing, embodied the worst-case scenario for the United States: it went beyond the original target for U.S. reductions but provided no mechanism for making developing countries reduce their emissions.

Opponents of climate change policies were appalled by the agreement. In a press conference before leaving Kyoto, Sen. Chuck Hagel, R-Neb., coauthor of the Byrd–Hagel Amendment, vowed that there was "no way, if the president signs this, that the vote in the United States Senate will even be close. We will kill this bill."[55] (The president ultimately signed the protocol but never submitted it to the Senate for ratification.)

Promoting the Kyoto Protocol Back Home

Rather than submitting the Kyoto Protocol for ratification, President Clinton instead proposed a five-year, $6.3 billion package of tax breaks and research spending in pursuit of the protocol's emissions reduction goals—measures that most experts regarded as too modest to have much impact on America's $500 billion fossil fuel-based economy. By executive order, Clinton also directed the federal government, the world's largest energy consumer, to reduce petroleum use in federally owned vehicles to 20 percent below 1990 levels by 2005 and reduce greenhouse gases from federal buildings by 30 percent by 2010. At the same time, the administration continued its campaign to persuade the public that the science underpinning the protocol was valid.

Even as the Clinton administration struggled to generate public support for domestic climate change policies, opponents geared up for a Senate ratification battle. Recognizing the political potency of the scientific consensus generated by the IPCC, by early spring 1998 a high-powered group including the American Petroleum Institute, Chevron, and Exxon had already planned a multimillion-dollar campaign to undermine that consensus. Aimed primarily at the public, the plan was to recruit a cadre of skeptical scientists and train them to convey their views persuasively to science writers, editors, columnists, and newspaper correspondents.[56]

In response to the campaign against climate change policy by industry and conservative activists, environmentalists undertook some new tactics of their own.

They approached businesses directly and tried to persuade them to reduce their emissions voluntarily—to save money, improve their environmental credentials, and gain a seat at the table when the inevitable regulations were enacted. In May 1998, for example, the Pew Charitable Trusts established the Pew Center for Global Climate Change, funded by a $5 million annual grant from the foundation and administered by Eileen Claussen, former deputy assistant secretary of state for environmental affairs. One of the center's objectives was to bolster the credibility of climate change science. To this end, in an ad published in *The New York Times*, the center's members—including American Electric Power, U.S. Generating Company, Maytag, Whirlpool, 3M, Toyota, Sunoco, United Technologies, Boeing, and Lockheed Martin—publicly accepted the views of most scientists that enough was known about the environmental impact of climate change to take steps to address the consequences, and they pledged to reduce their own emissions of greenhouse gases.[57] By autumn 1998 twenty major U.S. companies, including three electric power companies and two oil companies, had joined the Pew Center.

The Hague, 2000

This was the domestic context in which international talks to resolve outstanding issues from Kyoto resumed at the fourth COP in Buenos Aires, Argentina, which began in early November 1998. Perhaps it was not surprising, then, that little of substance was accomplished at the Buenos Aires meeting, and at the close of COP 4, the role of developing nations and the status of international emissions trading were still unresolved. In a promising advance on emissions trading, the G77 was no longer monolithic; China and India were leading the faction opposed to emissions limits on developing countries, but African and Latin American countries were showing interest in making emissions reductions in exchange for aid. The second point was stickier: U.S. negotiators were tenacious about emissions trading because they thought it might defuse domestic opposition, but Europeans and many developing nations objected that the United States was trying to buy its way out of reducing its own emissions.

As officials around the globe struggled to find solutions everyone could agree on, the scientific evidence on climate change continued to pour in. In March 1999 a study published in *Nature* concluded that the growing season in the northern temperate zone, from the sub-Arctic to the Mediterranean region, had lengthened by about eleven days since 1970. The authors attributed the shift to a rise in daily temperatures caused by the general warming of the climate. Their findings were consistent with a series of studies reported in 1996 and 1997 that detected early spring and a longer growing season in the northern hemisphere.[58]

Although the influx of worrisome scientific reports lent urgency to the climate change talks, COP 6 in The Hague in November 2000 foundered once again. The Hague conference was supposed to be the final meeting to establish greenhouse gas emissions reduction policies, and efforts to translate vague commitments into specific practices promised to be contentious.

At the same time, to placate its critics the administration continued to embrace the language of economic growth, efficiency, cost-effectiveness, and the primacy of markets. U.S. negotiators left for The Hague determined to obtain unlimited use of emissions-trading mechanisms and credit for land-use practices included in the agreement.

Critics were dubious about relying on forests to curb CO_2, pointing out that research suggested the role of forests and soils in sequestering CO_2 was not straightforward.[59] Through the final night and into the early morning, environmental groups helped the European delegation analyze a variety of formulas for calculating carbon equivalents attributable to forest protection. At 3 a.m., a small cadre of British, American, and European diplomats shook hands on a deal; the following day, however, the complex formula turned out to be unacceptable to many in the European Union. Jurgen Tritin, the German environment minister, explained that his country's opposition to forest credits was deeply rooted in his nation's values and derived from a sense that the United States and its collaborators were trying to get something for nothing.[60]

Observers were struck by the irony that previous talks had stumbled because of irreconcilable differences between developing and developed nations, or between environmentalists and industry, but The Hague negotiations fell apart primarily because of a schism within the environmental movement itself. While mainstream American environmental groups, as well as many climate scientists, supported the business-friendly U.S. solutions of emissions trading and forest conservation credits, hard-liners, such as Greenpeace, backed the German position. "We're better off with no deal than a bad deal," argued Bill Hare of Greenpeace moments after the negotiations ended.[61] The hard-line groups equated compromise with corruption and abdication to business interests but, ironically, found themselves allied with most of the business community in opposition to any treaty. For more moderate environmentalists, however, the failure to reach agreement at The Hague was particularly worrisome in light of the upcoming U.S. election. Michael Oppenheimer of Environmental Defense warned (presciently) that if George W. Bush became president, it would only become more difficult for American and European negotiators to find common ground.

The George W. Bush Presidency

In early 2001, a series of reports detailed the latest scientific understanding of the magnitude and likely impacts of climate change, and the overall tone was one of foreboding. Specifically, a new IPCC assessment confirmed that the global temperature had risen roughly one degree Fahrenheit in the twentieth century, blamed part of that increase on fossil fuel combustion, and cautioned that it represented the most rapid change in ten millennia. The IPCC suggested that earlier climate change estimates may have been conservative and that the Earth's climate could warm by as much as 10.5 degrees Fahrenheit by the end of the twenty-first century.[62] Then, in February, the IPCC released a report titled "Climate Change

2001: Impacts, Adaptations, and Vulnerability." Summarizing the work of 700 experts, the 1,000-page document concluded ominously that "projected climate changes during the 21st century have the potential to lead to future large-scale and possible irreversible changes in Earth systems," with "continental and global consequences." Among the likely outcomes were more "freak" weather conditions, such as cyclones, floods, and droughts; massive displacement of population in the most affected areas; greater risk of diseases like malaria; and extinction of entire species as their habitat disappeared. Over time, the report warned, global warming was also likely to cause large reductions in the Greenland and West Antarctic ice sheets and a substantial slowing of the circulation of warm water in the North Atlantic. Finally, changes in rainfall patterns due to climate change, combined with patterns of population growth, would likely lead to enormous pressure on water supplies.[63]

At conferences and in journals, researchers continued to document climate change effects that were already observable. At a meeting in San Francisco in February, scientists attributed the recently noted melting of equatorial glaciers in Africa and Peru to global warming. Scientists offered other evidence as well: thawing permafrost, delayed freezing, earlier breakup dates of river and lake ice, and longer growing seasons at mid- to high latitudes.[64] A study reported in *Nature* suggested yet another potential impact: droughts caused by global warming could prompt northern soils to release CO_2 into the air, speeding up changes in the climate.[65]

Despite mounting concern among scientists, the election of President Bush and Vice President Dick Cheney, both former oilmen, dimmed hopes for U.S. leadership on a climate change agreement. Bush took office saying that more research on climate change was needed before any policies were undertaken, and in March 2001 he vowed not to seek CO_2 emissions reductions, thereby defaulting on a campaign pledge to do so.

As if to put an exclamation point on the president's position, the administration's 2002 budget proposal cut spending on energy efficiency programs by 15 percent.[66] Moreover, Bush's energy plan, released in May 2001, emphasized loosening environmental regulations and developing new fossil fuel supplies. After meeting with the president in early April and failing to change his mind about the Kyoto Protocol, European leaders announced their intention to move forward with the treaty even without American leadership. According to the European Union's environmental commissioner Margot Wallstrom, "Other countries [were] reacting very strongly against the U.S."[67]

Recognizing his political vulnerability, Bush commissioned an expert panel, convened by the NAS National Research Council, that could either legitimate his views or provide cover for a reversal on the issue. The panel—which comprised eleven prominent atmospheric scientists, including Richard Lindzen of MIT—"reaffirmed the mainstream scientific view that the earth's atmosphere was getting warmer and that human activity was largely responsible."[68] Although the president subsequently conceded the scientific point, he continued to inflame Europeans by opposing the Kyoto Protocol and rejecting mandatory emissions limits.

Bush's resistance notwithstanding, in November 2001 negotiators representing 178 countries hammered out the details of the Kyoto Protocol, and many large industrial countries said they intended to ratify the agreement. Although it was only a first step, environmentalists were pleased. "The parties have reached complete agreement on what's an infraction, how you decide a case, and what are the penalties," said David Doniger of the Natural Resources Defense Council. "That's as good as it gets in international relations."[69] Furthermore, to the surprise of many observers, the European Union agreed to institute trading mechanisms for greenhouse gases—a move it hoped would gain the support of Canada, Japan, and Russia.

In hopes of appeasing his increasingly voluble critics, in February 2002 Bush announced a domestic plan that relied on $4.6 billion in tax credits over five years to encourage businesses to take voluntary measures that would reduce the "carbon intensity" (the ratio of greenhouse gases to gross domestic product) of the economy by 18 percent over ten years. In announcing the plan, the president made it clear that economic growth was his primary concern, in part because a thriving economy would provide the resources to invest in clean technologies. Eileen Claussen pointed out that the administration's carbon intensity targets were consistent with the trajectory the nation was already on and therefore represented little improvement. At the same time, Myron Ebell of the conservative Competitive Enterprise Institute warned that the administration's acknowledgment of climate change as a problem started the nation "down a dark path" toward mandatory emissions reductions.[70]

Climate Change Science Solidifies

The impacts of rising temperatures continued to manifest themselves more quickly than scientists had anticipated, particularly in the polar regions where nonlinear processes were causing ice to melt more rapidly than expected. For example, in February 2002 a 1,250-square-mile section of the Antarctic Peninsula's Larsen B ice shelf began splintering, and within a month it was gone. Studies of marine algae found in core sediment suggested that the Larsen B had been intact since it formed more than 10,000 years ago. Apparently, ice was flowing into the sea faster than scientists had expected: loss of coastal ice shelves was accelerating the flow of inland glaciers; in addition, water from summertime ponds percolating through cracks to the base was acting as a lubricant, facilitating the slide of glacial ice over the earth below.[71] The Arctic was showing the effects of global warming as well. In September 2003 researchers reported that the largest ice shelf in the Arctic had shattered, and a massive freshwater lake behind it had drained away.[72] Then, in fall 2004, a comprehensive four-year study of Arctic warming confirmed that the buildup of greenhouse gases was contributing to profound environmental changes in the region, including sharp retreats of glaciers and sea ice and thawing of permafrost.

In addition to detecting the consequences of climate change, scientists were refining their understanding of the interactions between carbon sources and sinks. For example, a study published in *Nature* in March 2004 suggested that changes in the Amazon rainforest appeared to make it less capable of absorbing CO_2 than once believed. The study revealed that trees were growing and dying at faster rates than twenty years earlier, and scientists attributed this change to rising CO_2 levels. Similarly, a study published in *Nature* in September 2004 described an experiment that confirmed the hypothesis that presumed carbon sinks, such as the Arctic tundra, may actually generate net CO_2 increases, as higher temperatures lead to more decomposition and leaching in the soil, the combination of which produces more CO_2 than new plants take up.[73]

Although media coverage of these scientific findings helped buttress the precautionary arguments of environmentalists, overall, the media—in their effort to attain "balanced" coverage of the issue—conveyed a sense of scientific disagreement that belied the underlying consensus.[74] In response, scientists tried to enhance their impact by publicizing the high level of agreement and concern in the scientific community. In October 2003 more than 1,000 scientists around the country signed a "state of climate science" letter affirming the claims of the IPCC and the National Research Council that anthropogenic climate change, driven by emissions of greenhouse gases, was already under way and was likely responsible for most of the observed warming of the past fifty years and that they expected the Earth to warm 2.5 degrees to 10.5 degrees Fahrenheit over the course of the twenty-first century, depending on future emissions levels and climate sensitivity. Adding to the chorus, in late December 2003 the prestigious American Geophysical Union followed the American Meteorological Society in adopting the official position that human-caused greenhouse gas emissions were causing global warming.

In hopes of persuading the public that U.S. citizens would be victims, not beneficiaries, of global warming, some activist scientists also undertook a campaign to portray its local impacts. For example, in fall 2004 the Union of Concerned Scientists released a city-by-city analysis for California that included dire forecasts for San Francisco and other major metropolitan areas. *Science* published a study in October 2004 suggesting that a warmer climate could exacerbate the severity and duration of western droughts.[75] And an EPA-commissioned study released in February 2005 concluded that by the end of the twenty-first century, global warming could raise sea levels enough that a heavy storm would send flood waters into Boston's downtown waterfront, the financial district, and much of the Back Bay.[76]

However, the accumulating scientific evidence made little impression on the Bush administration, which continued to argue that the unsubstantiated threat of global warming did not warrant action that would cripple the economy. In June 2002, the White House sent a climate report to the UN that acknowledged the role of fossil fuel burning in global warming but did not propose any major changes in U.S. policy. Some vocal Republicans in Congress backed the administration's view, and in August 2003 efforts by Sen. John McCain, R-Ariz., and Sen. Joseph

Lieberman, D-Conn., to force a vote on their bill to limit greenhouse gas emissions (S 139, the Climate Stewardship Act) prompted a furious resumption of the climate change debate. Representing the views of skeptics, Sen. James Inhofe, R-Okla., gave a two-hour speech on the Senate floor in which he said, "With all of the hysteria, all of the fear, all of the phony science, could it be that man-made global warming is the greatest hoax ever perpetrated on the American people? It sure sounds like it."[77] (The Senate ultimately defeated the McCain–Lieberman bill by a vote of 55 to 43. A second attempt, in June 2005, failed by an even larger margin, 60 to 38.) U.S. negotiators continued to impede international action as well: at the COP 10 meeting in December 2004, the U.S. delegation blocked efforts to undertake anything beyond limited, informal talks on ways to slow down global warming. In justifying the U.S. position, delegation leader Paula Dobriansky said, "Science tells us that we cannot say with any certainty what constitutes a dangerous level of warming, and therefore what level must be avoided."[78]

Despite continued obstruction by the U.S. government, in mid-February 2005 the Kyoto Protocol went into effect, thanks to Russia's ratification late in 2004. Europe began implementing a cap-and-trade system to meet its commitment to reduce CO_2 emissions 8 percent below 1990 levels by 2012. Back in the United States, even though the federal government refused to budge on the issue, a host of states took action to address climate change.

By 2005, twenty-eight states had plans to reduce their net greenhouse gas emissions. California, Minnesota, Oregon, New Jersey, Washington, Wisconsin, and the New England states had launched initiatives to curb their CO_2 output. Other states, including Texas and Colorado, had established requirements that electric utilities obtain a certain percentage of their portfolio from renewable energy sources—measures that, although not framed as climate change policies, promised substantial reductions in CO_2 emissions. Georgia, Nebraska, North Dakota, and Wyoming were investigating methods for sequestering carbon in agricultural soils by promoting no-till farming methods. Other state programs included fuel efficiency mandates for state fleets and tax credits for energy conservation and the purchase of fuel-efficient vehicles.[79] Local governments got on the bandwagon as well. In February 2005 Seattle mayor Greg Nickels announced a campaign to get U.S. cities to adopt the terms of the Kyoto Protocol. Nickels aimed to recruit 140 cities, to equal the 140 countries that had signed the treaty.[80] Cities signed on in droves, quickly exceeding Nickels's original goal, and signatories began crafting climate action plans that pledged to retrofit buildings to make them more energy efficient, improve public transportation, plant trees, and foster the installation of solar, wind, and other forms of renewable energy.[81]

Meanwhile, scientists' admonitions were becoming even more urgent. Atmospheric carbon concentrations continued to rise. In 2004 they reached 379 ppm, the highest level ever recorded. U.S. greenhouse gas emissions also continued to increase, albeit at a slackening pace. Furthermore, three countries—China, India, and the United States—were planning to build 850 new coal-fired plants, which together would pump up to five times as much CO_2 into the atmosphere as Kyoto aimed to reduce.[82] At the same time, it was beginning to appear that limiting global warming to below two degrees Celsius, and thereby avoiding the worst consequences of climate change, would require far more ambitious emissions reductions than scientists and policymakers had believed.

As the Bush administration's tenure drew to a close, the Supreme Court challenged its obstruction: in April 2007 the Court ruled that, contrary to the EPA's assertion, CO_2 and other greenhouse gases constitute air pollutants under the Clean Air Act, and the Bush administration broke the law in refusing to limit emissions of those gases (*Massachusetts v. EPA*). President Bush promptly issued an executive order calling on the EPA to begin writing regulations for greenhouse gas emissions, but when the agency sought to issue an "endangerment finding"—the first step in the regulatory process—the White House ordered it to retract the document. Finally, in July 2008, the EPA released a finding that emphasized the difficulties and costs of regulating greenhouse gases and drew no conclusions about whether global warming posed a threat to human health or public welfare.

The Obama Administration: The Greatest Threat to Our Future

The Obama administration made it known that climate change "poses a greater threat to our children, our planet, and future generations . . . and that no other country on Earth is better equipped to lead the world towards a solution."[83] In particular, President Obama pledged in 2009 that the United States would cut its greenhouse gas emissions 17 percent from 2005 levels by 2020 and 80 percent by 2050, having promised during the campaign to enact legislation that would cap U.S. greenhouse gas emissions and allow for trading of emission allowances—a cap-and-trade system (see Chapter 5).

Given that Obama was extraordinarily popular, there were strong Democratic majorities in both the House and Senate; the conservative movement appeared at least temporarily vanquished and the prospects for such legislation seemed favorable. In June 2009 the House made history by narrowly passing the nation's first-ever comprehensive energy and climate change bill. But numerous obstacles soon arose. Public support for climate change policies softened as the recession caused by the financial crisis of 2007 to 2008 dragged on, and a reinvigorated conservative-industry alliance capitalized on the delay created by the prolonged health care debate to discredit cap-and-trade as a policy solution. U.S. negotiators were forced to go empty-handed to international talks in Copenhagen, where they were supposed to develop a successor to the Kyoto Protocol; as a result, the meeting yielded only a nonbinding accord pledging the United States and other developed countries to submit economy-wide emissions targets by the end of 2010. The U.S. commitment was contingent on Congress passing legislation, however, and as the 2010 elections drew near it became apparent that, despite a concerted push by Sens. John Kerry, D-Mass., and Joseph Lieberman, I-Conn., the Senate would not reach agreement on a bill.[84]

Although foiled in Congress, Obama aggressively pursued greenhouse gas emissions reductions administratively. Through the $787 million stimulus package (the American Recovery and Reinvestment Act, or ARRA), the president funneled more than $80 million to energy efficiency projects administered by the Department of Energy. In addition, he ramped up efforts to control greenhouse gas emissions by the federal government, which is a major polluter. Most important, Obama's EPA began taking steps to regulate greenhouse gas emissions under the Clean Air Act: between 2009 and 2010 the agency issued an endangerment finding for CO_2, finalized an emissions-reduction rule for mobile sources, created a mandatory greenhouse gas registry for stationary sources, and wrote a "tailoring" rule that focused stationary-source permitting requirements on the largest industrial emitters (those with the potential to emit 75,000 tons or more). In late March 2012 the EPA unveiled the first-ever rules (often referred to as the Clean Power Plan Rule) to control CO_2 emissions from new power plants to 1,000 pounds per megawatt hour (MWh). (In September 2013 the agency released a revised version of the proposal that limited CO_2 emissions from coal-fired power plants to 1,100 pounds per MWh.) The new rule would effectively ban new coal plants, which generally emit about twice as much CO_2 as the proposed limit.

On August 3, 2015, the EPA finalized the Clean Power Plan Rule, establishing state-by-state targets with reductions of emissions by 32 percent below the 2005 levels by 2030. States were required to develop implementation plans to meet these targets and could adopt a variety of approaches. For example, states could invest in renewable energy to shift away from coal-fired power. If states failed to design their own approach, they would be forced to adopt a federal "model rule." (Not surprisingly, the Chamber of Commerce warned that the rule for existing power plants could lower the gross domestic product by $50 billion annually, and many Republicans labeled it a job killer and accused the president of waging a "war on coal."[85]) However, implementation of the rule was halted in February 2016. The Supreme Court placed a hold on the Clean Power Plan Rule until the D.C. Circuit Court of Appeals determined whether it was legally valid. Subsequently, the Trump administration has halted efforts altogether.

For some, President Obama's announcement in late 2014 was a turning point for the United States. The United States and China jointly committed to cutting their emissions. Under this agreement, the United States would reduce its emissions by 26 percent to 28 percent below 2005 levels by 2025, while China would stop its emissions from growing by 2030.[86] The administration hoped the move would help secure a new international climate change agreement in 2015. And in fact, the president's efforts transformed the United States' image: in a rare demonstration, at climate talks in Lima, Peru, in December 2014, attendees greeted U.S. delegates with cheers, applause, and praise. Delegates subsequently devised the Lima Accord, which does not include legally binding targets but rather requires each nation to enact domestic laws to cut CO_2 emissions and put forth a plan to do so well before the next meeting, in Paris in December 2015. Recognizing that there was no viable international enforcement mechanism, the deal relied on what political scientists call a "name-and-shame" (essentially, peer pressure) approach.[87]

As international negotiations proceeded, greenhouse gas emissions continued to rise, albeit at a slower rate than in the early 2000s: in 2012 they rose 2.1 percent over 2011, 2.3 percent in 2013 compared to 2012, and 2.5 percent in 2014 over 2013.[88] At this rate, the world would likely use up its "carbon budget"—calculated as the total all countries can emit without pushing global temperatures higher than two degrees Celsius above preindustrial levels—within thirty years. Given the steady rise in CO_2 emissions, it is hardly surprising that in 2014 the Earth's globally averaged surface temperature continued its long-term warming trend: the twenty warmest years in the historical record have all occurred in the past twenty years. In 2014 the combined land and ocean surface temperature was 1.24 degrees Fahrenheit above the twentieth-century average, making it the warmest year since records began in 1880.[89] The IPCC, which released its fifth climate science report in fall 2013, expressed with near certainty that human activity was the cause of most of the temperature increases in recent decades.

In the United States (and elsewhere), the impacts of climate change were already evident. In the western United States, severe wildfires were increasingly common, and forests were dying under assault from heat-loving insects, while the East and Midwest were experiencing more frequent torrential rains and flooding.[90] Severe weather was wreaking havoc with the nation's power-generation system: power plants were shutting down or reducing output because of shortages of cooling water; barges carrying coal and oil were being delayed by lower water levels in major waterways; floods and storms were inundating ports, refineries, pipelines, and rail yards; and powerful windstorms and raging wildfires were felling transformers and transmission lines.[91] Experts projected more havoc to come; for example, climate change was expected to pose sharp risks to the nation's food supply in coming decades. Based on a host of new research the IPCC predicted that rising temperatures could reduce crop production by as much as 2 percent each decade for the remainder of the twenty-first century.[92] According to the IPCC, the risks of climate change were so profound they could stall or even reverse generations of progress against poverty and hunger.[93] Such evidence helped President Obama be a crucial player at the Paris climate change conference in 2015.

Paris Climate Agreement and United States Policy

The Paris climate conference elevated international commitments to combat climate change on December 2015. Over 190 countries adopted the Paris Agreement, the most ambitious climate change agreement in history. The goal was to devise plans to decrease rising temperatures and mechanisms for developing countries and vulnerable cities. In order for the agreement to take effect, at least 55 countries representing at least 55 percent of global emissions needed to formally adopt it. On September 3, 2016, both China and the United States signed the Paris Agreement with additional countries to follow. By November 2016, the agreement was ratified with over 197 countries as signatories. The agreement requires countries to set their own targets to reduce greenhouse gas emissions by 2020. President Obama committed the United States to reducing carbon emissions by 26–28 percent by 2025.[94]

Although President Obama ardently positioned the United States to be an international leader to address climate change, this commitment was short lived. On June 1, 2017, President Trump announced the United States would withdraw from the 2015 Paris Agreement. The United States is now one of the few countries not committed to this international agreement. In December 2018, countries met to discuss how to reach targets in Poland, setting a tracking and reporting system.

OUTCOMES

President Trump is one of the most outspoken opponents of climate change. A month after taking office, he signed an executive order on March 28, 2017, to dismantle, halt, or slow down climate policies—this included the Clean Power Plan and new emission standards for vehicles.[95] Additionally, federal agencies were asked to remove climate change information from their websites. Most notably, the EPA's climate change website was completely removed, and when you visit the website it states, "This page is being updated." As a result, Gretchen Goldman, research director for the Center for Science and Democracy, released a statement, arguing, "Removing climate change resources from the EPA website is offensive and dangerous. At a time when Americans have lost their loved ones and their homes to floods and fires, are living without fresh water or electricity, and are experiencing multi-billion-dollar disasters exacerbated by climate change, this is not the time to impede public access to critical climate change information."[96]

Despite President Trump's animosity toward climate policy, two key climate reports sounded the alarm for lawmakers to act immediately. In October 2018, the IPCC released its annual comprehensive climate assessment report. In this report, the authors predicted that global warming is occurring more rapidly than expected. Thus, increased drought and severe weather are expected. As a result, famine, food shortages, and disease will affect countries across the globe. By way of comparison, in the United States, the fourth national climate assessment report was released in November 2018. This report blatantly points out that the United States is unprepared for the economic and social implications of climate change. In particular, upheaval will occur when our rising sea levels devastate coastal areas.[97]

As a result, others are forging ahead in the United States and internationally. California's former governor Jerry Brown and former New York City mayor Michael Bloomberg created America's Pledge Alliance. This organization provides a venue to share data-driven information and action at the local level. We Are Still In also emerged as an organization supportive of climate action. More than 2,700 mayors, business leaders, and governors pledged their support for international action toward climate change. Yet, most notably, are the twenty-one youth (ages 11–22) who filed a lawsuit claiming that the federal government violated their rights of due process to address pollution from the burning of fossil fuels. In June 2019, a federal district court heard arguments and a final decision is pending as we write this book.[98]

The results of the 2018 mid-term elections handed the majority to the Democratic Party in the House of Representatives. Representatives like Alexandria Ocasio-Cortez, D-N.Y., have organized to pass environmental legislation—the Green New Deal, for example. The goal is to provide a win-win for the environment and the economy, retraining the workforce in green jobs, while combating climate change. As Daniel Fiorino eloquently puts it, "[T]he United States has put the advantages of democracy on display: the vibrancy of state and local responses, the innovation in technology and investment from the private sector, the mobilization of voters and activists concerned about climate change. Indeed, a major strength of democratic governance is that periods of failure nationally may be countered in other ways."[99]

On the international stage, Greta Thunberg, a fifteen-year-old from Sweden, was one of the highlights of United Nations Climate negotiations (COP 24 Summit) in Poland (December 2018). Thunberg has inspired students across the globe to pressure lawmakers to stop talking and to take action. On Fridays she wanted to encourage others to walk out of school to demand that adults take action. Initially, others did not join her, but today she has sparked a global movement. Her message is simple and clear, "12 years left." More specifically, the international community has twelve years left to reach its 2030 climate goals.[100]

CONCLUSIONS

For decades domestic opposition has hampered efforts to build the groundswell of support necessary to force American national leaders to pursue climate change policies. Opposing greenhouse gas reductions is a powerful coalition of oil and coal producers and fossil-fuel-dependent industries that have deep pockets and strong, long-standing ties to elected officials. The relentless efforts by opponents of climate change policies to undermine the credibility of mainstream scientists and elevate the views of conservative economists illuminate the importance of problem definition in American politics—not only in bringing about policy change, but in preventing it. Environmentalists and their allies are at a disadvantage in the struggle to define climate change as a problem, not just because they have fewer resources, but also because they have a difficult case to make.

The climate system is complex and uncertain; the villains of the climate change story are ordinary Americans; and any crisis associated with climate change is likely to occur in the (politically) distant future. Recognizing their situation, proponents of climate change policies have adopted new approaches. They have lobbied at the state level, where the fossil fuel-based coalition is less entrenched. They have also tried to disrupt alliances among economic interests in the United States in a variety of ways, some of them confrontational and others more conciliatory. They have used boycotts and shareholder resolutions to raise business leaders' concern about their corporate image. In addition, they have formed partnerships with businesses to encourage voluntary greenhouse gas emissions reductions. Abetting their efforts is a growing chorus of scientists. As Princeton economist Robert Socolow observes, the experts who do ice core research or work with climate models have been unusually vocal, "going out of their way to say, 'Wake up!' This is not a good thing to be doing."[101]

Scientists' alarm about climate change, although unprecedented, has not resulted in legislation to curb greenhouse gas emissions—a clear indication of the extent to which the political impacts of science are mediated by advocates' ability to translate that science into a compelling political story. Having expended enormous resources in the effort to mobilize a winning coalition, environmentalists are faced with the dilemma of what to do next.

For environmental writer Bill McKibben, the answer lies in building a global social movement, which he has done with 350.org. That organization's campaign is rooted in the recognition that climate change is a profoundly moral issue and demands that institutions divest themselves of stocks and bonds that support fossil fuels. Financier Tom Steyer has adopted a more pragmatic approach: his group NextGen Climate supports political candidates who promise to make addressing climate change a priority, while exposing candidates who decline to do so. For activists of all stripes, the fundamental challenge remains to raise the salience of climate change so that elected officials feel sufficient heat that, unless they support policies that wean the country off fossil fuels, they will not be reelected. It remains to be seen what saliency will result from the U.S.'s Green New Deal and Sweden's Thunberg's pleas for adults to act.

QUESTIONS TO CONSIDER

- What will invoke lawmakers to pass climate policy in the United States?
- What do you think the prospects are for a shift in the U.S. position on policies to prevent or mitigate the damage from climate change, and why?
- Should Americans be concerned about the pace at which the U.S. government is responding to this issue, or is a go-slow approach preferable?

NOTES

1. Statement of Dr. James Hansen, director, NASA Goddard Institute for Space Studies, Greenhouse Effect and Global Climate Change, Hearing Before the Committee on Energy and Natural Resources, U.S. Senate, 100th Cong., 1st sess., on the Greenhouse Effect and Global Climate Change, Part 2 (June 23, 1988).
2. Jeff. Tollefson, "UN Climate Talks Keep Paris Accord Alive." *Nature*. Available from https://www.nature.com/articles/d41586-018-07824-w.
3. Norman J. Vig, "Introduction," in *The Global Environment: Institutions, Law, and Policy*, ed. Norman J. Vig and Regina S. Axelrod (Washington, D.C.: CQ Press, 1999).

4. Ian H. Rowlands, "Classical Theories of International Relations," in *International Relations and Global Climate Change*, ed. Urs Luterbacher and Detlef Sprinz (Cambridge, Mass.: The MIT Press, 2001), 43–65.

5. Stephen Krasner, quoted in Vig, "Introduction," 4.

6. Robert D. Putnam, "Diplomacy and Domestic Politics: The Logic of Two-Level Games," *International Organization* 42 (Summer 1988): 434.

7. See, for example, DeWitt John, *Civic Environmentalism: Alternatives to Regulation in States and Communities* (Washington, D.C.: CQ Press, 1994); Daniel A. Mazmanian and Michael E. Kraft, eds., *Toward Sustainable Communities: Transition and Transformation in Environmental Policy* (Cambridge, Mass.: The MIT Press, 1999).

8. Judith A. Layzer, "Deep Freeze: How Business Has Shaped the Legislative Debate on Climate Change," in *Business and Environmental Policy*, ed. Michael E. Kraft and Sheldon Kamieniecki (Cambridge, Mass.: The MIT Press, 2007), 93–125.

9. John Houghton, *Global Warming: The Complete Briefing*, 2d ed. (New York: Cambridge University Press, 1997).

10. Ian Rowlands, *The Politics of Global Atmospheric Change* (New York: Manchester University Press, 1995); Matthew Paterson, *Global Warming and Climate Politics* (New York: Routledge, 1996); Spencer R. Weart, *The Discovery of Global Warming*, rev. ed. (Cambridge, Mass.: Harvard University Press, 2008).

11. Roger Revelle and Hans E. Suess, "Carbon Dioxide Exchange Between Atmosphere and Ocean and the Question of an Increase of Atmospheric CO_2 During the Past Decade," *Tellus* 9 (1957): 18–27.

12. President's Science Advisory Committee, *Restoring the Quality of Our Environment: Report of the Environmental Pollution Panel* (Washington, D.C.: White House, 1965), 126–127.

13. National Academy of Sciences, *Carbon Dioxide and Climate: A Scientific Assessment* (Washington, D.C.: National Academy of Sciences, 1979).

14. Enduring cooperation among meteorologists began with the First International Meteorological Conference in 1853. Twenty years later, the International Meteorological Organization (IMO) was established. After World War II, the IMO turned into the World Meteorological Organization, and the latter began operating in 1951. See Paterson, *Global Warming*.

15. Quoted in William W. Kellogg, "Predictions of a Global Cooling," *Nature* (August 16, 1979): 615.

16. Stephen Seidel and Dale Keyes, *Can We Delay a Greenhouse Warming? The Effectiveness and Feasibility of Options to Slow a Build-Up of Carbon Dioxide in the Atmosphere* (Washington, D.C.: U.S. Environmental Protection Agency, September 1983).

17. World Meteorological Organization, *Report of the International Conference on the Assessment of the Role of Carbon Dioxide and of Other Greenhouse Gases in Climate Variations and Associated Impacts*, Villach, Austria, 9–15 October, WMO Publication No. 661 (Geneva: World Meteorological Organization, 1986).

18. Quoted in Michael Molitor, "The United Nations Climate Change Agreements," in Vig and Axelrod, *The Global Environment*, 221.

19. A framework convention is a broad but formal agreement; a protocol contains more concrete commitments. Both must be signed and ratified by participating nations.

20. Molitor, "The United Nations Climate Change Agreements."

21. Quoted in Paterson, *Global Warming*, 37.

22. Quoted in Michael Weisskopf, "Bush Says More Data on Warming Needed," *The Washington Post*, April 18, 1990, 1.

23. Quoted in Molitor, "The United Nations Climate Change Agreements."

24. Rowlands, *The Politics of Global Atmospheric Change*, 79.

25. John Hunt, "U.S. Stand on Global Warming Attacked," *Financial Times*, August 30, 1990, 4; Rowlands, *The Politics of Global Atmospheric Change*.

26. The complete text of the FCCC is available at http://www.unfccc.de.

27. Bert Bolin, "Report to the Ninth Session of the INC/FCCC" (Geneva: IPCC, February 7, 1994), 2.

28. IPCC Working Group I, *Climate Change* 1995*: The Science of Climate Change*, ed. J. T. Houghton et al. (New York: Cambridge University Press, 1996), 4.

29. Ibid.

30. Stuart Eizenstadt, Under Secretary of State for Economic, Business, and Agricultural Affairs, Statement before the Senate Foreign Relations Committee, 105th Cong., 2nd sess., February 11, 1998.

31. Quoted in Daniel Sarewitz and Roger A. Pielke Jr., "Breaking the Global-Warming Gridlock," *The Atlantic Monthly*, July 2000, 57.

32. Ross Gelbspan, *The Heat Is On: The High Stakes Battle over Earth's Threatened Climate* (New York: Addison-Wesley, 1997).

33. To estimate the influence of greenhouse gases in a changing climate, researchers run models for a few (simulated) decades and compare the statistics of the models' output to measures of the climate. Although observations of both past and present climate confirm many of the predictions of the prevailing models of climate change, at least some of the data used to validate the models are themselves model outputs. Moreover, the assumptions and data used to construct GCMs heavily influence their predictions, and those elements are themselves selected by scientists who already know what they expect to find. See Steve Rayner, "Predictions and Other Approaches to Climate Change Policy," in *Prediction: Science, Decision Making and the Future of Nature*, ed. Daniel Sarewitz, Roger A. Pielke Jr., and Radford Byerly Jr. (Washington, D.C.: Island Press, 2000), 269–296.

34. Michael L. Parsons, *The Truth Behind the Myth* (New York: Plenum Press, 1995); Patrick J. Michaels, *Sound and Fury: The Science and Politics of Global Warming* (Washington, D.C.: Cato Institute, 1992).

35. S. Fred Singer, "Benefits of Global Warming," *Society* 29 (March–April 1993): 33.

36. William K. Stevens, "At Hot Center of the Debate on Global Warming," *The New York Times*, August 6, 1996, C1.

37. Thomas Gale Moore, "Why Global Warming Would Be Good for You," *Public Interest* (Winter 1995): 83–99.

38. Thomas Gale Moore, *Climate of Fear: Why We Shouldn't Worry About Global Warming* (Washington, D.C.: Cato Institute, 1998), 1–2.

39. Warren T. Brookes, "The Global Warming Panic," *Forbes*, December 25, 1989, 98.

40. Quoted in Gale E. Christianson, *Greenhouse: The 200-Year Story of Global Warming* (New York: Penguin Books, 1999), 258.

41. U.S. Council of Economic Advisors, *Economic Report of the President* (Washington, D.C.: U.S. Government Printing Office, February 1990), 214.

42. Moore, *Climate of Fear*, 2.

43. Rowlands, *The Politics of Global Atmospheric Change*, 139.

44. William R. Cline, *Global Warming: The Economic Stakes* (Washington, D.C.: Institute for International Economics, 1992), 1–2.

45. Rowlands, *The Politics of Global Atmospheric Change*, 137.

46. John J. Fialka, "Clinton's Efforts to Curb Global Warming Draws Some Business Support, But It May Be Too Late," *The Wall Street Journal*, October 22, 1997, 24.

47. Craig Trumbo, "Longitudinal Modeling of Public Issues: An Application of the Agenda-Setting Process to the Issue of Global Warming," *Journalism & Mass Communication Monograph* 152 (August 1995).

48. Frank Loy, Under Secretary for Global Affairs, Statement Before the Committee on Foreign Relations and the Committee on Energy and Natural Resources, U.S. Senate, 106th Cong., 2nd sess., September 28, 2000.

49. Quoted in William K. Stevens, "Greenhouse Gas Issue: Haggling over Fairness," *The New York Times*, November 30, 1997, 6.

50. Quoted in Molitor, "The United Nations Climate Change Agreements," 219–220.

51. Ibid., 220.

52. John H. Cushman Jr., "U.S. Says Its Greenhouse Gas Emissions Are at Highest Rate in Years," *The New York Times*, October 21, 1997, 22.

53. John H. Cushman Jr., "Why the U.S. Fell Short of Ambitious Goals for Reducing Greenhouse Gases," *The New York Times*, October 20, 1997, 15.

54. According to journalist Eric Pooley, many senators who actually favored a binding commitment to reduce greenhouse gases believed that if the resolution were unanimous, its force would be blunted because observers could not possibly take it as a vote against the treaty. Their reasoning was clearly flawed. See Eric Pooley, *The Climate War: True Believers, Power Brokers, and the Fight to Save the Earth* (New York: Hyperion, 2010).

55. Quoted in James Bennet, "Warm Globe, Hot Politics," *The New York Times*, December 11, 1997, 1. The protocol takes effect once it is ratified by at least fifty-five nations; the terms become binding on an individual country only after its government ratifies the treaty.

56. John H. Cushman Jr., "Industrial Group Plans to Battle Climate Treaty," *The New York Times*, April 26, 1998, 1.

57. John H. Cushman Jr., "New Policy Center Seeks to Steer the Debate on Climate Change," *The New York Times*, May 8, 1998, 13.

58. William K. Stevens, "March May Soon Be Coming in Like a Lamb," *The New York Times*, March 2, 1999, 1.

59. By employing prudent land-use practices, the United States argued, parties can "sequester" CO_2—that is, they can store it in wood and soils, thereby preventing its release into the atmosphere. Critics resisted heavy reliance on forests to curb CO_2, however, pointing out that while forests currently do offset about one-quarter of the world's industrial CO_2 emissions, research at the Hadley Center in the United Kingdom indicated that many of the recently planted forests would, by the middle of the twenty-first century, begin releasing carbon back into the atmosphere. Moreover, they noted, during the same period, warming was likely to increase the amount of carbon released by soils, particularly in the peatland forests of the northern latitudes.

60. Andrew C. Revkin, "Treaty Talks Fail to Find Consensus in Global Warming," *The New York Times*, November 26, 2000, 1.

61. Quoted in Andrew C. Revkin, "Odd Culprits in Collapse of Climate Talks," *The New York Times*, November 28, 2000, C1.

62. "Climate Panel Reaffirms Major Warming Threat," *The New York Times*, January 23, 2001, D8.

63. "Paradise Lost? Global Warming Seen as Threat," *Houston Chronicle*, February 20, 2001, 1.

64. By 2001 Kilimanjaro had lost 82 percent of the ice cap it had in 1912. In the Alps, scientists estimate 90 percent of the ice volume of a century ago will be gone by 2025. See Eric Pianin, "U.N. Report Forecasts Crises Brought On by Global Warming," *The Washington Post*, February 20, 2001, 6.

65. James Glanz, "Droughts Might Speed Climate Change," *The New York Times*, January 11, 2001, 16.

66. Joseph Kahn, "Energy Efficiency Programs Are Set for Bush Budget Cut," *The New York Times*, April 5, 2001, 16.

67. Douglas Jehl, "U.S. Rebuffs European Plea Not to Abandon Climate Pact," *The New York Times*, April 4, 2001, 14.

68. Katharine Q. Seelye and Andrew C. Revkin, "Panel Tells Bush Global Warming Is Getting Worse," *The New York Times*, June 7, 2001, 1.

69. Quoted in Andrew C. Revkin, "Deal Breaks Impasse on Global Warming Treaty," *The New York Times*, November 11, 2001, Sec. 1, 1.

70. Quoted in "White House Plan Marks Turf Amid Already-Contentious Debate," *E&E Daily*, February 15, 2002.

71. By contrast, core sediment revealed that the Larsen A ice shelf, which disintegrated in the mid-1990s, had been open water 6,000 years ago. See Kenneth Chang, "The Melting (Freezing) of Antarctica," *The New York Times*, April 2, 2002, D1; Andrew C. Revkin, "Study of Antarctic Points to Rising Sea Levels," *The New York Times*, March 7, 2003, 8.

72. Usha Lee McFarling, "Arctic's Biggest Ice Shelf, a Sentinel of Climate Change, Cracks Apart," *Los Angeles Times*, September 23, 2003, Sec. 1, 3.

73. "CO_2 May Be Changing Structure of Amazon Rainforest—Study," *Greenwire*, March 11, 2004; "Study Finds 'Carbon Sinks' May Actually Generate More CO_2," *Greenwire*, September 24, 2004.

74. Maxwell T. Boykoff and Jules M. Boykoff, "Balance as Bias: Global Warming and the U.S. Prestige Press," *Global Environmental Change* 14: 125–136.

75. Andrew Freedman, "Warming May Lead to Western 'Megadroughts,' Study Says," *Greenwire*, October 8, 2004.

76. Susan Milligan, "Study Predicts City Flood Threat Due to Warming," *The Boston Globe*, February 15, 2005, 1.

77. Quoted in Andrew C. Revkin, "Politics Reasserts Itself in the Debate over Climate Change and Its Hazards," *The New York Times*, August 5, 2003, F2.

78. Quoted in Larry Rohter, "U.S. Waters Down Global Commitment to Curb Greenhouse Gases," *The New York Times*, December 19, 2004, 16.

79. Barry Rabe, *Statehouse and Greenhouse: The Emerging Politics of American Climate Change Policy* (Washington, D.C.: Brookings Institution Press, 2004).

80. "Cities Organizing to Reduce Greenhouse Gas Emissions," *Greenwire*, February 22, 2005.

81. By 2010 more than 1,000 cities had signed the Mayors' Climate Protection Agreement. Available at http://www.usmayors.org/climateprotection/revised.

82. Mark Clayton, "New Coal Plants Bury Kyoto," *The Christian Science Monitor*, December 23, 2004.

83. Obama White House. "A Historic Commitment to Protecting the Environment and Addressing the Impacts of Climate Change." National Archives and Records Administration, obamawhitehouse.archives.gov/the-record/climate.

84. For details on the Obama administration's legislative efforts, see Judith A. Layzer, "Cold Front: How the Recession Stalled Obama's Clean-Energy Agenda," in *Reaching for a New Deal: President Obama's Agenda and the Dynamics of U.S. Politics*, Project of the Russell Sage Foundation, 2010. Available at http://www.russellsage.org/research/working-group-obamas-policy-agenda. For the story of the Senate debacle, see Ryan Lizza, "As the World Burns," *The New Yorker*, October 11, 2010.

85. Coral Davenport, "Obama to Take Action to Slash Coal Pollution," *The New York Times*, June 2, 2014.

86. Beth Gardiner, "Testing the Limits of European Ambitions on Emissions," *The New York Times*, December 1, 2014.

87. Coral Davenport, "A Climate Accord Based on Global Peer Pressure," *The New York Times*, December 15, 2014.

88. Justin Gillis and David Jolly, "Slowdown in Carbon Emissions Worldwide, But Coal Burning Continues to Grow," *The New York Times*, November 19, 2013; Fiona Harvey, "Record CO_2 Emissions 'Committing World to Dangerous Climate Change,'" *The Guardian*, September 21, 2014.

89. LuAnn Dahlman, "Climate Change: Global Temperature," NOAA Climate.gov, January 16, 2015. Available at http://www.climate.gov/news-features/understanding-climate/climate-change-global-temperature.

90. Justin Gillis, "U.S. Climate Has Already Changed, Study Finds, Citing Heat and Floods," *The New York Times*, May 7, 2014.

91. John M. Broder, "U.S. Warns That Climate Change Will Cause More Energy Breakdowns," *The New York Times*, July 11, 2013.

92. Justin Gillis, "Climate Change Seen Posing Risks to Food Supplies," *The New York Times*, November 2, 2013.

93. Justin Gillis, "U.N. Panel Issues Its Starkest Warning Yet on Global Warming," *The New York Times*, November 3, 2014.

94. Stefan Becket. "Paris Climate Agreement: What You Need to Know," *CBS News*. June 1, 2017.

95. Trump White House. Presidential Executive Order on Promoting Energy Independence and Economic Growth, March 28, 2017.

96. Gretchen Goldman, "Removing Climate Change Resources from EPA Website 'Offensive and Dangerous,' Science Group Says." Union of Concerned Scientists. Washington. October 20, 2017.

97. Brad Plumer and Henry Fountain, "What's New in the Latest US Climate Assessment Report." November 23, 2018. Available at https://www.nytimes.com/2018/11/23/climate/highlights-climate-assessment.html.

98. Lawrence Hurly, "Supreme Court Rejects Trump Bid to Halt Climate Change Case." July 30, 2018. Available at https://www.reuters.com/article/us-usa-court-climate/supreme-court-rejects-trump-bid-to-halt-climate-change-case-idUSKBN1KK2FY

99. Fiorino, Daniel, *Can Democracy Handle Climate Change?* (Cambridge, Mass.: Polity Press, 2018).

100. John Sutter and Lawrence Davidson, "Teen Tells Adults They Are Not Mature Enough." December 17, 2018. Available from https://www.cnn.com/2018/12/16/world/greta-thunberg-cop24/index.html

101. Elizabeth Kolbert, "The Climate of Man—III," *The New Yorker*, May 9, 2005, 55.

CHAPTER THIRTEEN

CAPE WIND

If Not Here, Where? If Not Now, When?

In the summer of 2001 energy entrepreneur Jim Gordon unveiled a proposal to build 170 wind turbines in Nantucket Sound, off the coast of Massachusetts. According to its promoters, the development—known as Cape Wind—could furnish three-quarters of the energy used by the fifteen Barnstable County towns that comprise Cape Cod, as well as the nearby Elizabeth Islands, Martha's Vineyard, and Nantucket; what is more, it would do so without producing local air pollution or the greenhouse gases that cause global warming. In fact, proponents argued, the shallow waters and stiff breezes of the sound's Horseshoe Shoal made it one of the most hospitable sites for an offshore wind farm in the United States. Yet despite its promise of providing clean, safe energy, Gordon's proposal met with immediate and fierce resistance, and detractors capitalized on every procedural opportunity to resist it. As a result, nearly fifteen years after it was conceived, Cape Wind was all but dead. That said, because it was the first such project to be seriously considered, the travails of Cape Wind shaped the prospects for future offshore wind farms.

Many observers attribute the numerous delays in permitting Cape Wind to the objections of the influential Kennedy family and their wealthy and prominent allies, who simply want to protect the scenic view from their beachfront mansions.[1] In fact, although the Cape Wind saga features some unusually influential main characters, it is hardly unique: since the early 1990s local groups have thwarted efforts to site alternative energy projects—from solar and wind to geothermal developments—throughout the United States and Europe. And many of the detractors of such projects are neither wealthy nor politically powerful; they are simply determined. Critics have decried such resistance as the "not-in-my-backyard (NIMBY) syndrome," which they attribute to a selfish unwillingness to bear the costs of one's lifestyle. Scholars became intrigued by NIMBYism in the 1980s after local obstruction blocked efforts to site landfills, hazardous- and nuclear-waste disposal facilities, offshore-drilling facilities, and even homeless shelters and halfway houses.[2] Although many observers characterized such community activism as uninformed and parochial, a growing chorus of scholars described local resistance to unwanted land uses as a rational response to perceived risk or inequity, threats to community integrity, or an improper or arbitrary decision-making process.[3] Some argued that

NIMBYism actually *improved* public decision making by bringing to light factors that had been ignored in the project design or siting process.[4]

In the 2000s researchers began to investigate local resistance to wind farms in particular. They were puzzled because public opinion surveys invariably showed extremely high levels of support for wind energy, yet only a fraction of the available wind-power capacity was actually being commissioned. That disconnect reveals some ambivalence within the environmental community. In particular, alternative energy siting pits two important environmental goals—preserving highly valued landscapes and combating global warming—against one another.[5] More generally, it pits long-term, collective concerns against immediate, localized sacrifice: neighbors of a development are expected to subordinate their preferences and interests to the common good, even though their individual contribution to that societal goal will be imperceptible. Not surprisingly, the prospect of immediate, concrete, and concentrated loss arouses more intense reactions—and hence political mobilization—than does the possibility of uncertain and diffuse future benefits.[6]

In addition to exposing latent tensions within the environmental community, the Cape Wind case illuminates how both developers and their opponents take advantage of "veto points" in their efforts to translate their values and interests into policy. Scholars have long deployed the concept of veto points or veto players—those whose consent is necessary to change policy—to explain policy stability and change. In his book *Veto Players: How Political Institutions Work*, political scientist George Tsebelis proposes a formal theory elucidating why the more veto players a country has, the more stable policy is likely to be.[7] Others argue that it is the interaction between the number of veto points and the distribution of preferences of the officials that populate those veto points that determines how resistant the status quo is to policy change.[8]

Any major construction project involves obtaining numerous permits and approvals, each of which constitutes a potential veto point. At each veto point, opponents of the project typically raise a host of substantive concerns about the project's environmental and other impacts. Because Cape Wind was the first project of its kind in the United States, however, it raised additional concerns about the permitting process itself, such as who should be allowed to develop resources in publicly owned waters and under what conditions. Who has jurisdiction over development of the ocean? And should Cape Wind be allowed to proceed because of the urgency of addressing global warming, even in the absence of a regulatory framework? Such process questions, while often legitimate, also serve as a proxy for substantive concerns, and advocates on both sides of a proposed project raise them to delay decision making. Opponents hope that if they are able to postpone a final decision for long enough, sufficient opposition will form to derail the project, the cost of developing the project will become prohibitively high, or the developer will simply become discouraged.[9] Of course, developers use delay tactics as well: because they can pay representatives to attend a seemingly endless series of meetings, they can often outlast weary citizen volunteers.

The Cape Wind case also brings into sharp relief the different ways that proponents and opponents characterize solutions and specifically how they frame the costs and benefits of alternative energy projects. The capital costs of building a wind farm are high, particularly offshore: as of 2014, the price tag for a fully installed 500-megawatt (MW) offshore wind system was about $5,700 per kilowatt (kW), more than three times the approximately $1,750 per kW price of a land-based system.[10] In addition, the variability of wind means that integrating large blocks of wind capacity into an existing grid system can be expensive. Traditionally, utilities have built power systems around generating technologies that are predictable and "dispatchable"—that is, units that can be turned off and on and whose output can vary depending on changes in demand, such as coal-, oil-, or natural gas-based technologies. When the electrical grid is dependent on only small amounts of wind generation, the variations in wind output generally can be absorbed by the buffer capacity of the existing system. But when wind constitutes a large part of a system's total generating capacity—10 percent to 15 percent or more—utilities must incur additional costs to provide reliable backup for the wind turbines, and building such backup systems is expensive.[11] Wind-power advocates reject the assumption that baseload power plants are necessary for reliability; they argue that a new paradigm, in which demand and supply are managed in tandem, is taking hold.[12] Other, more objective analysts have weighed in as well with support for the thesis that integrating large amounts of intermittent power need not be more costly than using conventional fuels.[13]

Furthermore, as critics also point out, historically the financial viability of wind power has depended on government subsidies that are politically insecure. Two federal policies—the five-year depreciation schedule for renewable energy systems, enacted as part of the Economic Recovery Tax of 1981, and the production tax credit (PTC) put in place by the Energy Policy Act of 1992—create substantial tax breaks for renewable energy projects. The PTC has proven particularly effective. Worth 1.5 cents per kWh in 1992 and adjusted annually for inflation, it can decrease the cost of financing needed to build a project by 40 percent. But wind developers can never rely on it: between 1999 and 2004 the PTC expired on three separate occasions; it was extended in 1999, 2002, and 2004, and then extended again in 2005, 2006, 2008, 2009, and most recently in 2018.[14] There is widespread agreement that the "on-again, off-again" nature of the PTC has contributed to the boom-and-bust cycles characteristic of wind-power development to date.[15]

Defenders of offshore wind development respond that although the PTC artificially lowers the cost of wind energy, any price comparison between wind and more conventional fuels is nevertheless skewed in favor of the latter in several respects. First, the cost of generating energy from new wind-power plants is typically compared to that of running existing coal- and oil-fired or nuclear plants, which have already been largely or fully depreciated and—in the case of nuclear—have received substantial subsidies in the form of liability caps. Second, technologies to derive energy from solar, offshore wind, and other sources are at the early stages of development relative to fossil fuel technologies; costs are coming down as more generating units are installed and the technologies evolve.[16] Third, although construction and maintenance of wind turbines can be expensive, they are the only costs associated with power generation; unlike conventional fuels, whose prices are volatile and likely to increase over time, the cost of wind is stable and certain.

Most important, defenders note, the nonmonetary costs of extracting, transporting, and generating power from conventional fuels are not factored into price comparisons. Accidents such as the disaster in February 2014, in which Duke Energy spilled 39,000 tons of toxic coal ash into the Dan River in North Carolina, or the 2017 Big Bend Power Plant explosion in Florida, which killed two people, are reminders that using coal to generate electricity is not benign.[17] But the chronic, routine environmental consequences of fossil fuel use are actually more severe than those of the occasional accident. Coal from Appalachia is acquired at the expense of that region's mountaintops and valley streams, and mitigation measures do little to restore the local environment.[18] Burning coal produces a host of air pollutants—including mercury, sulfur dioxide, and nitrogen oxides—that are hazardous to human health and the environment; in 2009 the National Academy of Sciences (NAS) estimated that producing electricity from coal plants cost the United States $62 billion per year (or 3.2 cents per kWh) in hidden costs.[19] And, of course, coal combustion emits carbon dioxide (CO_2), the primary culprit in global warming. Obtaining natural gas, which has gained popularity as an electricity-generating fuel because it emits less CO_2 than coal, involves cutting well-site access roads into wilderness areas or through fragile coastal wetlands, fragmenting habitat and weakening shoreline defenses. In addition, natural gas is often obtained through hydraulic fracturing, a controversial process in which chemicals mixed with large quantities of water are injected into underground wells (see Chapter 14).[20] Another cost associated with thermoelectric power plants is less widely acknowledged: in 2005 they required 201 million gallons of water each day for cooling—amounting to 49 percent of all water use in the country.[21] Nuclear energy, often regarded as a clean alternative to fossil fuels because it produces no greenhouse gas emissions after plants are built, involves mining, transporting, and disposing of radioactive materials, as well as the risk of accidents. Cooling nuclear power plants also consumes enormous quantities of water.

BACKGROUND

In the 1990s growing awareness of the many environmental costs associated with fossil fuel-based electricity, as well as technological improvements in turbines and a variety of government policies, spurred renewed interest in wind energy. As land-based installations began to proliferate toward the end of the decade, developers realized that offshore wind, pioneered in Northern Europe, had the potential to provide massive quantities of energy. Because of its steady winds, which blow even on the hottest days of the summer, the coast of Massachusetts was considered one of the best places for wind energy in the nation. So it was not entirely preposterous when, in 2001, developers proposed Nantucket Sound as the location of the nation's first offshore wind farm. Nantucket Sound lies at the center of a triangle defined by the southern coast of Cape Cod and the islands of Martha's Vineyard and Nantucket. Created thousands of years ago by the retreating Laurentide Ice Sheet, Cape Cod is a low-lying, sandy peninsula that emerges from southeastern Massachusetts in an elegant curve. Between four and seven miles to the southwest lies the island of Martha's Vineyard, and thirty miles to the southeast lies Nantucket. Residents of the coastal Cape Cod towns of Falmouth, Mashpee, Barnstable, Yarmouth, Dennis, Harwich, and Chatham have views across the sound, as do those on the north coast of Nantucket and the towns of Edgartown, Oak Bluffs, and Tisbury on Martha's Vineyard (see Map 13.1).

Cape Cod and the islands are no strangers to wind power: in the early 1800s Cape Cod alone had more than a thousand working windmills, and they were a common site on Martha's Vineyard and Nantucket as well.[22] Windmills were an important source of power to pump water in rural America well into the twentieth century; they were also used to grind grain, saw wood, churn butter, and perform other tasks.[23]

Map 13.1 Cape Cod, Nantucket Sound, and Proposed Cape Wind Site

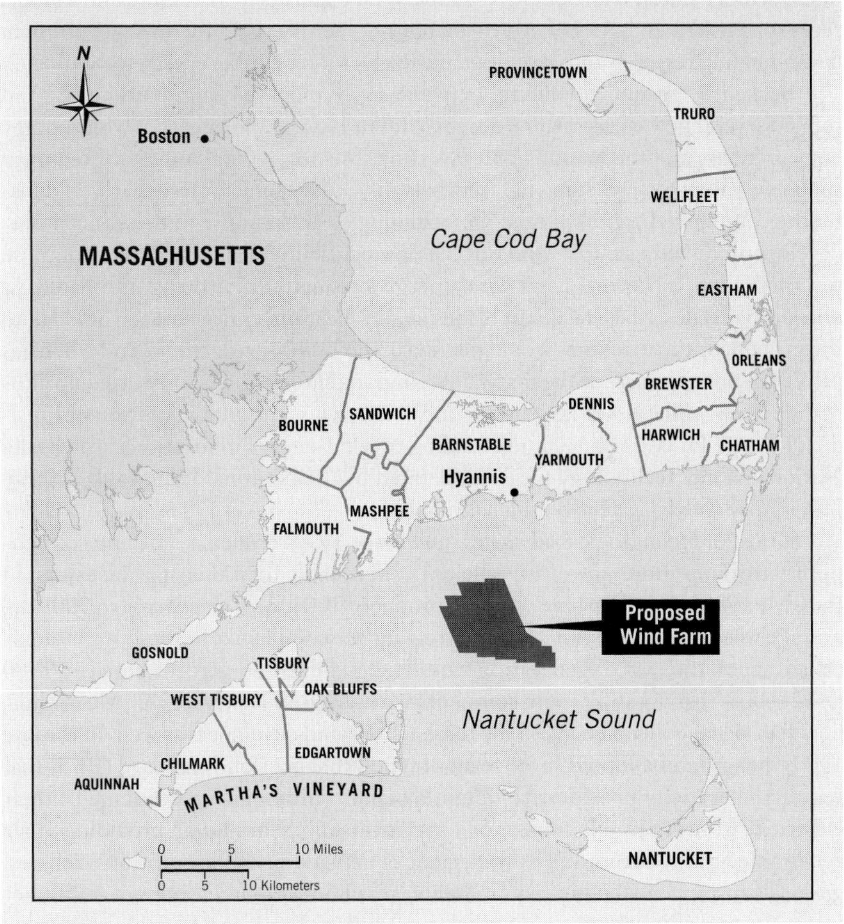

Source: Map and Cape Wind site compiled by author from ESRI, TeleAtlas, and AP.

By the early 1900s, however, large-scale, centralized power plants fueled by coal and oil were becoming the norm for generating electricity. During the 1930s and 1940s the federal government extended the electrical grid to the country's rural areas, hastening the demise of small-scale wind power. Then, between 1965 and 1975, plans to build nuclear power plants proliferated, thanks to a concerted push by the federal government. But skyrocketing construction costs and fears of an accident brought commissioning of new nuclear plants to a screeching halt in the late 1970s.[24]

Meanwhile, worries about the severe air pollution caused by coal-fired power plants breathed life into the nascent alternative energy business. Security concerns also played a role: in 1973 the United States experienced an oil shortage after the Organization of Arab Petroleum Exporting Countries (OAPEC)—a cartel comprising the major Middle Eastern oil-exporting states—placed an embargo on shipments to the West. Subsequently, OAPEC decided to limit supply to increase the income of its member states; crude oil prices in the West quadrupled as a result, and a global recession ensued. In hopes of averting a future "energy crisis," the U.S. government began funding research and development (R&D) for renewable energy technologies.

Backed by federal financing, between 1974 and 1981 engineers tested and refined several new wind-turbine designs. But in 1981, at the behest of wind-energy entrepreneurs and the banking and investment sector, federal policy shifted from supporting R&D to providing tax credits, loans, and loan guarantees that would foster the commercialization of existing technologies. In response to these incentives, developers erected a slew of wind turbines, particularly in California. Many of those turbines failed, however, giving wind energy a reputation for being unreliable. As one journalist describes it, "Giant blades would shear off, entire rotors would fall to the ground, or the machines would just shake themselves to death."[25] And when gas prices fell precipitously in the mid-1980s, investment in wind energy dried up altogether. Meanwhile, R&D languished until 1989, when the administration of President George H. W. Bush resumed funding of wind-energy innovation through the National Wind Technology Center, operated by the National Renewable Energy Laboratory (NREL), near Boulder, Colorado.

A host of technological advances during the 1990s ensued, rendering contemporary turbines more powerful, efficient, and reliable than their predecessors: in the early 1980s a wind turbine typically produced 100 kW of electricity; in 2003 the average was closer to 1.5 MW, a fifteenfold increase.[26] Thanks to these technological advances, the cost of generating wind energy fell by 80 percent between 1980 and 2000, so that it was far more competitive with coal and natural gas.[27] Meanwhile, new state-level policies changed the calculus for wind developers as well. In the late 1990s many states adopted some form of renewable portfolio standard (RPS) that required electricity providers to obtain a certain (minimum) percentage of their electricity from renewable sources by a specific date.[28] States began providing other renewable energy incentives as well, such as tax credits and exemptions, rebates, grants, loans, green labeling requirements, green power purchasing programs, net metering, fuel-mix and environmental disclosure policies, and tradable renewable certificates in the form of green tags or renewable energy credits.

Fueled by changing economics and a more favorable policy context, in 1999 the U.S. wind industry began a rapid expansion. By the early 2000s wind was the nation's fastest-growing energy source, increasing at a rate of 25 percent to 30 percent each year, according to the American Wind Energy Association (AWEA). The potential for wind in the United States was astounding: the Department of Energy (DOE) reported that only 0.6 percent of the land area in the lower forty-eight states would be needed to produce 560,000 million kWh—enough to supply more than 45 million average American households or nearly 200 million households if Americans consumed electricity at the same rate as Europeans.[29] Despite this promise, in 2000, the year before Cape Wind was proposed, the United States produced only 2,554 MW of wind energy out of a world production of 17,300 MW.[30] As of 2016, 6 percent of electricity in the United States is generated from wind, compared to 34 percent from natural gas, 30 percent from coal, 20 percent from nuclear, and 1 percent from oil.[31]

THE CASE

Even as wind energy was losing its luster in the United States during the 1980s, the northern European countries of Sweden, Denmark, Germany, and the Netherlands continued to pursue it. Recognizing that ocean breezes were stronger and more consistent than those on land and that ocean-based turbines could be larger and therefore more powerful than those assembled on land, the northern Europeans began looking seriously at offshore wind.[32] In the 1990s they installed small clusters of between four and twenty wind turbines in offshore waters. The first of these was built in 1991 near the village of Vindeby, Denmark, where developers placed eleven turbines with a combined capacity of 4.95 MW in shallow water a little more than one mile offshore.[33] By 2002 there were ten offshore wind farms operating in Northern Europe, with a combined generating capacity of 250 MW.[34]

Intrigued, the DOE began mapping U.S. offshore resources, and in 2004 the agency released a report showing that more than 900,000 MW of unharnessed wind capacity existed offshore within fifty miles of the nation's coasts—an amount roughly equivalent to the generating capacity of the country's existing power plants. The report pointed out that the winds off New England were the strongest in the United States and that wind speeds along virtually the entire Massachusetts coast would be "excellent" or "outstanding" for wind-generated electricity. Creating an incentive to exploit those wind resources, Massachusetts's RPS—enacted in 1997 as part of the Massachusetts Electric Utility Restructuring Act—required electricity suppliers to provide at least 1 percent of their electricity from renewable sources starting in 2003, increasing 0.5 percent annually and leveling off at 4 percent in 2009.[35]

The Cape Wind Proposal

Even before the DOE began its mapping project, a proposal to build the nation's first offshore wind farm gained notoriety. The Cape Wind idea originated with a small group of alternative energy entrepreneurs led by Jim Gordon, a successful

New England energy developer. The threesome called their new venture Cape Wind Associates LLC and made Gordon its president.[36] Cape Wind Associates began investigating the logistics of building a utility-scale wind farm in Nantucket Sound and soon settled on Horseshoe Shoal, a shallow bar about five miles from the Barnstable village of Hyannis and seven miles northwest of Nantucket Island. The site boasts several advantages: it receives strong, steady winds out of the southeast; because the water is relatively shallow—the average water depth is between forty and fifty feet—it is out of shipping lanes and not susceptible to crashing waves; and ferry routes to Martha's Vineyard and Nantucket skirt its edges—although passenger ferries sometimes tack into the area in bad weather. A quirk in the zoning makes Horseshoe Shoal advantageous as well: although almost completely surrounded by state waters that were, at the time, protected from energy development by the Massachusetts Ocean Sanctuaries Act, the heart of Nantucket Sound is federal water.[37]

The partners envisioned an array of 170 turbines that would be anchored to the ocean floor with steel piles and spaced one-third to one-half mile apart. With 263-foot-high carbine-steel columns and blades nearly 330 feet long, the turbines would rise 426 feet above the surface of the ocean. Two cables buried six feet under the seafloor would connect the turbines to each other and then travel about 6.5 miles through Nantucket Sound and Lewis Bay, making landfall in West Yarmouth. From there, a transmission line would run four miles under public streets to an NStar easement, where it would proceed another 1.9 miles to the Barnstable switching station.[38] The entire offshore facility would cover a twenty-eight-square-mile area and generate 420 MW, making it the world's largest operating wind farm. On average, it would yield about 168 MW, which Cape Wind Associates estimated would supply about three-quarters of the average total electric demand on Cape Cod and the islands.[39]

Having settled on a concept and a promising site, the company began conducting a more extensive investigation, monitoring bird-flight patterns, taking borings of the ocean bed, and searching the seafloor for shipwrecks and other archaeological relics. The main challenges they anticipated, though, were not environmental but financial and political. First, they would have to compete with existing nuclear and fossil fuel-based power, both of which were relatively inexpensive. Cape Wind would be eligible for the federal production tax credit, which at the time was worth about 1.7 cents per kWh. But the PTC could disappear before the project was fully permitted, wreaking havoc with its financing. A more formidable challenge was likely to be political: Cape Wind would alter the "viewscape" from parts of Cape Cod, Martha's Vineyard, and Nantucket, and some long-time residents were certain to object.

In late July 2001 *The Boston Globe* broke the story that developers were preparing to unveil a proposal for the nation's first offshore wind farm in Nantucket Sound. The article noted that the timing of Cape Wind's announcement was opportune: as usual, the summer of 2001 featured several days of smoggy weather that highlighted the environmental costs of Cape Cod's oil-fired Mirant Canal power plant, an old facility whose emissions consistently violated air pollution standards. On the other hand, the project faced some hurdles, particularly from residents concerned about what would be a significant change in the area's historical seascape. Journalist Jeffrey

Krasner quoted Jim Lentowski, executive director of the Nantucket Conservation Foundation, who observed, "Nantucket is a flat landscape with few trees, and anything visible that's near or far that looks like a technological intrusion in a natural area, people sort of wince at."[40] Because the proposed project lay in federal waters, it needed a variety of federal permits. It was also required to obtain state and local permits, particularly for the transmission lines that would carry electricity undersea from the wind farm to Yarmouth and then underground to Barnstable. In total, Cape Wind Associates was seeking some twenty permits or certificates, and each one offered opponents an opportunity to delay or block the project on substantive or procedural grounds, or both.

Opposition to Cape Wind Forms

Between the announcement of the proposal in 2001 and the end of 2002, vigorous opposition to the Cape Wind proposal arose. Recognizing that if they were going to have any impact they would need to be organized, opponents of the project formed the Alliance to Protect Nantucket Sound. Whereas the developers portrayed Cape Wind as a remedy for climate change, opponents characterized the wind farm itself as the problem. The alliance raised a variety of concerns, from impacts on birds and the marine environment to interference with aircraft flights and boat navigation. But the two objections that took particular hold concerned aesthetics—industrialization of a "pristine" landscape—and ownership—allowing a private developer to appropriate a public space with inadequate regulatory oversight. "A good portion of us who migrated to Cape Cod came to enjoy Nantucket Sound," explained Wayne Kurker, owner of the 180-slip Hyannis Marina and a cofounder of the alliance. "And if Nantucket Sound becomes an industrial, electrical generation area, then it's no longer the national treasure that people currently feel it is. We look at this as our wilderness, our national park."[41]

Isaac Rosen, the first executive director of the alliance, distributed flyers that captured the group's core objections to Cape Wind: "Private developers want to destroy one of the Cape's natural treasures and turn it into an industrial complex." To advance this perspective, the alliance began circulating computer-generated images of Nantucket Sound that depicted it as a major industrial site, with rows of turbines surrounded by construction barges and maintenance platforms. Voicing a refrain that would become commonplace in the coming years, retiree Sid Bennett told journalist Bob Wyss, "I think that building windmills is an excellent idea. I just don't think [Nantucket Sound] is a good place to put it."[42]

The alliance served as the mouthpiece for a host of disparate interests. A variety of local environmental groups expressed concerns about the toll of construction on marine life, noting that Nantucket Sound is a summer haven for endangered turtles, such as leatherbacks and loggerheads, and is frequented by harbor seals, porpoises, and humpback whales. The area's fishermen, many of them represented by the Massachusetts Fisherman's Partnership, pointed out that Horseshoe Shoal is a spawning ground for squid, flounder, sea and striped bass, and bluefish. And the Massachusetts

Audubon Society raised the possibility that migrating birds might be attracted to turbine warning lights or become disoriented by the towers during storms. Given that at least half the population of East Coast seabirds spends half the year on Nantucket Sound, this was not a trivial concern.[43]

For area merchants, the primary issue was the project's impact on tourism. The cape's tourism industry was worth $1.3 billion a year, according to the Cape Cod Chamber of Commerce, and was the region's largest employer. The Egan Foundation—prominent opponents of Cape Wind—funded a series of economic studies by the Beacon Hill Institute at Suffolk University, the first of which claimed that Cape Wind would cost Cape Cod's economy at least $64 million per year and 1,300 jobs because of reduced tourism.[44] Also weighing in were "expert sailors," who said that Cape Wind could jeopardize Nantucket Sound's standing as one of the premier sailing destinations in the world.[45] They contended that Cape Wind would disrupt sailing competitions in the area, both by presenting obstacles and by interfering with wind currents. Particularly vocal were organizers of the Figawi regatta, an annual event that attracts nearly 250 racing boats, as well as 3,000 visitors.

For their part, Cape Cod's commercial mariners and recreational sailors worried about the project's impact on navigation. The region's two major ferry lines, the Steamship Authority and Hy-Line Cruises, raised concerns about passenger ferries running between Martha's Vineyard, Nantucket, and the mainland, particularly their ability to divert from standard routes due to traffic or bad weather and increased traffic being squeezed into a smaller area as a result of the project. In July 2002 another party joined the fray: the union representing traffic controllers at a Cape Cod air force base charged that Cape Wind Associates had inaccurately depicted flight paths around the proposed wind farm. They subsequently argued that the wind turbines would pose a threat to small aircraft; about 1,000 aircraft per day crowd cape and islands airspace during peak summer months, the union said.

In addition to challenging the project's environmental, aesthetic, and navigational impacts, the opposition sought to discredit Cape Wind Associates' projections of long-term cost savings and suggested, instead, that Cape Wind was a financial boondoggle. In late July 2003 the alliance released a report prepared by a consultant that suggested Cape Wind could cost as much as $66 per MWh, compared to $42 per MWh for gas-fired power. Without the production tax credit, the cost would rise to $85 per MWh.[46] In another study funded by the Egan Foundation, the Beacon Hill Institute reported in March 2004 that the economic benefits of the wind farm (which it estimated at $753 million) would fall short of the costs to build and run it ($947 million).[47] An op-ed piece in *The Boston Globe* by two of the study's authors concluded that Cape Wind would be "a highly subsidized but socially inefficient enterprise."[48]

The alliance had unusual influence for a citizens' organization because it enjoyed the support of a host of wealthy denizens of Cape Cod and the islands.[49] Thanks to its well-heeled contributors, the alliance was able to wage a sophisticated and relentless campaign against Cape Wind; in its first fourteen months, the group raised $2 million in contributions—money it spent on TV, radio, and print ads, as well as yard signs, studies, and legal actions.[50] As the conflict dragged on, the alliance continued to have a formidable fundraising capacity: it took in $4.8 million in 2004 and at least $3.7 million in 2005, with 94 percent of its money coming from 93 "major donors" who gave $20,000 or more.[51]

In promoting their position Cape Wind's critics benefited not just from their ability to tap wealthy donors; they also had the support of influential public figures. Although he was not a member of the alliance, a tacit supporter of the opposition was the veteran Democratic Massachusetts senator, Edward M. (Ted) Kennedy, who had consistently supported subsidies for alternative energy development but whose front porch would look out over the proposed wind farm. Rather than citing the impact on his property, Senator Kennedy emphasized the lack of federal oversight: along with Rep. William Delahunt, D-Mass., and Attorney General Tom Reilly, Kennedy argued that allowing a private developer to build a massive wind farm in the absence of any kind of regulatory framework would be inappropriate and that the benefits and costs of such a project needed to be fully explored before granting permission to build it. Shortly after Cape Wind was proposed, Kennedy inserted a little-noticed amendment into the federal energy bill that called for an NAS study to assess existing federal authority over the development of wind, solar, and ocean-based energy on the outer continental shelf and make suggestions for tighter regulatory review by federal agencies.[52] (The amendment failed.)

Another high-profile detractor of Cape Wind was Ted Kennedy's nephew, Robert F. Kennedy Jr., whose entrance into the debate flummoxed some environmentalists. Kennedy, the founder of the environmental group Riverkeeper, insisted that he was an ardent supporter of offshore wind-energy production but that Nantucket Sound was simply not an appropriate location. It was, he said, one of the state's most important assets and some people's "only access to wilderness."[53] "I am all for wind power," Kennedy insisted in a radio debate with Jim Gordon. But "[t]he costs . . . on the people of this region are so huge, . . . the diminishment to property values, the diminishment to marinas, to businesses. . . . People go to the Cape because they want to connect themselves with the history and the culture."[54] In an op-ed piece in *The New York Times*, Kennedy compared Nantucket Sound to Yosemite and disparaged Cape Wind as a government-subsidized industrial boondoggle.

Further enhancing the clout of Cape Wind's opponents, in late January 2003 the venerable newsman Walter Cronkite, who had a home on Martha's Vineyard, appeared in a TV ad commissioned by the alliance. In late August 2003 Cronkite reconsidered his opposition to Cape Wind and asked the alliance to pull the ads he had made.[55] In the meantime, however, another high-profile supporter had weighed in: Pulitzer prize-winning author David McCullough, who had lived on Martha's Vineyard for thirty years. McCullough's radio ad decried the "scheme" to build "a sprawling industrial factory" in the middle of "one of the most beautiful unspoiled places in America."[56] McCullough rejected the charge of NIMBYism, pointing out that he had no view of the sound from his home. He added that he had been involved in lots of other fights to preserve beautiful places, including the Manassas National Battlefield in northern Virginia.

Cape Wind's Defenders Fight Back

Cape Wind had vocal supporters of its own, albeit none as eminent as the Kennedys, Cronkite, or McCullough. Gordon himself was the most compelling spokesperson for the project. Gordon began any discussion of Cape Wind by touting the project's environmental benefits: he pointed out that Cape Wind could offset more than 1 million tons of CO_2, noting that to produce as much power as Cape Wind, an electricity-generating plant would have to burn 85 million gallons of oil or 500,000 tons of coal.[57] With respect to the project's impact on birds, Gordon claimed the turbines would be designed to minimize bird kills, so they would be no more lethal than plate-glass windows. Oil spills and global warming pose a bigger threat to the avian world, he pointed out.[58] Gordon mustered other reasons for the project as well, most often energy security. After the September 11, 2001, terrorist attacks, he pointed out, "We have an inexhaustible supply of wind off the coast of Massachusetts, and no cartel can economically manipulate or cut off that supply. We have to reduce our reliance on foreign oil," he added. "The geopolitical price is just too high to pay."[59]

In response to criticisms of the project's visual impacts, Gordon said: "There are a lot of people who look at wind turbines and see them as a study in power and grace and a visual testimony to us working with nature. So there are a lot of people who enjoy looking at wind turbines. For those who don't, we've significantly reduced the visual impact" by taking them offshore.[60] For Cape Wind Associates and its supporters, Denmark offered a successful example of offshore wind near population centers with strong local buy-in. Of particular interest was the Middelgrunden wind farm in Copenhagen harbor, comprising forty turbines with a capacity of 40 MW. For many, Middelgrunden was a visually pleasing addition to the city. As one journalist described the view, "From the harbor in Copenhagen, you see a wall of 200-foot windmills that dominates the horizon, their rotors silently spinning and glinting in the sunshine as sailboats and fishing trawlers glide past."[61]

Gordon noted that studies had found no damage to local economies from wind turbines; in fact, Denmark found that tourists came specifically to see the wind farms. And a federally funded review of 25,000 real-estate transactions by the Renewable Energy Policy Project found no drop in property values near wind projects in the United States. As for the decision to locate the country's first major wind farm in Nantucket Sound, he argued that Horseshoe Shoal was an ideal site. "If we can't do it here," he said, "we can't do it anywhere"—a brilliant formulation, given that he could hardly have picked a site that was more challenging politically.[62] Of the opposition, Gordon was dismissive, saying, "These fat cats with waterfront estates go to cocktail parties and claim they're all for renewable energy. But when it comes to their own views, it is pure nimbyism."[63]

Gordon shrewdly added another dimension to his arsenal of arguments for Cape Wind: regional employment. Early on, Gordon made clear that he would like to build the turbines locally. One site he was considering in June 2002 was the dormant Fore River Shipyard in Quincy, Massachusetts, which already had a skilled workforce to draw from, as well as the huge cranes and steel fabricating buildings needed for production, proximity to the ocean, and a rail line linked to cross-country tracks.[64] In early April 2003 Cape Wind Associates released a study conducted by the consulting firm Global Insight that claimed Cape Wind could generate at least 597 full-time jobs and $6.1 million in revenue during construction. According to the authors, the project could generate up to 1,013 jobs and tax revenues could reach $10 million before the turbines became operational. Another 154 jobs should be created once the plant was running, and annual tax revenue was estimated at $460,000.[65] "This is going to be the largest economic-development project in Southern New England," Gordon claimed.[66]

For many, the employment argument was compelling: in late April 2003 hundreds of fishermen, maritime workers, and union members endorsed Cape Wind for the first time.[67] In February 2005, after Cape Wind Associates selected General Electric (GE) as its turbine manufacturer, blue-collar workers at GE's Riverworks plant in Lynn, Massachusetts, passed out petitions and met with lawmakers to convey their support for the project. (Cape Wind Associates subsequently chose a similar Siemens turbine that would be manufactured in Europe.) Also moved by job claims was the Republican governor of Rhode Island, Donald Carcieri. He wrote to the administration of President George W. Bush, asking it to approve permits for Cape Wind because the project would create jobs in Quonset, Rhode Island, a possible construction site.

Although they withheld final judgment until the environmental review process was complete, several national environmental groups—including the Natural Resources Defense Council, Clean Water Action, and the Union of Concerned Scientists—as well as the New England-based Conservation Law Foundation pressed for Cape Wind to get a fair hearing. Greenpeace USA campaigned actively for the project from its inception: its workers walked Cape Cod beaches preaching green energy to sunbathers; in addition, it paid for prowind energy spots on cable, toured parades and fairs across the cape with a solar-powered truck, and hosted a speech by an Arctic explorer warning about the hazards of climate change. Bolstering Cape Wind's efforts to demonstrate it had local support was a local group, Clean Power Now. Led by Matt Palmer, a power-plant engineer who previously had worked for Gordon, the group claimed more than 100 members in June 2003—although one contributor accounted for most of the group's revenue. Another local group, the Coalition for Buzzards Bay, mobilized in support of Cape Wind after an April 2003 oil spill leaked nearly 100,000 gallons of oil into Buzzards Bay, coating ninety-three miles of beaches and highlighting the costs of conventional energy.

The Tortuous Permitting Process Begins

Although battles have erupted in many places where alternative energy projects are proposed, the struggle over Cape Wind was unusually prolonged and garnered extensive media coverage, both locally and nationally. The project's opponents raised both substantive and procedural objections to the project, seeking to capitalize on every possible veto point. Although Cape Wind garnered substantial support at the federal level, at the state level it faced an uphill battle because the state's Republican governor, Mitt Romney, strongly opposed it.

Federal Permits and the Draft Environmental Impact Statement. The first serious skirmish between Cape Wind Associates and its opponents came in the summer of 2002, when a ten-member coalition—including the Ocean Conservancy, the Earth Institute, the International Wildlife Coalition, the Humane Society, and the Massachusetts Society for the Prevention of Cruelty to Animals—called for an independent environmental study before Cape Wind Associates could gain approval for a 197-foot test tower that would gather data on wind and water conditions in Horseshoe Shoal. The coalition argued that state and federal reviews of the tower's environmental impacts were superficial and, in any case, were tainted because the applicant paid for them.[68] Despite these complaints, on August 19, 2002, the U.S. Army Corps of Engineers (Corps)—which had jurisdiction over the project—approved plans for the $2 million tower.[69]

The alliance immediately challenged the Corps' authority to issue the permit for the tower, claiming in federal court that the Corps had violated one of its own regulations, a requirement that developers control the land or water they want to build on. Another set of plaintiffs, the Ten Taxpayer Citizens group and several fishing and boating concerns, filed suit in state court.[70] But U.S. District Court Judge Joseph Tauro was unconvinced by opponents' arguments and allowed work on the tower to proceed, and by mid-December 2002 the tower was complete and data collection had begun.

Even as opponents appealed the test-tower decision (unsuccessfully), the Corps began the public "scoping" process for a combined environmental impact statement/environmental impact report (EIS/EIR) for the wind farm itself.[71] (The EIS would satisfy federal requirements; the EIR would satisfy state requirements.) Among the questions raised were these: What sorts of hazards will the turbines pose for birds, particularly endangered species? How will the towers affect commercial fishing vessels and recreational boaters? Will they disturb radio and TV frequencies used for navigation? And what will they look like from shore? As part of the EIS process, the Corps asked Cape Wind Associates to investigate reasonable alternatives to building the wind farm, as well as alternative locations to Horseshoe Shoal. Eventually, the Corps required a full analysis of six alternative sites—three in shallow water, one in deep water, and two smaller projects—as well as one land-based alternative and a no-build option.[72]

On January 22, 2003, the *Cape Cod Times* reported that Cape Wind Associates had scaled back its proposal by 23 percent, from 170 towers to 130, because efficient new turbines would allow them to generate the same amount of power with fewer structures. (Each of the new turbines would produce 3.6 MW of power, up from the 2.7 MW per turbine originally planned.) Reducing the number of turbines enabled Cape Wind Associates to move the entire complex about a half-mile further from the shores of Cape Cod, Martha's Vineyard, and Nantucket. The towers would also be slightly smaller, rising 417 feet above sea level, rather than 426 feet; the diameter of the turbines at sea level would shrink from 20 feet to either 16.75 feet or 18 feet, depending on the water depth; and the footprint of the entire project on Horseshoe Shoal would be reduced from twenty-eight square miles to twenty-four square miles. The alliance was unmoved by the new proposal. "The number doesn't make any difference," said associate director John Donelan. "It's the location we have a problem with and the total lack of federal standards governing wind power projects."[73]

Although they lacked formal veto power, local politicians were creative in their efforts to obstruct Cape Wind. In mid-October 2003 Attorney General Thomas Reilly, one of the most vocal opponents of the project, sent a letter urging the Corps to halt its review of the project until questions about its legal basis were fully addressed. Reilly also appealed to Rep. Barbara Cubin, R-Wyo., chair of the U.S. House Energy and Mineral Resources subcommittee, to add more teeth to a federal energy bill she was proposing that would require leases for "nonextractive" uses of the outer continental shelf. Reilly wanted the bill to include additional environmental safeguards, mandates for meaningful state input, and a competitive bidding process for developers. Not to be outdone, Representative Delahunt vowed to do "whatever is necessary to protect Nantucket Sound from industrialization."[74] To this end, he revived dormant legislation to make Nantucket Sound a national marine sanctuary, a designation that could block Cape Wind or any other development on the seabed.[75] In early February 2003 Delahunt's proposal got a boost when the Center for Coastal Studies in Provincetown released a study concluding that Nantucket Sound was a valuable coastal resource that warranted preservation. The sound "remains a pristine and tremendously productive ecosystem worthy of environmental conservation and protection," the group wrote.[76] (Delahunt's legislation failed.)

The Public Relations Battle Heats Up. With the controversy surrounding Cape Wind becoming more fractious by the day, the Massachusetts Technology Collaborative, a quasi-public agency that runs the state's Renewable Energy Trust, held six meetings from October 2002 to March 2003 on Cape Wind, ostensibly to serve as an "honest broker" that could produce some sort of consensus. Despite the efforts of the technology collaborative to calm the debate, during the summer of 2003 the battle between Cape Wind's allies and opponents became more vitriolic. Many journalists and their editors depicted the struggle over Cape Wind as a battle between selfish elites and an altruistic developer and Nantucket Sound as an "exclusive summertime playground." Among the headlines about the controversy were these: "Rich and Famous Lining Up to Protect Nantucket Sound" and "Celebrities in a Huff Over Cape Wind Farm." Even the *Times* of London picked up this theme, in an article titled "Wind Farm Threatens Kennedys' Playground." The *Times* noted ironically that many of the mansions overlooking Nantucket Sound bore signs declaring "Nantucket Sound is not for Sale."[77] In a particularly scathing article in *The New York Times Magazine*, journalist Elinor Burkett pointedly ridiculed the opposition to Cape Wind. Burkett was emphatic that this was no David-and-Goliath story: the alliance included some of the wealthiest and most influential businesspeople in the United States, many of them with ties to fossil fuel companies; its attorneys and lobbyists were employed by high-powered D.C. firms.[78]

Caricatures notwithstanding, Cape Wind presented an agonizing choice for many in the land- and wildlife-conservation community, and the project sharply divided residents of the Cape and Islands region. A poll commissioned by Cape Wind Associates in the fall of 2002 found that 55 percent of Cape and Island voters supported the wind farm project, while 35 percent were opposed and 11 percent were undecided.[79] But another poll, commissioned by the *Cape Cod Times* and the WCAI/WNAN radio stations, found that of the 80 percent of respondents who actually answered the question, 55.1 percent opposed the project, while 44.9 percent supported it.[80] A poll taken by the University of Delaware in early 2005 yielded similar results: 56 percent of respondents from the Cape and Islands opposed the project.[81] In 2005, four years into the debate, the *Cape Cod Times* and WCAI commissioned a third poll, which revealed a dead-even split between supporters and opponents: 39 percent opposed the project, 37 percent backed it, and 24 percent were uncertain.[82] Polls consistently showed that, as with other wind farm siting projects, Massachusetts residents as a whole were far more likely to support Cape Wind than were nearby residents, and those who lived nearest the Cape Wind site were the most likely to oppose it.

The Draft EIS Is Finally Released. After a series of delays that frustrated proponents of Cape Wind, in November 2004 the Corps released its 4,000-page draft environmental impact statement (DEIS) for the project. The report asserted that Cape Wind would do little or no permanent harm to fish, birds, and the seafloor and would not drive down property values on the cape and islands. The draft estimated the turbines could kill as many as 364 birds each year, a number it claimed was unlikely to affect endangered species or specific populations of birds.[83] Effects on fish and shellfish populations were likely to be felt only during construction. Navigation would not become more hazardous, although boaters would have to take extra care in the vicinity of Horseshoe Shoal. The report acknowledged that the wind farm would mar scenic views in a small number of historic areas, including the Kennedy compound and the Nantucket Historic District. But overall, it predicted the wind farm would create jobs and increase tourism, and it would yield public health benefits worth $53 million annually. Several major environmental groups took the occasion to praise the Corps, an agency they usually regarded as a foe, and to endorse the Cape Wind project formally. But the alliance largely dismissed the Corps' findings, pointing out that Cape Wind had paid the $13 million cost of the report (although the Corps chose the experts who conducted the analyses), and vowed to conduct its own analysis.

The release of the DEIS triggered a lengthy public comment period, during which the Corps hosted four public hearings: three on the cape and islands and one in Cambridge. At the first public meeting, at Martha's Vineyard Regional High School, participants delivered a series of short speeches one after the other, and little new ground was broken. A subsequent meeting, held at the Massachusetts Institute of Technology (MIT), was more heavily tilted in favor of Cape Wind and featured more entertainment: a half-dozen protestors showed up wearing captain's hats or ball gowns, mocking opponents of Cape Wind as wealthy and self-interested. The protestors chanted, "Save our Sound. Save our Sound. Especially the view from my compound." A man calling himself Preston Cabot Peabody III held a sign that read, "Global Warming: A Longer Yachting Season."

While the Corps reviewed and crafted responses to some 5,000 comments on the DEIS in spring 2005, opponents of Cape Wind looked for ways to derail the project at the state level. For example, Governor Romney sought to expand the state by twelve square miles, thereby gaining jurisdiction over the wind farm. Contending that a recently discovered rock outcrop off the coast of Yarmouth was part of Massachusetts, the Romney administration adjusted its three-mile boundary further into the sound, putting ten proposed turbines within state rather than federal waters. Cape Wind responded by relocating the turbines, thereby moving the project one-half mile further from Point Gammon, the area closest to the development. (The reconfigured proposal also shifted another twenty turbines to accommodate concerns about commercial fishing, navigation, and archaeological interests.) In addition, Governor Romney proposed legislation to zone state water like private land, in an effort to protect the coast and oceans from offshore drilling and commercial development.[84]

Despite Romney's efforts to impede it, Cape Wind continued to advance. In May 2005 the state's Energy Facilities Siting Board (EFSB), which is charged with ensuring the state has reliable energy supplies with minimal impact on the environment and at the lowest possible cost, voted five to two to approve the laying of underwater transmission lines between Cape Wind and the Cape Cod shoreline. (The alliance appealed the decision, arguing that the board had made several procedural and substantive errors.[85]) Then, in late July 2005, President Bush signed a comprehensive energy bill that renewed the PTC for two years and set up a mechanism by which the federal government could charge renewable energy projects for the use of federal waters. The new law also included language that allowed Cape Wind to avoid competitive bidding for Horseshoe Shoal, a grandfathering provision justified by the fact that Cape Wind had applied before the new rules were in place.

Although favorable to Cape Wind in most respects, the energy law also cheered Cape Wind opponents because it transferred lead federal authority for offshore wind projects, including Cape Wind, from the Corps to the Minerals Management Service (MMS) in the Department of the Interior. The shift in responsibility promised more delays, which opponents of Cape Wind hoped to use to frustrate the developers into giving up.[86] (The MMS planned to issue a new DEIS by May 2006, hold another round of public hearings over the summer of 2006, and issue a final decision by the fall.) Meanwhile, as the MMS geared up for its review, the maneuvering in Congress continued. In mid-September 2005 House Transportation and Infrastructure Committee Chair Don Young, R-Alaska, inserted into the $8.7 billion Coast Guard authorization bill an amendment calling for the commandant of the Coast Guard to furnish a written opinion on whether any proposed offshore wind farm would interfere with navigation. When that amendment failed, Young tried another tack: in conference committee on the bill, he offered language that would prohibit wind turbines within 1.5 miles of shipping and ferry lanes, on the grounds that wind turbines could interfere with shipping and radio transmissions.[87] The provision was clearly aimed at Cape Wind, whose turbines would stand within 1,500 feet of the shipping lane to the south of Horseshoe Shoal and within one nautical mile of the ferry route to the east.

Young's amendment received a deluge of unfavorable attention from newspaper editors along the East Coast, including *The New York Times*, and was eventually dropped. But in early April 2006 Sen. Ted Stevens, R-Alaska, asked the conference committee to consider language that would allow state leaders to veto a project if they believed it would obstruct navigation. Again, the provision was narrowly written and clearly targeted Cape Wind. This measure drew abundant unfavorable publicity as well, and *The Boston Globe* wrote a fierce editorial echoing Gordon's characterization of his detractors:

> The country's most important renewable energy project is in danger of being sandbagged in Congress. If Congress accepts the bill with the veto power for Romney, it would be a victory for the project's well-heeled opponents on the Cape and Islands, who have funded the lobbying campaign waged against Cape Wind in the backrooms of Congress. . . . Congress should not let NIMBY property owners on the Cape and Islands deal a body blow to the most promising plan yet for steering the country away from uncontrolled global warming and dependence on foreign oil.[88]

Despite these criticisms, the measure ended up garnering enough support to be attached to the final bill, which had to be approved by the House and Senate, prompting a furious lobbying effort by Cape Wind supporters. Although Cape Wind Associates howled about the "special-interest provision" tucked into legislation behind closed doors in response to lobbying, critics pointed out that the project itself had benefited from such provisions.[89] Suspecting that Stevens had attached the measure at the behest of Senator Kennedy, Greenpeace began running a TV ad in six Northeast states (although not Massachusetts) mocking Kennedy's efforts. The spot featured a giant cartoon caricature of Kennedy standing shin-deep in Nantucket Sound and hitting windmills with a mallet as they popped out of the water. The ad urged viewers to call their senators and tell them, "It's not about the view from Ted Kennedy's porch. It's about a vision of America's clean energy future."

Important figures from the New England region weighed in against the measure as well. In early May 2006 ISO (Independent System Operator) New England, the nonprofit organization that runs New England's power grid and oversees its wholesale electric markets, urged Congress not to block Cape Wind, pointing to the region's "perilous overreliance on natural gas."[90] And in mid-May sixty-nine state lawmakers, many of whom ordinarily were loyal to Kennedy, sent Congress a letter urging legislators to oppose the amendment. (Only one representative from the cape and islands—staunch Cape Wind supporter Matthew Patrick—signed the letter.) Finally, in late May 2006, when it appeared that the amendment could scuttle the entire bill, Kennedy dropped his insistence on veto power for the Massachusetts governor and instead accepted a provision that the Coast Guard commandant be given a veto over Cape Wind, after taking into consideration threats to navigation and public safety.

Cape Wind Gains Ground Despite Setbacks

Legislative maneuvering aside, momentum seemed to shift decisively in Cape Wind's favor in 2006. In late March 2006 the Massachusetts Audubon Society gave its preliminary blessing to the wind farm, saying studies showed that turbine blades were unlikely to cause significant harm to birds. Also fortuitous for Cape Wind was the release in early April of a Danish study—the world's most comprehensive analysis of offshore wind farms to date—concluding that the world's two largest offshore projects had little or no environmental impacts. Based on eight years of data, the report's executive summary stated, "Danish experience from the past 15 years shows that offshore wind farms, if placed right, can be engineered and operated without significant damage to the marine environment and vulnerable species. . . . [U]nder the right conditions, even big wind farms pose low risks to birds, mammals, and fish."[91] Among the key findings were that the underwater turbine bases acted like artificial reefs and increased the number of benthic organisms, fish populations neither increased nor decreased, and marine mammals left the area during construction but largely returned after construction ceased.[92] The report also surveyed Danish public opinion and found it generally positive—although, as in the United States, opinion was more supportive nationally than locally; even after learning that environmental impacts were minimal, local residents still wanted future projects to be placed further offshore and out of sight.

The election of Democratic Governor Deval Patrick in early November 2006 was yet another auspicious event for Cape Wind. As a candidate, Patrick had come out in support of the project. "We can't keep saying we understand the problem of energy cost and supply and the problem of global warming," he argued, "and then refuse to act when we have an opportunity to do so." Patrick did not explicitly disparage the project's opponents. But he infuriated them when he said, "I think one of the things a governor can do through his public leadership is to . . . try to explain to people including those who oppose it in good faith that there may be some modest sacrifice that's required of us for the common good."[93] Once elected, Patrick enhanced Cape Wind's prospects by appointing Ian Bowles, a former Cape Cod resident and a Cape Wind supporter, as secretary of energy and environmental affairs.

Although the tide seemed to be turning in favor of Cape Wind, major veto points remained. The Pentagon released a report in late 2006 urging further study of the impact of Cape Wind on military radar systems. The report raised concern that the 2004 Air Force study, which found that Cape Wind posed "no threat"[94] to missile-detecting radar, was oversimplified and technically flawed. The Federal Aviation Administration (FAA) was also planning to take a third look at Cape Wind. And the Coast Guard was scheduled to issue a report in December specifying terms and conditions necessary to ensure navigational safety, as required in its 2006 appropriations legislation.

Obtaining Permits From the State

In late February 2007 Cape Wind Associates filed its final 5,407-page environmental impact report (FEIR) with the state's Executive Office of Environmental Affairs. The report included another set of modifications Cape Wind Associates had made to its proposal, most of which reflected new technological developments: the turbines would be 5 percent taller and 7 percent more efficient than those originally proposed, and blade tips would reach 440 feet rather than 417 feet from the ocean surface. The maximum output of 3.6 MW apiece would remain the same, but the new windmills would capture more wind energy on days with light wind, increasing their total output. Cape Wind Associates also reduced the number of aviation lights from 260 to 57, making the turbines less visible at night. The alliance immediately demanded that the state's thirty-day comment period on the FEIR be extended while pointing out that the report did not address how Cape Wind would be dismantled or who would pay for decommissioning it at the end of its twenty-year lifetime. The alliance also expressed skepticism about the document's claims with respect to birds and fishing. And it contended that the final EIS improperly dismissed a potential threat to national security from the project's interference with military radar systems.

In mid-March the Cape Cod Commission, the regional planning agency for Barnstable County, issued a report that echoed the Alliance's criticisms of Cape Wind's FEIR. Like the alliance, the planning agency wanted more information on decommissioning; it also requested additional detail on the impact of turbines on commercial fishing and the viability of potential alternative sites. The commission strongly recommended that the state demand more information from Cape Wind Associates before accepting the report.[95] (Once the state accepted the FEIR, the commission would have forty-five days to review Cape Wind as a Development of Regional Impact.) But, to the dismay of Cape Wind opponents, Massachusetts's energy and environmental chief Bowles limited the scope of his review to the potential impacts from the installation of underwater transmission cables, deferring more detailed consideration of the project's larger impacts to ongoing federal review, and in late March he asserted that Cape Wind's FEIR passed muster with the state.[96]

The next battleground, therefore, was review by the Cape Cod Commission. Dissatisfied with the amount of information provided to its staff by Cape Wind Associates, the commission unanimously denied the permit. In response, Cape Wind Associates asked the EFSB to overrule the commission's permit denial and issue a composite permit that would cover the Development of Regional Impact permit as well as the eight remaining state and local permits needed to build Cape Wind. If the siting board granted this request, the only outstanding state-level permission Cape Wind Associates would need for siting purposes would be from the Office of Coastal Zone Management, which had to determine whether the project was consistent with the state's overall coastal-zone management plan. Such a strategy would dramatically reduce the number of veto points available to the opposition; rather than being able to appeal to the state Department of Environmental Protection and conservation commissions in Barnstable and Yarmouth, it would only have recourse to the Supreme Judicial Court.

Local officials reacted strongly to the possibility that the siting board might override the Cape Cod Commission: the Edgartown (Martha's Vineyard) selectmen weighed in against Cape Wind, while the Martha's Vineyard Commission planned to file its own request to intervene on procedural grounds (it opposed the precedent of overruling local control). Officials from the town of Barnstable filed a complaint in Barnstable County Superior Court the week of April 11 claiming that the commission had exclusive jurisdiction over the transmission cables and that the siting board did not have the authority to review the commission's denial.[97] After holding hearings on the matter, in late July the siting board ruled that it *was* entitled to review the commission's permit denial, that it would *not* remand the case to the commission, and that its own review (like that of energy secretary Bowles) would be limited to the transmission lines—decisions that were all favorable to Cape Wind.

In May 2009, after more than a year of deliberation and fact finding, the siting board voted unanimously to approve the composite permit for Cape Wind, a major milestone for the project. The alliance promptly announced that it would appeal the decision to the Massachusetts Supreme Judicial Court. The issue was no longer Cape Wind, according to Alliance Executive Director Audra Parker. It was the precedent that would be set if the project were approved. If the state could override the wishes of communities and the regional planning body, it could do so on other issues as well; the subversion of local control, she said, would affect everyone in the state.[98]

While state and local agencies were debating Cape Wind's fate, the Massachusetts Legislature passed several laws that further improved the prospects for the project. First, in 2008, the legislature passed the Green Communities Act, which included two critical provisions: one that increased the amount of renewable energy to be sold by electricity retailers over time (the RPS) to 15 percent by 2020 and one that required electricity-distribution companies to enter into long-term contracts with renewable energy developers to help them get financing for their projects. Second, the Global Warming Solutions Act called for a reduction in Massachusetts's greenhouse gas emissions over time. Third, the Oceans Act allowed the siting of "appropriate scale" offshore renewable-energy facilities in state waters. Fourth, in 2016, the Act to Promote Energy Diversity required utility companies to procure a combined 1,600 megawatts of electricity from offshore wind farms within ten years.[99]

Moving Toward a Federal Permit

Running in parallel with the state process was Cape Wind's effort to garner a set of federal permits. After repeated delays, in early January 2008 the MMS issued its 2,000-page DEIS for Cape Wind. Like its predecessor written by the Corps, the new DEIS claimed that the project would have no lasting adverse impacts on wildlife, navigation, fishing, tourism, or recreation. On the other hand, it recognized that there would be a moderate impact on "visual resources" from the shore and a major one for boats in proximity to the turbines.[100] And it acknowledged that the project "represents a large, manmade feature in the natural landscape of Nantucket Sound that would be viewed by many people in numerous shoreline areas used for recreation." Apart from its visual impacts, the most significant negative identified was a potentially moderate impact on coastal and marine birds once the turbines were up and running.

Although the DEIS was generally favorable, the issue of air traffic remained unresolved: the FAA had issued a "notice of presumed hazard" for the project because it would hinder air-traffic control radar systems.[101] In addition, the Steamship Authority, Hy-Line Cruises, and the National Passenger Vessel Association were concerned about the interference with their boats' radar systems; at a minimum, the Steamship Authority complained, the ferry lines might have to alter their routes, which would add to their fuel costs. Fishermen were particularly exercised about the report, which they said had gotten the fishing situation on Horseshoe Shoal completely wrong. According to the fishermen, the DEIS had conch landings maximized at 300,000 pounds and in decline since 2000, which was false. In fact, they said, the value of the fishery was increasing and was the largest by volume on the cape and islands. The vineyard alone was landing between 2 million and 3 million pounds per year, primarily from the Horseshoe Shoal area. In addition to conch, the fishermen added, there were squid, sea bass, flounder, and fluke on Horseshoe Shoal. (An economic analysis conducted for the Massachusetts Fisherman's Partnership concluded that Cape Wind would cost the region's fishing industry between $8 million and $13 million.[102]) Fishermen feared that pile driving for each turbine as well as dredging six-foot trenches to lay wires between the turbines and from the turbines to the mainland would have devastating effects on the fishery.

In addition to the fishermen, the Mashpee and Aquinnah Wampanoag tribes were deeply troubled by the DEIS.[103] The Wampanoag, whose name means People of the First Light, pointed out that the 440-foot turbines would hinder their traditional spiritual practices, which rely on an unobstructed view of the eastern horizon. The tribes also feared that construction of Cape Wind would obliterate sacred burial sites.[104] Working with the alliance to publicize their concerns, the tribes argued vehemently throughout the process that the MMS did not engage in adequate government-to-government consultation, as required by their status as sovereign nations.

While the MMS considered these objections, it convened another round of public hearings in Hyannis/Yarmouth, Nantucket, Martha's Vineyard, and Boston. Although the public meetings, held in mid-March 2009, revealed little new information, the testimony did suggest an increasing emphasis among Cape Wind opponents on tribal and historic concerns.[105] Cheryl Andrew-Maltais, chair of the Wampanoag tribe of Gay Head (Aquinnah), read a statement from a consortium of twenty-five federally recognized tribes, urging the MMS to "deny the permitting of such a devastating and destructive experiment, which will adversely affect and destroy the essence of tranquility, sanctity and spirituality of this Sacred Place for all time."[106] There was also a twist: shortly before the Boston hearing, a company called Blue H USA LLC announced a plan to build a 120-turbine, floating, deep-water wind-energy farm twenty-three miles off the coast of Martha's Vineyard and forty-five miles from New Bedford. Blue H's parent company had launched its first offshore floating wind turbine, an 80 kW demonstration project, off the southeastern coast of Italy in December; it planned to move their turbine about ten miles offshore to water more than 350 feet deep. Blue H teamed up with Cape Wind's opponents in touting its project as an alternative to Cape Wind, though in fact no one knew whether a floating wind farm was viable.[107] Although Blue H quickly receded into the background, the company's proposal lent credence to the notion that alternative locations and technologies might provide a solution to the seemingly intractable Cape Wind debate.

Finally, after more delays, on January 16, 2009, Cape Wind cleared its most significant hurdle by garnering a favorable review from the MMS. Much like the draft, the final EIS found mostly "negligible" and "minor" effects of the project on the environment, public safety, and fisheries, and "moderate" impacts on some bird species, marine mammals, visual resources, marine traffic, and marine radar. Appended to the document was an opinion by the Fish and Wildlife Service (FWS) that Cape Wind would not "jeopardize the continued existence" of piping plovers or roseate terns, both of which are protected under the Endangered Species Act. In addition, the Coast Guard commander responsible for the waters off southeastern New England certified that Cape Wind posed no significant problem for marine radar.[108] In response to public comments on the draft, the final EIS had increased the number of historic views that would be adversely affected from three to twenty-eight, and it acknowledged that the project would hamper "traditional religious and ceremonial practices" of the Aquinnah and Mashpee Wampanoag.

The final EIS was not the last word, however; the MMS could issue a lease for the project only after it produced a "record of decision." To do that, it had to await the outcome of the consultation process required by Section 106 of the National Historic Preservation Act. Interior secretary Ken Salazar promised to intervene personally in negotiations between Cape Wind Associates and the Wampanoags, and if the two sides could not reach an agreement, he would make the decision himself.[109] Meanwhile, the tribes were seeking to have the National Park Service designate all of Nantucket Sound as cultural property and list it on the National Register of Historic Places—a move that would further complicate Cape Wind's efforts to garner permits. The Cape Wind record of decision also awaited a ruling by the FAA on interference with airplane radar, a statement from the Coast Guard on marine radar, and the results of an investigation into the environmental review process by the inspector general for the interior department requested by Senator Kennedy and the alliance.[110]

For Cape Wind Associates the roller coaster ride continued when, in early January 2010, the National Park Service announced that Nantucket Sound *was* eligible for listing on the national register as both a "traditional cultural property" and a "historic and archaeological property" based on its association with the ancient and historic period of exploration by Native Americans. Adding to Cape Wind's woes, in mid-February the FAA released a report saying that Cape Wind *would* interfere with air-traffic control radar systems. But the agency suggested a solution: Cape Wind Associates could pay for improvements to one of the three air-traffic radar systems covering Nantucket Sound, at a cost of about $1.7 million; if that upgrade was unsuccessful, then a completely new system, costing $12 million to $15 million, might be needed.

Other news was more hopeful for the developers: Cape Wind got new support from unexpected quarters. In early December 2009 the University of Delaware released a report showing for the first time that more than half of the people living on the Cape and Islands (57 percent of respondents) supported Cape Wind.[111] Then, in early April 2010, seventy-eight Massachusetts lawmakers—almost half the state legislature—urged Secretary Salazar to approve Cape Wind "as soon as possible." (Again, Representative Matthew Patrick was the lone signatory for the cape and islands delegation.) And in mid-April the Association to Preserve Cape Cod threw its support behind Cape Wind; the forty-year-old organization's seventeen-member board of directors voted unanimously to support the project after hearing from both sides and reviewing the final EIS.

OUTCOMES

With the release of the final EIS, the environmental arguments against Cape Wind seemed to lose traction, and the alliance returned to what it hoped would be its strongest case of all: the claim that Cape Wind was not cost-effective and would raise electricity rates. By 2006 Bill Koch, cochair of the alliance, had emerged as the high-profile spokesperson for opposition to Cape Wind on financial grounds; as a successful energy developer himself, Koch had some credibility on this issue. In an op-ed piece in *The Wall Street Journal* in late May 2006, he claimed that the arguments over Cape Wind had been missing a crucial element: "a rational analysis of the Cape Wind project's effects on the supply of and demand for electricity in New England." He concluded:

> When you do the math, it is clear that every other form of power generation would be cheaper to build, produce more electricity at a consistent rate and save consumers more money. When you consider the costs and risks of an offshore wind farm, and the fact that New England does not need more power, the project becomes nonsensical, a giant boondoggle for the benefit of one developer.[112]

To bolster its position, in January 2010 the alliance released a poll, conducted by the University of Massachusetts Dartmouth Center for Policy Analysis, which found that support for a wind farm in Nantucket Sound dropped if customers' electric bills went up by $100 per year or more.[113] "Cape Wind's oversized costs do not represent a reasonable return on the public investment," Joseph P. Kennedy III wrote in a letter to the *Cape Cod Times* in February 2010.[114] Representative Delahunt added, "This will be the most expensive and heavily subsidized offshore wind farm in the country, at over $2.5 billion, with power costs to the region that will be at least double [the current rates]."[115] Cape Wind Associates responded to these charges by commissioning its own analysis: on February 10 it released a study by the Charles River Associates that claimed Cape Wind would save New England $4.6 billion in wholesale electric costs over twenty-five years. But critics pointed out that the study assumed the implementation of a federal cap-and-trade program and a carbon cost of $30 per ton in 2013 and $60 per ton in 2030. It also assumed that the PTC would be retained, which was unlikely in the event of carbon regulations.

As debates raged over Cape Wind's costs, a final decision from the interior department loomed. When Secretary Salazar's March 1, 2010, deadline arrived, the Wampanoag tribes and Cape Wind Associates were still at odds. Salazar immediately terminated the consultation process and instead requested a final recommendation from the Advisory Council on Historic Preservation (ACHP), a federal agency that promotes the preservation of historic resources. One month later, the ACHP panel recommended that Salazar "not approve the project," having determined that Cape Wind's impact on thirty-four historic properties would be "pervasive, destructive, and, in the instance of seabed construction, permanent." The damage, the panel said, "cannot be adequately mitigated at the proposed site." The ACHP suggested that "the selection of nearby alternatives might result in far fewer adverse effects to historic properties, and holds the possibility that those effects could be acceptably minimized or mitigated."[116] The ACHP's judgment notwithstanding, on April 28 Salazar approved Cape Wind, hastening to add that he was requiring the company to modify the turbines' color and configuration to mitigate their visual impacts.[117] In May the FAA determined the project posed "no hazard" to planes or radar, after Cape Wind agreed to its terms, and in early August it rejected a request for a review of that decision. In late October the secretary signed a twenty-eight-year lease with Cape Wind Associates. Six groups immediately filed suit, arguing the project would "exact a terrible toll" on federally protected migratory birds and possibly whales.[118]

Meanwhile, the march to finalizing approval for Cape Wind proceeded at the state level. A power-purchase agreement was essential for Cape Wind to persuade investors there would be customers for its power, and in spring 2010 National Grid and Cape Wind Associates struck a tentative deal in which National Grid would buy half the power from Cape Wind for 20.7 cents per kWh, plus a 3.5 percent annual increase—significantly higher than the general rate of 8.1 cents per kWh the utility was paying other suppliers in Massachusetts.[119] In late July the state's attorney general intervened to reduce the starting price by 10 percent, to 18.7 cents per kWh, which was lower but still well above the price of other sources. The Department of Public Utilities (DPU) spent the summer of 2010 reviewing the power-purchase agreement, holding three hearings across the state to gather public comments on the proposed deal. In the midst of those hearings, Walmart weighed in to argue that Cape Wind was not cost-effective and that the above-market cost of power would be unfairly distributed among all of National Grid's distribution customers rather than just those who buy their power from the utility.[120] The Associated Industries of Massachusetts, representing some 6,000 companies, also objected vociferously to the prospect of higher costs. Cape Wind Associates responded that its project was necessary to help the state meet its RPS goals and that its environmental and job-creation benefits outweighed the additional cost. These claims were fleshed out in the formal evidentiary hearings, which were held in September and October and involved testimony under oath by a series of fact and expert witnesses.[121] On November 22 the DPU approved the fifteen-year power-purchase agreement, having concluded that the benefits of the Cape Wind project outweighed its costs. Still, Cape Wind's path was not clear; the wind farm faced lawsuits and appeals of the DPU decision by the alliance and other organizations. Moreover, Cape Wind had missed the deadline for the most lucrative federal incentive: the ITC, an up to 30 percent cash grant that could have been worth hundreds of millions of dollars.

Then, for a brief period, the construction of Cape Wind seemed imminent. In March 2012 NStar agreed to pay 18.7 cents per kWh for 27.5 percent of Cape Wind's total output, in return for being allowed to merge with Connecticut-based Northeast Utilities. In late November the DPU approved the fifteen-year contract. In January 2013, as part of the "fiscal cliff" deal, Congress extended the PTC and the ITC so that projects that began construction in 2013 could apply for these incentives. In mid-June 2013 a large Danish pension fund pledged $200 million in financing, and for the next year Cape Wind continued to shore up its financing. The project would have to start construction by the end of 2014 to meet the terms of its power contract with utilities, however, and that turned out to be the main stumbling block: in January 2015 both National Grid and NStar terminated their contracts with Cape Wind, saying the company had missed the December 31 deadline to begin construction.[122] In late January Cape Wind terminated contracts to buy land and facilities in Falmouth, Massachusetts, and in Rhode Island, the latest sign that the project was never going to come to fruition. And in November 2017, Cape Wind submitted a notification to the Bureau of Ocean Energy Management that it is surrendering its leasing and ceasing development.

While Cape Wind struggled, the federal government moved ahead with its new process for issuing offshore wind farm leases. On November 24, 2010, the interior department announced that it would begin conducting environmental assessments on proposed "Wind Energy Areas" in early 2011 and would start issuing offshore development leases within a year. Between 2013 and 2017 the BOEM awarded eleven competitive wind energy leases off the Atlantic coast: two offshore Massachusetts/Rhode Island, two offshore Maryland, two offshore Massachusetts, two offshore New Jersey, one offshore New York, one offshore North Carolina, and another offshore Virginia. As of spring 2019 only one project has been completed: in December 2016, Deepwater Wind began generating electricity off the coast of Rhode Island, with five wind turbines able to provide 30 MW of power.

CONCLUSIONS

While U.S. developers and prospective financiers watched the Cape Wind project struggle, in other countries offshore, wind was taking off. According to the European

Wind Energy Association (EWEA), as of 2015 there were 128.8 gigawatts (GW) of installed wind energy capacity in the European Union—approximately 120.6 GW onshore and just over 8 GW offshore. In 2014 Europe connected 408 offshore wind turbines in nine wind farms and one demonstration project, for a combined capacity totaling 1,483 MW. Once completed, the twelve offshore projects currently under construction will increase installed capacity by 2.9 GW, bringing the cumulative capacity in Europe to 10.9 GW.[123] In 2011 China put forth a plan to build 5,000 MW of offshore wind turbines in four years. As of mid-2014, however, only 10 percent of that capacity was in place.[124]

In the United States, although Cape Wind appeared to be doomed, and other offshore projects were slow to materialize, land-based wind power was experiencing explosive growth: between 2001, when Cape Wind was proposed, and the end of 2009, installed capacity went from 2.39 GW to more than 35 GW, with 10 GW added in 2009 alone;[125] the only source growing faster was natural gas. The DOE reports that as of 2013 there were more than 61 GW of wind-generating capacity installed—and that wind power accounted for 4.5 percent of electricity end-use demand, up from 1.5 percent in 2008.[126] And the land-based market continued to boom: developers installed another 4.85 GW of capacity in 2014, according to the AWEA.[127] The industry's phenomenal growth notwithstanding, there were numerous wind-energy siting controversies in the United States throughout the 2000s. Everywhere conflict arose, the refrain was, "We support wind power, just not here."[128]

Although it is tempting to label such sentiment as NIMBYism and hypocrisy, doing so obscures the legitimate concerns raised by critics of wind power. For example, in some locations the impact of turbines on birds and bats may be substantial. More than 1,000 raptors are killed by wind-power facilities in Northern California each year—a fact that many experts attribute to the number of facilities (there are more than 5,000), the type of turbines used, and the presence of abundant raptor prey in the region. New wind farms proposed for ridgelines in the Appalachian Mountains could harm already declining populations of songbirds, which tend to fly lower than seabirds and raptors. Bats face even more hazards because they seem to be attracted to turbines. A study of bat mortality at a forty-four-turbine facility in West Virginia revealed that in a six-week period, the turbines killed an estimated 1,364 to 1,980 bats—34 per turbine. If West Virginia were to build the more than 400 turbines currently permitted, up to 14,500 bats could be killed during a six-week migratory period alone.[129] The U.S. Government Accountability Office (GAO) concluded that the impact of wind-power facilities on wildlife varies by region and by species, but a great deal is not known about the threat wind poses or about the flyways or populations of bird species; as a result, it is extremely difficult to generalize about the impact of wind power on birds.

Adam Davis, John Rogers, and Peter Frumhoff contend that what creates real distress for wildlife and conservation advocates is the prospect of immediate and tangible damage to habitat and wildlife. As one interviewee told the Union of Concerned Scientists, "You can't see how many birds die from climate change, but you can go to the base of a turbine and see dead birds there."[130] And concerns about wildlife impacts are not limited to wind-power facilities. In the January 2008 issue of *Scientific American*, Ken Zweibel, James Mason, and Vasilis Fthenakis propose a "solar grand plan" to harness the abundant solar energy of the American Southwest. They estimated that at least 250,000 square miles of land in the Southwest was suitable for solar-power plants; just a fraction of that—only 10,000 square miles by some estimates, or 9 percent of degraded federal lands in Nevada—would be sufficient to supply all U.S. electricity demand at current rates of consumption.[131] But Andrew Silva, aide to San Bernardino, California, county supervisor Brad Mitzfelt, cautioned, "[T]he Southwest is not a big, flat, empty parking lot. It is a thriving, sensitive, fragile ecosystem."[132] The Mojave Desert, which already hosts nine concentrating solar-power plants, is also the home of the desert tortoise, listed as threatened under the Endangered Species Act.

Perhaps even more intractable than wildlife-based concerns are aesthetic objections that arise out of attachment to a place. On Sunday, March 7, 2010, the *Cape Cod Times*, which remained staunchly opposed to Cape Wind, ran an editorial that exemplified this sentiment. "With a footprint larger than Manhattan, with turbines each the size of the Statue of Liberty, this industrial project is out of place in an area that borders marine sanctuaries on all sides. Nantucket Sound is a national treasure and it must be protected at all costs."[133] Two weeks later an op-ed in the *Cape Cod Times* reflected on the final public hearing on the final EIS held by the federal government. The author, North Cairn, said that the two sides would never reach agreement because

> at its heart the irresolvable conflict arises out of two different approaches to history, not to mention two different experiences of what Nantucket Sound is and means. The Cape Wind project presents the Sound as mere location, the locus of a resource to be exploited for commercial and industrial purposes. Those who want the Sound reserved from development hold it as "place," a living presence that carries the authority of individual and collective memory.[134]

Kert Davies, a research director with Greenpeace, warned that "[p]eople should be cautious of thinking that protecting the viewshed protects the place. The truth is that every model shows that the cape and islands and Nantucket are going to be drastically different places if climate change continues apace."[135] But as historian Robert Righter explained, "Familiarity with a place generates attachment [to], and indeed love, of that landscape. With love comes a sense of stewardship and a determination to protect the land *as it is*" (emphasis added).[136] Ironically, it is precisely this place attachment that policymakers have sought to harness with "place-based" strategies that aim to address the growing challenge posed by diffuse, nonpoint-source environmental problems.

Notwithstanding deeply held differences, most experts believe that using properly designed stakeholder processes can help resolve alternative energy siting controversies; in fact, early and effective public engagement is the near-universal prescription for averting or mitigating such conflicts. According to political scientists Susan Hunter and Kevin Leyden, "[I]f governmental agencies or facility developers would address community concerns adequately through open meetings, then public concern would dissipate."[137] Survey evidence confirms that the earlier, more open, and participatory the public process, the greater the likelihood of public support for a project.[138]

One element of such participatory processes is joint fact-finding (JFF), in which stakeholders work with experts to develop the informational foundations for a decision.[139] But even a well-designed JFF process cannot alleviate the need to make value-based judgments in the face of uncertainty. A forum seeking to bring the two sides in the Cape Wind debate together in late September 2007 encountered just such a situation. Organized under the auspices of the Vineyard Haven library lecture and workshop series, the meeting aimed to establish the factual basis for discussion. Not surprisingly, however, beyond the basics of the project—the number of turbines, the length of the transmission lines—there was little agreement to be found; the two sides differed on the effects of the project on birds and fisheries, the viability of deep-water wind, the extent of navigation hazards, and other issues. As with the forum held by the Massachusetts Technology Collaborative five years earlier, the organizers found that the unresolved questions concerned willingness to accept uncertain risk, which varies irreducibly from one person to the next.

Other remedies for contentious siting disputes have been proposed as well. Some advocates have suggested federal preemption or at least a streamlined permitting process to ensure that local resistance does not derail projects with important public benefits. But many people find the idea of overriding local concerns unappealing and undemocratic and are reluctant to sanction such approaches. Still others have proposed a reverse Dutch auction, in which communities bid for a facility and the price rises until a town or county comes forward with an environmentally acceptable location. In theory, such a process would reveal the price people in the United States are willing to pay to avoid having undesirable land uses in their own backyard.[140] Such an approach might be of limited utility for wind-power projects, however, given that the choice of sites is physically constrained and multiple jurisdictions are typically affected.

A more effective way to mitigate the negative response to alternative energy siting may be to involve the affected community in ownership; in Cape Wind and other cases, distrust of private developers has figured prominently among the reasons for resistance.[141] Careful analysis suggests that the Danes were able to build offshore wind not only because they paid attention to the concerns of the local community but also because they made the first offshore wind farm a community-owned enterprise: the Middelgrunden Offshore Wind Farm Cooperative is controlled by 8,500 individual investors, who own half the project, and the municipal utility provider, which owns the other half. According to Jens Larsen, the developer of Middelgrunden, "The main reason for our success is local ownership. . . . If it is just a big company doing a wind farm, the community doesn't feel it owns it. And then they will resist. And once you encounter resistance, you are going to have to work harder and harder against the negativity."[142]

Even if JFF, federal preemption, or community ownership can sometimes succeed in mitigating siting disputes, ideally nations would alleviate case-by-case skirmishes through comprehensive planning; given the scale of land-use impacts implied by a massive conversion to renewables, comprehensive planning appears essential. Admittedly, Massachusetts's ocean-planning process in 2009 encountered difficulties of its own. The state's draft ocean-management plan, released in the summer of 2009, allowed as many as 166 turbines in two sites off Martha's Vineyard. According to state officials, scientific analysis identified the two areas, which presented a minimum of conflict with other valued resources. Yet the affected communities immediately mobilized to challenge the designations, raising both substantive and procedural concerns, and began exploring ways to thwart the plan.[143] That said, the United Kingdom and other countries have successfully established comprehensive planning processes in which appropriate sites are identified and rights to lease them are publicly auctioned. In 2010 the United States adopted a similar approach. Given that there was no such comprehensive planning system in place in 2001, when Cape Wind was proposed, choosing to build the nation's first offshore wind farm somewhere less controversial than Nantucket Sound might have yielded faster, more encouraging results.

QUESTIONS TO CONSIDER

- Historically, environmentalists have capitalized on procedural reviews to block or delay projects they did not like. How should regulators deal with siting and zoning controversies over installing alternative energy, given that time is of the essence? Do you think regulatory "streamlining" is a good idea for "environmentally beneficial" projects, such as alternative energy projects? If so, who gets to decide which projects qualify?
- Why, in your view, have some countries been able to move ahead with offshore wind, while the United States has not?
- What do you think is at the heart of the Cape Wind debate, and what might the developer have done differently to produce a different outcome?
- The Cape Wind debate uniquely pits environmentalists against other environmentalists. How can we reconcile two differing priorities (preservation versus clean energy)?

NOTES

1. Journalists Wendy Williams and Robert Whitcomb make this argument in their book, *Cape Wind: Money, Celebrity, Class, Politics, and the Battle for Our Energy Future* (New York: Public Affairs, 2007), a spirited but deeply biased account of the controversy.
2. Michael O'Hare, Lawrence Bacow, and Deborah Sanderson, *Facility Siting and Public Opposition* (New York: Van Nostrand Reinhold, 1988); Howard R. Kunreuther, W. H.

Desvouges, and Paul Slovic, "Nevada's Predicament: Public Perceptions of Risk from the Proposed Nuclear Waste Repository," *Environment* 30 (October 1988): 16–20; Michael E. Kraft and Bruce B. Clary, "Citizen Participation and the NIMBY Syndrome: Public Response to Radioactive Waste Disposal," *Western Political Quarterly* 44 (1991): 299–328; Barry G. Rabe, *Beyond NIMBY: Hazardous Waste Siting in Canada and the United States* (Washington, D.C.: Brookings Institution, 1994); Kent E. Portney, "Allaying the NIMBY Syndrome: The Potential for Compensation in Hazardous Waste Treatment Facility Siting," *Hazardous Waste* 1 (Fall 1984): 411–421; and Kent E. Portney, *Siting Hazardous Waste Treatment Facilities: The NIMBY Syndrome* (New York: Auburn House, 1991).

3. Denis J. Brion, *Essential Industry and the NIMBY Phenomenon* (New York: Quorum Books, 1991); Desmond M. Connor, "Breaking Through the 'Nimby' Syndrome," *Civil Engineering* 58 (December 1988): 69–71; William R. Freudenburg and Susan K. Pastor, "NIMBYs and LULUs: Stalking the Syndromes," *Journal of Social Issues* 48 (1992): 39–61; Kraft and Clary, "Citizen Participation"; Susan Hunter and Kevin M. Leyden, "Beyond NIMBY: Explaining Opposition to Hazardous Waste Facilities," *Policy Studies Journal* 23 (Winter 1995): 601–619; Kristy E. Michaud, Juliet E. Carlisle, and Eric R.A.N. Smith, "Nimbyism vs. Environmentalism in Energy Development Attitudes," *Environmental Politics* 17 (2008): 20–39.

4. See, for example, Gregory E. McAvoy, *Controlling Technocracy: Citizen Rationality and the NIMBY Syndrome* (Washington, D.C.: Georgetown University Press, 1991).

5. See Charles R. Warren, Carolyn Lumsden, Simone O'Dowd, and Richard V. Birnie, "Green on Green: Public Perceptions of Wind Power in Scotland and Ireland," *Journal of Environmental Planning and Management* 48 (November 2005): 853–875; Maarten Wolsink, "Wind Power and the NIMBY-Myth: Institutional Capacity and the Limited Significance of Public Support," *Renewable Energy* 21 (2000): 49–64. The landscape impact of wind farms is exacerbated by the fact that some of the areas with the best wind and proximity to population centers are highly valued for their scenic qualities and ecological sensitivity. That said, much of the best wind resource in the United States is in the Great Plains or far offshore; in both places, the main issue is the need for long and costly transmission lines to bring the power to population centers.

6. James Q. Wilson, ed., *The Politics of Regulation* (New York: Basic Books, 1980).

7. George Tsebelis, *Veto Players: How Political Institutions Work* (Princeton, N.J.: Princeton University Press, 2002).

8. Thomas H. Hammond, "Veto Points, Policy Preferences, and Bureaucratic Autonomy," in *Politics, Policy, and Organizations: Frontiers in the Scientific Study of Bureaucracy*, ed. George A. Krause and Kenneth J. Maier (Ann Arbor: University of Michigan Press, 2003), 73–103.

9. Public relations consultant Robert Kahn (2000) quotes a California official who observed, "The best way to stop a project . . . is to litigate an environmental document. The delays alone are enough to kill the project or cause its costs to rise to prohibitive levels." See Robert P. Kahn, "Siting Struggles: The Unique Challenge of Permitting Renewable Energy Power Plants," *Electricity Journal* 13 (March 2000): 21–33. For developers, time is of the essence because they have to spend capital up front in anticipation of an eventual return. Moreover, the capital costs of a construction project tend to rise over time.

10. Navigant, "Offshore Wind Market and Economic Analysis: 2014 Annual Market Assessment," September 8, 2014. Available at http://energy.gov/sites/prod/files/2014/09/f18/2014%20Navigant%20Offshore%20Wind%20Market%20%26%20Economic%20Analysis.pdf; U.S. Department of Energy, "2013 Wind Technologies Market Report," August 2014. Available at http://energy.gov/sites/prod/files/2014/08/f18/2013%20Wind%20Technologies%20Market%20Report_1.pdf. Electricity-generating capacity, measured in watts, is an expression of instantaneous power output. A kilowatt (kW) is one thousand watts. One thousand kW equals one megawatt (MW), and one thousand MW equals one gigawatt (GW).

11. Jeffrey Logan and Stan Mark Kaplan, "Wind Power in the United States: Technology, Economic, and Policy Issues," CRS Report for Congress, June 20, 2008, RL34546.

12. See American Wind Energy Association, "Wind Power Outlook 2010." Available at http://www.awea.org/pubs/documents/Outlook_2010.pdf.

13. See, for example, European Climate Foundation, Roadmap 2005. Available at http://www.roadmap2050.eu/

14. NC Clean Energy Technology Center. (2018, February 28). Renewable Electricity Production Tax Credit (PTC). Retrieved from http://programs.dsireusa.org/system/program/detail/734.

15. Union of Concerned Scientists, "Production Tax Credit for Renewable Energy" (2015). Available at http://www.ucsusa.org/clean_energy/smart-energy-solutions/increase-renewables/production-tax-credit-for.html#.VRc0Ad6qf8s.

16. Onshore wind is a mature technology, but researchers believe the cost of offshore wind power may fall as much as 30 percent to 50 percent as technologies improve. See Carolyn S. Kaplan, "Congress, the Courts, and the Army Corps: Siting the First Offshore Wind Farm in the United States," *Boston College Environmental Affairs Law Review* 31 (2004): 177–220.

17. Trip Gabriel, "Ash Spill Shows How Watchdog Was Defanged," *The New York Times*, February 8, 2014.

18. M. A. Palmer et al., "Mountaintop Mining Consequences," *Science* 327 (January 8): 2010, 148–149.

19. National Research Council, *Hidden Costs of Energy: Underpriced Consequences of Energy Production and Use* (Washington, D.C.: National Academies Press, 2009).

20. Margot Roosevelt, "Gulf Oil Spill Worsens—But What About the Safety of Gas Fracking?" *Los Angeles Times*, June 18, 2010. The oil and gas industry estimates that 90 percent of the more than 450,000 operating gas wells in the United States rely on hydraulic fracturing. See Zeller, "Far from Gulf, Due Diligence."

21. U.S. Geological Survey, "Thermoelectric Power Use," The USGS Water Science School. Available at water.usgs.gov/edu/wapt.html.

22. Pamela Ferdinand, "Windmills on the Water Create Storm on Cape Cod," *The Washington Post*, August 20, 2002, 3.

23. Robert W. Righter, *Wind Energy in America: A History* (Norman: University of Oklahoma Press, 1996).

24. Interestingly, though, the 1980s was the period of most nuclear capacity additions, as plants built in the 1970s came online.

25. Jeffrey Krasner, "Newer Technology Renews Interest in Wind Power," *The Boston Globe, August 6, 2001, C2.*

26. Jack Coleman, "Boston Wind-Farm Proponents Scale Back Proposal," *Cape Cod Times*, January 22, 2003. A megawatt is roughly equal to the electricity consumed in 1,000 U.S. homes.

27. Scott Disavino, "New England's EMI Plans 420 MW Nantucket Wind Farm," Reuters News, October 30, 2001; Bob Wyss, "The Winds of Change," *Providence Journal*, October 14, 2001, G1.

28. As of 2010, twenty-four states and the District of Columbia had RPSs; together, these states accounted for more than half of the electricity sales in the United States The DOE provides a map of states with RPSs and a table showing the particular policy in each state. Available at http://apps1.eere.energy.gov/states/maps/renewable_portfolio_states.cfm.

29. Martin J. Pasqualetti, Paul Gipe, and Robert W. Righter, "A Landscape of Power," in *Wind Power in View*, ed. Martin J. Pasqualetti, Paul Gipe, and Robert W. Righter (New York: Academic Press, 2002), 3–16.

30. Wyss, "The Winds of Change."

31. US Energy Information Administration. (2017, May 10). "Electricity Explained: Electricity in the United States." Retrieved from https://www.eia.gov/energyexplained/index.cfm?page=electricity_in_the_united_states

32. Most experts also contend it is easier to assemble a large turbine in water than on land. Manufacturers can build the enormous blades in a vast shipyard on the coast and then transport them to the site by barge, rather than by truck over busy highways.

33. Pasqualetti, Gipe, and Righter, *Wind Power in View*, 115–132.

34. Kaplan, "Congress, the Courts, and the Army Corps."

35. This requirement was weakened by the provision of an "alternative compliance payment" that allowed Massachusetts retail electricity providers to pay a fixed penalty in lieu of purchasing qualified energy. All the New England RPSs have a similar provision. In most cases, they have met their RPSs this way, rather than by purchasing energy from renewable energy providers. Michael T. Hogan, Power Programme Director for the European Climate Foundation, personal communication, August 2010.

36. Cape Wind Associates LLC was a joint venture between Gordon's firm, Energy Management Inc., and Wind Management Inc., a subsidiary of the European-based UPC. See Karen Lee Ziner, "Offshore Harvest of Wind Is Proposed for Cape Cod," *The New York Times*, April 16, 2002, F3.

37. In 1986, in *United States v. Maine*, a federal court rejected Massachusetts's claim that the center of Nantucket Sound was state "internal waters."

38. NStar is a Massachusetts-based, investor-owned electric and gas utility. In 2015 the company changed its name to Eversource Energy.

39. Originally NStar estimated the average electricity demand for the cape and islands at 350 MW. But in late 2002 it revised that estimate to 230 MW, at which point Cape Wind could claim to furnish three-quarters of the electricity consumed by the cape and islands. For comparison, the Pilgrim nuclear power plant in Plymouth, Massachusetts, generates about 630 MW.

40. Jeffrey Krasner, "Offshore Wind Farm Blows into Cape View," *The Boston Globe*, July 28, 2001, 1.

41. Quoted in Ziner, "Offshore Harvest of Wind."

42. Quoted in Wyss, "The Winds of Change."

43. Ziner, "Offshore Harvest of Wind."

44. Scott Allen, "Study Funded by Foe Says Wind Turbines to Hurt Cape Tourism," *The Boston Globe*, October 28, 2003, B3.

45. John Leaning, "Expert Sailors in Hyannis, Mass., Say Wind Farm Would Affect Area's Allure," *Cape Cod Times*, January 8, 2003.

46. Cosmo Macero, "Tax Credit Powering Windmills," *Boston Herald*, July 28, 2003, 21.

47. In mid-May 2006 the Beacon Hill Institute released a third study, again funded by the Egan Foundation, claiming that the developers of Cape Wind would reap a 25 percent return on equity, or $139 million in profits, largely as a result of taxpayer-supported subsidies. See Jay Fitzgerald, "Taxpayer to Fund Windfall," *Boston Herald*, May 15, 2006, 27. This study seemed to contradict the 2004 study that contended Cape Wind would not be economically feasible.

48. Jonathan Haughton and David G. Tuerck, "Offshore Wind Power Still Isn't Worth the Cost," *The Boston Globe*, April 3, 2004.

49. For example, Doug Yearley—formerly an executive with Phelps Dodge Corp., the nation's second-biggest mining company, and owner of a $3.2 million home in Osterville—was the organization's first president. Other prominent members included Christy Mihos, a successful convenience-store owner and former vice chairman of the Massachusetts Turnpike Authority; Bill Koch, founder of the Oxbow Group and an America's Cup sailing winner; Paul Fireman, former Reebock CEO; and Michael Egan, son of the founder of EMC Corp.

50. Ethan Zindler, "Cape Cod, Mass., Environmentalists, Energy Firm Spar over Wind-Turbine Farm," *Cape Cod Times*, July 24, 2003.

51. Michael Levenson, "Cape Project's Foes Tap Big Spenders," *The Boston Globe*, April 1, 2006, 1.

52. Andrew Miga, "Pols Tilt at Windmills," *Boston Herald*, May 14, 2002, 3.

53. Quoted in Pam Belluck, "A Wind Power Plan Stirs Debate in Massachusetts," *The New York Times*, March 2, 2003, F3.

54. Quoted in Elinor Burkett, "A Mighty Wind," *The New York Times*, June 15, 2003, 6.

55. Cronkite changed his position after meeting with Gordon; he said he had taken his initial position before fully researching the project and that the alliance had exploited him in making him the spokesperson for its cause. Jay Fitzgerald, "Cronkite Changes Tune on Cape Wind Project," *Boston Herald*, August 29, 2003, 3. Just days before changing his position, Cronkite's ambivalence was apparent in a column he had written for the *Cape Cod Times* deploring the aesthetic impact of the turbines, as well as their impact on wildlife and birds. "I find myself a nimbyist," he wrote, "abashedly, and perhaps only temporarily."

56. Quoted in Jennifer Peter, "Rich and Famous Lining Up to Protect Nantucket Sound," Associated Press Newswires, August 5, 2003.

57. Wyss, "The Winds of Change."

58. Anthony Flint, "Bird Safety a Factor in Debate over Cape Wind Farm," *The Boston Globe*, September 1, 2003, B1.

59. Quoted in Wyss, "The Winds of Change." Given that oil is a minimal contributor to the nation's electricity supply, this argument is somewhat specious; that said, the stability of the constrained local grid in the area of Cape Cod and the islands is unusually reliant on an oil-fired power plant (Canal). Hogan, personal communication.

60. Ziner, "Offshore Harvest of Wind."

61. Charles M. Sennott, "Denmark's Windmills Flourish as Cape Cod Power Project Stalls," *The Boston Globe*, September 22, 2003, 1. As Cape Wind detractors point out, however, there are important differences between Cape Wind and Middelgrunden. The latter is much smaller—one square mile versus twenty-five square miles, 20 turbines versus 130, and turbine heights of 200 feet versus well above 400 feet. Moreover, Middelgrunden is located in Copenhagen's industrial harbor, on a shoal that was used as a dumping area between 1980 and the early 1990s, when the wind farm was built. Kate Dineen, former assistant director, Alliance to Protect Nantucket Sound, personal communication, August 2010.

62. Quoted in Andrew Miga, "Ted K Joining Fight Against Energy Plan," *Boston Herald*, May 14, 2002, 3.

63. Ibid.

64. Elizabeth W. Crowley, "New Idea Blows into Quincy," *The Patriot Ledger* (Quincy), June 15, 2002.

65. Donna Goodison, "Study: Cape Plan Could Be Benefit to Area Economy," *Boston Herald*, April 4, 2003, 33.

66. Quoted in Mark Reynolds, "In the Wind," *Providence Journal*, June 15, 2003, F1.

67. Cynthia Roy, "Unions, Fishermen Back Wind Farm for Nantucket Sound," *The Boston Globe*, April 25, 2003, B3.

68. Environmental studies are typically paid for by permit applicants; this is regarded as appropriate, since the alternative is for taxpayers to fund studies that would benefit private developers.

69. The Rivers and Harbors Act requires applicants to obtain a permit from the Corps for the installation of a structure in U.S. navigable waters. The Outer Continental Shelf Lands Act (OCSLA) extends the Corps' jurisdiction to include the submerged lands seaward of state and coastal waters between 3 and 200 nautical miles offshore.

70. In that case, Worcester attorney John Spillane charged that the Corps had overstepped its authority in issuing the permit because the site was more than three miles from shore, and the project developer had no legal interest in or ownership of the seafloor where the tower would be secured. Spillane also argued the tower was unnecessary, since all the information Cape Wind Associates needed was available from the Coast Guard and Woods Hole Oceanographic Institute. Finally, he argued that the permit was illegal because Nantucket Sound was protected under the state's Cape and Islands Sanctuary Law.

71. An EIS is mandatory for any major project that involves a federal permit under the National Environmental Policy Act (NEPA). The state requires an EIR under the Massachusetts Environmental Policy Act (MEPA). In this case, the EIS/EIR also served as the environmental assessment on a Development of Regional Impact (DRI) for the Cape Cod Commission, which had oversight over the project as well. As the lead federal agency the Corps was responsible for producing a single document to serve as a joint EIS/EIR/DRI, soliciting the input of all the relevant federal and state agencies in the process.

72. The Corps selected the comparison sites from among the seventeen it considered based on five criteria: the amount of wind power, the ease of connecting to the New England power grid, the available area on land or water, engineering challenges, and legal constraints.

73. Quoted in Jack Coleman, "Boston Wind-Farm Proponents Scale Back Proposal," *Cape Cod Times*, January 22, 2003.

74. Quoted in Donna Goodison, "Report, Pol Picks Nature over Power," *Boston Herald*, February 11, 2003, 30.

75. Reilly and Delahunt were also drawing up separate federal legislation to address siting, bidding, compensation, and state input for wind projects. Their efforts to attach legislation to the House energy bill failed, and the stand-alone bill stalled in Cubin's subcommittee. Other politicians opposed to Cape Wind engaged in legislative machinations as well. Sen. Lamar Alexander, R-Tenn., in consultation with Senator Kennedy, was drafting an amendment to the energy bill working its way through Congress that would require local or state governments to approve such projects. Working at the state level, in early December Democratic State Senator Robert O'Leary of Barnstable proposed a bill requiring state environmental officials to draw up a coastal zoning map to determine where wind farms would be appropriate before they are proposed, and Democratic State Representative Demetrius Atsalis of Hyannis filed a bill that would prohibit any "ocean-based energy project" unless the state environmental affairs office determined it would pose no environmental, aesthetic, or safety hazards.

76. "Coastal Center Calls for Conservation, Study of Nantucket Sound," Associated Press, February 10, 2003.

77. James Bone, "Wind Farm Threatens Kennedy's Playground," *The Times* (London), July 5, 2003, 15.

78. Burkett, "A Mighty Wind."

79. John Leaning, "Cape Cod, Mass.-Area Poll Finds Majority Approval for Wind-Farm," *Cape Cod Times*, October 3, 2002.

80. "Poll: Cape, Island Residents Oppose Wind Farm," Associated Press Newswires, March 4, 2004.

81. Patrick Cassidy, "Poll: Most Cape Codders Favor Wind Farm," *Cape Cod Times*, December 3, 2009.

82. Jon Chesto, "Poll Finds Even Split over Cape Wind Plan," *Quincy Patriot Ledger*, June 14, 2005.

83. As journalist Beth Daley pointed out, however, the threat to birds was highly uncertain. The Corps was working with reports that offered very different estimates of how many birds flew through the area and potential fatalities: one based on aerial and boat surveys, the other based on radar data. Complicating matters, the number of *acceptable* bird deaths is always a judgment call. See Beth Daley, "Report on Possible Risks from Wind Farm Fuels Ire," *The Boston Globe*, October 17, 2004, B1.

84. The bill grew out of the work of a task force Romney had convened in the spring of 2003, which had called for a comprehensive Ocean Resources Management Act that would institute stronger environmental protection while streamlining the statutes governing the use and protection of the state's oceans.

85. On December 18, 2005, the Supreme Judicial Court upheld the decision by the EFSB.

86. By this time, Cape Wind Associates had twelve full-time employees, two offices, lawyers, and a public relations firm; it was spending about $250,000 each month, or $8,000 a day, to keep its operation afloat. See Stephanie Ebbert, "Foes' Tactics Slow Advance of Cape Wind Farm Proposal," *The Boston Globe*, December 6, 2004, B3.

87. For comparison, oil rigs can be situated within 1,500 feet of a shipping channel.

88. "An Ill Wind from Congress," *The Boston Globe*, April 7, 2006, A16.

89. For example, after a provision was added to the 2005 Energy Policy Act exempting Cape Wind from competitive bidding requirements, the alliance took out a full-page ad in the D.C. magazine *Roll Call* charging that Cape Wind was the beneficiary of a "sweetheart deal" for exclusive rights to build on Nantucket Sound. Cape Wind supporters pointed out that grandfathering the project was fair given that it had been in the works years before the proposed rule changes.

90. Stephanie Ebbert, "Grid Operator Urges Congress Not to Block Wind Farm Proposal," *The Boston Globe*, May 4, 2006. In mid-October 2005 ISO New England had granted approval for Cape Wind to tie into the region's electricity grid.

91. "Danish Offshore Wind: Key Environmental Issues," DONG Energy, Vattenfall, Danish Energy Authority and the Danish Forest and Nature Agency, November 2006, p. 8. Available at http://ec.europa.eu/ourcoast/download.cfm?fileID=975.
92. Benthic organisms live on the seafloor and are integral to the marine food web.
93. Quoted in Stephanie Ebbert, "Patrick Supports Wind Farm Plan," *The Boston Globe*, October 18, 2005, B7.
94. Cited in U.S. Department of Energy, "Environmental Impact Statement for the Proposed Cape Wind Energy Project, Nantucket Sound, Offshore of Massachusetts, Final Environmental Impact Statement," December 2012, pp. 5–229. Available at http://energy.gov/sites/prod/files/DOE-EIS-0470-Cape_Wind_FEIS_2012.pdf.
95. The Cape Cod Commission was formed in 1990 to slow the dizzying pace of development the region experienced during the 1980s. It wields substantial authority over development decisions across the cape's fifteen towns and often exerts influence beyond its traditional jurisdiction of large-scale developments. It has a reputation for trying to protect the region's natural resources and impeding development.
96. Cape Wind opponents proceeded to file suit against Bowles in Barnstable Superior Court, complaining that he had abdicated his responsibility to protect Cape Cod's waters when he accepted the Cape Wind FEIR and that he should have demanded more information from Cape Wind Associates before signing off on the report. Lawyer John Spillane, who filed the suit, said he hoped the complaint would stall state and local permitting for the project and eventually kill it. See David Scharfenberg, "Foes of Wind Project File Suit Against Massachusetts Environmental Officials," *Cape Cod Times*, May 26, 2007. In June, however, Barnstable Superior Court Judge Robert Kane ruled in favor of Cape Wind Associates on most of the company's requests to dismiss five complaints in the case.
97. In April 2009 Barnstable Superior Court Judge Robert Rufo threw out the lawsuit filed by the town of Barnstable because the EFSB had not yet finished its review of Cape Wind's "super permit" request.
98. While the EFSB was deliberating, several more decisions broke in Cape Wind's favor. First, during the spring of 2008, the Department of Environmental Protection quietly issued two permits: a water quality certificate that set conditions for the installation of transmission lines and a Chapter 91 waterways license that permitted construction on filled tidelands and lands that are property of the state—also necessary for the transmission lines. Second, in January 2009, the state Office of Coastal Zone Management issued a pair of favorable determinations for Cape Wind. The alliance and the Town of Barnstable immediately filed two lawsuits challenging the office's findings.
99. C. Harvey (2016, August 8). "Massachusetts Just Gave a Huge Boost to the Offshore Wind Industry." *The Washington Post*. Retrieved from https://www.washingtonpost.com/news/energy-environment/wp/2016/08/08/at-long-last-the-u-s-offshore-wind-industry-is-ramping-up/?utm_term=.2800ba4b3519.
100. The DEIS rated impacts on a 4-point scale from negligible through minor (short-term/capable of mitigation), moderate (unavoidable, irreversible, but not a threat

to the viability of the resource), and major (irreversible and damaging, regardless of mitigation). The report also distinguished between impacts during the construction and operation phases of the project.

101. In addition, the Barnstable Municipal Airport, Martha's Vineyard Airport, and Nantucket Memorial Airport were all on record as opposing Cape Wind, as was the National Aircraft Owners and Pilots Association.

102. Joshua Wiersma, "An Economic Analysis of Mobile Gear Fishing Within the Proposed Wind Energy Generation Facility Site on Horseshoe Shoal in Nantucket Sound." Paper prepared for the Massachusetts Fisherman's Partnership. Available at www.fishermenspartnership.org/pdf/HorseshoeShoalEconomicReport.pdf.

103. The Wampanoag tribe of Gay Head (Aquinnah) on Martha's Vineyard gained formal federal recognition in 1987, and the Mashpee Wampanoag received such recognition in 2007.

104. Archaeological excavations under Nantucket Sound had unearthed evidence of a submerged forest six feet under the mud, but no signs of Native American camps or other human relics. The tribes questioned the integrity of these excavations.

105. Journalist Stephanie Ebbert commented in *The Boston Globe* that she knew the debate had gone on too long when a half-dozen faux oil sheiks stalked into the gymnasium at the University of Massachusetts, Boston, and hardly anyone raised an eyebrow. See Stephanie Ebbert, "A Long-Winded Debate," *The Boston Globe*, March 14, 2008, B1.

106. "Obstructed Sunrise," *Cape Cod Times*, April 21, 2008.

107. Blue H modeled its technology after deep-sea oil rigs; its floating tension-leg turbine platforms are chained steel or concrete anchors on the seafloor. Peter Sclavounos, a mechanical engineer at MIT, suspected that Blue H's technology would not be economical in the rough waters off New England, where thirty- to forty-foot waves are not uncommon. See Peter Fairley, "Wind Power That Floats," *Technology Review*, April 2, 2008.

108. At the behest of Democratic Minnesota Congressman James Oberstar, chair of the House Committee on Transportation and Infrastructure, the Coast Guard agreed to delay its recommendation, thereby pushing back the release of the final EIS. Journalists speculated he did so at the request of Senator Kennedy.

109. Complicating matters, in a letter sent to Secretary Salazar on February 9, 2010, Jeffrey Madison—a former member of the Aquinnah tribe's tribal council and an attorney at the law firm Wynn & Wynn, hired in January 2010 by Cape Wind Associates—claimed that the idea that the wind farm would harm the tribe's cultural tradition was a "fabrication" invented by opponents "who wish to derail the project." Quoted in Patrick Cassidy, "Tribe Members: Hot Air Driving Wind Farm Alarm," *Cape Cod Times*, February 20, 2010. Madison attached a petition signed by eight other tribe members, including Beverly Wright, a former chairman of the tribe and current commissioner on the state Commission on Indian Affairs.

110. On February 3, 2010, the interior department's inspector general released a report saying that while officials at several federal agencies—including the FWS and the

Coast Guard—felt they were "unnecessarily rushed" to review the Cape Wind proposal before President Bush left office, they did not believe the expedited timeline affected their conclusions. See Patrick Cassidy, "Wind Farm Review 'Rushed,'" *Cape Cod Times*, February 4, 2010.

111. Patrick Cassidy, "Poll: Most Cape Codders Favor Wind Farm," *Cape Cod Times*, December 3, 2009.

112. William Koch, "Tilting at Windmills," *The Wall Street Journal*, May 22, 2006, 12.

113. Patrick Cassidy, "Poll Gauges Willingness to Pay for Wind Power," *Cape Cod Times*, January 23, 2010.

114. Quoted in Tom Zeller, "Reaping Power from Ocean Breezes," *The New York Times*, April 27, 2010, B1.

115. Quoted in Valerie Vande Panne, "Cape Wind: It's Complicated," *The Boston Phoenix*, May 7, 2010. Cape Wind opponents emphasized the cost to purchase wholesale power from Cape Wind rather than the final impact on ratepayers' bills; because electricity from Cape Wind would constitute only a fraction of the region's supply, electricity rates would increase by a relatively small amount.

116. Advisory Council on Historic Preservation, Comments of the Advisory Council on Historic Preservation on the Proposed Authorization by the Minerals Management Service for Cape Wind Associates, LLC to Construct the Cape Wind Energy Project on Horseshoe Shoal in Nantucket Sound, Massachusetts, April 2, 2010, B1.

117. In particular, he asked Cape Wind Associates to move the wind farm further from Nantucket and reduce its breadth. He also asked that the turbines be painted off-white to reduce the contrast with the sea and sky, while remaining visible to birds. And he said the lights should be turned off during the day and dimmed more at night than originally planned. (Oddly, Salazar also took credit for reducing the number of turbines from 170 to 130, a change that had been made years earlier.)

118. Quoted in Beth Daley, "6 Groups File First Suit to Halt Wind Farm," *The Boston Globe*, June 26, 2010.

119. Cape Wind critics charged that even this quoted price was artificially low because it assumed Cape Wind would get both the PTC and the investment tax credit, when it would be allowed to take only one. National Grid said it was assuming the project would get the more lucrative ITC. If it did not get the PTC, the price would go up to 22.8 cents per kWh. If it did not get the ITC, the price would go up to 23.5 cents per kWh.

120. The Green Communities Act allows the utility to spread the cost of the contract with Cape Wind among its distribution customers.

121. Contrary to the wishes of the Patrick administration, the DPU allowed the alliance, many other power generators, and some big customers (including Walmart) to be "intervenors" in the formal proceeding.

122. Jim Gordon had declined to put up financial collateral to extend the deadline. Instead, he sent a letter to the utilities asking them not to terminate the contracts, citing relentless litigation by the alliance, a phenomenon he tried to portray as "force

majeure." That term is ordinarily reserved for unforeseen events, however, and litigation over Cape Wind was anything but that. See Jim O'Sullivan, "Two Utilities Opt Out of Cape Wind," *The Boston Globe*, January 7, 2015.

123. European Wind Energy Association, "Statistics," n.d. Available at http://www.ewea.org/statistics/.

124. Anon., "China Three Years Late on Installing Offshore Wind Farms," *Bloomberg Business*, July 16, 2014.

125. American Wind Energy Association, *AWEA U.S. Wind Industry Annual Market Report, Year Ending* 2009 (2010). Available at http://awea.files.cms-plus.com/File Downloads/pdfs/Annual-Market-Report-Year-Ending-2009(1).pdf.

126. U.S. Department of Energy, "Wind Vision: A New Era for Wind Power in the United States," March 2015. Available at http://www.energy.gov/windvision.

127. Diane Cardwell, "Offshore Wind Farm Leases Draw Few Bids from Wary Industry," *The New York Times*, January 29, 2015.

128. These examples are from Belluck, "A Wind Power Plan Stirs Debate"; Meredith Goad, "Debate over Wind Turbines Heats Up," *Portland Press Herald*, September 28, 2003, 1; Michael Levenson, "Buzzards Bay Wind Farm Proposed," *The Boston Globe*, May 24, 2006, 1; and John Rather, "When Nimby Extends Offshore," *The New York Times*, January 29, 2006, 14LI-1.

129. U.S. Government Accountability Office, *Wind Power*. The GAO notes that humans cause the death of millions of birds and bats each year. The FWS estimates that some of the leading causes of bird mortality each year are collisions with building windows, which cause 97 million to 976 million bird deaths; collisions with communication towers, which cause 4 million to 50 million bird deaths; poisoning from pesticides, which kills at least 72 million birds; and attacks by domestic and feral cats, which kill hundreds of millions of birds. In addition, human activity that causes habitat loss and fragmentation destroys untold numbers of birds.

130. Quoted in Adam Davis, John Rogers, and Peter Frumoff, "Putting Wind to Work," *Planning* 74 (October 2008), 40.

131. Ken Zweibel, James Mason, and Vasilis Fthenakis, "Solar Grand Plan," *Scientific American* 298 (January 2008): 64–73. Michael Hogan points out that this finding puts into relief one of wind power's major disadvantages relative to solar: its much larger spatial impact. Hogan, personal communication.

132. Quoted in Allen Best, "Solar Power's Friends and Enemies," *Planning* (October 2008).

133. "Sound Objections," *Cape Cod Times*, March 7, 2010.

134. North Cairn, "The Sound, a Precious Place," *Cape Cod Times*, March 28, 2010.

135. Quoted in Zeller, "Reaping Power from Ocean Breezes."

136. "Exoskeletal Outer-Space Creations," in *Wind Power in View*, 37.

137. Hunter and Leyden, "Beyond NIMBY."

138. Charles R. Warren, Carolyn Lumsden, Simone O'Dowd, and Richard V. Birnie, "Green on Green: Public Perceptions of Wind Power in Scotland and Ireland," *Journal of Environmental Planning and Management* 48 (November 2005): 853–875.

139. Clinton J. Andrews, *Humble Analysis: The Practice of Joint Fact Finding* (Westport, Conn.: Praeger, 2002).

140. Herbert Inhaber, "Of LULUs, NIMBYs, and NIMTOOs," *Public Interest* 107 (Spring 1992): 52–63.

141. Willett Kempton, Jeremy Firestone, Jonathan Lilley, Tracey Rouleau, and Phillip Whitaker, "The Offshore Wind Power Debate: Views from Cape Cod," *Coastal Management* 33 (2005): 119–149.

142. Quoted in Charles M. Sennott, "Denmark's Windmills Flourish as Cape Cod Power Project Stalls," *The Boston Globe*, September 22, 2003, 1.

143. Mike Seccombe, "Plan Okays Turbines in Vineyard Waters," *Vineyard Gazette*, July 3, 2009.

CHAPTER FOURTEEN

FRACKING WARS

Local and State Responses to Unconventional Shale Gas Development

In August 2010, Dryden, New York, banned the development of shale gas using a process colloquially known as "fracking."[1] In June 2014, New York's highest court ruled that Dryden had the authority to institute such a ban under the state's constitution. Less than a year later Governor Andrew Cuomo shocked the nation when he announced a statewide ban on fracking, making New York the first state with significant shale reserves to adopt such a ban.[2] This series of decisions was especially noteworthy given that fracking promised to revitalize a region of New York that has struggled economically. Elsewhere, including neighboring Pennsylvania, fracking has pumped millions of dollars into rural economies, and state-level politicians have strongly supported the shale gas industry in its struggles with local activists. But in New York, antifracking advocates succeeded in making sure that concerns about impacts on the environment, health, and quality of life dominated the debate from the outset. These sentiments also prevailed in nearby Vermont and Maryland where statewide fracking was also banned.

This case shines a spotlight on how federalism works in the United States. Federalism refers to a political organization in which smaller units of government join to form a larger whole, with political authority distributed among the units. As of 2018, the U.S. federal government had opted not to regulate most aspects of fracking, although advocates continued to debate whether it ought to take a more aggressive role. In the meantime, the states were by default the primary overseers of shale gas development. Proponents offer a host of arguments for devolving regulatory authority to the states.[3] They point out that states can devise policies that are tailored to local conditions and more responsive than centralized policies to the demands of the people. States can serve as "laboratories of democracy," experimenting with different approaches to similar problems. Critics argue that decentralization makes it more difficult to address the cumulative impacts of industrial activity, as well as impacts that cross state lines. They contend that relying on states to regulate such activity is likely to exacerbate unequal levels of protection across the states because many state governments lack the administrative capacity to regulate complex industrial processes. Moreover, they fear that competition to attract industry will provoke a "race to the bottom." Industry itself often prefers regulatory centralization on the grounds that it is more efficient to comply with a single, national policy than with dozens of state programs.[4]

Also supporting the notion that politics, not principle, lies behind the federalism debate is the contest over what some scholars have called second-order federalism, referring to the devolution of authority by states to localities. In many states, localities are seeking to control their own experience with fracking, in the process running into resistance from state government, which has an interest in promoting the orderly exploitation of its natural resources and in the economic growth that accompanies that development. Localities are "creatures of the state," their authority granted by state constitutions, statutes, and sometimes court decisions.[5] Historically states have granted localities some measure of autonomy, particularly with respect to zoning and land-use issues. But the extent of municipalities' home-rule authority varies widely from state to state.[6]

One challenge facing any level of government that seeks to regulate fracking is the dearth of unbiased information on the actual impacts of the practice. Most of what information seekers encounter on this topic is "advocacy science," in which proponents or opponents of fracking enlist the evidence that supports their preexisting position, often leaving out caveats and discussions of uncertainty, as well as studies that yield results contrary to their beliefs. Numerous university reports have come under scrutiny because of their authors' (often undisclosed) ties to industry.[7] In some cases, researchers have released their results before peer review in hopes of influencing the political debate.[8] Sincere efforts to discern the impacts of fracking are also hampered by the industry's unwillingness to allow access to well sites or to disclose information about its practices, citing proprietary considerations. Further impeding information-gathering efforts are incomplete and disorganized files in state regulatory offices.[9] In the face of inconclusive, preliminary, and often unreliable information, policymakers must choose between precautionary and risk-tolerant positions, depending on the balance of political forces. For many, the promise of jobs and tax revenue is sufficiently alluring that it overwhelms environmental and safety worries. That inclination is often institutionalized in state-level commissions, whose members are drawn heavily from industry; in the state agencies whose staff write permits for oil and gas development; and in longstanding, industry-friendly resource development laws. Reinforcing a proindustry disposition is pressure from two sources: landowners who receive signing bonuses and royalty payments in exchange for mineral rights, as well as payments from pipeline companies for rights of way; and oil and gas industry lobbyists, who threaten to leave the state if regulations become too onerous.

BACKGROUND

In the 1980s and 1990s, Mitchell Energy, a company run by petroleum engineer, geologist, and billionaire George Mitchell, began experimenting with the use of hydraulic fracturing to free natural gas from shale. Along the way, Mitchell received help from the government. The Department of Energy gave him technical help and some financing; he also received federal tax credits. In 2001, Mitchell sold his company to the Devon Energy Corporation for $3.5 billion. In 2003, according to energy expert Daniel Yergin, Devon engineers linked hydraulic fracturing to horizontal drilling, though it took another five years before the combined process now commonly known as fracking took off—mainly because the rising price of natural gas made the practice profitable.[10]

Hydraulic fracturing entails injecting massive quantities of fracking fluid deep into the ground at high pressures to fracture rock. Fracking fluid contains millions of gallons of water, as well as chemicals that serve a variety of purposes, including eliminating bacteria and preventing corrosion. Fracking fluid also contains sand that serves as a "proppant," holding open fractures so that gas can flow into the production well. Horizontal drilling allows operators to drill multiple wells, and therefore reach more energy, from a single pad. Combining the two processes requires using much more fluid and higher pressures than a vertical well; it also makes shale gas development extraordinarily productive and efficient (see Figure 14.1).

With demand for (and hence prices of) natural gas rising, fracking took off in the late 2000s, and by 2012, according to the EPA, at least twenty-four states had wells that were fracking shale to recover natural gas.[11] By 2013, shale gas accounted for 40 percent of total U.S. gas production.[12] The rapid spread of fracking created thousands of jobs in states rich in shale, particularly Colorado and Texas. The fracking boom also caused a radical shift in natural gas markets. Once notoriously volatile, by 2011 natural gas prices appeared likely to be stable and relatively low for the foreseeable future thanks to large known shale-gas reserves. Low and stable prices, along with increasingly strict federal regulations on power plants, led utilities to shift their investments from coal-fired power plants to gas-fired electricity generation. Low, stable natural gas prices also prompted the return to the United States of manufacturing, particularly steel, petrochemical, and fertilizer producers.[13]

Even as it bolstered rural economies and lowered electricity prices, fracking prompted environmentalists and local communities to raise questions about pollution. Critics of fracking were alarmed about its potential impact on water supplies, a topic that was fraught with uncertainty and about which there was a lively scientific debate. Cases of contamination have been documented in states such as Ohio, Pennsylvania, and Wyoming. Scientists continue to investigate precise causes.

There are numerous points in the fracking process when damage to water supplies can occur. Clearing land and building a well pad, which occupies several acres, can cause erosion and sedimentation that harms surface water.[14] If a well is not properly cased and cemented, it can leak methane during drilling, sending mud and fluids into underground water supplies; alternatively, gas can migrate through

Figure 14.1 Steps in Unconventional Shale Gas Development

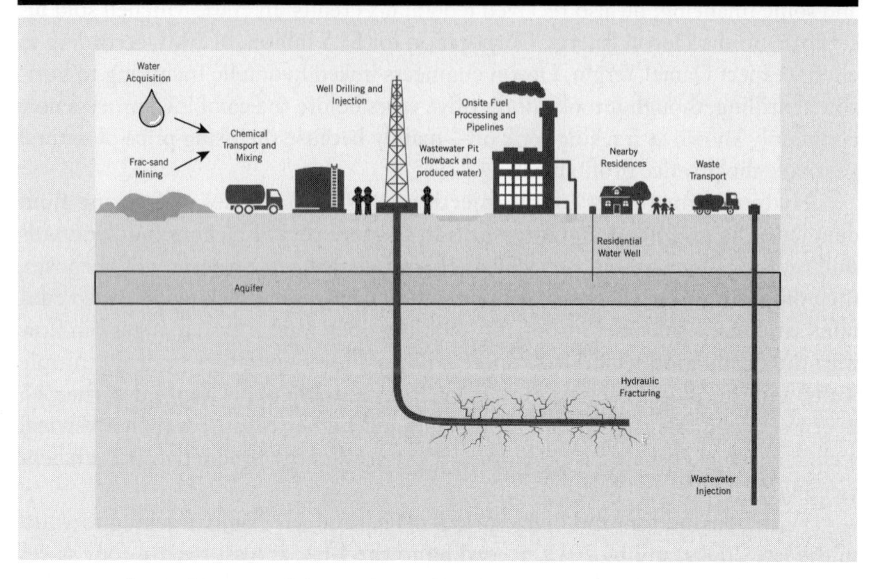

Source: Gretchen Goldman et al., "Toward an Evidence-Based Fracking Debate: Science, Democracy, and Community Right to Know in Unconventional Oil and Gas Development" (Cambridge Mass.: The Center for Science and Democracy at the Union of Concerned Scientists, 2013), https://www.ucsusa.org/center-for-science-and-democracy/toward-an-evidence-based-fracking-debate.html#.XEnrXlxKiUl

abandoned wells and possibly even through naturally occurring pathways. Liquid wastes generated during the drilling process and stored in pits or tanks on-site can leach or spill and pollute surface or underground water. Once a well is fracked, developers are left with "flowback" water, the 10 percent to 40 percent of fracking fluids that return to the surface; they also must dispose of the well's "produced" water, which is underground when drilling begins but can rise to the surface during or after the fracking process. Injection of wastewater into underground wells has caused minor earthquakes, while disposal into rivers has caused excessive loads of "total dissolved solids" and radioactivity.[15] Moreover, a single well can use between 3 million and 12 million gallons of water—the average fracked well requires between 2 million and 4 million gallons—volumes that in some locations can threaten local water supplies and harm aquatic life.[16]

Critics became worried about high concentrations of local air pollutants as well. Several elements of the fracking process produce air pollution: the trucks that bring water and equipment to the site, the drilling process, fracturing rig motors and other well-site equipment; processing units on-site, such as machines that separate water from oil and gas; compressor units that send gas through pipelines; and storage tanks and flowback water, both of which emit volatile organic compounds.[17] In addition, fracking causes emissions of methane, a greenhouse gas that is released throughout the life cycle of natural gas. Proponents have touted natural gas as a "bridge" fuel to a future dominated by renewables because combustion of natural gas results in fewer carbon dioxide emissions than combustion of coal or oil. Importantly, they note that, in the absence of large-scale storage capacity, gas-fired capacity is needed to compensate for the variability and intermittency of renewables. They point out that, according to the U.S. Energy Information Agency (EIA), U.S. carbon emissions fell by about 12 percent between 2007 and 2012, a decline the agency attributed mainly to a widespread shift from coal to natural gas for electricity generation.[18] But some scientists have questioned whether fracking is really so benign for the climate. Specifically, they argue that while combustion of natural gas results in fewer carbon dioxide emissions, fugitive releases of methane—which is far more potent in its heat-trapping potential than carbon dioxide in the short run—negate the climate benefits of natural gas.[19] (Between 2010 and 2014 a scientific debate raged over the extent and impact of fugitive methane emissions from fracking; the most conclusive studies to date suggest that in terms of greenhouse gas emissions, natural gas performs better than coal and oil for generating electricity.[20]) Skeptics also fear that the availability of cheap natural gas will slow investment in both energy efficiency and renewables; a slew of modeling studies conducted in 2013 and 2014 suggests those fears are well founded.[21]

Although concerns about water and air pollution are serious, the most universally galvanizing aspect of fracking has been its impact on rural landscapes and lifestyles. Fracking generates major increases in truck traffic; a single well requires between 900 and 1,300 round trips by trucks hauling equipment, water, sand, chemicals, and flowback.[22] In addition to disrupting rural peace and damaging country roads, these trucks can create hazardous driving conditions.[23] Furthermore, draining a shale gas reserve requires a grid of wells and related infrastructure across an entire region, all of which generate noise, light pollution, and odors and industrialize the landscape. The rapid pace of well construction has led to population increases that strain many communities' social safety nets with increased rates of crime and other social problems.[24]

Despite the rapid spread of fracking and the pervasiveness of local concern, attempts by the federal government to intervene have been halted. In 2015, the Obama administration attempted to finalize a rule that would ensure that industry discloses the chemicals it uses in the fracking process. These efforts were halted by a Wyoming federal judge in 2016. Moreover, in 2017, Senator Casey, D-Pa., introduced the Fracturing Responsibility and Awareness of Chemicals Act (aka, FRAC Act). The purpose of the bill was to amend the Safe Drinking Water Act to allow the U.S. EPA to oversee enforcement of fracking processes.[25]

Nevertheless, fracking is exempt from a host of federal laws, as we explain here. In 2005, Congress granted oil and gas producers an exemption from the requirements of the Safe Drinking Water Act: the Energy Policy Act amended the definition of "underground injection" so that fracking operators can legally inject anything but diesel into the ground as part of the fracking process without obtaining an underground injection control (UIC) permit.[26] (Operators must obtain UIC permits to inject wastewater from fracking into underground wells, however.) In 1978 the EPA opted not to regulate oil and gas exploration and production wastes under Subtitle C of the Resource Conservation and Recovery Act (RCRA), which governs the disposal of hazardous waste; Congress codified that exemption in the 1980 amendments to RCRA, and in 1988 the EPA issued a regulatory determination that oil and gas waste would not be treated as hazardous.[27] Fracking is also exempt from the Emergency Planning and Community Right-to-Know Act (EPCRA), which requires industries to submit a Toxic Chemical Release Form describing the toxic chemicals used in their industrial processes and how they are disposed of.[28]

Congress considered a slew of laws to regulate the practice. The most prominent of these was the FRAC Act, which would have eliminated the oil and gas industry's exemption from the Safe Drinking Water Act and forced companies to disclose the chemicals they were injecting into the ground. Despite strong support from environmental groups, none of the bills debated by Congress between 2008 and 2018 stood a serious chance of passing. In an effort to spur action on the issue, in 2010 the House Committee on Energy and Commerce launched an investigation into the health impacts of fracking. The committee's report, issued in 2011, concluded that leading companies were using fracking fluids that contained 750 compounds, of which more than 650 were known human carcinogens or air pollutants.[29] Congress subsequently directed the EPA to fund a peer-reviewed study of the relationship between fracking and groundwater. In response to intense pressure from industry, however, the EPA significantly narrowed the scope of that study; moreover, industry refused to allow EPA scientists to collect baseline water quality data, impeding efforts to conduct prospective studies, which are more definitive than retrospective ones.[30] In addition to weakening the study, industry obstruction delayed its completion: although initially due in 2012, the study was not released until 2016.

The Obama administration was more engaged with issues raised by fracking than Congress. In May 2011, energy secretary Steven Chu appointed a seven-member panel of scientific and environmental experts to make recommendations on how fracking could be done more safely and cleanly. The Natural Gas Subcommittee issued its recommendations in August. The report called for better tracking and more careful disposal of wastewater from wells, stricter standards on air pollution and greenhouse gas emissions associated with drilling, and the creation of a federal database so the public could better monitor drilling operations. The report also called for companies to eliminate diesel fuel from their fracking fluids and to disclose the full list of ingredients in those fluids. It was unclear who was supposed to enact these measures.

In October 2011, in response to intense pressure from environmental groups, the EPA announced that it would set national standards for the disposal of fracking wastewater. The standards focused on wastewater going to publicly owned wastewater treatment plants and so do not apply to most western states, where fracking waste is injected underground, buried, or stored in surface ponds.[31] Then, in April 2012—in response to litigation by two environmental groups—the EPA issued rules requiring new facilities to capture up to 95 percent of the volatile organic compounds (VOCs) emitted by fracked wells. These rules require "green completions," which involve capturing greenhouse gases and other toxic emissions from wells, compressors, storage sites, and pipelines. After complaints that the rule was costly and unnecessary, however, the EPA granted the industry more than two years to comply, thereby lowering their costs significantly.[32] The EPA estimates that implementation of this rule, which took effect in January 2015, will result in a 26 percent reduction in emissions from the natural gas sector.[33] President Trump has since rescinded these endeavors.

In March 2015, the Obama administration finalized rules for fracking on its 700 million acres of public land. The interior department estimates that 90 percent of the 3,400 wells drilled each year on public land use fracking.[34] In May 2012, the BLM released draft regulations that required public disclosure of chemicals used, established new guidelines for casing wells and testing well integrity, and contained requirements for submission of water management plans. In a concession to industry, the agency proposed allowing companies to reveal the composition of fracking fluids only after they had completed drilling—an about-face from the agency's original proposal, which would have required disclosure of the chemicals thirty days before a well could be started. That retreat followed a series of meetings at the White House after the original regulation was proposed in February. Lobbyists representing oil industry trade associations and major individual producers met with officials from the OMB, arguing that the additional paperwork would slow the permitting process and jeopardize trade secrets.

The OMB subsequently reworked the rule—although industry remained dissatisfied.[35] In mid-2013, after receiving 177,000 comments on its proposed rules, the administration issued revised regulations that contained even more concessions to industry. For example, the new version allowed oil and gas companies to keep some components of their fracking fluid secret and allowed them to conduct well-integrity tests on one representative well rather than all wells in a field whose geology and well-construction techniques were similar.[36] The final rules, issued in March 2015, require companies to publicly disclose the chemicals used within thirty days of completing fracking operations, via the industry-run website, FracFocus. The rules also include provisions governing inspection of wells and storage of chemicals around well sites. The Independent Petroleum Association of America immediately filed suit, calling the regulations "a reaction to unsubstantiated concerns."[37] These rules never went into effect. This is because President Trump argued that they placed an undue economic burden on energy development. Moreover, interior secretary Ryan Zinke's *Promoting Energy Independence and Economic Growth* (March 29, 2017) stated that fracking fluids were already disclosed, that the Obama administration rule is unnecessary, and that, therefore, it should be rescinded.[38]

Although the federal government has been hesitant to constrain the oil and gas industry, states where fracking is occurring have moved quickly to regulate the practice, although they have done so in very different ways.[39] Some have delegated independent authority to their environmental regulatory agencies, while others regulate fracking through their oil and gas commissions' well-permitting processes. Whatever regulatory entity has primary responsibility, states have numerous policy options. They can try to entice industry with accelerated permit processes, low severance tax rates, drilling incentives, and limited public disclosure programs.[40] Alternatively, they can take a more restrictive approach, imposing requirements for comprehensive disclosure of fracking chemicals; ensuring wells are adequately lined with casing and properly cemented; strictly regulating the storage and disposal of oil and gas wastes; creating stringent rules governing the prevention and reporting of spills; limiting the sources and quantity of water that can be withdrawn for fracking and requiring testing of water supplies before, during, and after fracking; issuing rules controlling air pollution from oil and gas operations; regulating the location of wells relative to protected resources, the construction of roads, and the restoration of well sites; and requiring bonding insurance for well sites to ensure they are properly restored.

THE CASE

With the practice spreading at lightning speed, by the late 2000s fracking had transformed the economies of oil- and gas-rich states and localities. Northeast Pennsylvania, for example, experienced a gas rush in response to discoveries in the Marcellus Shale, which stretches from New York to Tennessee and is one of the world's biggest natural gas fields (see Map 14.1). The public became aware of the potential of the Marcellus Shale in 2009, when Terry Engelder released his estimate that the formation contained at least 489 trillion cubic feet of technically recoverable gas.[41] But oil and gas companies had begun exploiting gas reservoirs in the Marcellus Shale four years earlier, with the "Renz" well in Washington County.[42] After a small number of initial wells were drilled, rapid expansion ensued, with 478 permits approved in 2008, 1,984 in 2009, and 3,314 in 2010. Total natural gas production in Pennsylvania reached 1.3 trillion cubic feet in 2011, more than six times the level in 2008.[43]

The influx of wells across rural Pennsylvania prompted citizens to mobilize in opposition, which in turn spurred growing media attention to fracking.[44] In particular, the experience of one northeast Pennsylvania town, Dimock, and subsequent coverage of its woes in the documentary film *Gasland*, had a disproportionate influence on public perceptions of fracking. Plans to frack in Dimock began in 2006,

Map 14.1 Marcellus Shale

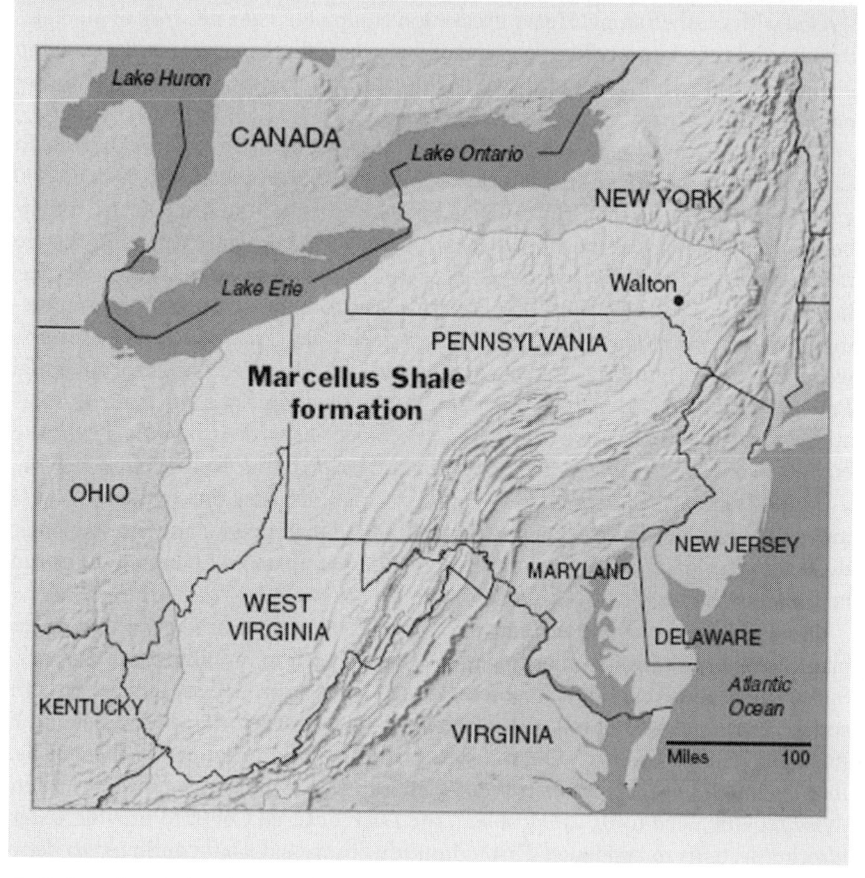

Credit: Penn State University Marcellus Center for Outreach and Research

inspired by the encouraging results of seismic surveys and a test well. Over the next two years, the Cabot Oil & Gas Corporation of Houston secured the rights to eighty-seven square miles in Susquehanna County, and in 2008 the president and CEO of Cabot—undaunted by the region's mountainous, forested terrain and narrow, winding roads—revealed to investors his intention to begin fracking there. As Wilber notes, when they signed their leases in 2006 and 2007, the residents of Dimock lacked the information that would later become available on the economics, environmental consequences, and other aspects of fracking. Moreover, Dimock sat in the midst of a northern Appalachian countryside that had always relied on natural resource extraction to support its economy and was experiencing high rates of poverty. So it was not surprising that many residents leapt at the chance to sign mineral leases. As a result, by fall 2008, "noise from well pad construction and drilling echoed through the hills and hollows of Dimock. . . . Flatbeds hauled rigs, generators, tanks, and massive pieces of plumbing to fields and woodlots cleared and leveled with crushed stone. Heavy excavation equipment cut a network of pipelines through fields and woods."[45] By the end of the year, Cabot's operations in the region included twenty wells, most of them in Dimock, producing 13 million cubic feet of natural gas per day.

It was construction-related accidents that first set off alarm bells in Dimock. In early July 2008 a truck knocked over a storage tank, spilling between 600 and 800 gallons of diesel fuel; other truck-related accidents soon followed. Then, in September, Julie and Craig Sautner found their tap water was brown, a dramatic change that occurred just as Cabot completed a gas well less than 1,000 feet away from their home. Soon several other Dimock residents near drilling operations began reporting problems with their well water. On New Year's Day 2009 Norma Fiorentino's well exploded, prompting an investigation by the Department of Environmental Protection (DEP). And in September 2009 Cabot had to suspend its fracking operations for three weeks after causing three spills of hazardous materials over the course of nine days. In November, as the investigation of the explosion dragged on, a group of Dimock residents filed suit against Cabot, accusing the company of causing methane to bleed into residents' water wells and then fraudulently asserting that the drilling process used in fracking could not contaminate groundwater and posed no threat to the people who lived there.[46]

Eventually, the DEP concluded that Cabot's poorly constructed wells were to blame for the high levels of methane in the water and air in various parts of Dimock. Specifically, it said that Cabot's contractors had failed to properly seal the wells with concrete; as a result, natural gas migrated upward through the voids outside the steel casing that lined the wells and into shallow aquifers, then into private wells. In November 2009 regulators ordered Cabot to repair the faulty gas wells. When those repairs failed to fix the problem, the DEP ordered Cabot to provide water filtering systems to residents. Throughout this process, Cabot continued to deny responsibility for contaminating residents' wells, although it did hire a contractor to deliver water to households whose wells had been disconnected. Finally, in December 2010, the DEP announced a deal with Cabot to pay $4.1 million to nineteen Dimock families whose water wells were contaminated with methane. The settlement was reached after outgoing DEP Secretary John Hanger agreed to drop a plan to force Cabot to build an $11.8 million water pipeline. Cabot also agreed to pay a fine of $500,000. In mid-October 2011 the DEP told Cabot it could stop water deliveries to homeowners whose wells had been contaminated. And in August 2012 the state gave Cabot the green light to resume well completion and production in Dimock—although not to begin new wells.[47]

The uproar in Dimock—which was reported on by *Vanity Fair, Discover, Forbes, 60 Minutes, The New York Times*, CNN, and NBC *Nightly News*[48]—piqued the interest of aspiring documentary filmmaker Josh Fox, who grew up in northern Pennsylvania. His film *Gasland* claimed that over its lifetime a well could be fracked up to eighteen times, requiring up to 7 million gallons of water and some 596 (undisclosed) chemicals; the life of each well could also involve at least 1,150 truck trips to haul water, sand, and chemicals, including between 400 and 600 tanker truck trips.[49] The film contained interviews with several distraught Dimock residents and featured a scene in which a Colorado resident lit his tap water on fire. Industry vociferously decried *Gasland*, pointing to inaccuracies and distortions in its depiction of fracking; more balanced reviewers also found fault with Fox's depiction.[50] Nevertheless, activists in New York recognized the film's potential to mobilize the public across the ideological spectrum, and they began inviting Dimock residents to speak at their events alongside high-profile fracking opponents.

Weighing the Pros and Cons of Fracking in New York

Like Susquehanna County, the Southern Tier of New York State, just over the Pennsylvania border, also sits atop the Marcellus Shale. In hopes of locking up drilling rights there, companies began approaching landowners in the Southern Tier around 2006, focusing on owners of large tracts of land, many of them farmers or descendants of farmers, and promising them a way to earn large sums with minimal risk.[51] In 2008 a group of 300 landowners who controlled 50,000 acres in Broome County struck a major deal with XTO Energy for $90 million—a figure that subsequently grew to $110 million as more landowners signed on. The terms included $2,411 per acre for leasing rights for five years and the same amount for a three-year extension. If prospectors hit gas, the landowner would get 15 percent of the royalties. After the so-called Deposit deal was announced, a gold-rush mentality took hold, and lease offers in Broome County towns soared.[52] As alarming reports began filtering out of Dimock, however, some upstate New York residents organized local groups to oppose fracking. Although New York was no stranger to oil and gas drilling—there were 13,000 active wells in the state as of 2010—fracking was new and posed poorly understood threats.[53] Concerned residents began to crowd auditoriums where the Department of Environmental Conservation (DEC) had organized public meetings, asking pointed questions about how the agency would manage the impacts of fracking, given a declining budget and inadequate staffing. Initially, representatives of the DEC were reassuring. But the more they dismissed residents' worries, the more agitated and skeptical residents became. As the possibility of fracking gained visibility, supporters joined forces as well, creating the Binghamton-based Joint Landowners Coalition of New York (JLCNY), an amalgamation of thirty-seven landowners' groups whose members collectively owned 800,000 acres across fourteen counties. As Carol Robinson, an unemployed pipeline worker from Steuben County and a member of JLCNY, explained, "We need to put the money in the hands of the people who own the land and the mineral rights. Not the politicians. It's our land, our royalties, and our lives. I am an American. We are the people. If we the people don't protect our rights, we have no one to blame but ourselves."[54] In July 2008, in an effort to make the state's permitting requirements more appropriate for fracking, then-governor David Paterson signed an industry-backed bill to streamline the permitting process by extending the state's uniform well-spacing system to unconventional shale gas development. At the same time, however, Paterson charged the DEC with completing a comprehensive review of the impacts of fracking before it could issue drilling permits—a Supplemental Generic Environmental Impact Statement (SGEIS).[55] (In the meantime, applicants could go through an ad hoc review that involved preparing a complete environmental impact statement for each well, something most were loath to do.) In September 2009 the DEC issued an 800-page SGEIS recommending that fracking be allowed with extensive safeguards in place. The proposed regulations did not ban drilling near sensitive watersheds, including the one-million-acre watershed that supplies drinking water to New York City and surrounding counties, as many activists had urged. Instead, they set forth strict rules for where wells could be drilled and required companies to disclose the chemicals in fracking fluids.

Antifracking activists seized on the process that accompanied the drafting of the SGEIS—public hearings and an extended comment period—to overwhelm the DEC with criticism of and concerns about fracking. Bolstering their efforts were findings by Walter Hang, head of an environmental research firm based in Ithaca, that were reported in the Binghamton *Press & Sun-Bulletin* in November 2009. Hang discovered 270 DEC files documenting wastewater spills, well contamination episodes, methane migration, and ecological damage related to gas production in the state dating back to 1979.[56] Ultimately, the draft SGEIS drew 14,000 comments, most of them complaining that the proposed rules were not restrictive enough. Weighing in just days before the comment period closed, the regional EPA office expressed major worries about the impact of fracking on public health and the environment. The agency expressed particular concern about the regional water supply, air quality, wastewater treatment, and the fate of radioactive materials dredged up during the drilling process.[57] Although state officials had anticipated releasing final regulations on fracking by the end of 2010, as the year drew to a close, no revised SGEIS was in sight. Instead, after vetoing legislation imposing a broad ban on fracking, on December 13, 2010, Governor Paterson issued an executive order imposing a fracking moratorium through July 1, 2011, to allow time for further environmental review and public comment—a major accomplishment for fracking opponents, who had more time for analysis and mobilization. (Meanwhile, industry was not sitting on its hands. Ten companies or trade groups that lobbied on fracking and other issues of concern to the natural gas industry spent $4.5 million lobbying in Albany between 2009 and 2011, according to the New York Public Interest Research Group.[58])

Complicating matters, in mid-May 2009, in the midst of New York's state-level review, the multi state Delaware River Basin Commission (DRBC) announced that natural gas producers would have to apply for commission approval before drilling in shale formations that lie within the Delaware River basin.[59] (The Marcellus Shale underlies an area that includes the Delaware River's headwaters. The Delaware River provides drinking water for 15 million people, from Philadelphia to New York. Portions of the river, whose watershed sustains a thriving ecosystem, have been designated "wild and scenic."[60]) Then, in July 2010, the commission declared a moratorium on drilling gas wells, even exploratory wells, in the upper Delaware watershed until it could approve new drilling regulations. In late 2010 the commission proposed new rules for the four-state watershed, but the moratorium was to remain in place until the rules were finalized after a 90-day comment period. The proposed rules gave companies incentives to drill away from forested areas and stream banks, required wastewater to be held in tanks rather than open ponds, prohibited burying drilling refuse on-site, and instituted a $125,000 per well fee as financial assurance for the plugging and restoration of the well site. By the end of 2011, however, overwhelmed by the 58,000 comments it received on the proposed regulations, the DRBC still had not finalized them.[61]

While the state and DRBC agonized, a vigorous debate over fracking in New York State erupted. Anthony Ingraffea, an engineering professor at Cornell who boasted decades of experience with oil and gas development, became a vocal antifracking advocate. In videos posted on YouTube and in more than fifty public presentations over the course of 2010, he described the process of drilling a fracking well and what could go wrong. He portrayed industry public relations campaigns as inaccurate and insincere and argued that large-scale gas drilling was incompatible with the economy of upstate New York, which relies heavily on agriculture, tourism, and recreation.[62] In response, Thomas West, a lawyer who represented the gas industry, focused on the plight of landowners, saying, "You're talking about people who are barely eking out an existence suddenly getting a huge source of revenue, which is huge for an economically challenged part of the state."[63] West pointed out that industry could spend its capital elsewhere, in states with less onerous regulatory climates. "New York has to be careful how it deals with these issues so it doesn't send a negative signal to the industry that chills the interest in the play in New York," he warned.[64] Scott Kurkoski, a lawyer for the Joint Landowners Coalition of New York, echoed West's assertion that the failure to develop gas would be an economic hardship to landowners, pointing out that drilling in Pennsylvania had created 44,000 jobs and had a $1.8 billion impact. Susan Christopherson, an economic geographer at Cornell, countered that although two-thirds of the short-term economic gains from fracking would accrue to upstate landowners, most of the jobs and long-term economic benefits would go elsewhere.[65]

With industry increasingly on the defensive, the antifracking movement gained momentum, thanks in part to a host of celebrities—including Natalie Merchant, Pete Seeger, Mark Ruffalo, Robert Kennedy Jr., Yoko Ono, Sean Lennon, Susan Sarandon, Margot Kidder, and Paul McCartney—who made appearances at antifracking rallies. The movement also gained credibility from the support of a group of professionals led by investor Chip Northrup. That team included Lou Allstadt, a retired Mobil vice president, Brian Brock, a geologist, and Jerry Acton, a retired systems engineer for Lockheed Martin.[66] Northrup's team emphasized that the terrain in New York was very different from that of the West and far more inhospitable to fracking—more vulnerable to groundwater contamination, in particular. Northrup's group also conducted analyses showing that the natural gas potential of the Southern Tier was well below industry estimates. And it touted the work of Elizabeth Radow, a lawyer who warned of risks to homeowners' property values, mortgages, and insurance.[67] Finally, as promised, in July 2011 the DEC released the details of proposed permit conditions and regulations for unconventional shale gas development. By this time there was a new governor, Democrat Andrew Cuomo, whose inclination was to allow fracking as long as the state imposed tight oversight.[68] Consistent with those preferences, the DEC recommended that the practice be permitted on private lands "under rigorous and effective controls"; for example, no permits would be granted for drilling within 500 feet of a private drinking water well or a domestic use spring or within 2,000 feet of a public reservoir or drinking water supply without further study. In addition, Cuomo's DEC commissioner, Joseph Martens, expressed support for a ban on fracking in the New York and Syracuse watersheds and in all primary aquifers statewide. Martens assured the public that the agency's proposals reflected lessons learned from fracking in states like Pennsylvania.

But environmentalists complained that the DEC's proposed rules came up short on issues like mapping buffer zones where drilling would be banned. They also pointed out that, although the state would require companies to disclose to state officials the chemicals and formulas they used, even those considered proprietary, the rules did not prohibit the use of toxic chemicals like benzene. Environmentalists also voiced concern about proposed rules governing the disposal of drilling waste. The draft called for drillers to submit a disposal plan for state approval before a permit was issued and for the state to adopt a tracking system to monitor the pathway taken by all liquid and solid wastes. Critics wanted the state to go a step further and designate the waste as hazardous material.[69] In hopes of defusing the controversy, the governor appointed a twelve-member High Volume Hydraulic Fracturing Advisory Committee and charged it with recommending oversight mechanisms for the DEC.

Finally, in early September 2011, the DEC released a complete, revised SGEIS (known as the rdSGEIS). Included in the document was an economic study prepared by the consulting firm Ecology and Environment, which argued that fracking could create up to 37,000 jobs and generate between $31 million and $185 million each year in additional state income taxes for New York at the peak of well development. It acknowledged, however, that communities on or near the Marcellus Shale would pay a heavy price for the economic bonanza. They would have to endure large-scale industrial activity, heavy truck traffic, and higher spending on police and fire protection, as well as rising home prices in response to the influx of new workers. With the state's population divided on the issue of fracking, this report promised to intensify the debate.[70] Sure enough, by the time the public comment period for the rdSGEIS closed in January 2012, the DEC had received 80,000 public comments, the most it had ever received on an environmental issue.[71] Most of those comments were critical of fracking and the proposed regulations. But some were supportive, including one from a member of the JLCNY that offered a comprehensive rejection of the antifracking case:

> Throughout the four years this issue has been debated, the forgotten voice has been the upstate landowner. There seems to be two sides to the issue: the person who uses energy, but doesn't feel comfortable living

near its development, and the industry that coaxes it from the ground. The former considers himself an environmentalist. I disagree. The true environmentalists are the farmers and other upstaters who have spent a lifetime taking care of their land, paying taxes, and braving the hardships that often come with rural life. NY uses more natural gas than any other state except CA and TX. But it imports 95 percent of it.

Many of the natural gas issue opponents acknowledge that natural gas is better than coal and oil, but feel we can get the gas from somewhere else. And I've heard folks say that since there already is enough gas coming out of PA's ground, there's no need for us to put up with truck traffic through the Catskills. Everyone has a right to their opinion, but what's wrong with this debate is that instead of rolling up their sleeves and finding the best environmental protections we can enact, the anti-natgas folks perpetuate misinformation—such as the famed flaming faucet—which was proven before it even appeared in Gasland to be caused by naturally occurring methane. The best chance we have at protecting our environment and saving the people too—those that have been without economic opportunity for decades—is to have earnest, informed landowners serve as stewards of their own properties with airtight, environmentally sound leases.

The Joint Landowners Coalition of NY (JLCNY) has done just that. Spending thousands of hours studying this issues [*sic*] in the SGEIS, they have provided two dozen pages of highly informed and technical input into the most effective ways to prevent gas migration, which setbacks make the most sense, and ways to reduce truck traffic. They have taken a pragmatic but earnest look at protecting grasslands and birds; at the best recommendations for handling fluids and protecting water; and verified for themselves the characteristics of the most secure well pads. The farmers I know wouldn't risk their land or their way of life for anything. They don't have to. We can have safe natural gas development in NY, just as the thirty or so other states do. We can be responsible for producing the energy we use. And that's greener than importing it.[72]

In mid-January 2012, as the agency was once again sifting through comments, the EPA regional office weighed in, posting on its website dozens of suggested revisions to New York's plans for fracking regulation. The recommendations, which touched on most of the issues raised by environmentalists, focused on the disposal of potentially radioactive wastewater and the safety record of recycled wastewater, the importance of predrilling tests of residential water wells, the adequacy of proposed buffer zones around selected water supply wells, and the need to evaluate the environmental impact of new gas pipelines. In addition, EPA regional director Judith Enck (formerly New York's deputy secretary for the environment) again raised questions about whether the DEC, which had been dealing with staff cuts in recent years, was prepared to oversee natural gas drilling.[73] Meanwhile, the governor's fracking advisory panel seemed irrelevant: its January 11 meeting was canceled abruptly, and the rescheduled January 26 meeting was also canceled. The panel never reconvened—another indication Cuomo was wavering on the subject of fracking.[74]

Enacting a Ban in Dryden

In August 2011, as the state was wrestling with how to devise rules that would satisfy both proponents and opponents of fracking, the town of Dryden (population 14,500) approved a change in its zoning ordinance making oil and gas activities, including fracking, "prohibited uses" of land within the town's ninety-four square miles. The Anschutz Corporation, which held most of the leases within the town, promptly sued. Although Dryden was not the first locality to enact such a ban, nor the first where a ban was subject to legal challenge, the stakes in Dryden were particularly high because, unlike many towns that had enacted such bans, Dryden actually sits atop significant natural gas reserves: it is in Tompkins County, one of two New York counties where the Marcellus Shale formation juts out from Pennsylvania.[75] The process that led to the Dryden ban started in the spring of 2009, when town officials began to engage with the Tompkins County Council of Governments, a voluntary regional organization built around issue-focused working groups, including a gas drilling task force.[76] The task force was particularly interested in the experience with fracking of towns to the south, in Pennsylvania.

As town officials initiated their investigation, a handful of citizens energized by concern about the possibility of fracking in their neighborhoods created the Dryden Resources Awareness Coalition (DRAC). Among the founders of DRAC was Marie McRae, a long-time farmer who was first approached by a representative of the oil and gas industry in 2007. Over the next year, companies had approached McRae six more times, telling her they would drill whether or not she signed the lease—but that signing was the only way she could protect her farm. (companies commonly argue that by not signing, landowners are punishing their neighbors, will be excluded from a good deal, and will be confronted with nearby oil and gas development anyway.[77])

In May 2009, McRae bowed to the pressure and signed a five-year lease on her thirteen acres, receiving $1,200 for doing so. "I felt really backed into a corner," she told journalist Kate Sheppard.[78] Other DRAC founders were Deborah and Joan Cipolla-Dennis, who had also been approached by a company but refused to sign the proffered lease. Some who joined DRAC did so after attending educational meetings around the county hosted by the county council or by scientists at Cornell.[79] Like the county council's task force, the members of DRAC agreed that they needed to find out more about the experience of fracking in nearby Pennsylvania.[80] DRAC's leaders soon learned that two lawyers from a nearby town were investigating New York State law and had made a promising discovery. Lawyers David and Helen Slottje realized that under New York's Oil, Gas and Solutions Mining Law, municipalities could not regulate fracking. But under their home-rule authority, they could use local zoning laws to prohibit oil and gas activities altogether.

In 2010, following the Slottjes' advice, DRAC members began collecting signatures from their fellow residents on a petition asking the town board to develop an industrial noise ordinance and make a firm statement against drilling for natural gas. Since their goal was to convey widespread unease about fracking, they solicited support from residents across the political spectrum, many of whom were primarily worried about the ancillary impacts of fracking, like truck traffic, air pollution, and increased demand for local services.[81] After the board received the petition containing 1,600 signatures—"enough to make or break an election for a town board member"[82]—it held a series of public meetings over a period of four to five months. Although the majority of attendees supported strong restrictions on fracking, there was opposition to the move as well. As Town Supervisor Mary Ann Sumner recounts, "It was pretty damn passionate on both sides. And everybody [was] scared. . . . The pro-ban folks [were] scared that their lives [were] going to be changed, and the anti-ban folks [were] scared that their rights [were] being diminished somehow." Although the townspeople were divided, DRAC founder Marie McRae sympathized with her critics. "A lot of farmers around here are land rich and money poor," she explained. "So they had a lot of acres that could be drilled, theoretically, and they signed that lease with the thought that they could be a millionaire."[83]

Dryden officials spent about a year determining their legal position with respect to fracking. They also weighed the strength of the opposition: about one-third of the town's land was already leased, and a ban would mean that residents who had signed those leases would not see the royalty revenues they had anticipated. In addition, local officials were cognizant that the town was home to a sizable property rights group. Ultimately, the town supervisor, attorney, and planning director decided that an outright ban would be more legally defensible than, while achieving the same results as, a complex set of zoning amendments that would permit some industrial activity but exclude oil and gas drilling. When the issue was put to a vote in August, the town board unanimously passed a series of zoning amendments that prohibited oil and gas exploration, extraction, storage, treatment, disposal, and "support activities." Voters tacitly endorsed the board's decision in the fall of 2012, when they reelected Town Supervisor Mary Ann Sumner by nearly 60 percent.[84] The Anschutz Corporation, which had leased 22,200 acres in Dryden for a total of $4.7 million, immediately filed suit to challenge the town's ban.[85] The plaintiffs argued that Dryden's ordinance was precluded by New York's Environmental Conservation Law, which includes an express preemption provision that says, "The provisions of [Article 23 governing mineral resources] shall supersede all local laws or ordinances related to the regulation of the oil, gas and solution mining industries; but shall not supersede local government jurisdiction over local roads or the rights of local governments under the real property tax law."[86] So the question before the courts was whether Dryden's ban "related to the regulation of" natural gas extraction.

In February 2012, a New York State supreme court ruled in favor of Dryden.[87] In essence, the judge concluded that state law precluded municipalities from regulating the "how" of fracking but not the "where."[88] In May 2013 the four judges that heard the case for the appellate court agreed unanimously that state law "does not preempt, either expressly or impliedly, a municipality's power to enact a local zoning ordinance banning all activities related to the exploration for, and the production or storage of, natural gas and petroleum within its borders."[89] The ruling of the intermediate court prompted a third and final round of litigation.

In June 2014 the State Court of Appeals upheld (by a vote of five to two) the lower courts' rulings, with support from both liberal and conservative judges. Thomas West, the attorney for Norse Energy, which became the plaintiff after Anschutz declared bankruptcy, had tried to persuade the court that Dryden—as well as the town of Middlefield, which had also enacted a fracking ban—had exceeded its authority, adding that a ruling in favor of the towns would "have a very chilling effect" and would be "very hard for operators to justify spending hundreds of millions of dollars to come in and not have regulatory certainty."[90] Unmoved, Judge Victoria A. Graffeo, who was appointed by Republican Governor George Pataki, wrote, "We do not lightly presume preemption where the preeminent power of a locality to regulate land use is at stake. Rather, we will invalidate a local zoning law only where there is a 'clear expression of legislative intent to preempt local control over land use.'" She added, "At the heart of these cases lies the relationship between the State and its local government subdivisions, and their respective exercise of legislative power. These appeals are not about whether hydrofracking is beneficial or detrimental to the economy, environment or energy needs of New York, and we pass no judgment on its merits."[91] Rather, she and the four judges who joined her said the language and history of the state oil and gas law indicated no intent to exercise its power to supersede local home rule. Industry groups condemned the decision. Frank Macchiarola, executive vice president for government affairs at America's Natural Gas Alliance, told journalist Steven Mufson that a patchwork of local regimes was problematic for companies and that state-level regulation was preferable for a number of reasons, one of which was the expertise brought to bear by the State Department of Environmental Conservation, versus that of the local or county council.[92]

The State Enacts a Ban of Its Own

The fracking bans enacted by Dryden—and eventually more than 100 other New York towns—eased the pressure on Governor Cuomo to move quickly on the issue, while also signaling strong antifracking sentiment within the Southern Tier. The Cuomo administration remained tight-lipped about its plans for fracking throughout 2012 and 2013, but observers speculated that the governor's preference was for a plan to allow limited fracking in the Southern Tier, where it had been heavily promoted as a source of economic revival for depressed communities. Under Cuomo's proposed plan, permits would be issued only in those Southern Tier communities that supported the technology. According to journalist Danny

Hakim, with this approach Cuomo was "trying to acknowledge the economic needs of the rural upstate areas, while also honoring the opposition expressed in some communities, and limiting the ire of environmentalists."[93] But Sandra Steingraber, founder of New Yorkers Against Fracking, decried the governor's idea. "Sending a polluting industry into our most economically impoverished communities is a violation of environmental justice," she said. "Partitioning our state into frack and no-frack zones based on economic desperation is a shameful idea."[94] While the governor delayed, the antifracking forces continued to gain steam. Although some environmentalists and community activists started out open to the idea of fracking with safeguards, as time wore on momentum built for complete opposition, with national groups like the Natural Resources Defense Council and the Sierra Club gradually shifting their positions.[95] A key focus of the antifracking groups in 2012 was persuading the state to conduct a comprehensive public health assessment.[96] To that end, hundreds of doctors, scientists, and environmentalists had sent a letter to then-Department of Health (DOH) Commissioner Dr. Nirav Shah asking that the DOH partner with the DEC on the SGEIS process, based on growing evidence of threats to public health.[97] Subsequently, physicians with close personal ties to the governor weighed in as well. In September 2012 the Cuomo administration agreed to postpone a decision on fracking until it had conducted a thorough review of the potential public health impacts—another signal that the administration's position was evolving.[98]

In late November 2012, the DEC released an ostensibly final draft of its fracking regulations. Despite allowing just thirty days for public comment, over the holiday season, the agency received more than 200,000 comments, almost all of them once again opposed to the proposed rules. Over the next year, the antifracking forces were relentless; they hounded the governor at every opportunity, with a growing chorus of celebrities lending their voices to the movement and taking direct aim at the governor.[99] At one point, the governor told Susan Arbetter, host of *Capitol Pressroom*, that the antifracking campaign was the most effective political action campaign he had seen.[100] As 2014 wore on, a strong primary challenge from a previously unknown candidate who opposed fracking, combined with falling natural gas prices, numerous local bans, and the fact that many leases had expired during the drawn-out process of developing rules, decisively altered the governor's political calculus.[101]

Finally, in mid-December 2014, Governor Cuomo announced that the state would ban fracking altogether because of concerns about health risks. Cuomo acknowledged that his decision would leave the economic problems of Southern Tier counties unaddressed. But he explained, "I've never had anyone say to me, 'I believe fracking is great.' Not a single person in those communities. What I get is, 'I have no alternative but fracking.'"[102] Cuomo announced the ban at the same time as he presented the state's health study, which concluded that there were "significant uncertainties about the kinds of adverse health outcomes that may be associated with [fracking], the likelihood of the occurrence of adverse health outcomes and the effectiveness of some of the mitigation measures in reducing or preventing environmental impacts which could adversely impact public health."[103] Among the specific impacts identified by the study were the following:

- Methane from gas wells has leaked into the atmosphere. Other dangerous gases like benzene and volatile organic compounds are also found in the air near fracking wells.
- Children born in areas with fracking are more likely to be underweight and to have congenital defects of the heart and neural tube.
- Communities with fracking experience overburdened roads, more accidents, and greater crime and other social problems.
- Methane leaking from gas wells contributes to climate change.
- Fracking can induce small-scale earthquakes.

In justifying the ban, Cuomo went to considerable lengths to portray it as a "decision made by experts objectively weighing the facts," saying, "I am not a scientist. I'm not an environmental expert. I'm not a health expert. I'm a lawyer. I'm not a doctor. I'm not an environmentalist. . . . So let's bring the emotion down, and let's ask the qualified experts what their opinion is."[104] For his part, acting State Health Commissioner Dr. Howard A. Zucker said the investigation had revealed "significant public health risks" associated with fracking. He adopted a precautionary rationale, saying there was insufficient evidence to affirm the safety of the practice. And he explained that his view boiled down to a simple question: Would he want his family to live in a community where fracking was occurring? His answer was no. "We cannot afford to make a mistake," he said. "The potential risks are too great. In fact, they are not even fully known."[105]

Environmentalists lauded Cuomo's decision, maintaining the rhetoric of objectivity. "Governor Cuomo repeatedly stated that he would follow the science when making a decision on fracking" said Adrienne Esposito, executive director of the Albany-based Citizens Campaign for the Environment. "Today he has been proven a man of his word by making a conclusive decision based on scientific facts and clear evidence."[106] But industry representatives were disgusted. "For the time-honored natural gas industry, 'the new New York' is not open for business," complained Brad Gill, director of the Independent Oil and Gas Association of New York. "Thousands of families just had their property rights extinguished," said Karen Moreau of the New York State Petroleum Council.[107] Although the JLCNY threatened a "takings" lawsuit and members of the legislature quickly proposed bills to override the governor's action, these actions were largely symbolic; veteran observers of the battle did not expect profracking forces to mount an effective response. As Tom West explained, "I don't know of anybody . . . who cares enough about New York now to invest the money in that kind of legal fight. We've been shut down for six years. Most of the industry has left. They're not going to invest to try and turn around a political decision when they have plenty of other shale resources in other states."[108]

OUTCOMES

New York's story is just one of many; for better or worse, abdication by the U.S. federal government has resulted in wide variation in the treatment of fracking among the states. States vary in terms of whether or not they regulate a particular aspect of fracking, the mechanism used to regulate (rules, performance standards, inducements, or other means), the stringency of the regulation, and the extent of enforcement. As of 2013 (when the report was last revised), thirty-one states had actual or potential shale gas production. Of those, nine states had no laws requiring disclosure of chemicals used, while seventeen did have chemical disclosure laws—although comprehensive disclosure of all chemicals used or produced onsite was the exception, not the rule. Only six states required disclosure before drilling occurred; the rest required it only after drilling was completed. And in some states access to the information was limited. (Most relied on FracFocus, a database maintained by the Ground Water Protection Council and Interstate Oil and Gas Compact Commission that was launched in 2011.) Similarly, a majority of states that allowed fracking did not have laws requiring water quality monitoring near drilling sites. The laws on water quality monitoring that did exist were often ineffective, relying on voluntary action, for example, or requiring baseline or ongoing monitoring but not both. Laws on air quality monitoring were rare as well, with states most commonly regulating venting or flaring of gas.[109]

States also vary in their treatment of municipalities seeking to limit or ban fracking within their borders. In Pennsylvania, for instance, Republican Governor Tom Corbett worked with the state's Republican legislature to pass the Unconventional Gas Well Impact Fee Act, known as Act 13. Although Act 13 strengthened many state environmental regulations for unconventional wells, it also dramatically expanded state preemption of local control.[110] Specifically, the act provided that any local ordinance regulating oil and gas activity must "authorize oil and gas operations, other than activities at impoundment areas, compressor stations and processing plants, as a permitted use in all zoning districts," except for residential areas. It also required municipalities to "authorize natural gas compressor stations as a permitted use in agricultural and industrial zoning districts and as a conditional use in all other zoning districts" as long as the compressor stations meet state standards. Another section of the law called for statewide rules on oil and gas to preempt local zoning rules.

By 2012, however, seven Pennsylvania municipalities had already banned fracking within their borders, and several of these municipalities joined forces with the Delaware Riverkeeper Network to challenge this expansion of state authority. Specifically, they argued that Act 13 denied local governments the ability to use zoning powers to protect natural resources granted under the Municipal Planning Code.[111] In July 2012 the Pennsylvania Commonwealth Court struck down by a four-to-three vote many of the statute's preemption provisions. Specifically, the court found that some provisions violated due process by forcing municipalities to "violate their comprehensive plans for growth and development."[112] Governor Corbett appealed the ruling to the State Supreme Court. But in December 2013 that court ruled by a four-to-two margin that two provisions of Act 13—which it described as a "blanket accommodation of industry and development"—violated the Environmental Rights Amendment to the Pennsylvania constitution.[113] That amendment guarantees Pennsylvanians the right to "clean air, pure water, and to the preservation of the natural, scenic, historic and esthetic values of the environment." In January 2014 the Corbett administration asked the court to reconsider its decision, but it declined to do so.

Even states whose officials are not particularly receptive to the exercise of local control have had to respond to such uprisings. In Colorado, for example, fracking of the Niobrara shale has caused wells to proliferate in Front Range Urban communities that are unaccustomed to industrial activity. In response, between mid-2011 and 2012 several Colorado municipalities and counties adopted policies that restricted the issuance of drilling permits. The Colorado Oil and Gas Conservation Commission (COGCC) responded by threatening legal action against localities that enacted bans or regulatory restrictions. By contrast, despite being a supporter of fracking, Governor John Hickenlooper adopted a conciliatory stance aimed at tamping down local activism. He began in 2011 by working with the COGCC to devise one of the nation's strictest chemical disclosure rules for fracking fluids. In his State of the State address in January 2012, he emphasized the need to work cooperatively with local governments while also stressing the importance of regulatory consistency. He then established a twelve-member task force to coordinate state and local activity on fracking. The task force was to address issues such as setback requirements, floodplain restrictions, noise abatement, air quality, traffic, financial assurance, and the protection of wildlife and livestock. The task force did not propose any changes in the division of authority between state and local governments, instead calling for "collaboration and coordination."

One concession made by the COGCC in response to a task force recommendation was to adopt a rule allowing a city or county to designate a well-site inspector and forward evidence of violations to COGCC staff.[114] Efforts by the governor and the COGCC to quell local resistance were only partly successful, however. The Front Range city of Longmont placed on its ballot for the fall 2012 election an initiative to ban fracking within city limits. Despite a costly campaign by industry, which outspent supporters of the initiative by thirty to one, nearly 60 percent of Longmont voters supported the ban.[115] The Colorado Oil and Gas Association, an industry trade group, promptly filed suit challenging the ban. For its part, the COGCC responded to local unrest by enacting strict new rules on oil and gas development that included a new groundwater testing requirement for water sources within a half mile of new wells, as well as more restrictive setback requirements that lengthened the distance between drilling sites and occupied structures. Despite these concessions, the relationship between state regulators and local officials remained uneasy. In fall 2013 four more cities passed measures calling for bans or long moratoria on fracking, thereby even further increasing the pressure on the state to enact restrictive rules.

In February 2014 Colorado's Air Quality Control Commission adopted rules requiring oil and gas companies to find and fix methane leaks, as well as to install technology that captures 95 percent of VOCs and methane. Governor Hickenlooper said both the rules and the collaborative process that produced them should reassure people about the state's willingness to impose controls on fracking.[116] Then, in a compromise aimed at avoiding additional local bans, in fall 2014 Governor Hickenlooper established a new, twenty-one member panel to draft legislative protections for residents who lived near existing or proposed oil and gas drilling sites. (While that panel deliberated, the Colorado courts threw out several local bans, ruling them unconstitutional.)

The main reason states are working so hard to accommodate local opposition and avert pressure to ban fracking altogether, as New York did, is the desire to benefit from the "shale-gas revolution," which has created jobs, reduced consumer costs, and stimulated growth. According to an industry-sponsored study released in early December 2011 by IHS Global Insight, natural gas drilling supported more than 600,000 jobs nationwide in 2010, about one-quarter of them in the gas industry, another one-third in "indirect" jobs supplying the industry, and the remainder "induced" by further spending throughout the economy. By the end of 2011 shale gas accounted for 34 percent of domestic gas production and was expected to account for 60 percent by 2035. The study projected that natural gas, which averaged $6.73 per thousand cubic feet between 2000 and 2008, would cost an average of $4.79 through 2035. In the absence of shale gas, the price of natural gas would be between $10 and $15 per thousand cubic feet, and meeting domestic demand would require importing gas from overseas.[117] Proponents of lax regulations also pointed out that, as fracking lowered the price of domestic natural gas, some manufacturing industries—particularly producers of steel, petrochemicals, and fertilizer—were returning to the United States. On the other hand, those persistently low prices signaled a glut of natural gas: by March 2018, the U.S. benchmark price of natural gas was $2.69 per million Btus and production was at 2,400 billion cubic feet for the month of January 2018.[118] Jobs in the oil and gas sector were disappearing fast. Prices in the United States were so low that companies were pushing for licenses to export liquefied natural gas—a move that not only environmentalists but also domestic manufacturers opposed because it would negate the price advantage that lured them home in the first place.[119]

Further undermining any prospects for long-term economic benefits were the declining projections of shale gas reserves. A 2014 analysis of the major shale plays in the United States indicates that the EIA's forecasts for the medium and long term were wildly optimistic. (This was not the first time: in 2012 the EIA lowered its estimate of natural gas in the United States by 40 percent—from 827,000 cubic feet in its 2011 report to 482,000 cubic feet—and in the Marcellus Shale by 66 percent—from 401,000 cubic feet in the 2011 report to 141,000 cubic feet.[120]) After analyzing a wealth of data, geologist J. David Hughes concluded that shale gas production from the top seven plays will likely peak before 2020. Barring major new discoveries on the scale of the Marcellus, production will be far below the EIA's forecast by 2040.[121]

The most significant consequence of state efforts to mollify locals has been the passage of restrictive laws and regulations. Although the industry has strongly opposed such constraints on its activities, some innovative technologies and practices have emerged in response to either regulation or the prospect of regulation or liability. For example, wastewater recycling took hold in Pennsylvania after alarming reports about radioactivity in wastewater that was being taken to the state's wastewater treatment plants. Because of the state's geology, unground injection of drilling wastewater was infeasible; therefore, in 2008 and 2009 oil and gas companies sent about half the waste they generated to public sewage treatment plants, disposing of the rest directly into rivers or trucking it to New York or West Virginia. It soon became clear that total dissolved solids in fracking wastewater were overwhelming the capacity of several Pennsylvania waterways to absorb those pollutants. So, following a series of critical media reports and complaints from public water suppliers, the DEP wrote strict new discharge standards that took effect in 2011 over drillers' objections.[122]

Then, in April 2011, regulators asked gas drillers in the Marcellus to voluntarily stop sending wastewater to fifteen treatment plants.[123] Between them, these moves led to a dramatic increase in wastewater recycling in Pennsylvania. By contrast, Texas has been slow to adopt wastewater recycling because despite its scarcity, fresh water is inexpensive, and disposal wells are prevalent. (Texas is home to about 7,500 active disposal wells.[124]) Also, in hopes of allaying concerns about excessive water use, fracking operations in Canada and some U.S. states have begun to employ fracking gel made of liquefied propane, which exits the well as vapor after use and can be collected for reuse. Unlike water, propane does not return to the surface carrying chemicals, salts, and radioactivity. But there are several obstacles to the widespread adoption of this process, which was developed by a small Canadian company named GasFrac. Water is cheap, and the infrastructure for using it is in place; by contrast, switching to propane involves higher up-front costs for the propane itself and for the infrastructure to capture and reuse it. (Propane is also explosive and must be handled carefully, using special equipment.) Representatives of oil and gas producers also cite a lack of published data on the effectiveness of gel fracking as an obstacle to adopting it.[125]

CONCLUSIONS

In a federal system like that of the United States, advocates argue over the appropriate political level at which to make decisions. Ideally, local communities would be empowered to act on issues that affect them exclusively, while larger political units—the states or the federal government—would address problems whose consequences cross political boundaries. This case makes abundantly clear, however, that decisions about which level of government ought to address a particular problem have little to do with principle and much to do with politics. For many fracking-related issues, it arguably makes more sense to adopt a single, national rule that protects all citizens equally than to allow dozens of disparate approaches. Why, for example, should each state have a different rule for the disclosure of chemicals in fracking fluid? Why, for that matter, should companies be allowed to conceal the composition of that fluid in the name of trade secrets, if there is any possible threat to public health? Yet, the federal government has largely abdicated responsibility for fracking regulation in response to political pressure from the oil and gas industry, which has capitalized on political opportunities to gain exemptions from the nation's main environmental laws. Similarly, it is political factors, rather than geographic, geological, hydrological, or environmental ones, that account for variation in states' regulatory approaches. And it is political considerations that shape states' responses to localities that seek to ban or restrict fracking within their borders—decisions that may fall within their traditional authority over land use but can easily result in a patchwork of well-to-do communities and "sacrifice zones."

With most fracking regulation occurring at the state level, many environmental advocates are working with citizen groups to either defend local fracking bans or enhance the stringency and efficacy of state-level regulations. Only in New York have antifracking activists gained enough momentum to persuade state officials to enact a statewide fracking ban. One key to their victory was the persistence with which they pursued Governor Cuomo personally; a second was the use of a canny legal strategy to defend local bans, which in turn put swaths of potentially valuable land off limits to fracking, while also communicating local grievances to state officials; and a third was their ability to exploit the regulatory process to delay rulemaking until economic conditions had shifted.

Now that fracking is prohibited, however, New York State is left with the question of how it will obtain the massive amounts of energy its residents demand, while also moving away from fossil fuels. As commentator James Conca points out, natural gas is the largest supplier of electricity in New York State, followed by nuclear power and hydropower. Preventing local natural gas production means increasing reliance on other energy sources or increasing pipeline delivery to the state, which is also controversial.[126] State officials are also left with the quandary of economic development in rural New York, where struggling counties and landowners had banked on fracking (as well as on a casino, which the governor also ruled out) to resuscitate their finances.

Across the United States, rural communities have long been subject to the boom-and-bust cycles that accompany natural resource exploitation (e.g., North Dakota). Each time a new opportunity arises, local boosters are eager to seize it, despite overwhelming evidence that the economic benefits are short-lived and most of them do not accrue to local residents. Fracking is just the latest iteration of this phenomenon. A more visionary strategy for rural economic development—perhaps one built around the construction of renewable energy and storage capacity, as well as around sustainable agriculture—might simultaneously nurture rural economies, while also building support for a future that does not depend on fossil fuels.

QUESTIONS TO CONSIDER

- Do you think relegating responsibility for regulating shale gas development to the states increases the likelihood that economic development considerations will trump environmental concerns—particularly long-term, global ones like climate change? Why, or why not?
- Why might the abundant natural gas supplies provided by fracking slow the transition to renewables, and what, if anything, should government do to avert that outcome?
- How should state officials approach fracking if their goal is to reap the economic benefits of shale gas development while also protecting their residents from harm?
- Many assert that a single federal regulation would simplify the debate and processes of fracking. Who would benefit more from a sweeping federal rule: the environmentalists, industry, or both?

NOTES

1. The oil and gas industry insists it has been "fracking" for decades without serious problems. Contemporary use of the term, however, denotes a process that combines hydraulic fracturing and horizontal drilling to gain access to deposits of natural gas (and oil) within shale formations deep underground. More precise but cumbersome terms for this process include high-volume hydro fracking and unconventional shale gas development.
2. In 2012 Vermont banned fracking, but that state has negligible shale gas reserves.
3. The agreed-on principle at the root of the normative debate is that the jurisdiction of the responsible institution should be aligned with the scope of the problem. The debate, then, typically revolves around the nature of the problem itself.
4. Michael Burger, "Fracking and Federalism Choice," *University of Pennsylvania Law Review Online* 150 (2013). Available at http://www.pennlawreview.com/responses/2-2013/Burger.pdf; David B. Spence, "Federalism, Regulatory Lags, and the Political Economy of

Energy Production," *University of Pennsylvania Law Review* 161 (2013): 431–508; Barbara Warner and Jennifer Shapiro, "Fractured, Fragmented Federalism: A Study in Fracking Regulatory Policy," *Publius: Journal of Federalism* 43,3 (2013): 474–496.

5. Ann O'M. Bowman and Richard C. Kearney, "Second-Order Devolution: Data and Doubt," *Publius* 41,4 (2013): 563–585. Dillon's Rule is the nineteenth-century legal precedent according to which any legal authority wielded by local officials is derived from the state constitution but may be extended by subsequent legislative enactments.

6. Shaun A. Goho, "Municipalities and Hydraulic Fracturing: Trends in State Preemption," *Journal of Planning and Environmental Law* 64,7 (July 2012): 3–9.

7. See, for example, Mireya Navarro, "Institute's Gas Drilling Report Leads to Claims of Bias and Concern for a University's Image," *The New York Times*, June 12, 2012; Andrew Revkin, "When Agendas Meet Science in the Gas Drilling Fight," *The New York Times*, July 23, 2012.

8. See, for example, Andrew Revkin, "When Publicity Precedes Peer Review in the Fight Over Gas Impacts," *The New York Times*, July 25, 2012.

9. See Tom Wilber, *Under the Surface: Fracking, Fortunes, and the Fate of the Marcellus Shale* (Ithaca: Cornell University Press, 2012).

10. Wald, Matthew, "Sluggish Economy Curtails Prospects for Building Nuclear Reactors," *The New York Times*, October 11, 2010; Douglas Martin, "George Mitchell, 94, a Pioneer in Hydraulic Fracturing," *The New York Times*, July 27, 2013; Andrew C. Revkin, "Daniel Yergin on George Mitchell's Shale Energy Innovations and Concerns," *The New York Times* (Dot Earth Blog), July 29, 2013.

11. Warner and Shapiro, "Fractured, Fragmented Federalism." According to Warner and Shapiro, as of 2012 Colorado had the largest number of gas wells, at 3,850, many of them extracting coal-bed methane; next was Texas with 3,454 wells. Pennsylvania was third at 1,825.

12. U.S. Energy Information Administration, "Shale Gas Provides Largest Share of U.S. Natural Gas Production in 2013," *Today in Energy* (Website), November 25, 2014. Available at http://www.eia.gov/todayinenergy/detail.cfm?id=18951.

13. Jim Motavalli, "Natural Gas Signals a 'Manufacturing Renaissance,'" *The New York Times*, April 11, 2012.

14. Unconventional plays require well pads that are four to six times larger than conventional, single-acre pads. The larger size is necessary for the higher number of tank farms, reservoirs, mixing areas, waste pits, pipes, equipment, and trucks needed to drill and stimulate a horizontal well. See Wilber, *Under the Surface.*

15. The level of total dissolved solids (TDS) is a measure of all organic and inorganic substances in water that scientists use to gauge the quality of freshwater systems.

16. Gretchen Goldman et al., *Toward an Evidence-Based Fracking Debate: Science, Democracy, and the Community Right to Know in Unconventional Oil and Gas Development* (Cambridge, Mass.: Union of Concerned Scientists, October 2013). Available at http://www.ucsusa.org/assets/documents/center-for-science-and-democracy/fracking-report-full.pdf; Alan Krupnick, Hal Gordon, and Sheila Olmstead, Overview of

"Pathways to Dialogue: What the Experts Say About the Environmental Risks of Shale Gas Development" (Washington, D.C.: Resources for the Future, 2013). Available at http://www.rff.org/Documents/RFF-Rpt-PathwaystoDialogue_Overview.pdf; Hannah Wiseman, "The Capacity of States to Govern Shale Gas Development Risks," *Environmental Science & Technology* 48,15 (2014): 8376–8387.

17. Goldman et al., *Toward an Evidence-Based Fracking Debate;* Wiseman, "The Capacity of States to Govern."

18. Stanley Reed, "Cracking the Energy Puzzles of the 21st Century," *The New York Times*, October 15, 2013. Natural gas supplied just under one-third of U.S. electricity in 2012, up from 16 percent in 2000, while coal-fired electricity dropped from 50 percent to about 36 percent over the same period. The CO2 Scorecard Group, which is funded by a company that sells energy-efficiency services to large corporations and governments, argues that aggressive energy efficiency efforts led to the decline in energy use in the United States and are therefore primarily responsible for the decline in carbon emissions. See Maria Gallucci, "Efficiency Drove U.S. Emissions Decline, Not Natural Gas, Study Says," *Inside Climate News*, July 30, 2013. Available at http://insideclimatenews.org/news/20130730/efficiency-drove-us-emissions-decline-not-natural-gas-study-says.

19. A 2011 report by Howarth et al. estimated that between 3.6 percent and 7.9 percent of methane was leaking into the atmosphere at various points in the shale gas production life cycle. The authors concluded that natural gas was therefore not a bridge to a more sustainable future but, as coauthor Anthony Ingraffea put it, "a gangplank to a warm future." Other scientists challenged both the data on which Howarth et al. based their estimated leakage rates and their assertions about the relative global warming potential of methane and carbon dioxide (CO_2). They also questioned the authors' gas-to-coal comparison, noting that, because gas-powered electricity generation is far more efficient than coal-based generation, a gigajoule of gas produces a lot more electricity than a gigajoule of coal. Perhaps most important from the perspective of defenders of natural gas, the short lifetime of methane in the atmosphere means that the climate impact of methane emissions disappears within twenty years of ceasing those emissions, whereas CO_2 accumulates in the atmosphere practically indefinitely, so CO_2 emissions are, for all intents and purposes, irreversible. See Tom Zeller Jr., "Poking Holes in a Green Image," *The New York Times*, April 12, 2011; Andrew C. Revkin, "On Shale Gas, Warming and Whiplash," *The New York Times* (Dot Earth Blog), January 6, 2012; Andrew C. Revkin, "Two Climate Analysts Fault Gas Leaks, but Not as a Big Warming Threat," *The New York Times* (Dot Earth Blog), August 1, 2013; Justin Gillis, "Picking Lesser of Two Climate Evils," *The New York Times*, July 8, 2014.

20. A 2014 review of twenty years of technical literature on natural gas emissions in the United States and Canada found that (1) measurements at all scales showed that official inventories consistently underestimated actual methane emissions, with natural gas and oil sectors as important contributors; (2) many independent experiments suggest that a small number of "superemitters" could be responsible for a large fraction of the leakage; (3) recent regional atmospheric studies with very high emissions rates were unlikely to be representative of typical natural gas system leakage rates; and (4) assessments using

100-year impact indicators showed that systemwide leakage is unlikely to be large enough to negate climate benefits of coal-to-natural-gas substitution. The authors conclude, "If natural gas is to be a 'bridge' to a more sustainable energy future, it is a bridge that must be traversed carefully: Diligence will be required to ensure that leakage rates are low enough to achieve sustainability goals." See A. R. Brandt et al., "Methane Leaks from North American Natural Gas Systems," *Science* 343,6172 (February 2014): 733–735.

21. Matthew Carr, "Low Natural Gas Prices Hamper U.S. Energy Efficiency, IEA Says," *Bloomburg Business*, October 16, 2013; Energy Modeling Forum, "Changing the Game?: Emissions and Market Implications of New Natural Gas Supplies," Stanford University, EMF Report 26, vol. 1 (September 2013); Christine Shearer, John Bistline, Mason Inman, and Steven J. Davis, "The Effect of Natural Gas Supply on US Renewable Energy and CO2 Emissions," *Environmental Research Letters* 9 (2014).

22. Wilber, *Under the Surface*, 82.

23. Ian Urbina, "Deadliest Danger Isn't at the Rig But On the Road," *The New York Times*, May 15, 2012.

24. Goldman et al., *Toward an Evidence-Based Fracking Debate*; Wilber, *Under the Surface*.

25. FRAC Act. Available from: https://www.congress.gov/bill/115th-congress/senate-bill/865.

26. Burger, "Fracking and Federalism Choice." This provision was designed to invalidate an Eleventh Circuit Court decision made in 1997 that forced the State of Alabama to regulate fracking fluids under the Safe Drinking Water Act. Environmentalists had hoped to use that decision to force other states to adopt regulations under the act. Congress justified the exemption on the grounds of a 2004 study whose findings, according to EPA whistle-blower West Wilson, were strongly influenced by pressure from industry and its political allies. (Five of the seven members of the study's peer review panel were current or former employees of the oil and gas industry.) See Goldman et al., *Toward an Evidence-Based Debate;* Theo Stein, "Group: Extraction Process Threatens Water," *The Denver Post*, April 14, 2005; Ian Urbina, "Pressure Stifles Efforts to Police Drilling for Gas," *The New York Times*, March 4, 2011.

27. Specifically, in December 1978, the EPA proposed hazardous waste management standards that included requirements for several types of "special wastes" that the EPA believed were lower in toxicity than other wastes regulated as hazardous under RCRA. Those included "gas and oil drilling muds and oil production brines." In amendments to the act passed in 1980, Congress expanded the exemption to include "drilling fluids, produced water, and other wastes associated with the exploration, development, or production of crude oil or natural gas" and codified the exemption pending a regulatory determination by the EPA. In 1988 the EPA issued a regulatory determination stating that control of exploration and production wastes under RCRA Subtitle C (hazardous waste) regulations was not warranted. See Environmental Protection Agency, "Exemption of Oil and Gas Exploration and Production Wastes from Federal Hazardous Waste Regulations," EPA530-K-01-004, October 2002. Available at http://www.epa.gov/osw/nonhaz/industrial/special/oil/oil-gas.pdf. Journalist Ian Urbina notes that EPA researchers had, in fact, concluded that some oil and gas waste was hazardous and should be tightly controlled; but according to Carla Greathouse, the author of the 1987 study on which the 1988 decision was based, EPA

officials altered her findings in response to pressure from the White House Office of Legal Counsel. See Urbina, "Pressure Stifles Efforts to Police Drilling."

28. Federal law does require fracking operations to file material safety data sheets with local governments for each hazardous chemical present at a site. See Spence, "Federalism, Regulatory Lags, and the Political Economy of Energy Production."

29. Warner and Shapiro, "Fractured, Fragmented Federalism."

30. Initially, researchers planned to consider the dangers of toxic fumes released during drilling, the impact of drilling waste on the food chain, and the risks of radioactive waste to workers. Earlier planning documents also called for study of the risks of contaminated runoff from landfills where drilling waste is disposed and included plans to model whether rivers can adequately dilute hazardous wastewater from gas wells. According to agency scientists and consultants, these topics were eliminated in response to pressure from industry and its allies in Congress and justified by time and budgetary constraints. See Ian Urbina, "Pressure Stifles Efforts to Police Drilling for Gas," *The New York Times*, March 4, 2011. For details on industry obstruction of EPA data collection efforts, see Neela Banerjee, "Can Fracking Pollute Drinking Water? Don't Ask the EPA," *InsideClimate News*, March 2, 2015.

31. Bruce Finley, "Oil, Gas Companies Begin to Reveal Fracking Chemicals," *The Denver Post*, April 24, 2011.

32. John M. Broder, "U.S. Caps Emissions in Drilling for Fuel," *The New York Times*, April 19, 2012.

33. Spence, "Federalism, Regulatory Lags, and the Political Economy of Energy Production."

34. John M. Broder, "New Fracking Rules Proposed for U.S. Land," *The New York Times*, May 17, 2013.

35. John M. Broder, "New Rule Requires Disclosure of Drilling Chemicals, but Only After Their Use," *The New York Times*, May 5, 2012.

36. Broder, "New Fracking Rules Proposed."

37. Quoted in Coral Davenport, "New Federal Rules Are Set for Fracking," *The New York Times*, March 20, 2015.

38. Bureau of Land Management. BLM Rescinds Hydraulic Fracturing Rule. December 28, 2017. Available at https://www.blm.gov/press-release/blm-rescinds-rule-hydraulic-fracturing.

39. Barry G. Rabe, "Shale Play Politics: The Intergovernmental Odyssey of American Shale Governance," *Environmental Science & Technology* 49,15 (2014): 8369–8375; Richardson et al., *The State of State Shale Gas Regulation;* Warner and Shapiro, "Fractured, Fragmented Federalism.

40. A severance tax is a fee assessed on the extraction of a nonrenewable resource on privately owned land (or where mineral rights are privately owned).

41. See Anon., "How Much Natural Gas Can the Marcellus Shale Produce?" Penn State Extension (Website), February 5, 2012. Estimates of the amount of the technically and

economically recoverable oil and gas are notoriously unreliable. Technological advances can dramatically change what is deemed technically recoverable. What is economically recoverable depends on a variety of factors, including technology but also the value of the resource on the market. That figure, in turn, is influenced by supply, demand, and the availability of infrastructure. All of these are affected by regulation, which is unknown. See Tom Wilber, "NY Shale Prospects Dim Six Years After Leasing Frenzy," Shale Gas Review (Blog), May 22, 2014.

42. Goho, "Municipalities and Hydraulic Fracturing."

43. Wilber, *Under the Surface;* Goho, "Municipalities and Hydraulic Fracturing."

44. Charles Davis, "Substate Federalism and Fracking Policies: Does State Regulatory Authority Trump Local Land Use Autonomy," *Environmental Science & Technology* 48,15 (2014): 8397–8403.

45. Wilber, *Under the Surface*, 75.

46. Wilber, *Under the Surface;* Abrahm Lustgarten, "Pa. Residents Sue Gas Driller for Contamination, Health Concerns," ProPublica, November 20, 2009. Available at http://www.propublica.org/article/pa-residents-sue-gas-driller-for-contamina tion-health-concerns-1120; Jad Mouawad and Clifford Krauss, "Dark Side of a Natural Gas Boom," *The New York Times*, December 8, 2009.

47. Anon., "Pa., Cabot Reach Settlement Over Methane Contamination," *Greenwire*, December 16, 2010; Clifford Krauss and Tom Zeller Jr., "When a Rig Moves in Next Door," *The New York Times*, November 7, 2010; Andrew Maykuth, "Susquehanna Residents Wary of Gas-Drilling Operation," *The Philadelphia Inquirer*, December 13, 2009; Andrew Maykuth, "Pa. Fines Chesapeake Energy Corp. $1.1 Million For Drilling Violation," *The Philadelphia Inquirer*, May 18, 2011; Andrew Maykuth, "Long Fight Over Fracking Still Divides Pa. Town," *The Philadelphia Inquirer*, August 26, 2012. In November 2011 the EPA began a three-month project to test dozens of water wells in Dimock, in the meantime paying to deliver fresh water to four homes where the agency had discovered high levels of manganese, sodium, and arsenic. In December the agency told Dimock residents it could find no evidence that their water wells posed an immediate health threat to water users. And in May 2012 the agency announced that its test results of well-water samples did not reveal elevated levels of contaminants related to fracking. See Mike Soraghan, "Pa. Town's Water Not 'An Immediate Threat'—EPA," *E&ENewsPM*, December 2, 2011.

48. Wilber, *Under the Surface*.

49. Warner and Shapiro, "Fractured, Fragmented Federalism."

50. See, for example, Energy in Depth, "Debunking Gasland." Available at energyindepth/wp-content/uploads/2010/06/Debunking-GasLand.pdf. For a more balanced review, see Mike Soraghan, "Ground Truthing Academy Award Nominee 'Gasland,'" *The New York Times*, February 4, 2011.

51. Wilber, *Under the Surface*.

52. Tom Wilber, "How Fracking Got Stopped in New York," Shale Gas Review (Blog), December 20.

53. Peter Applebome, "Before a 'Yes' on Drilling, Smart Moves," *The New York Times*, June 17, 2010.

54. Quoted in Wilber, *Under the Surface*, 147.

55. The SGEIS amended a 1992 GEIS that had been drafted with the aim of expediting oil and gas permitting by relieving operators of the requirement to prepare an environmental impact statement for each well under the state's Environmental Quality Review Act.

56. Wilber, "How Fracking Got Stopped."

57. Jad Mouawad, "State Issues Rules on Upstate Natural Gas Drilling Near City's Water," *The New York Times*, October 1, 2009; Mireya Navarro, "E.P.A., Concerned Over Gas Drilling, Questions New York State's Plans," *The New York Times*, December 31, 2009.

58. Danny Hakim, "Cuomo Proposal Would Restrict Gas Drilling to Struggling Region," *The New York Times*, June 14, 2012.

59. Congress created the DRBC in 1961 to resolve decades of conflict over water supplies between New York, Pennsylvania, New Jersey, and Delaware. The commission's powers trump those of its member states when it comes to the 13,539-square-mile Delaware River basin. See Mike Soraghan, "Obscure Regulator Hits Brakes on Northeast Shale Rush," *Greenwire*, September 13, 2010.

60. Sandy Bauers, "Shale-Gas Regulation Near River Divides Pa.," *The Philadelphia Inquirer*, April 3, 2011.

61. Sandy Bauers, "Suit Seeks Study of Effects of Gas Drilling on Delaware River Basin," *The Philadelphia Inquirer*, June 1, 2011.

62. Applebome, "Before a 'Yes' on Drilling."

63. Quoted in Peter Applebome, "The Light Is Green, and Yellow, on Drilling," *The New York Times*, July 27, 2008.

64. Quoted in Peter Applebome, "Putting Water Ahead of Natural Gas," *The New York Times*, August 10, 2008.

65. Applebome, "Before a 'Yes' on Drilling"; Peter Applebome, "On Drilling, Pleasing Both Sides," *The New York Times*, December 13, 2010.

66. Wilber, "NY Shale Prospects Dim."

67. Chip Northrup, "A Brief History of the New York State Frack Ban Movement," *No Fracking Way* (Blog), December 21, 2014. Available at http://www.nofrackingway.us/2014/12/21/a-brief-history-of-the-new-york-frack-ban-movement/.

68. Danny Hakim and Nicholas Confessore, "Cuomo Moving To End a Freeze On Gas Drilling," *The New York Times*, July 1, 2011.

69. Mireya Navarro, "Latest Drilling Rules Draw Objections," *The New York Times*, July 15, 2011.

70. Mireya Navarro, "Report Outlines Rewards and Risks of Upstate Natural Gas Drilling," *The New York Times*, September 8, 2011.

71. Mireya Navarro, "State Plans Health Review As It Weighs Gas Drilling," *The New York Times*, September 21, 2012.

72. Quoted in Andrew C. Revkin, "Beyond Hype, A Closer Look at New York's Choice on Shale Gas," *The New York Times* (Dot Earth Blog), January 23, 2012.

73. Mireya Navarro, "Evaluating Feedback on Fracking Rule," *The New York Times*, January 13, 2012; Brian Nearing, "EPA Questions Fracking Study," *Albany Times Union*, January 12, 2012.

74. Mireya Navarro, "New York's Fracking Deliberations Inch Along," *The New York Times*, January 27, 2012.

75. Jesse McKinley, "Town's Fracking Fight May Decide Future of the Oil and Gas Industry in New York," *The New York Times*, October 24, 2013; Steven Mufson, "How Two Small New York Towns Have Shaken Up the National Fight Over Fracking," *The Washington Post*, July 2, 2013.

76. Jessica C. Agatstein, "Localities and Their Natural Gas: Stories of Problem Diffusion, State Preemption, and Local Government Capacity," Master of City Planning thesis, Department of Urban Studies and Planning, Massachusetts Institute of Technology, 2012.

77. Under New York's "compulsory integration" law, passed in 2005, gas companies can drill under land without the owner's consent if they have leases in most of the surrounding area. Owners without leases would still get royalties of 12.5 percent on gas extracted from beneath their property, but they would not get the other protections (and financial benefits) available to leaseholders. See Mireya Navarro, "At Odds Over Land, Money and Gas," *The New York Times*, November 28, 2009.

78. Kate Sheppard, "This Town Took on Fracking and Won," *Mother Jones*, May 8, 2013.

79. Agatstein, "Localities and Their Natural Gas."

80. Sheppard, "This Town Took on Fracking."

81. Sheppard, "This Town Took on Fracking."

82. Marie McRae, quoted in Agatstein, "Localities and Their Natural Gas."

83. Quoted in Sheppard, "This Town Took on Fracking."

84. Agatstein, "Localities and Their Natural Gas."

85. Sheppard, "This Town Took on Fracking."

86. Quoted in Goho, "Municipalities and Hydraulic Fracturing."

87. In the New York State court system, cases are originated in the supreme courts, appealed to the intermediate appellate courts, and are finally resolved by the Court of Appeals.

88. Goho, "Municipalities and Hydraulic Fracturing."

89. Quoted in Sheppard, "This Town Took on Fracking."

90. Quoted in Mufson, "How Two Small New York Towns Have Shaken Up the National Fight."

91. Quoted in Mufson, "How Two Small New York Towns Have Shaken Up the National Fight."

92. Mufson, "How Two Small New York Towns Have Shaken Up the National Fight."

93. Hakim, "Cuomo Proposal Would Restrict Gas Drilling."

94. Quoted in Hakim, "Cuomo Proposal Would Restrict Gas Drilling."

95. Peter Applebome, "Drilling Critics Face a Divide Over the Goal of Their Fight," *The New York Times*, January 10, 2012.

96. Peter Iwanowicz, executive director, Environmental Advocates of New York, personal communication, February 8, 2015.

97. Bill Huston, "A Timeline of Fracking in New York," Bill Huston's Blog. Available at http://www.williamahuston.blogspot.com/2013/02/a-timeline-of-fracking-in- ny.html.

98. Mireya Navarro, "State Plans Health Review As It Weighs Gas Drilling," *The New York Times*, September 21, 2012.

99. Mireya Navarro, "Celebrities Join to Oppose Upstate Gas Drilling," *The New York Times*, August 23, 2012.

100. Tom Wilber, "Cuomo's Choice to Ban Fracking Driven by Science, Politics," *Shale Gas Review* (Blog), December 17, 2014.

101. A full two-thirds of the potential 12 million acres containing natural gas in the Marcellus Shale in New York would be off limits anyway because of local bans, being part of the watersheds of New York City or Syracuse, being part of an aquifer, or having gas that was located within 2,000 feet of the surface. See Brian Nearing, "Citing Perils, State Bans Fracking," *Albany Times Union*, December 18, 2014. A report by the League of Women Voters released in April 2014 concluded that fracking in New York would not be economically viable below $6 per MMBtu, and even at that price, production would be modest. A price of $8 per MMBtu would encourage more production but not the bonanza many people were expecting in 2008. See Wilber, "NY Shale Prospects Dim."

102. Quoted in Thomas Kaplan, "Citing Health Risks, Cuomo Bans Fracking in New York State," *The New York Times*, December 17, 2014.

103. Quoted in Nearing, "Citing Perils."

104. Quoted in Kaplan, "Citing Health Risks."

105. Quoted in Kaplan, "Citing Health Risks."

106. Quoted in Nearing, "Citing Perils."

107. Quoted in Kaplan, "Citing Health Risks."

108. Quoted in Chip Northrup, "What the Frack Happens Now New York?" *No Fracking Way* (Blog), December 20, 2014.

109. Goldman et al., *Toward an Evidence-Based Fracking Debate*; Richardson et al., *The State of State Shale Gas Regulation*.

110. Under Pennsylvania's Home Rule Chapter and Optional Plans Law, local officials are empowered to make policy and regulatory decisions, as long as those policies don't conflict with prior policies enacted by the state's general assembly. Initially, policymakers

dealt with fracking under the Pennsylvania Oil and Gas Act (POGA), administered by the Bureau of Oil and Gas Management within the DEP. That law expressly preempted any local ordinance that conflicted with its substantive provisions. (It did, however, allow for locally crafted rules under the Municipalities Planning Code [MPC] or the Floodplain Management Act.) In a pair of decisions issued on the same day in 2009, the Pennsylvania Supreme Court interpreted POGA to prohibit ordinances that "imposed conditions, requirements, or limitations on the same features of oil and gas activities regulated by the Act" but to allow ordinances that "sought only to control the location of wells consistent with established zoning principles." See Goho, "Municipalities and Hydraulic Fracturing."

111. Davis, "Substate Federalism and Fracking Policies"

112. Quoted in Hannah H. Wiseman, "Pennsylvania's Supreme Court Affirms Unconstitutionality of Parts of Pennsylvania's Act 13," Environmental Law Prof Blog, December 19, 2013. Available at http://lawprofessors.typepad.com/environmental_law/2013/12/pennsylvania-supreme-court-affirms-unconstitutionality-of-parts-of-pennsylvanias-act-13.html.

113. Quoted in Barry G. Rabe, "Shale Play Politics: The Intergovernmental Odyssey of American Shale Governance," *Environmental Science & Technology* 49,15 (2014): 8369–8375.

114. Davis, "Substate Federalism and Fracking Policies"; Goho, "Municipalities and Hydraulic Fracturing."

115. Davis, "Substate Federalism and Fracking Policies"; Mufson, "How Two Small New York Towns Have Shaken Up the National Fight."

116. Mike Lee, "Colo.'s New Air Rules Dampen the Push to Ban Fracking—Governor," *EnergyWire*, March 6, 2014.

117. Andrew Maykuth, "Industry Study Touts Large Economic Impact of Shale-Gas Drilling," *The Philadelphia Inquirer*, December 6, 2011.

118. US EIA. U.S. Dry Natural Gas Production (Billon Cubic Feet) [Table]. March 30, 2018. Retrieved from https://www.eia.gov/dnav/ng/hist/n9070us1m.htm

119. Clifford Krauss, "Reversal of Fortune for U.S. Gas," *The New York Times*, January 5, 2013; Matthew Wald, "Would Exporting the Natural Gas Surplus Help the Economy, or Hurt?" *The New York Times* (Green Blogs), January 11, 2013.

120. Ian Urbina, "New Report by Agency Lowers Estimates of Natural Gas in U.S.," *The New York Times*, January 29, 2012.

121. David J. Hughes, Drilling Deeper: A Reality Check on U.S. Government Forecasts for a Lasting Tight Oil & Shale Gas Boom (Santa Rosa, Calif.: Post Carbon Institute, 2014). Available at http://www.postcarbon.org/publications/drillingdeeper/.

122. Andrew Maykuth, "Shale Drillers Tout Recycling As Option for Wastewater," *The Philadelphia Inquirer*, March 23, 2011. Those standards—updates to Chapter 95 of Pennsylvania's Clean Streams law—limited the discharge of drilling wastewater effluent to TDS levels of 500 milligrams per liter—the federal and state standard for drinking water. See Wilber, *Under the Surface*.

123. Andrew Maykuth, "Pa. Drillers Told to Stop Sending Wastewater to Treatment Plants," *The Philadelphia Inquirer*, April 20, 2011.

124. Jim Malewitz and Neena Satija, "In Oil and Gas Country, Water Recycling Can Be an Extremely Hard Sell," *The New York Times*, November 22, 2013.

125. Anthony Brino and Brian Nearing, "New Waterless Fracking Method Avoids Pollution Problems, But Drillers Slow to Embrace It," *Inside Climate News*, November 6, 2011. Available at http://insideclimatenews.org/news/20111104/gasfrac-pro pane-natural-gas-drilling-hydraulic-fracturing-fracking-drinking-water-marcellus-shale-new-york.

126. James Conca, "New York Fracking Ban Contrary to State's Energy Future," *Forbes*, December 27, 2014.

CHAPTER FIFTEEN

MAKING TRADE-OFFS

Urban Sprawl and the Evolving System of Growth Management in Portland, Oregon

It can be said that the story of growth management in Portland is really two stories: the first describes the evolution of Portland's unique growth management experiment; the second describes both the sources and implications of the backlash against it. At the heart of this case lies the issue of sprawl. After World War II, federal programs—particularly the establishment of the federal highway system and the Federal Housing Administration's insurance program for long-term housing construction and purchase loans, which favored single-family homes over apartment projects—promoted sprawling, low-density development.[1] Reinforcing the trend toward suburbanization was the introduction of shopping malls that could be reached only by car and the exodus of manufacturing from cities in response to the availability of trucking as an option for transporting freight. Then, beginning in the 1950s and intensifying in the 1960s and 1970s, middle-class white Americans fled the cities in response to a wave of migration of blacks from the South and, later, school desegregation efforts. With the out-migration of prosperous families and businesses, urban tax bases shrank, making it difficult for municipal governments to provide services to their hard-pressed residents. By the 1980s these patterns had become entrenched, and America's older cities were dotted with pockets of extreme poverty and crime, even as the suburbs that surrounded them thrived.

By the latter two decades of the twentieth century, however, those suburbs—as well as rural counties—were choking on new development. As they became disillusioned by the environmental, aesthetic, and social costs of new growth, local officials adopted regulations on lot coverage, floor-area ratio, the number of unrelated people who may live together, and even buildings' architectural details—all aimed at promoting low-density development and separation of land uses.[2] These measures in turn forced development even farther out in a "leapfrog" process that placed corporate offices out of commuting range of city dwellers. In response, people moved still farther from the city and its nearby suburbs, spreading the urbanizing area even more.

Its detractors point out that, aside from its aesthetic drawbacks, sprawl has serious environmental costs: it eliminates wetlands and forests, alters water courses and runoff patterns, disrupts scenic vistas, destroys and fragments wildlife habitat, and damages local air and water quality. Sprawling development also generates more energy use, and hence more greenhouse gas emissions, than more compact development. Transportation-related energy use is lower in compact cities whose residents are more inclined (and able) to walk, bicycle, or take public transit.[3] Building energy use is also lower in compact cities primarily because residents tend to live in smaller units with shared walls, reducing the amount of energy needed for heating and cooling.[4] Moreover, because power lines are shorter it is more efficient to transmit energy in more compact cities; in fact, sufficient density creates the possibility of adopting district energy systems that are far more efficient than the conventional grid. In addition to its environmental costs, sprawl has severe economic and social consequences: communities become more segregated by income, and urban disinvestment reduces opportunities for the poor.

Public concern about sprawl originated in the early 1970s, as the impacts of the rapid postwar suburbanization manifested themselves. In response, some states adopted policies to redirect public and private investment away from infrastructure and development that spreads out from already-built areas. Hawaii enacted the first state-level growth management law in the 1960s, and in the next three decades nine states—Vermont (1970 and 1988), Florida (1972, 1984, and 1986), Oregon (1973), New Jersey (1986), Maine (1988), Rhode Island (1988), Georgia (1989), Washington (1990), and Maryland (1992 and 1997)—passed laws that aim to regulate the rate and location of growth.[5] Early growth management policies included constraints on the intensity of development through zoning limitations on subdivisions, design and capacity standards for lots and buildings, requirements for adequate impact fees, reductions in the supply of land open for development, and restrictions on where development is permitted.[6] In the 1990s the term *smart growth* came into vogue to describe a more proactive and comprehensive growth management approach designed specifically as an antidote to sprawl. Among the crucial elements of smart growth are creating walkable cities; establishing fast, efficient public transportation systems; and locating dense, mixed-use development at transit nodes (see Box 15.1). More recently, cities have begun promoting compact urban form in the name of *urban sustainability*—a term that comprises efforts to enhance cities' contribution to societal efforts to become more environmentally sustainable.[7]

Not everyone agrees, however, with either the sprawl diagnosis or the smart-growth prescription. Planning scholars Marcial Echenique and his coauthors challenge the notion that compact development is the urban form most able to accommodate growth in an environmentally sustainable fashion. They also point to the potential negative consequences of compactness—including higher housing costs, crowding, and congestion.[8] Planning Professor Robert Bruegmann argues that sprawl simply reflects people's growing ability to satisfy their desire for space, privacy, and mobility.[9] Libertarian Randal O'Toole goes even further, offering a scathing indictment of planning in general and growth management in particular. Smart growth, he says, is "based on the design fallacy, the idea that urban design shapes human behavior."[10] Like Bruegmann, O'Toole dismisses the notion that Americans actually want to live in dense cities, disputes the claim that sprawl consumes excessive amounts of land, and strongly opposes "coercive" policies formulated by elites who want to impose their preferences on others.

But planning scholar Jonathan Levine contends the causes of sprawl are more complex than antiregulatory advocates acknowledge. He argues that municipal regulations impede compact, mixed-use development, even where there is public demand for it; therefore, reform should seek to eliminate such regulatory impediments, not create new mandates. His argument rests on the recognition that sprawl is not an organic by-product of consumers' free choice. Rather, ubiquitous municipal land-use regulations strongly constrain market processes. By treating municipal zoning as a market force, he notes, defenders of sprawl short-circuit public debate over the appropriate scope of local action. In reality, efforts to mitigate sprawl by removing regulatory impediments to compact development provide consumers with greater choice. Thus, he concludes, "Where compact, mixed-use development is no longer zoned out, people who prefer these alternatives would be able to get more of what they want in their transportation and land-use environment."[11]

This case goes beyond theoretical debates over policies aimed at preventing sprawl, however; it also concerns the implementation of those policies. Critics' main complaint about Portland is that rigid bureaucrats administer rules with insufficient

BOX 15.1 SMART GROWTH GOALS AND PRINCIPLES

- Preserve, if not advance, public goods such as air, water, and landscapes.
    - Prevent further expansion of the urban fringe.
    - Use a systems approach to environmental planning.
    - Preserve contiguous areas of high-quality habitat.
    - Design to conserve energy.
- Minimize, if not prevent, adverse land-use impacts (effects on others).
    - Prevent negative externalities among land uses.
    - Separate auto-related land uses from pedestrian-oriented uses.
- Maximize positive land-use impacts; for example, neighborhood schools create synergy.
    - Achieve a job/housing balance within three to five miles of development.
    - Design a street network with multiple connections and direct routes.
    - Provide networks for pedestrians and bicyclists that is as good as the network for motorists.

(Continued)

(Continued)

- o Incorporate transit-oriented design features.
- o Achieve an average net residential density of six to seven units per acre (clustering with open space).

- Minimize public fiscal costs.
  - o Minimize cost per unit of development to provide public facilities and services.
  - o Channel development into areas that are already disturbed.
- Maximize social equity.
  - o Balance jobs and housing within small areas; provide accessibility to work, shopping, leisure; offer socioeconomic balance within neighborhoods.
  - o Provide for affordable single-family and multifamily homes for low- and moderate-income households.
  - o Provide life-cycle housing.

Source: Adapted from Arthur C. Nelson, "How Do We Know Smart Growth When We See It?" in *Smart Growth: Form and Consequences*, ed. Terry S. Szold and Armando Carbonell (Washington, D.C.: Island Press, 2002), 83–101.

flexibility. But defenders respond that only the strict application of clear standards will contain the powerful forces in favor of sprawling development. Political scientists have long bemoaned the trade-offs between flexibility and accountability. Proponents of flexibility point out that it allows administrators to respond to new information and capitalize on emerging opportunities.[12] But the more flexible a regulatory regime, the less predictable its result and the more tempting it is to engage in unprincipled compromise, self-dealing, and even agency capture (see Chapter 12)—particularly in an urban context, in which development interests and their allies wield disproportionate power.[13] Moreover, as noted elsewhere in this volume (see Chapter 12, for example), strictly enforced rules tend to yield results more reliably than instruments that rely primarily on voluntary cooperation by the private sector.

In Portland, direct democracy in the form of ballot initiatives has been a key mechanism for resolving disputes between those who favor growth management and those who want to fend off or dismantle such policies. Twenty-four states allow voters who gather enough signatures to place a measure on the ballot for voters' consideration. Adopted during the Progressive Era as a means to curb patronage and ensure government responsiveness, use of the ballot initiative has surged in recent decades. Scholars have demonstrated that ballot initiatives have direct effects on policy and politics—by instituting term limits or tax cuts, for example, or by placing an issue on the political agenda. Other scholars have amassed evidence to support the claim that direct democracy leads to outcomes that are closer to what the public wants and that information cues enable voters to make reasoned choices, even though they often lack extensive knowledge of the issues at stake.[14] It has proven more difficult to demonstrate conclusively that well-heeled interests exert undue influence on the initiative process.[15] Similarly, evidence of the hypothesized procedural impacts of direct democracy—secondary effects that include enhanced turnout, voter knowledge, and political efficacy—is elusive.[16] The Portland case provides fodder for these scholarly debates; in particular, it shows how ballot initiatives can capture a genuine public sentiment, influence both the political agenda and the content of public policy, and—even when ultimately overturned—shape the behavior of government officials responsible for implementing policy.

BACKGROUND

Journalist David Goldberg depicts Portland as "among the nation's most livable cities, lacking the blight of so many central cities while offering a broad array of vibrant urban and suburban neighborhoods." He attributes this outcome to the fact that "citizens of all stripes have a say in making the plans everyone will live by."[17] As Carl Abbott observes, however, Portland's contemporary image contrasts sharply with its persona in the 1950s and 1960s.[18] According to Henry Richmond, cofounder of the environmental group 1000 Friends of Oregon, prior to 1970 Portland was anything but vibrant: "There was no street life. The Willamette River was polluted. There weren't a lot of jobs. . . . Public transportation was poor."[19] Similarly, community activist Steven Reed Johnson describes Portland in the 1950s as a "strikingly dull and derivative city, only a restaurant or two above a logging town."[20] The city, he says, was run by white men, and the rare instances of citizen involvement consisted of discussions among well-known elites.

Portland's politics underwent a dramatic transformation during the late 1960s and early 1970s, however. Between 1969 and 1973, the average age of Portland City Council members dropped by nearly fifteen years, as civic leadership shifted from men in their sixties and seventies to men and women in their thirties.[21] During the same period, the number of advocacy organizations in the city rose from 31 to 184.[22] The first outward sign of this "revolution" was a protest on August 19, 1969, when about 250 people gathered for a picnic in the median strip of Harbor Drive, a freeway that separated the city from the waterfront. The goal of the protest was to generate support for a proposal (that eventually succeeded) to tear down the highway and replace it with a waterfront park. A second indication of the sea change under way in Portland was the 1972 release—after three years of collaborative work by business owners, planners, and citizens—of the Portland Downtown Plan. The plan rejected conventional urban redevelopment orthodoxy and instead responded to overwhelming public sentiment in favor of "a pedestrian atmosphere with interesting and active streets."[23] It provided for new parks and plazas, high-density retail and office corridors, improved mass transit, and pedestrian-oriented street design.

Environmental concern and a growing willingness to adopt innovative approaches extended beyond Portland's city limits. By the early 1970s the state of

Oregon had acquired a reputation as an environmental leader by cleaning up the Willamette River, banning flip-top beverage containers, requiring a deposit on beer and soft drink cans and bottles, and setting aside 1 percent of its highway revenues for bike path construction. As planning scholar John DeGrove explains,

> Oregonians seemed to feel that the state enjoyed a special place in the world by virtue of its natural features and that these should be protected. The beauties of a wild and rugged coast, the pastoral valley stretching for a hundred miles between mountain ranges, and the rugged sagebrush and timber areas of eastern Oregon all were seen by many residents of the state as a precious and valuable heritage that made the state different from most other places.[24]

Denizens of the 100-mile-long Willamette Valley, which runs from Portland to Eugene and is home to 70 percent of the state's population, were especially disconcerted by the symptoms of sprawl and determined not to repeat the mistakes made by California, where a natural paradise was rapidly disappearing. For some Oregonians, rampant coastal development was particularly disturbing: throughout the 1960s developers erected condominiums, high rises, and amusement parks along the seacoast, prompting Mark Hatfield, the Republican governor, to dub the coastal Lincoln City "the 20 miserable miles."[25] Others worried that the loss of rural land presaged the demise of Oregon's rich agricultural heritage.

Alarmed by the appearance of new houses encroaching on farmland, in 1971 newly elected Republican legislator and dairy farmer Hector MacPherson of Linn County became a champion of statewide planning. He joined forces with Portland Democrat Ted Hallock and Republican Governor Tom McCall in a campaign to link Oregon's increasingly salient problems to their preferred solution: a new state planning law. While MacPherson and Hallock recruited potential allies inside and outside the legislature, McCall used his formidable rhetorical powers to mobilize the public. In 1972 the governor made national news with a speech in which he linked victims to villains in a simple causal story, declaiming,

> There is a shameless threat to our environment . . . and to the whole quality of life—[that threat] is the unfettered despoiling of the landscape. Sagebrush subdivisions, coastal "condomania," and the ravenous rampage of suburbia in the Willamette Valley all threaten to mock Oregon's status as the environmental model for the nation. We are in dire need of a state land use policy, new subdivision laws, and new standards for planning and zoning by cities and counties. The interests of Oregon for today and in the future must be protected from the grasping wastrels of the land.[26]

With overwhelming support from the Willamette Valley, in 1973 the legislature passed Senate Bill (SB) 100, a land-use planning law that required each of Oregon's 241 cities and 36 counties to prepare a comprehensive plan that was consistent with a set of statewide goals and provided legal support for zoning and other local regulations.[27] In addition, the law created a new entity, the seven-member Land Conservation and Development Commission (LCDC), as well as an administrative agency (confusingly named the Department of Land Conservation and Development, or DLCD) staffed with professionals to serve it. The law also established a process by which the commission would monitor local plans for accordance with state goals and gave the LCDC the authority to revise nonconforming plans. To formulate the state's planning goals, the newly established LCDC embarked on a massive statewide citizen engagement initiative, mailing out more than 100,000 invitations for residents to attend workshops in thirty-five locations. Approximately 10,000 people participated in these workshops, simultaneously helping devise the system's goals while getting a "crash course" in land-use planning.[28]

In late 1974, with the public process complete, the LCDC adopted fourteen statewide planning goals (later expanded to nineteen), several of which were particularly critical to metropolitan planning: Goal 5 emphasized setting aside open space; Goal 10 promoted the provision of affordable housing; Goal 11 aimed to ensure the orderly development of public facilities and services; and Goal 12 involved the development of regional transportation systems. But two goals in particular formed the backbone of the state's growth management system: Goal 3, which promoted farmland preservation by requiring counties to zone productive land exclusively for agricultural use, and Goal 14, which directed local governments to establish urban growth boundaries to separate rural from "urbanizable" land. As journalist Rebecca Clarren explains, the system's founders hoped that by establishing such ground rules they could create an "ideal Oregon, a place where small towns could remain small, urban areas would hum with efficiency, and farms and forests would surround cities in great ribbons of green."[29]

THE CASE

In 1975 the LCDC began implementing its new planning framework and immediately encountered pockets of resistance, particularly in rural areas. This reaction was not surprising: forty-nine of the sixty Willamette Valley legislators had supported SB 100, but only nine of the thirty legislators from rural counties had voted to adopt it; moreover, according to Randal O'Toole, "nearly all the effects of state land-use planning rules fell on rural landowners, who were denied the right to use their land for things other than traditionally rural purposes."[30] In 1976 and again in 1978 the law's detractors tried to repeal it using ballot initiatives, but voters defeated the measures, 57 percent to 43 percent and 61 percent to 39 percent, respectively. In addition to trying to roll back the law, rural officials simply refused to implement it. By 1981 about 1.6 million acres of prime farmland in the Willamette Valley had been zoned exclusively for farm use, but a study by 1000 Friends of Oregon revealed that this designation carried little weight with local officials: of the 1,046 actions taken by twelve county governments dealing with applications for residential development in farm zones, 90 percent were approved, and 81 percent of those were "illegal."[31] In 1982 opponents took yet another stab at repealing SB 100 through the initiative process, failing yet again after an appeal from the terminally ill former governor McCall. A fourth effort, in 1984, did not even garner enough signatures to get on the ballot.

A sluggish economy in the 1980s alleviated some of the pressure to relax SB 100 and gave the LCDC some breathing room to work on enhancing local implementation and increasing the law's legitimacy in rural areas. During this period planners laid the foundations of the Portland metropolitan area's growth management system. They altered the region's transportation options dramatically by ripping up the waterfront freeway and replacing it with a park, tearing down parking garages, ramping up bus service, and building light rail. In addition, they established an urban growth boundary (UGB) and worked to coordinate local comprehensive planning by communities inside and outside of it. All of this was done in consultation with the region's citizens through a variety of mechanisms designed to both incorporate their views and garner their support.

Beginning in the late 1980s, however, a booming regional economy increased the demand for housing and prompted building and real estate interests to begin pressing policymakers to expand the growth boundary and relax development rules. Over the course of the 1990s, even as the fruits of Portland's growth management system manifested themselves in the form of a highly livable urban metropolis, a full-fledged campaign aimed at undermining support for it was taking shape. Ultimately, that movement led to the passage of a ballot initiative that substantially loosened the government's control over development in metropolitan Portland and the state as a whole.

Regional Planning in the Portland Metropolitan Area

To carry out its responsibilities under SB 100, the Portland metropolitan area created a unique entity: the country's only elected regional government, the Metropolitan Service District (known as Metro), which oversees land-use matters for three counties and twenty-four municipalities. The seeds of Metro were sown in 1957, with the establishment of the Metropolitan Planning Commission, whose purpose was to get a share of federal funds for regional planning. Ten years later that organization evolved into the Columbia Region Association of Governments, a weak and underfunded coalition of five counties and thirty-one municipalities. In May 1978, Portland-area voters approved the merger of this association and the existing Metropolitan Service District (which was established in the early 1970s to manage solid waste disposal and the Washington Park Zoo) into the new entity, Metro, whose jurisdiction covers 460 square miles in three metropolitan Portland counties: Clackamas, Multnomah, and Washington. Metro assumed responsibility for establishing and managing the growth boundary for the region's twenty-four cities and the urbanized parts of those counties. In 1979 voters elected the first Metro Council, consisting of twelve members chosen from districts and an executive officer elected at large.

Metro's first major task was to designate the Portland region's growth boundary, and in 1979 it drew a line around a 364-mile (232,000-acre) territory. During most of the 1980s, when growth pressure was minimal, Metro did not engage in much planning for development either within or outside the newly established boundary; rather, it focused on transportation planning, selecting sites for landfills, managing the solid waste and recycling system, and approving and coordinating local comprehensive plans. But in the late 1980s, as the economy rebounded, challenges began to emerge that demanded a response from Metro. According to planning scholar Ethan Seltzer, inside the growth boundary communities were experiencing the symptoms of sprawl, from traffic congestion to disappearing open space. Outside the boundary, speculation on rural land was ubiquitous, and high-end mansions were springing up in rural residential zones. Furthermore, there was little clarity about where the boundary might expand, creating uncertainty for farmers, elected officials, and local service providers. A study commissioned by the state and conducted by ECONorthwest, a land-use planning and economic consulting firm, confirmed that inappropriate development was occurring both inside and outside Portland's growth boundary.

To address these concerns, Metro began crafting a set of regional urban growth goals and objectives (RUGGOs) that would underpin an overall regional growth management plan. In late 1991, following an intensive public participation process, Metro adopted its new RUGGOs. The first major goal spelled out how regional planning would be done, when Metro would exercise its powers, and what role citizens, jurisdictions, and organized interests would play in the planning process. To advance this goal, Metro established a Committee for Citizen Involvement and a Regional Policy Advisory Committee. The second major goal concerned urban design and incorporated specific objectives for the built and natural environments and growth management. Parallel to the RUGGO process, Metro developed the Regional Land Information System, a geographical information system that links a wide range of public records to a land parcel base map. The objective was to provide a database to support land-use planning and management throughout the region. In particular, Metro hoped the database would enable planners to develop overlays linking landscape characteristics to environmental planning and would facilitate citizen involvement by making possible real-time depictions of land-use policy choices.[32]

In an apparent affirmation of Metro's movement toward more aggressive regional planning, in 1990 statewide voters approved a constitutional amendment allowing the agency to function under a home rule charter, and two years later voters in the metropolitan area adopted such a charter.[33] The charter retained the independently elected executive, reduced the number of councilors from twelve to seven, added an elected auditor (for accountability), established a standing system of policy advisory committees (to professionalize the council), reiterated Metro's broad mandate, clarified and strengthened its planning powers, and declared urban growth management to be the agency's primary responsibility.[34] In 2000, voters revised the charter slightly to provide for six councilors elected by districts, a council president elected at large, and an appointed administrator.

As its 1992 charter was being debated, Metro was undertaking an even more ambitious endeavor, the Region 2040 Planning Project, to give concrete form to the vague conceptions articulated in the RUGGOs. More specifically, the new project sought to specify the extent and location of future growth-boundary expansion; identify the major components of the regional transportation system; designate a hierarchy of places, from downtown Portland to the outlying neighborhoods; and integrate a system of green spaces into the urban region.[35] As part of the Region 2040 project, Metro made an explicit effort to engage citizens: it held public hearings and workshops, publicized the effort on cable TV and the news media, mailed more than 25,000 newsletters to households, and made hundreds of presentations to local governments and civic organizations.

After this exhaustive public process, Metro established a set of alternative scenarios for the region's development. The agency then disseminated more than a half million copies of a publication outlining the trade-offs involved in different strategies, prompting 17,000 citizen comments and suggestions. From these responses planners gleaned that participating citizens generally favored holding the existing growth boundary in place, retaining the emphasis on green space preservation and transit-oriented development, reducing traffic and encouraging alternative travel modes, increasing density within the growth boundary, and continuing public education and dialogue.[36] Consistent with these preferences, the final Metro 2040 Growth Concept that emerged in December 1994 called for very little UGB expansion in the ensuing fifty years and instead encouraged higher density, mixed-used centers around light-rail stations. Following a series of public hearings, Metro formally adopted the 2040 Growth Concept, and in December 1996 the state acknowledged its consistency with applicable land-use laws, goals, and rules.

Metro's planning push culminated in 1997 with the adoption of the Regional Framework Plan. The threefold purpose of the plan was to allocate a projected 500,000 new residents by 2017, unify all of Metro's planning activities, and address ten regional planning concerns—such as management and adjustment of the growth boundary, the transportation system, and green spaces—in ways consistent with the Region 2040 Growth Concept. In essence, the plan was a composite of regional functional plans that would be implemented through a combination of local actions and Metro initiatives.[37]

Transportation Planning

Achieving the quality-of-life goals articulated in Metro's planning documents rested heavily on improving the region's transportation system and in particular on decreasing residents' dependence on cars. In 1991 the LCDC adopted the controversial Oregon Transportation Planning Rule (TPR) to try to get people out of their cars. The TPR set an ambitious statewide goal of reducing the number of per capita vehicle miles traveled (VMT) by 10 percent in twenty years and 20 percent in thirty years. To instigate movement toward this goal, the TPR mandated that local governments provide bikeways and pedestrian paths, designate land for transit-oriented development, and require builders to link large-scale development with transit. It also directed local governments and metropolitan planning organizations to produce their own transportation plans within four years.

The TPR was largely the product of a sustained campaign by 1000 Friends of Oregon and other environmental groups to integrate land use and transportation and create a compact, dense urban area that would be hospitable to bicyclists and pedestrians, rather than cars.[38] But local officials' attempts to implement the TPR immediately provoked hostility from big-box retailers, homebuilders, and commercial developers, who bitterly opposed efforts to reduce the number of parking spaces and dictate street connectivity. In response to these complaints, the LCDC relaxed the deadline for adopting ordinances to comply with the TPR and convened a group of stakeholders to amend the rule. That process yielded a more flexible approach and scaled back building orientation and street connectivity requirements.[39]

Despite the LCDC's adjustments to accommodate critics, a scheduled review of the TPR in the mid-1990s revealed continuing reluctance by some planners to use VMT per capita as a measure of reliance on cars and strong opposition among builders to mechanisms such as regulating the parking supply. The consulting firm that conducted the review—Parsons, Brinckerhoff, Quade, and Douglas—recommended tempering the TPR's VMT-reduction goal and further increasing the rule's flexibility by shifting the emphasis from regulating the supply of parking spaces to adjusting the price and location of parking.[40] In 1998, in partial conformance with this advice, the LCDC decreased the initial VMT target from a 10 percent reduction to 5 percent for all the metropolitan planning organizations in the state except Metro, for which it retained a slightly higher goal. Moreover, the rule allowed planners to develop alternative approaches to reducing automobile reliance in place of the VMT target—although any alternative had to be approved by the LCDC.[41]

In 2000 Metro issued an updated version of its TPR that reflected changes in the state's plan. Metro's plan took advantage of the TPR's flexibility to propose an alternative to a VMT standard: it set targets for the percentage of trips taken by individuals traveling by bicycle or transit, on foot, or by carpool (as opposed to traveling alone in a car), ranging from 42 percent to 67 percent for various parts of central Portland. Although it established performance standards against which to measure progress, Metro allowed local jurisdictions to figure out how they would meet the targets. Metro's plan also recommended doubling the existing transit service to accommodate an expected 89 percent increase in ridership by 2020. And it included a regional parking policy, which required cities and counties to regulate the amount of free, off-street parking for most land uses. Finally, Metro required local governments to implement its Green Streets program, which was designed to improve regional and local street design and connectivity, and minimize the environmental impacts of transportation choices (particularly the impact on stream habitat for endangered salmon and steelhead) by adding landscaped areas to capture and filter stormwater. Although Metro's approach remained deeply controversial in the metropolitan region, environmentalists applauded the council's efforts to integrate transportation and land use and to reduce the metropolitan area's dependence on the automobile.[42]

The city of Portland completed its own transportation plan in 2002. Unlike the region's other cities, Portland decided to retain the 10 percent parking reduction target set in the original TPR. The city was well positioned to achieve this goal thanks to thirty years of investment in an elaborate system of bus, light-rail, bicycle, and pedestrian transit. In 1969 the legislature had begun the process by establishing the Tri-County Metropolitan Transportation District (Tri-Met), which immediately took over the failing Rose City Transit system and a year later subsumed the suburban bus lines. On taking office in 1973, Mayor Neil Goldschmidt had made transit a core element of his urban revitalization strategy by shifting investment from highways to public transportation.

Consistent with its new emphasis, in 1975 the city canceled plans to build the Mount Hood Freeway, a five-mile connector highway between I-5 in the center of Portland and I-205 through the eastern suburbs. Three years later the city completed the transit mall, a central feature of the Downtown Plan that dedicated two north–south streets through the heart of downtown to buses. To encourage ridership Tri-Met made buses free along the mall and everywhere else within the city's inner freeway loop. In 1986 the city inaugurated its fifteen-mile Eastside light-rail line (the Metropolitan Area Express, or MAX), and it opened the eighteen-mile Westside line in 1998. Although voters balked at financing a north–south line, a spur line to the airport began running in 2001, thanks to a substantial contribution from the Bechtel Corporation, which owned an industrial park along the line. And builders completed a scaled-down six-mile North Portland line in 2004. The city also built the first modern streetcar system in North America, connecting downtown and Portland State University to the Pearl District and trendy Northwest neighborhoods. To complement its efforts at reducing auto dependence through mass transit, the city also adopted a Bicycle Master Plan in 1996 and a Pedestrian Master Plan in 1998. By 2001 the city had more than doubled the amount of bikeways (off-street paths, on-street lanes, and bicycle boulevards) from 111 miles to 228 miles.[43]

Development Rules

In addition to improving residents' transportation options, the Portland region's smart growth approach involved combining infill and density in urban areas with strict protection of open space and rural land rich in natural resources. As the growth management regime evolved, the number of regulations burgeoned, and development rules became more complex and detailed. For example, to protect rural areas, Metro issued regulations that prevented a developer from building a destination resort on prime forestland, within three miles of high-value farmland, or anywhere in the 300,000-acre Columbia River Gorge National Scenic Area. Planners devised numerous restrictions on homebuilding in rural areas as well: they divided rural property into eight categories based on characteristics such as soil depth and quality, slope, and erosion, and zoned about 16 million acres of farmland for exclusive farm use. Any house built on high-value farmland had to qualify as a farm dwelling, a temporary hardship shelter, or a home for relatives helping out on the farm; in addition, as of 1994 the property must have yielded $80,000 in annual gross farm income over the previous three-year period. Such measures reduced the market value of agricultural land dramatically: in 2000 ECONorthwest estimated that residentially zoned land inside the growth boundary was worth about $150,000 an acre, while property outside the boundary zoned for farm or forest sold for about $5,000 per acre.[44] But farmers and owners of forestland got enormous property tax breaks in return; farmland was assessed at as little as 0.5 percent of the value of land where development was encouraged.[45]

Outside the growth boundary, developers had to show their land was worth little for forestry or farming and was easily supplied with services and that comparable land was not available inside the boundary but inside the UGB; the burden of proof rested on opponents of development. For example, the LCDC issued housing rules that required every jurisdiction to zone at least half its vacant residential land for attached single-family homes or apartments, and it prohibited suburbs from adopting exclusionary zoning to block construction of affordable housing. To encourage dense development, Metro established aggressive housing targets for jurisdictions within the growth boundary. And to attract development near transit stops, Metro and other taxing districts, as well as many individual jurisdictions, created tax abatements and other incentives. As a result of these policies, in 1998 and 1999 half of the new housing starts in the metropolitan area were apartments and attached dwellings, up from 35 percent between 1992 and 1995.[46]

In addition to putting in place measures to encourage infill, the city of Portland enacted a variety of rules that aimed to maintain neighborhood aesthetics and protect natural resources. In the early 1990s city planners devised a system of environmental overlays designed to prevent development in areas with the highest resource values and to ensure environmentally sensitive development in conservation zones. As the decade wore on, the city enacted standards that required developers to landscape foundations and preserve or plant trees when building new housing in single-family zones. Street standards became progressively tougher, so that by the late 1990s every new Portland street had to have sidewalks, a tree strip, and street trees. Then, in 1999, the city council voted unanimously to outlaw the new garage-front "snout houses" that had proliferated in response to the city's efforts to promote infill. The new rules specified that a garage could not dominate the front of the house, the main entrance had to be close to the street, and the street-facing side of the house had to meet specific standards for window and door space. The aim, planners said, was to ensure that houses made "more connection to the public realm."[47]

The Results of Planning

As a result of its formidable urban planning regime, by the late 1990s Portland was defying many of the trends occurring nationally. While development was gobbling up farmland and forestland around the country, Oregon's losses were modest.

According to agriculture department figures, even at the height of the economic boom of the 1990s, Oregon yielded rural land to development at a slower rate than the national average.[48] And while most of the nation's major metropolitan areas were spreading out during the 1970s and 1980s, Portland remained compact. Metropolitan Cleveland's population declined by 8 percent between 1970 and 1990, but its urban land area increased by a third. Cities whose populations increased spread at an even faster rate. Metro Chicago's population grew 4 percent between 1970 and 1990, but its metropolitan area expanded 46 percent; during the same period, metropolitan Los Angeles's population grew 45 percent, and its developed land area expanded 300 percent.[49] Metropolitan Denver, which is often compared to Portland, expanded its developed area from 300 to 530 square miles between 1970 and 1997, a 77 percent increase; by contrast, Portland's 364-square-mile-developed metropolitan area grew by fewer than five square miles between 1979 (when the growth boundary was put in place) and 1997.[50]

Portland's pattern of housing development was also unusual compared to other U.S. cities. In 1999 the Brookings Institution reported that Portland was one of the few places in the nation where new housing construction was increasing faster within the central city than in the larger metropolitan region.[51] In addition, Portland was building smaller, denser housing units. When Metro drew the growth boundary in 1979, lot size for the average new single-family detached house in the metro area was 13,200 square feet. By the mid-1990s it was down to 8,700 square feet. Although the median size lot for single-family homes sold in the United States in 1990 exceeded 9,000 square feet, Metro encouraged builders to average 6,700 square feet or build multifamily units.[52] Furthermore, in 1985 attached homes accounted for 3 percent of units in the city of Portland; a decade later that figure was 12 percent.[53] In 1994 Portland was building new houses at a density of five dwellings per acre; by 1998 that figure was eight dwellings per acre—higher than the 2040 Plan target.[54] Even as it increased density and infill, Portland managed to save oases of green throughout the city: between the 1995 passage of a $135.6 million bond initiative and 2001, Metro acquired nearly 8,000 acres of green space.[55]

Portland's downtown businesses were thriving as well, apparently unharmed by the city's concerted effort to limit driving. Between 1973 and 1993 the amount of office space in downtown Portland nearly tripled, and the number of downtown jobs increased by 50 percent. Nevertheless, the volume of cars entering Portland's downtown remained the same.[56] Rather than driving, many workers relied on mass transit: by the mid-1990s Tri-Met's radial bus and rail system carried 30 percent to 35 percent of the city's commuters into downtown Portland, roughly twice the proportion in comparable cities like Phoenix or Salt Lake City.[57] Furthermore, according to the 2000 Census, Portland ranked in the top five among sixty-four U.S. cities with a population of 250,000 or more in the percentage of workers who commute by bicycle. Counts of cyclists crossing the main bridge with bicycle access into downtown went up 143 percent from 1991 to 2001, outpacing population growth for that period. The 2000 Census also showed that the Portland metropolitan area had more people walking to work than in most regions of a similar size: over 5 percent of workers in the city of Portland got to their jobs on foot.[58] In short, "Downtown Portland's landscaped streets, light rail system, innovative parks, and meticulously designed new buildings . . . transformed what was once a mediocre business district into a thriving regional center."[59]

The combination of dense housing, efficient public transportation, and a lively downtown made Portland a national model of smart growth and New Urbanism.[60] Beginning in the mid-1970s and continuing through the 1990s, Portland consistently made lists of the top ten places to live and the best-planned cities in the United States. In 1975 the prestigious Midwest Research Institute rated 243 cities on their economic, political, environmental, social, and health and education qualities, and Portland was the only city rated as "outstanding" in all categories. In spring 1980 the *Chicago Tribune* profiled a dozen American cities and described Portland as "largely free of graffiti, vandalism, litter, street gangs, Mafiosi, racial tensions, desolate slums, choking pollution, crooked politicians, colonies of freaks, phony social climbers, and those legions of harassed men whose blood pressure rises whenever the Dow Jones average dips a few points."[61]

The accolades continued to roll in over the next two decades, as a spate of books cited Portland as a city worth imitating, and complimentary articles abounded. Writing in 1997 in *The Denver Post*, journalist Alan Katz said,

> Swirling with street life and commerce, Portland is best seen at afternoon rush hour, when office workers pour out of buildings and into taverns and coffeehouses. Buses roar up and down the narrow streets by the dozen. Light-rail trains whisk by, carrying commuters to the east-side suburbs. What's particularly striking about central Portland is its residential and retail density—unlike downtown Denver, which is pockmarked with surface parking lots and boarded up buildings. Portland's density creates a lively retail climate, and the presence of pedestrians makes the downtown streets feel safer.[62]

And after *Money* magazine named Portland the "best place to live in the U.S." in 2000, a British visitor wrote admiringly,

> Portland is a delight. Walking through the city's bustling streets on a sunny day, it is easy to forget that one is in the most car-dominated country on the planet. The streets are lined with trees ablaze in green and gold. Sculpture and fountains distract the eye and please the soul. Safe and direct pedestrian crossings at every junction make walking easy and quick for people in a hurry. For those with time to spare, benches, street cafes and pocket parks provide places for reflection and relaxation. The traffic is slow and light. The pavements are wide and clean. There is not a guard railing to be seen.[63]

Institutionalizing Civic Engagement in Portland

A host of observers attributed the longevity of Portland's unusually stringent growth management regime to the participation and buy-in of the city's engaged citizenry. The number of advocacy groups in Portland, which increased sixfold in the early 1970s, continued to grow—reaching 222 in 1985 and 402 in 1999.[64] Neighborhood associations, which originated in Portland in the 1940s, thrived under Mayor Goldschmidt after he created the Office of Neighborhood Associations in 1974. This office provided funding and technical training to members and thereby "legitimized direct democratic action at the grassroots level by allowing neighbors to directly influence city policies."[65] In addition the city created a host of citizen advisory committees and task forces: the number of citizen advisory committees grew from twenty-seven in 1960 to fifty-six in 1972; the number of task forces jumped from five in 1960 to twenty-five in 1972. By 1986 there were twenty-three bureau advisory committees that engaged citizens in the everyday business of city agencies. As longtime Portland activist Steven Johnson explains, the city "created an open-door policy that changed the expectation of citizens' relationship to their government."[66]

According to political scientist Robert Putnam, the responsiveness of Portland's leaders to the surge in citizen activism produced a virtuous circle of "call and response" that resulted in "a positive epidemic of civic engagement." As Putnam explains, in most other American cities, after a great deal of political activity in the 1960s and early 1970s, "the aging baby boomers left the streets, discarded their placards, gave up on politics, and slumped onto the couch to watch television. In Portland, by contrast, the ranks of civic activists steadily expanded as Portlanders experienced a resurgence of exuberant participation in local affairs."[67] Abbott says the result was an unusually inclusive and civil political dialogue:

> The openness of civic life in Portland is the basis for an emphasis on team play. Public life takes place around a big table. Some of the seats are reserved for elected officials and heavy hitters from the business community. But anyone can sit in who accepts the rules (politeness is important) and knows how to phrase ideas in the language of middle class policy discussion (the goal is to do "what's good for the city"). Once an organization or interest group is at the table, or on the team, it has an opportunity to shape policy outcomes.[68]

As Johnson notes, however, during the 1990s the tone of citizen activism changed, and some of Portland's civic engagement became directed at stopping growth control and preventing the increases in density prescribed by Metro. For example, in Oak Grove, a blue-collar suburb with large lots, homeowners successfully resisted a plan for a dense, transit-oriented "town center." Similarly, leaders in the Multnomah Village section of southwest Portland blanketed the streets with signs opposing what they said was city officials' heavy-handed attempt to increase the area's population by 20 percent—changes that would deprive single parents of an affordable place with yard space for their kids. Resisters in these and other low-density neighborhoods claimed they were being targeted, while wealthy areas escaped planners' push for density.[69] Meanwhile, residents in the more affluent suburbs, such as Milwaukie, West Linn, and Tigard, feared the environmental costs and socioeconomic diversification associated with compact growth. West Linn City Council member John Jackley complained about Metro planners' disdain for suburbs and their apparent desire to impose their "urban village" concept on unwilling residents. Similarly, the mayor of Tigard defended the large-lot, upscale developments and "lifestyle opportunities" that Tigard provided.[70]

Trouble in Paradise: Measures 7 and 37

Neighborhood resistance to density increases was only one manifestation of festering divisions over the state's growth management program. In 1998, 53 percent of Portland-area voters rejected a new light-rail line—a major shift from the 1990 vote, when 75 percent of voters approved the previous light-rail line.[71] More ominously, as Oregon's economy boomed through the 1990s, a coalition of homebuilders, real estate brokers, the Oregon Association of Counties, Safeway and other large retail companies, as well as the property rights group, Oregonians in Action, began to complain bitterly about Portland's growth boundary—and particularly the rules limiting rural development. Bill Moshofsky, leader of Oregonians in Action, captured the essence of the coalition's complaint, saying,

> Instead of a planning tool, the urban growth boundaries have become a Berlin Wall to restrict growth to a high-density model. The whole thing is, they want Oregon to look like Europe. The screws have been tightened year by year. . . . Property rights have been swept under the rug. And the high density is going to reduce the quality of life. People will be crowded together, living on small lots. The land supply will be so limited that housing costs will become unaffordable.[72]

To garner support for their position among the public, growth management opponents claimed that Portland's boundary had raised home prices—more precisely, that the price of a buildable acre of land inside the growth boundary had tripled in the two years prior to 1996 and that land prices had gone from 19 percent below the national average in 1985 to 6 percent above it in 1994. Defenders of the growth boundary responded that high demand was the primary cause of the price increase, noting that land prices were rising in many western cities that lacked growth boundaries. They observed that Portland house prices rose 26 percent between 1992 and 1996, while Denver's increased 44 percent during the same period; house prices rose 25 percent between 1994 and 1996 in Salt Lake City, compared to Portland's 19 percent.[73] Planning scholar Deborah Howe suggested that those who attributed Portland's house price increases to market constraints should explain the 80 percent increase in values reported by homeowners during the 1970s, before the growth boundary was established.[74] Not surprisingly, building interests and property rights advocates were not persuaded.

Throughout the mid- to late 1990s Democrats in the state legislature fended off bills aimed at dismantling elements of the state's land-use planning system.[75] After several failed legislative efforts to rewrite the planning laws to accommodate developers' concerns and defuse attacks, in 2000 the conflict came to a head with the passage of Measure 7, a ballot initiative written by Oregonians in Action that required state and local governments to compensate landowners whose property values were affected by regulation or, if they could not find the money, to waive the rules. The Oregon Supreme Court struck down Measure 7 in 2002 on the grounds that it contained multiple constitutional changes that should have been voted on separately. But the measure's approval by 53 percent of voters, including both rural and urban residents, sent a chill through the state's environmental and planning community. Although polls by 1000 Friends of Oregon and the League of Conservation Voters consistently showed support at a rate of two to one for the state's land-use planning system, the proponents of Measure 7 clearly had tapped into some latent dissatisfaction with the implementation of that system. As journalist Rebecca Clarren commented,

> While Oregon's land-use system was born with huge public support, over half the state's current population wasn't around when it was created. For many newcomers, the regulations are confusing and restrictive. Even many longtime Oregonians say that a system that began with clear goals has grown into a bureaucratic snarl, and that the rules have become onerous and arbitrary.[76]

In hopes of defusing conflict and reducing the impetus to overhaul the system, in early 2002 Metro took 18,638 acres of land—an area equal to more than 7 percent of the existing boundary—and designated it as "urban reserves" eligible to be incorporated into the growth boundary. Metro also addressed complaints about a shortage of industrial land by identifying a small number of "regionally significant industrial areas," which would be reserved for industries that provided "family wage" jobs. Opponents of growth management were not mollified, however, and they became even more determined after legislative efforts to craft a compromise fell apart over how to finance a landowner compensation system.

In 2004, Oregonians in Action spearheaded another initiative campaign, and this time they carefully crafted a measure to avoid a constitutional challenge. Measure 37 created a state law that allowed property owners to petition to be excused from any state or local land-use rules put in place after they bought their land or be paid for any decrease in their land's market value as a result of those rules. The measure gave the government 180 days to make a decision or the rule would be waived automatically. As the battle over Measure 37 heated up before the November 2004 vote, the rhetorical battle lines were sharply drawn. Proponents offered a handful of anecdotes that emphasized the unfairness of the planning system and the burdens its regulations imposed on ordinary folks:

- According to one story, Barbara and Gene Prete bought twenty acres in 1990, envisioning a horse barn and farmhouse in the shadow of the mountains. When they got ready to move from Los Angeles to Sisters in 1995, however, they got bad news: a new Oregon rule required them to produce $80,000 in farm goods a year if they wanted to build their house. The Pretes became chief petitioners for Measure 37, which they said would bring "justice" to property owners.[77]
- According to a second story, Connie and Steve Bradley were unable to rebuild the run-down house on their family's property because it had been abandoned for more than a year and so reverted to its farming designation. Connie Bradley said she trusted Oregonians to preserve their landscape and liked Measure 37 because it put power back into the hands of the people (and took it out of the hands of bureaucrats).[78]
- By far the most effective story featured Dorothy English, a ninety-one-year-old widow who recorded a radio ad that was played constantly in the weeks leading up to the election. English recounted a series of battles with the state over subdividing twenty acres of land, which she bought in 1953, into residential lots. Her punch line was, "I've always been fighting the government, and I'm not going to stop."[79]

Defenders of the growth management system, led by 1000 Friends of Oregon, responded to these anecdotes by saying that Measure 37 would create uncertainty, bankrupt local governments, and dismantle the state's planning system. They raised the specter of factories and big-box stores mixed in with houses and farms. Their message apparently fell flat, however; although opponents of Measure 37 outspent proponents four to one, the initiative passed with 61 percent of the vote—even carrying Multnomah County, the "epicenter of planning," by a slim majority.[80] Explanations for the measure's passage abounded. Some said voters did not appreciate its full implications; others pointed to its straightforward and appealing title ("Governments Must Pay Owners, or Forgo Enforcement, When Certain Land Use Restrictions Reduce Property Values"), the power of the anecdotes told during the campaign, and the opposition's failure to convey the success of the state's planning system. Some observers said that the sheer comprehensiveness of Oregon's growth management system made it especially vulnerable. The puzzling fact remained that, when polled, Oregonians said they either supported or strongly supported land-use regulations in the same proportions by which voters passed Measure 37.[81]

By mid-August 2005 more than 2,000 Measure 37 claims were in the pipeline across the state.[82] Washington County was overwhelmed with claims, which were coming in at a rate of between 50 and 60 a month. In Clackamas County filings shot up in April, May, and June, as county commissioners steadily waived zoning regulations in response to claims. The majority of the claims were for relatively small developments, from one to a handful of homes. But even those cumulatively threatened the very death-by-a-thousand-cuts result that Oregon's planning framework was devised to avoid. And some claims promised much more substantial short-run impacts. One proposal in Clackamas County for a subdivision of fifty-two houses that was approved in late March threatened to undo years of planning for the rolling hills and farmland of the Stafford Triangle between Tualatin, Lake Oswego, and West Linn. Another, in scenic Hood River County, would convert 210 acres of pear orchard land into housing. Farmer John Benton, who was asking for $57 million or permission to build as many as 800 houses, told listeners at a hearing on his application that Oregon legislators should "quit trying to be social engineers and let the market forces and the good people of this state realize their potential."[83]

There was some variation in the process cities and counties used to handle Measure 37 claims, but in the vast majority of cases officials felt they had little choice but to approve nearly every claim. In some instances they supported the applications, but in many they felt hamstrung by Measure 37's clear language and lacked the resources to compensate landowners rather than waive the rules. Some county commissioners also bemoaned the fact that the law contained no provisions for involving neighbors in the process of deciding the appropriate uses of property, a dramatic reversal of Oregon's usual approach.

Although environmentalists were alarmed by the increasing number of Measure 37 claims, county approval was just the first step. Most proposals had to clear state review, and the state's early rulings—which denied some requests and attached strings to others—suggested that development under Measure 37 would not be as straightforward as applicants had hoped. Most claimants also had to get land-use and building approvals typical of any subdivision, and developments outside the growth boundary needed clearance for new wells and septic systems. Complicating matters even further, banks were skittish about financing Measure 37 properties because it was unclear whether development rights were transferable to new owners. (The state attorney general issued an advisory letter in February saying that in his opinion, any change in zoning under Measure 37 was *not* transferable.)

As local and state officials grappled with claims, the legislature worked frantically on revising the law. The central questions were how to set up a compensation system for landowners and whether to make development rights transferable. The bill that got the most interest from both Democrats and Republicans established a compensation system in which applicants who got permission to develop their land would have to pay the property tax that had been forgiven when the land was zoned for agriculture, forestry, or habitat protection, and that money would be used to pay off other claimants. The bill also limited the potential for large-scale subdivisions in exchange for making small-scale development easier. Even this proposal foundered, however. Fundamental philosophical differences over the importance of ensuring private property rights versus limiting development divided legislators, and the highly visible controversy surrounding the issue discouraged compromise. The 2005 legislative session ended without a resolution, and the issue remained to be decided case by case within individual jurisdictions. Meanwhile, Dave Hunnicutt—executive director of Oregonians in Action and a prime mover behind Measure 37—was crafting initiatives that targeted suburban housing density requirements and restrictions on rural road construction.

As Oregonians struggled to implement Measure 37, its impact spread well beyond the state's borders. Emboldened by the measure's success, property rights activists in other western states began planning ballot initiatives of their own. In Washington State, rural property owners mobilized to challenge the state's Growth Management Act and the regulations counties had adopted to comply with it. Hopeful advocates from Alaska, California, Colorado, Idaho, Maine, North Carolina, South Carolina, and Wisconsin contacted Oregonians in Action for strategic advice on getting similar measures passed.[84]

Property rights activists' euphoria was short-lived, however. In mid-October 2005, in response to litigation brought by 1000 Friends of Oregon and several agricultural groups, Judge Mary James of the Marion Circuit Court ruled that Measure 37 violated the state constitution because it granted special privileges and immunities, impaired the legislative body's plenary power, suspended laws, and violated the separation of powers. The judge also ruled that the measure violated substantive and procedural due process guaranteed by the U.S. Constitution. Vowing to appeal the ruling, Hunnicutt called it "another kick in the face to the citizens of [the] state who want nothing more than protection of their right to use their land as they could when they bought it, to Dorothy English and Gene and Barbara Prete, and to all those people who have bought property over the years, only to have the State of Oregon change the land use rules and take away their rights."[85]

OUTCOMES

With Measure 37 in legal limbo, a coalition of environmental groups, farmers, and Democratic legislators devised Measure 49, which the 2007 legislature referred to the voters on a party-line vote. Measure 49 drastically revised Measure 37: it prohibited the use of claims for commercial and industrial development and strictly limited how many houses most claimants could build. More precisely, it offered Measure 37 claimants three options: an "express route" to develop one to three houses; a "conditional" path that allowed building four to ten houses, if claimants could prove by appraisal that land-use laws devalued their property by a sufficient amount; or a designation of "vested rights," meaning claimants could complete a project beyond the scope of what otherwise would be allowed under Measure 49 because they had spent a significant amount of money and done substantial work before the law changed. Backers said Measure 49 would allow the "little guy" to build a home or two but would prevent rampant sprawl development. In November 2007 voters passed Measure 49 by a margin nearly identical to that of Measure 37.[86]

By May 2008 about 90 percent of the 7,500 claimants under Measure 37 who responded to the state had chosen the express option, and Measure 49 seemed to have resolved most of the issues associated with Measure 37.[87] Several dozen vested-rights cases remained unresolved, however, and they involved the largest proposed developments. The state left vested-rights decisions to the counties, and they inevitably ended up in court. Two rulings issued in November 2008 alarmed environmentalists: one allowed landowners to finish subdivisions started under Measure 37, and the other said that Measure 37 waivers amounted to binding contracts that could not be undone by Measure 49. But in July 2010 the Ninth Circuit Court of Appeals overturned the latter decision. Ultimately, about 4,600 claimants filed for development rights under Measure 49, and between 80 percent and 85 percent of those claims were approved.[88] By fall 2010, however, little construction had occurred—partly because of delays in the approval process but also because of a stagnant housing market.

While the Department of Land Conservation and Development grappled with express claims, and vested-rights cases wended their way through the courts, the state undertook a new scan of its land-use planning system. In 2005, the governor and legislature had convened the Oregon Land Use Task Force, known as the "Big Look," and charged it with making recommendations for reform to the legislature. In 2009, after more than three years of stop-and-go work, the task force concluded that Oregon's land-use system was overly complicated, relied too heavily on regulatory sticks rather than inducement carrots, and had a top-down, one-size-fits-all approach to managing growth. The group recommended allowing Oregon counties to seek regional rezoning (as opposed to retaining state control over rezoning) of unproductive or compromised farm, forest, and resource land. They also called for replacing the state's nineteen land-use planning goals with four broad (and relatively vacuous but easily communicated) principles: provide a healthy environment, sustain a prosperous economy, ensure a desirable quality of life, and provide fairness and equity to Oregonians. In summer 2009 the Oregon legislature approved and the governor signed House Bill 2229, based on the task force's recommendations.

The combination of Measure 49 and HB 2229 quelled some of the controversy over Oregon's land-use system, but debates continued to arise; in particular, critics challenged the 2010 growth management plan adopted by Metro and Portland-area counties. In March 2014, shortly after the Oregon Court of Appeals reversed and remanded land designations in that plan, the legislature passed House Bill 4078, known as the "grand bargain," to resolve legal uncertainty over the regional plan. But disgruntled senators cited the need for legislative intervention as evidence that reforming the state's land-use system should be a priority for the 2015 legislature. "It's so bureaucratic, it's so messed up, it's so broken that we have to have special pieces of legislation to get us out of the court cases," complained Republican Senator Larry George, once a supporter of Measure 37.[89]

CONCLUSIONS

The debate over Oregon's growth management regime in general, and Measure 37 in particular, captures many of the most important elements of contemporary environmental politics. Competing activists are divided by genuine value differences and have engaged in a fierce struggle to persuade the public to view the good life, and the benefits and costs of government intervention, as they do. For proponents of Measure 37 and similar measures, sprawl is simply a reflection of Americans' desire for a house with a yard, and government policies to address it threaten individuals' private property rights and, by extension, their freedom to pursue a better life. They believe property owners have a right to a return on their investments and that the housing market, not lines on a map, should determine where people live.[90] The measure's opponents, by contrast, challenge the notion that sprawl is a result of free choice and market forces, noting that existing land-use regulations shape consumers' options. They also believe that sprawl jeopardizes not only the environment but the very connections to a community that make life worthwhile. For them, some restrictions on individual freedom are worth the collective benefits. As Robert Liberty, former president of 1000 Friends of Oregon and a member of the Metro board, says, "Quality of life is something that is shared. A golf course is not. A four-car garage is not. One of the best things about the planning process is that it makes a better community for everyone, regardless of income."[91] Given this divide, it is not surprising that, just as fans of growth management cite Portland as an exemplar, critics tout it as a prime example of overbearing and unrepresentative government run amok.

The views of the general public are much more ambiguous, however, and are therefore susceptible to the way arguments are framed. Many environmentalists believe that the defenders of the status quo in Oregon made tactical mistakes in characterizing their position but that the state's residents fundamentally support growth management. They point out that backers of Measure 37 chose their words carefully; they did not challenge the system itself but rather criticized the rules, which they called irrational and unfair, and those responsible for implementing those rules, whom they described as elitist and out of touch with the people. This strategy enabled them to play on the public's ambivalence: many Oregonians seem to want to avoid the effects of sprawl without having to comply with the rules government has employed to achieve that result.

Public opinion is probably even more complicated than this depiction suggests, however. Notwithstanding the fervent convictions of smart growth advocates, it is unclear how many people really dislike sprawl and even less obvious that the public will readily accept density as an alternative. Certainly, a vocal coalition opposing growth control, led by Samuel Staley of the Reason Public Policy Institute, challenges the idea that middle-class Americans will (or should) trade their lawns, gardens, and privacy for housing in dense urban neighborhoods. As journalist Tim Ferguson writes, "A nation in love with truck-size sport utility vehicles is unlikely to embrace the housing equivalent of an Escort."[92] Whereas antisprawl advocates blame misguided government policies, such as traditional zoning laws and highway subsidies, opponents of growth management insist that it is Americans' passionate desire for a spread-out, car-centered way of life that has shaped the American landscape.[93] Economist J. Thomas Black of the Urban Land Institute adds that changes in the workforce complicate housing decisions: jobs are often in different suburbs, and a close-in address does not work; moreover, downtown housing becomes more expensive once renewal/gentrification begins and the economy is booming.[94]

Despite challenges, environmentalists and other smart growth advocates remain hopeful, and the results of the Measure 49 vote in Oregon, as well as other votes taken around the country, seem to bear out some of their optimism. In the 2004 elections, voters in red and blue states alike approved smart growth measures and voted to tax themselves to expand transit options and set aside open space. Nationally, 80 percent of the more than thirty transit funding measures passed, for a total value of $40 billion. Voters in 111 communities in twenty-five states passed ballot measures to invest $2.4 billion in protecting parks and open spaces, a success rate of 76 percent, according to the Trust for Public Land. Even in Colorado, Montana, and Utah, candidates ran and won on smart growth platforms. David Goldberg argues, "Though it was little noted in the political coverage, the desire to address growth issues was one of the few arenas that transcended partisanship in this era of polarized discourse."[95]

A decade later, the 2014 election saw similar results: with just a handful of exceptions, voters across the United States passed statewide, countywide, and citywide measures in support of smart growth, open space preservation, and improvements in transportation and water infrastructure. Voters also rejected measures that would impede or roll back compact development.[96] The question becomes if President Trump's anti-environmental rhetoric will impact smart growth plans at the state and municipal levels.

QUESTIONS TO CONSIDER

- What are the strengths and weaknesses of using ballot initiatives to decide the fate of Oregon's land-use planning system?
- Portland's population continues to increase. What approach would you take to evaluate whether communities and existing land inside the growth boundary have enough room for people over the next 20 years?
- What might supporters of the growth management system in Oregon have done differently if they wanted to head off a property rights challenge?
- Do you think it is reasonable for cities, towns, and regions to impose requirements and restrictions on its citizens to avert sprawl? Why, or why not?

NOTES

1. Oliver Gillham, *The Limitless City* (Washington, D.C.: Island Press, 2002).
2. Jonathan Levine, *Zoned Out: Regulation, Markets, and Choices in Transportation and Metropolitan Land-Use* (Washington, D.C.: Resources for the Future, 2006).
3. Jose A. Gomez-Ibanez and Nancy Humphrey, "Driving and the Built Environment: The Effects of Compact Development on Motorized Travel, Energy Use, and CO_2 Emissions," *TR News* 268, May–June, 2010, 24–28; Reid Ewing and Robert Cervero, "Travel and the Built Environment," *Journal of the American Planning Association* 76 (2010): 265–294.
4. Reid Ewing and Fang Rong, "The Impact of Urban Form on U.S. Residential Energy Use," *Housing Policy Debate* 19,1 (2010): 1–30. Although compact cities typically experience a greater urban heat island effect than more spread-out ones, and therefore expend more energy on cooling in hot months, that effect is outweighed by the energy-saving impacts of smaller units and shared walls.
5. The growth management literature distinguishes between policies that aim to control—that is, limit—growth and those that aim to manage—that is, shape but not limit—growth. I use the term "growth management" in this chapter because virtually no U.S. policies actually seek to impose limits on growth. See Gabor Zovanyi, *Growth Management for a Sustainable Future: Ecological Sustainability as the New Growth Management Focus for the 21st Century* (Westport, Conn.: Praeger, 1998).
6. Wim Wievel, Joseph Persky, and Mark Senzik, "Private Benefits and Public Costs: Policies to Address Suburban Sprawl," *Policy Studies Journal* 27 (1999): 96–114.
7. Raquel Pinderhughes, *Alternative Urban Futures* (Lanham, MD: Rowman & Littlefield, 2004); Sadhu Aufochs Johnston, Steven S. Nicholas, and Julia Parzen, *The Guide to Greening Cities* (Washington, D.C.: Island Press, 2013).
8. Marcial H. Echenique, Anthony J. Hargreaves, Gordon Mitchell, and Anil Namdeo, "Growing Cities Sustainably," *Journal of the American Planning Association* 78,2 (2012): 121–137. Their analysis assumes, however, that policies that foster compactness influence only the location of residential growth, not employment. And they argue that we can achieve greater reductions in greenhouse gas emissions through shifts to electric cars and public transport—while failing to acknowledge that both of these fare better in more compact settings.
9. Robert Bruegmann, *Sprawl: A Compact History* (Chicago: University of Chicago Press, 2005).
10. Randal O'Toole, *The Best-Laid Plans: How Government Planning Harms Your Quality of Life, Your Pocketbook, and Your Future* (Washington, D.C.: Cato Institute, 2007), 93.
11. Levine, *Zoned Out*, 3.
12. See, for example, Daniel Fiorino, "Flexibility," in *Environmental Governance Reconsidered: Challenges, Choices, and Opportunities*, ed. Robert F. Durant, Daniel J. Fiorino, and Rosemary O'Leary (Cambridge, Mass.: The MIT Press, 2004): 393–425.

13. John R. Logan and Harvey L. Molotch, "The City as a Growth Machine," in *Urban Fortunes: The Political Economy of a Place* (Berkeley: University of California Press, 1987), 50–98.

14. Arthur Lupia, "Shortcuts versus Encyclopedias: Information and Voting Behavior in California Insurance Reform Elections," *American Political Science Review* 88 (March 1994): 63–76; Sean Bowler and Todd Donovan, *Demanding Choices: Opinion, Voting and Direct Democracy* (Ann Arbor: University of Michigan Press, 1998); J. G. Matsusaka, *For the Many or the Few: The Initiative, Public Policy, and American Democracy* (Chicago: University of Chicago Press, 2004).

15. Elisabeth R. Gerber, *The Populist Paradox: Interest Group Influence and the Promise of Direct Legislation* (Princeton, N.J.: Princeton University Press, 1999); David S. Broder, *Democracy Derailed: Initiative Campaigns and the Power of Money* (New York: Harcourt, 2000); Richard J. Ellis, *Democratic Delusions: The Initiative Process in America* (Lawrence: University of Kansas Press, 2002).

16. See Daniel A. Smith and Caroline J. Tolbert, *Educated by Initiative: The Effects of Direct Democracy on Citizens and Political Organizations in the American States* (Ann Arbor: University of Michigan Press, 2004); Joshua J. Dyck and Edward L. Lascher "Direct Democracy and Political Efficacy Reconsidered," *Political Behavior* 31 (2009): 401–427; Daniel Schlozman and Ian Yohai, "How Initiatives Don't Always Make Citizens: Ballot Initiatives in the American States, 1978–2004," *Political Behavior* 30 (2008): 469–489. Smith and Tolbert argue that ballot initiatives produce important secondary effects. But Dyck and Lascher find little support, either logical or empirical, for secondary impacts of ballot initiatives. Schlozman and Yohai find limited support for the claim that ballot initiatives increase turnout and political knowledge; like Dyck and Lascher, they discern no effects on political efficacy.

17. David Goldberg, "The Wisdom of Growth," *The Sacramento Bee*, December 12, 2004, E1.

18. Carl Abbott, *Portland: Planning, Politics, and Growth in a Twentieth-Century City* (Lincoln: University of Nebraska Press, 1983).

19. Quoted in Alan Katz, "Building the Future," *The Denver Post*, February 9, 1997, 1.

20. Steven Reed Johnson, "The Myth and Reality of Portland's Engaged Citizenry and Process-Oriented Governance," in *The Portland Edge*, ed. Connie P. Ozawa (Washington, D.C.: Island Press, 2004), 103.

21. Abbott, *Portland: Planning, Politics, and Growth*.

22. Johnson, "The Myth and Reality."

23. Abbott, *Portland: Planning, Politics, and Growth*, 220.

24. John M. DeGrove, *Land, Growth & Politics* (Washington, D.C.: Planners Press, 1984), 235.

25. Rebecca Clarren, "Planning's Poster Child Grows Up," *High Country News*, November 25, 2002.

26. Quoted in DeGrove, *Land, Growth & Politics*, 237.

27. In 1969 Oregon had adopted SB 10, a precursor to SB 100 that mandated that all counties and cities develop land-use plans that met ten state goals, but few local governments had taken steps to implement the law, and the state had not enforced it.

28. Johnson, "The Myth and Reality."

29. Clarren, "Planning's Poster Child Grows Up."

30. O'Toole, *The Best-Laid Plans*, 87.

31. The approvals were "illegal" in that they did not provide sufficient findings of fact to allow a judgment as to whether the exception was justified. See DeGrove, *Land, Growth & Politics.*

32. Ethan Seltzer, "It's Not an Experiment: Regional Planning at Metro, 1990 to the Present," in *The Portland Edge*, 35–60; John M. DeGrove, *The New Frontier for Land Policy: Planning and Growth Management in the States* (Cambridge, Mass.: Lincoln Institute of Land Policy, 1992).

33. Home rule is the power of a substate-level administrative unit to govern itself.

34. Seltzer, "It's Not an Experiment."

35. Ibid.

36. Ibid.

37. Ibid. A functional plan addresses a narrow set of concerns associated with an issue of regional significance. Although Metro's charter gave it functional planning authority, the agency never used it prior to the mid-1990s.

38. Sy Adler and Jennifer Dill, "The Evolution of Transportation Planning in the Portland Metropolitan Area," in *The Portland Edge*, 230–256. Randal O'Toole argues that 1000 Friends glossed over weaknesses in the study it commissioned, titled Land Use, Transportation and Air Quality (LUTRAQ), in hopes of persuading Metro to adopt land-use policies that would reduce auto dependence. O'Toole derides transportation planning in general and Metro's efforts in particular, which he argues have dramatically worsened Portland-area congestion. See O'Toole, *The Best-Laid Plans.*

39. Adler and Dill, "The Evolution of Transportation Planning."

40. Parsons et al. recommended the less ambitious VMT goal in part because even Metro's sophisticated transportation planning effort did not promise to achieve that goal.

41. Adler and Dill, "The Evolution of Transportation Planning."

42. Ibid.

43. Ibid.; Carl Abbott, *Greater Portland: Urban Life and Landscape in the Pacific Northwest* (Philadelphia: University of Pennsylvania Press, 2001), 5..

44. R. Gregory Nokes, "Study Sees Billions in Boundary Costs if Measure Passes," *The Oregonian,* October 17, 2000, 1.

45. Felicity Barringer, "Property Rights Law May Alter Oregon Landscape," *The New York Times,* November 26, 2004, 1.

46. Abbott, *Greater Portland.*

47. Timothy Egan, "In Portland, Houses Are Friendly. Or Else," *The New York Times,* April 20, 2000, F1.

48. Hal Bernton, "Oregon Slows Sprawl to a Crawl," *The Oregonian*, December 8, 1999, 1.

49. John Turner and Jason Rylander, "Land Use: The Forgotten Agenda," in *Thinking Ecologically*, ed. Marion R. Chertow and Daniel C. Esty (New Haven, Conn.: Yale University Press, 1997), 60.

50. Katz, "Building the Future."

51. Gordon Oliver, "City Outpaces Region in Growth of New Housing," *The Oregonian*, December 14, 1999, 1.

52. Alan Ehrenhalt, "The Great Wall of Portland," *Governing*, May 1997, 20; Tim Ferguson, "Down with the Burbs! Back to the City!" *Forbes*, May 5, 1997, 142–152.

53. Ehrenhalt, "The Great Wall."

54. Abbott, *Greater Portland*.

55. Seltzer, "It's Not an Experiment."

56. Marcia D. Lowe, "Alternatives to Sprawl: Shaping Tomorrow's Cities," *The Futurist*, July–August 1992, 28–34.

57. Abbott, *Greater Portland*.

58. Adler and Dill, "The Evolution of Transportation Planning."

59. Jonathan Barnett, "Shaping Our Cities: It's Your Call," *Planning*, December 1995, 13.

60. New Urbanism is a movement initiated by architects that aims to restore many of the principles of traditional urban design, particularly the emphasis on pedestrian movement and comfort, instead of vehicular mobility: front doors rather than garages face the street; on-street parking is allowed; tightly spaced houses with small front yards abut the sidewalk; neighborhood shopping is within walking distance; public parks and squares are designed to foster neighborly gatherings. In short, "The multiple components of daily life—workplace, residence, shopping, public space, public institutions—are integrated, after 50 years of sorting, separating, and segregating functions and people." See Roberta Brandes Gratz and Norman Mintz, *Cities Back from the Edge: New Life for Downtown* (Washington, D.C.: Island Press, 1998), 327–328.

61. Paul Gapp, "Portland: Most Livable—'But Please Don't Stay,'" *Chicago Tribune*, April 2, 1980, 1.

62. Alan Katz, "Developing the Future," *The Denver Post*, February 10, 1997, 1.

63. Ben Plowden, "A City Ablaze in Green and Gold," *New Statesman*, January 8, 2001.

64. Johnson, "The Myth and Reality"; Robert D. Putnam, Lewis M. Feldstein, and Don Cohen, *Better Together: Restoring the American Community* (New York: Simon & Schuster, 2003).

65. Johnson, "The Myth and Reality," 109.

66. Ibid., 110.

67. Putnam et al., *Better Together*, 244.

68. Abbott, *Greater Portland*, 151.

69. Johnson, "The Myth and Reality"; Ferguson, "Down with the Burbs!"

70. Abbott, *Greater Portland.*

71. O'Toole, *The Best-Laid Plans.*

72. Quoted in Katz, "Developing the Future."

73. Ehrenhalt, "The Great Wall"; Katz, "Developing the Future."

74. Deborah Howe, "The Reality of Portland's Housing Market," in *The Portland Edge*, 184–205.

75. John M. DeGrove, *Planning Policy and Politics: Smart Growth and the States* (Cambridge, Mass.: Lincoln Institute of Land Policy, 2005).

76. Clarren, "Planning's Poster Child Grows Up."

77. Laura Oppenheimer, "Initiative Reprises Land Battle," *The Oregonian*, September 20, 2004, B1.

78. Laura Oppenheimer, "Property Compensation Fight: People on Both Sides," *The Oregonian*, October 7, 2004, 4.

79. Quoted in Barringer, "Property Rights Law."

80. The measure won in all thirty-five of the state's counties except Benton County, which encompasses Corvallis and Oregon State University. See Laura Oppenheimer, "Land-Use Laws on Turf That Is Uncharted," *The Oregonian*, November 4, 2004, 1.

81. Laura Oppenheimer and James Mayer, "Poll: Balance Rights, Land Use," *The Oregonian*, April 21, 2005, C1. Respondents in a poll by the Oregon Business Association and Portland State University's Institute of Portland Metropolitan Studies said they value property rights more than farmland, the environment, or wildlife habitat, and 60 percent chose individual rights over responsibility to their community. On the other hand, two-thirds said growth management made Oregon a more desirable place to live, and respondents strongly favored planning over market-based decisions and wanted to protect land for future needs instead of using it now for homes and businesses.

82. Laura Oppenheimer, "Battle Intensifies over Oregon's Property Rights Law," *The Oregonian*, August 13, 2005, 1.

83. Quoted in Blaine Harden, "Anti-Sprawl Laws, Property Rights Collide in Oregon," *The Washington Post*, February 28, 2005, 1.

84. Barringer, "Property Rights Law."

85. "Marion County Judge Overturns Measure 37," press release, Oregonians in Action, October 14, 2005.

86. Eric Mortenson, "Rhetoric May Cool as Measure 49 Settles In," *The Oregonian*, April 20, 2008, W67.

87. Eric Mortenson, "Trying to Beat the System," *The Oregonian*, May 28, 2008, 4. Mortenson reports that 1,258 property owners did not respond at all to the state, which gave claimants ninety days to choose one of three options.

88. Eric Mortenson, "Measure 49 Housing Boom a Bust," *The Oregonian*, September 2, 2010.

89. Quoted in Christian Gason, "Land-Use 'Grand Bargain' Heading to Governor's Desk, But Oregon Growth Debate Far From Settled," *OregonLive*, March 4, 2014; Doug Burkhardt, "Legislature Oks Land Use Grand Bargain," *Portland Tribune*, March 5, 2014.

90. James Mayer, "Planners Brace for Growth Battle," *The Oregonian*, August 12, 1990, B1.

91. Quoted in Barringer, "Property Rights Law."

92. Ferguson, "Down with the Burbs!"

93. Christopher R. Conte, "The Boys of Sprawl," *Governing*, May 2000, 28.

94. Ferguson, "Down with the Burbs!"

95. Goldberg, "The Wisdom of Growth."

96. Alex Dodds, "Voters Strongly Support Smart Growth Measures on Election Day 2014," Smart Growth America, November 5, 2014. Available at http://www.smartgrowthamerica.org/2014/11/05/voters-strongly-support-smart-growth-measures-on-election-day-2014/.

CHAPTER SIXTEEN

POST-KATRINA

Lessons From a Disaster

Since Hurricane Katrina in 2005, the United States has encountered numerous and devastating hurricanes (e.g., Sandy, Ike, Rita, Maria, Michael). Yet, the story of Hurricane Katrina is important because it not only is one of the costliest and deadliest U.S. hurricanes to date, but it also examines our ability to recover from a disaster in the face of a rapidly changing climate. As geographer James Mitchell observes, Hurricane Katrina was a truly exceptional event by almost any measure—from the size of the affected population to the degree to which buildings and infrastructure were destroyed or rendered unusable to the range and scale of the economic costs.[1] But the overwhelming impacts of the storm also made manifest the risks facing many Americans, as well as others around the world. According to the National Oceanic and Atmospheric Administration (NOAA), more than half of the U.S. population lives in 673 coastal counties, up from 39 percent in 1970. Coastal areas have always been subject to hurricanes, earthquakes, and tsunamis; they suffer from the effects of chronic processes as well, including erosion, subsidence, and saltwater intrusion. But with scientists forecasting rising sea levels and more severe storms as climate change proceeds, urban areas on the coast are even more exposed to hazards than they have been historically.[2]

In the late twentieth century a growing number of Americans moved to areas routinely affected by hurricanes, forest fires, earthquakes, and floods. In seeking to make such areas safer for human habitation, government policies such as subsidized flood insurance, wildlife suppression and firefighting, construction of levees, and federal relief payments fostered, rather than discouraged, dangerous settlement patterns.[3] As Paul Farmer, executive director of the American Planning Association, explains, the government's message when it comes to disasters consistently has been, "We will help you build where you shouldn't, we'll rescue you when things go wrong, and then we'll help you rebuild again in the same place."[4] Although improved building techniques, forecasting technology, and evacuation planning mitigated disaster-related losses for a time, the movement of masses of people into disaster-prone areas in the latter decades of the twentieth century has reversed that trend.[5]

Also contributing to the devastation wrought by Hurricane Katrina are entrenched poverty and racial disparities that have long bedeviled New Orleans. Although the storm affected black and white, rich and poor, the vast majority of residents hit hardest by the storm were poor and African Americans. Those same people were the least able to evacuate, the last to be allowed back into the city to inspect the damage and retrieve their belongings, and the least well equipped to rebuild. This is a common pattern, as indicated by environmental justice experts.[6] Sociologists Robert Bullard and Beverly Wright argue that, in general, "Race tracks closely with social vulnerability and the geography of environmental risk."[7] Geographers Susan Cutter and Christopher Emrich observe that physical vulnerability correlates strongly with social vulnerability, defined not just by one's age, income, and race but also by the extent of one's social network and access to health care and emergency-response personnel.[8] Race and poverty affect not only the severity of damage from disasters but relief efforts as well.

Further compounding the damage caused by Katrina and the flooding that followed was the extraordinarily incompetent response by every level of government. Unfortunately, few disaster experts were surprised by the inadequate public-sector response. Geographer Rutherford Platt points out that the "Byzantine" federal disaster-management system relies heavily on a variety of partnerships with state and local governments, nongovernmental organizations, and the private sector.[9] The Department of Homeland Security, created after September 11, 2001, undermined those already tenuous partnerships while giving natural disasters short shrift, instead directing most of the resources of the U.S. Federal Emergency Management Agency (FEMA) to counterterrorism activity.[10] For their part, local governments historically have been reluctant to prepare for major disasters. Municipal officials devote most of their energies to addressing immediate concerns for which there are active constituencies, such as schools, roads, and crime.

Finally, efforts to rebuild New Orleans illustrate the challenges associated with trying to modify past practices in an effort to enhance a city's resilience, while respecting the wishes of citizens, who typically want to restore the place that existed before. For urban planners, New Orleans' experience illustrates a familiar set of tensions between bottom-up and top-down problem solving. Most planners believe that collaborating with stakeholders yields opportunities for mutual learning among citizens and officials and therefore results in more effective plans. But citizens' desire to return to normalcy can impose severe constraints on efforts to introduce ecological sensitivity to the planning process. Moreover, even a plan that reflects citizens' wishes can be thwarted by a lack of resources and capacity for effective implementation.

Meanwhile, under conditions of scarce resources, those who are wealthier can rebuild first, while bureaucratic requirements designed to ensure accountability dampen the energy of even the most determined low-income rebuilders. Nonetheless, Hurricane Katrina "was not an equal opportunity storm."[11] The effects of years of discriminatory practices expose the institutionalized racism that is also evident in the environmental injustice's wrought by Hurricane Katrina.

BACKGROUND

Established in 1718 by the French governor of Louisiana, Jean-Baptiste Le Moyne de Bienville, New Orleans originally occupied a crescent of high ground between the Mississippi River and the brackish, 630-square-mile Lake Pontchartrain (see Map 16.1).[12] Bienville and subsequent settlers encountered an alluvial plain built over thousands of years, as the Mississippi River deposited vast amounts of sediment during storms and annual spring floods. The drying sand and silt compressed but was replenished each year by new infusions. As a result of this ongoing process, southern Louisiana had gradually grown into a vast expanse of marshland interspersed with bayous.[13] Along the Mississippi River, natural levees rose ten to fifteen feet above sea level, while the banks of Lake Pontchartrain stood only a few feet or even mere inches above the bay.[14] Early settlers made modest changes to this delta landscape: they built their houses on relatively high ground, often on stilts, and they erected small riverfront levees to protect themselves from periodic flooding. They were not entirely successful: the Mississippi River inundated New Orleans in 1719, 1735, 1785, 1791, and 1799.[15]

Aggressive efforts to confine the Mississippi River over the next two centuries dramatically changed the region's geography, however. In 1803 the United States acquired New Orleans as part of the Louisiana Purchase, and American farmers began moving to the region to build plantations along the bayous. In the mid-1800s wealthy property owners created levee districts to spread the costs of maintaining and reinforcing levees that would contain the region's waterways, but floods intermittently overwhelmed these paltry defenses. Finally, in 1879, Congress commissioned the Army Corps of Engineers (Corps) to build more substantial levees along the entire lower Mississippi in hopes of mitigating floods. This more extensive levee system had the effect of raising the level of floodwaters, however, often with disastrous consequences.[16]

After the great flood of 1927, Congress authorized the Mississippi River and Tributaries Project, in which the federal government assumed the entire cost of erecting levees, spillways, and other structures from Cairo, Illinois, southward to protect against an 800-year flood. Upon completing this project, the Corps boasted, "We harnessed it, straightened it, regularized it, shackled it."[17] But this massive construction project had unanticipated consequences: over time, the dams and levees on the Mississippi and its tributaries, particularly the Missouri River, captured 60 percent to 70 percent of the 400 million cubic yards of sediment that for thousands of years had flowed to the delta and built up the region's marshes and barrier islands. In addition, to prevent the formation of sandbars that impeded navigation, engineers straightened the last miles of the Mississippi by building parallel 2.5-mile jetties, which carried whatever sediment did make it to the river's end over the continental shelf and out into the deep ocean. Without sediment to nourish its marshes or rebuild its barrier islands, the entire delta began to subside and erode.

Another process got under way in the early 1900s that further altered New Orleans' geography. Public officials and private entrepreneurs were intent on draining water from the city after heavy rainstorms, a regular feature of the city's weather, and in 1899 voters approved a comprehensive plan to do so.[18] During the first forty years of the twentieth century, the Sewer and Water Department installed a network of drainage canals and pumps that successfully lowered the water table throughout the city.[19] To keep the canals dry enough to capture water during storms, officials pumped them continuously, in the process causing the city to subside. The process

Map 16.1 New Orleans and Hurricane Katrina's Storm Surge

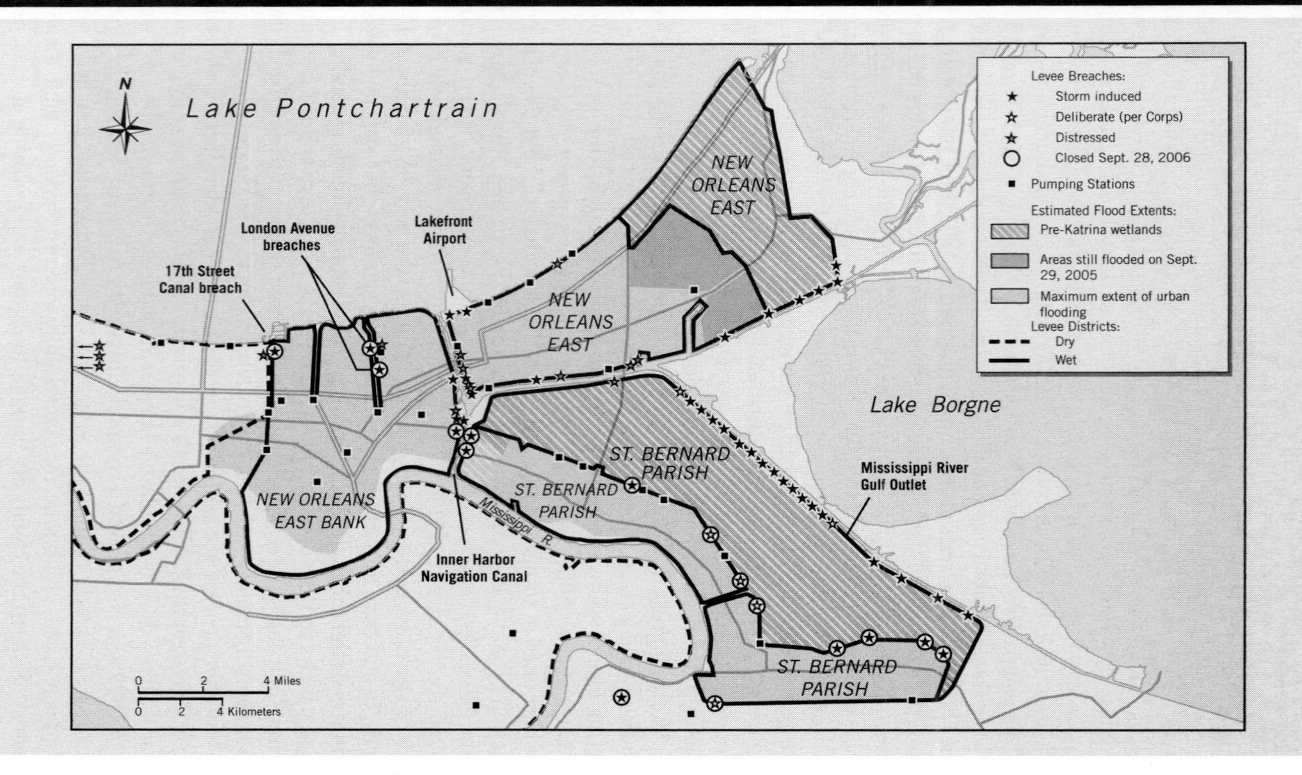

Source: U.S. Geological Survey and Army Corps of Engineers.

was self-perpetuating: as swampy areas were drained, the reclaimed land sank; the more the city sank, the more it flooded, and the deeper the canals and more pumping that was needed to keep it dry. The drainage system also enticed people into harm's way, while giving them a false sense of security.[20]

The dredging of thousands of miles of channels for oil and gas and shipping aggravated the subsidence and erosion of the wetlands between New Orleans and the gulf. Beginning in the mid-1900s, fossil fuel companies dredged hundreds of miles of navigation channels and pipeline canals through the southern Louisiana marsh. Those canals not only caused massive coastal erosion, they also transported saltwater inland, where it killed the cypress swamps and the grasses and bottomwood forests of the interior marshes. Local development interests contributed to the damage as well: at the behest of the Port of New Orleans, the Corps built a series of shipping channels to facilitate commerce.[21] Completed in 1923, the 5.5-mile Inner Harbor Navigation Canal, universally known as the Industrial Canal, connected the Mississippi River and Lake Pontchartrain; it also allowed for docks that were shielded from the fluctuations of the unruly Mississippi. An even larger channel, the Gulf Intracoastal Waterway, was built in the 1930s.

The most controversial shipping channel of all was the $62 million Mississippi River–Gulf Outlet, or MR-GO, known locally as "Mr. Go." Justified as a way to allow freighters easier access to New Orleans' inner harbor from the gulf, MR-GO cut a path directly through the unspoiled marshes of St. Bernard Parish.[22] But at thirty-six feet deep, MR-GO was too shallow for the deep-draft container ships that were coming into use by the time it was completed in 1968. (This was a recurring problem for the Corps in New Orleans: by the time its projects were built, they were often obsolete.) Within a decade, traffic along MR-GO began to decline and by 2004 averaged just one vessel per day—15 percent of all Port of New Orleans traffic. Although not heavily used, MR-GO had dramatic environmental consequences. Originally dredged to 650 feet across, the channel eroded rapidly, reaching 2,000 feet wide in places by the early 2000s. With devastating efficiency, MR-GO facilitated saltwater intrusion that destroyed 28,000 acres of marsh and caused substantial changes to another 30,000 acres;[23] it also provided a pathway into eastern New Orleans and St. Bernard Parish for hurricane storm surges. Detractors had warned of all these outcomes, but they had been overruled by development interests.[24]

The cumulative result of these processes was that Louisiana's marshes vanished, while the land that remained began to sink. Coastal marshland disappeared at a rate of twenty-five to thirty-five square miles a year during the latter half of the twentieth century. In total, more than 1,900 square miles of Louisiana's coastal marshland disappeared between 1930 and 2000.[25] Those vast marshes had once served to blunt storm surges because hurricanes weaken as they travel over the irregular surfaces of the land; a commonly cited estimate is that every four miles of marsh reduces storm surge by as much as a foot. Meanwhile, New Orleans itself was sinking by about one-quarter inch per year. As a consequence, by 2005 nearly half of the city was below sea level—although the most densely populated areas were well above sea level.[26] Observers warned that in a severe storm the city would flood, and it would be hard to remove the water trapped behind the levees.

New Orleans' vulnerability was made manifest in 1965, when Hurricane Betsy struck the Gulf Coast, inundating parts of the city under eight feet of water and causing $1 billion in damages. The Corps' response to Hurricane Betsy was to propose engineering an even more extensive levee system, to be completed in the early 1980s, that would protect the city from future Category 3 hurricanes.[27] An important selling point was that the new system would facilitate continued urbanization of the region: protection of existing development accounted for just 21 percent of the benefits used to justify the proposed $80 million Lake Pontchartrain and Vicinity Hurricane Protection Project, while 79 percent were to come from the new development that would be feasible given the additional protection.[28] In 1965 Congress authorized the Lake Pontchartrain project as part of the Flood Control Act.

Three years later Congress passed the National Flood Insurance Act, which enabled households and businesses to insure their property against floods—something most private insurers refused to do. Under the new flood insurance program, homeowners in areas designated as flood prone—that is, with a 1 percent or more chance of catastrophic flooding in a given year—are required to buy policies from insurance companies; the government pays for flood damage with federal funds collected from homeowner premiums. Insurance was supposed to go only to property owners in communities with floodplain-management laws that were enforced—so that, ostensibly, insurance was exchanged for a commitment to reduce vulnerability. In an effort to increase the program's coverage, in 1973 Congress made insurance mandatory for anyone who took out a mortgage from a federally regulated lender to buy property in a flood zone, adding penalties in 1994 for lenders that did not comply. To entice more people into the program, Congress also limited the amount premiums could rise in a single year.

In combination, the enhanced levee system and the availability of flood insurance facilitated explosive growth into the wetlands of eastern Orleans Parish and Jefferson Parish. Critics warned, however, that these new developments were at serious risk in the event of a major storm. First of all, in designing the Lake Pontchartrain and Vicinity Project the Corps had used a "standard project hurricane" that was based on a mix of characteristics of past storms, all of which were relatively mild; the agency had concluded that stronger protection would be "cost prohibitive."[29] Moreover, as the completion date for the project slipped and costs escalated, the Corps made compromises in its execution: engineers focused on fortifying existing levees and built flood walls in places where land acquisition would have been too costly.[30] Meanwhile, no accommodation was made for the fact that the levees themselves were sinking.[31] As a result, according to computer simulations done in the 1990s, the 350-mile "system" of levees and flood walls that surrounded the New Orleans metropolitan area was capable of protecting against a fast-moving Category 3 storm at best. After Hurricane Georges gave the city a near miss in 1998, Congress authorized the Corps to begin studying ways of bolstering the city's defenses against a Category 5 hurricane. But that work moved slowly; by 2004 the Corps had just completed its preliminary study.

Hurricane Georges also prompted scientists, engineers, federal agencies, and the region's politicians to converge on a blueprint for restoring coastal Louisiana. At $14 billion, "Coast 2050" was the most expensive restoration plan yet proposed in the United States. Its central elements were rebuilding the marshes and reconnecting the barrier islands, both of which would, in theory, protect the coast from storm surges. More specifically, the plan featured several key projects. First, at critical spots along the Mississippi River, engineers would build diversions to allow suspended sediments to wash down through the marshes toward the gulf. A second project involved taking 500 million cubic yards of sand from Ship Shoal to rebuild the southern barrier islands and cutting a channel in the neck of the river delta about halfway down. This would enable the Corps to stop dredging the southern end of the river and would allow the mouth of the river to fill with sediment that would eventually flow to the west and rebuild the barrier islands. A third project entailed building a new port and closing MR-GO. And a fourth consisted of building a pair of gates on the narrow straits on Lake Pontchartrain's eastern edge where it connects to the gulf. Those gates could be lowered during storms but otherwise would remain open to allow tidal flushing.

Congress refused to fund the ambitious coastal restoration plan, however, so the state had to rely on the meager $50 million annually provided under the Coastal Wetlands Planning, Protection, and Restoration Act of 1990, known as the Breaux Act.[32] Although few projects were actually built using Breaux Act money, one of the projects described in "Coast 2050" that did become a reality was the Davis Pond Diversion, a dam that opens and closes to allow water (and sediment) to flow into 33,000 acres of wetlands, oyster beds, and fishing grounds. That project mimicked the Caernarvon Freshwater Diversion Structure, near MR-GO, which releases 8,000 cubic feet per second of Mississippi River water in an effort to preserve 16,000 acres of marsh. Completed in 1991 at a cost of $26 million, the Caernarvon project illuminated not only the benefits of restoration but also the pitfalls. In 1994 oyster farmers in the area, who had paid $2 per acre for fifteen-year claims, filed a class action suit against the state, arguing that the project had reduced the value of their leases. In December 2000 a local jury awarded five of the farmers $48 million in damages; applied across the entire class the award added up to $1.3 billion. Although the verdict eventually was overturned by the Louisiana Supreme Court, the legal wrangling unnerved backers of restoration. To shield itself from further liability, the state practically shut down the Caernarvon diversion and stopped work on fifteen other restoration projects while the litigation was pending.

In short, over the course of the twentieth century, development in and around New Orleans left the city in a precarious situation, and efforts to bolster its defenses yielded negligible improvements while facilitating further development in flood-prone areas. As a result of both subsidence and migration patterns, whereas only 48 percent of New Orleans residents were below sea level in 1960, when the city's population peaked at 627,535, by 2000, 62 percent of New Orleans residents lived below sea level.[33] That New Orleans was vulnerable despite its hurricane-protection system was widely known. Journalist Elizabeth Kolbert notes that "Katrina was probably the most comprehensively predicted disaster in American history."[34] At an annual meeting shortly after the September 2001 terrorist attacks, the nation's disaster scientists and emergency planners warned that a major hurricane would destroy New Orleans. In October 2001 journalist Mark Fischetti published an article in *Scientific American* that sought to publicize experts' concerns. He argued that "[i]f a big, slow-moving hurricane crossed the Gulf of Mexico on the right track, it would drive a sea surge that would drown New Orleans under 20 feet of water. . . . New Orleans is a disaster waiting to happen."[35] In December Eric Berger wrote in the *Houston Chronicle* about New Orleans' dire prospects in the face of a major hurricane, predicting that hundreds of thousands would be left homeless. Then, in early summer 2002, *The Times-Picayune* ran a sobering series by journalists John McQuaid and Mark Schleifstein suggesting that the levees around New Orleans would breach in a serious storm, with "apocalyptic" consequences.

Disaster management officials struggled to devise ways of responding to the severe hurricane that many believed was inevitable. In July 2004, FEMA and the Louisiana Office of Homeland Security and Emergency Preparedness conducted a week-long simulation in which New Orleans was hit by a hypothetical Hurricane Pam, a slow-moving hurricane preceded by twenty inches of rain. The scenarios used to construct the simulation did not count on the levees failing; nevertheless, the exercise predicted ten to twenty feet of water in some parts of the city, the evacuation of 1 million people, and the need to rescue 100,000 more who would remain behind. Although it generated some useful insights, as well as a momentary burst of media attention, the exercise was never translated into a workable plan—a victim of budget cuts. Then, just months before Katrina struck, a report by the Corps on the region's hurricane-protection plan identified weaknesses in the levee system, which was decades behind schedule, and expressed concern about a $71 million cut in the fiscal year 2005 budget of the New Orleans District.[36] "Continuing land loss and settlement of land in the project area may have impacted the ability of the project to withstand the design storm," the agency warned.[37]

THE CASE

Despite the warnings and planning exercises, New Orleans was woefully underprepared for Hurricane Katrina and its aftermath. As a result, days after the hurricane struck, thousands were still stranded in a city that was largely under water, enduring temperatures hovering in the muggy 90s, and with limited access to food, water, or medical supplies. Subsequent analyses revealed that shoddy engineering beforehand and poor coordination among federal, state, and local governments after the fact transformed a severe storm into a disaster. Even as they struggled to untangle the causes of the disaster and provide relief to victims, Louisiana officials seized the opportunity to promote both a more protective levee system and an ambitious coastal restoration plan that scientists believed would enhance the region's resilience in the face of storms. But skeptics wondered whether shoring up a city in such a perilous place made sense, and Congress and the administration of President George W. Bush were reluctant to endorse the costly project. Meanwhile, local officials struggled to plan the monumental task of rebuilding, a process fraught with racial and economic tensions.

Disaster Strikes

Katrina began its life early in the week of August 22, 2005 as a tropical depression off the Bahamas. By August 25 it was officially a hurricane. By late Friday, August 26, Katrina had gained strength and, after killing nine people and knocking out electricity in south Florida, it was on track to slam into the Gulf Coast. At that point, forecasters at the National Hurricane Center in Miami were predicting it would hit southeast Louisiana on Monday, August 29, as a Category 4 storm with top winds of 132 miles per hour; they warned that if the storm moved through New Orleans the city could see storm surges of eighteen to twenty-two feet.[38] Ominously, forecasters expected the storm to pass over the "loop current," a 200-foot-deep swath of 90-degree-Fahrenheit tropical seawater floating in the Gulf of Mexico. If it did, it was sure to intensify.

The normally placid Max Mayfield, director of the National Hurricane Center, called dozens of federal, state, and local officials to transmit an urgent message. "This is the 'Big One,'" he told them. "I'm as sure as I can be."[39] A computer model devised by the Louisiana State University (LSU) Hurricane Center late Saturday suggested that the New Orleans metro area could see flooding on the scale of Hurricane Betsy, with a storm surge of as much as sixteen feet moving up MR-GO, topping levees, (see Map 16.1). According to the model, high water from Lake Pontchartrain would also flood over levees to the north. (The model did not account for waves that could overtop the levees along the lake's south shore.[40]) "All indications [were] that this [was] absolutely worst-case scenario," said Ivor van Heerden, deputy director of the LSU Hurricane Center.[41] According to an unusually explicit alert sent by the National Weather Service, the New Orleans metropolitan area would experience blown-out windows, airborne debris, power outages, and uprooted trees.

State and local officials seemed to take the threat seriously. The governor's office held a conference call with emergency preparedness directors from Louisiana parishes at 5 p.m. on Friday to update them on the forecast and review state plans. At 11 p.m. Democratic Governor Kathleen Babineaux Blanco declared a state of emergency. The following day, after speaking with Max Mayfield, the governor ordered a mandatory evacuation of all low-lying areas, and at 5 p.m. New Orleans Mayor C. Ray Nagin declared a state of emergency and issued a voluntary evacuation order.[42] (According to *Times-Picayune* reporter Bruce Nolan, Nagin was hesitant to issue a mandatory evacuation order because of the possibility that hotels and businesses would sue the city for lost trade—a charge Nagin vehemently denied.[43]) An hour before the mayor's declaration, state police activated the state's contraflow plan, which allows traffic to use both sides of I-55, I-59, and I-10 to leave the city.

Federal officials appeared to be responding to the hurricane center's increasingly strident warnings as well. On Saturday, August 27, from his vacation home in Texas, President Bush declared a state of emergency for the Gulf Coast, authorizing the Department of Homeland Security and FEMA to "coordinate all disaster relief efforts which have the purpose of alleviating the hardship and suffering caused by the emergency on the local population." On Sunday, after being told by the National Hurricane Center that Katrina's storm surge was likely to top the levees, FEMA Director Michael Brown convened a videoconference with disaster-management officials in Louisiana, Mississippi, and Alabama, as well as the president and Homeland Security Secretary Michael Chertoff. While awaiting information from the state about precisely where to deploy supplies and specialized personnel, Brown proceeded to have generators, tarps, and stockpiles of water, ice, and ready-to-eat meals delivered to bases around the Gulf Coast. He also dispatched twenty-three medical assistance teams and seven search-and-rescue teams to the region.[44]

At 1 a.m. on Sunday, August 28, as predicted, Katrina was declared a Category 4 hurricane, with sustained winds of more than 140 mph. Six hours later it was upgraded to a "potentially catastrophic" Category 5 storm with sustained winds above 155 mph and a storm surge of fifteen to twenty feet topped by large, dangerous waves. At that time, it was the strongest hurricane ever recorded in the gulf.[45] Finally, at 10 a.m., Mayor Nagin ordered a mandatory evacuation and warned residents that floodwaters could top the levees. For those who could not evacuate (prequalified special-needs residents), the city provided transportation on Regional Transit Authority (RTA) buses to the Superdome, a covered sports stadium and the only building in the city designed to withstand a severe hurricane. Municipal officials publicized the twelve pickup points on TV and radio and by shouting through megaphones on the streets.[46] Those heading to the Superdome were told to bring enough food, water, and medicine to last five days—a requirement that disaster management experts generally regard as unrealistic.

The vast majority of the city's 485,000 residents heeded the orders to evacuate, and traffic was at a crawl leaving the city late Saturday and into Sunday. By 3 p.m. on Sunday, about 10,000 people had taken shelter at the Superdome. An estimated 100,000 residents remained in their homes, however.[47] Some stayed because, having survived Hurricane Betsy, they were confident they could ride out another storm. Others were worried they would have nothing to come back to if they did not stay and protect their property. Still others were resigned to whatever fate the storm dished out. But many simply lacked the resources to evacuate: the storm came at the end of the month, when the city's poorest were out of cash and so were unable to pay for gas, food, or hotel rooms, and those who were native-born residents typically did not have relatives in nearby states who could take them in. In any case, approximately 51,000 New Orleans residents (28 percent of the adult population) did not have cars.[48] Local officials knew that the least mobile residents lived in some of the most flood-prone parts of town; they were aware that water rescues were likely.[49]

By the time the 460-mile-wide Hurricane Katrina made landfall at Buras, Louisiana, at 6:10 a.m. Monday, it had been downgraded from a Category 5 to a

Category 4 storm; nevertheless, it was bearing down on New Orleans packing 127-mph winds and pushing a storm surge of up to twenty-eight feet.[50] For eight straight hours, wind and heavy rains lashed the city, uprooting trees and tearing roofs and siding off houses. At 11:37 a.m., the National Weather Service issued an advisory, saying, "Widespread flooding will continue. . . . Those seeking refuge in attics or rooftops are strongly urged to take the necessary tools for survival."[51] By midafternoon, however, it appeared to many in New Orleans as though the worst was over: as of 9 a.m. the eye of the storm had passed 100 miles to the east, most city streets were dry, and newspapers around the country crafted headlines to the effect that "New Orleans Dodged a Bullet." But the sigh of relief was premature; in fact, water had begun rising throughout the city early that morning, and by late afternoon it was painfully clear that the levees surrounding the city had been breached.[52] Houses in the Lower Ninth Ward were inundated, and many residents had climbed onto their roofs to escape the rising waters.

By Tuesday, August 30, the city was a disaster zone. Engineers were struggling to repair massive breaks in the levees that separated New Orleans from Lake Pontchartrain, dropping 3,000-pound sandbags into a 300-foot-and-growing gap in the flood wall along the 17th Street Canal and two more on the London Avenue Canal. (This tactic had little impact on the breaches.) In the meantime, thanks to the combination of the storm surge and heavy rainfall, the surface of Lake Pontchartrain was nine feet above sea level, and water was pouring through breaches. Levees had also failed along the 80-year-old Industrial Canal, funneling water into the Lower Ninth Ward, New Orleans East, and St. Bernard Parish. The floodwaters had caused the city's twenty-two pump stations to fail, making it impossible to drain water out of the streets; as a result, 80 percent of the city was submerged, with water levels in some neighborhoods twenty feet deep and rising as fast as three inches per hour. Only a narrow band containing the French Quarter and parts of Uptown—the same strip that was settled by Bienville almost 300 years earlier—remained dry.

The official response to the news of widespread flooding was chaotic. Late in the day on Monday President Bush had declared a major disaster in Louisiana, thereby making federal funds available for relief and recovery. FEMA chief Brown had arrived at the Emergency Operations Center in Baton Rouge, but FEMA search-and-rescue teams had difficulty getting into flooded areas. The U.S. Coast Guard was deploying helicopters to pluck people from rooftops, while personnel from the Louisiana Department of Wildlife and Fisheries conducted rescue operations by water. National Guard and wildlife officials ferried people to the Superdome in trucks. But the scale of the devastation quickly overwhelmed the capabilities of government response teams, which were severely hampered by an inability to communicate with one another: federal, state, and local officials were using incompatible equipment; different agencies used different radio frequencies; and power, cable, and telephone service had been knocked out.

As officials struggled to respond, conditions in the city deteriorated. Inside the Superdome, where some 23,000 people had taken shelter, the heat and humidity were stifling (power had gone out even before the storm hit on Monday morning), toilets overflowed, and food and water were scarce. As the heat inside soared above 100 degrees, the stench became overpowering; the elderly, sick, disabled, drug addicted, and mentally ill began to break down.[53] Late on Tuesday, Governor Blanco announced that everyone needed to be evacuated from the city, including those in the Superdome, but there were not enough buses on hand to carry out her order. As the promised buses repeatedly failed to appear, despair mounted.

Throughout the rest of New Orleans, electricity was sporadic and looting was widespread. Although there was scattered theft of luxury items and reports of armed gangs looting the Walmart in the Lower Garden District, most of the stealing was done by people desperate for food and water.[54] Nevertheless, media coverage treated black and white looters differently, focusing on the former and exacerbating racial tension. Reports of widespread theft and violence caused FEMA to hold back rescue workers who might be imperiled. More worrisome from the perspective of those trying to evacuate trapped residents was the fact that in various parts of the city, including at the Charity Hospital, sniper shots had been fired.[55] Lacking the resources to maintain order and fearing for their own safety, some New Orleans police officers fled, while others joined in the looting.[56]

By Wednesday, August 31, New Orleans had descended into chaos. Besieged rescuers—including citizen volunteers from surrounding parishes and, eventually, other states—were dropping people off anywhere that was dry. Overpasses, parking lots, and highway ramps became scenes of intense suffering, as people—many of them elderly and disabled—waited for help in the sweltering heat without food, water, or medical supplies. Fires broke out in empty buildings, and without water pressure firefighters were unable to respond, so a smoky pall hung over the city. Heightening concerns about the safety of rescue workers, Police Superintendent P. Edwin Compass III told journalists stories of gangs attacking tourists and beating and raping them in the streets—stories he later confessed were exaggerated.[57] Stranded travelers continued to arrive at the Superdome; after being turned away, they went to the Morial Convention Center, where frustrated evacuees had broken in and set up camp. (Eventually, some 22,000 evacuees were camped in and around the convention center awaiting transportation out of the city.) Like the Superdome, the convention center lacked working toilets, clean water, or electricity. Moreover, because it had been occupied spontaneously, there was inadequate security and people had not been searched for weapons, so many observers portrayed it as dangerous. Desperate to restore order, Governor Blanco made an urgent appeal to President Bush for federal troops. She also ordered the city's remaining police officers to stop engaging in search and rescue and focus on keeping the peace.

Late on Wednesday, Mayor Nagin reiterated the governor's order for a total evacuation of the city, but—after a frantic search—state officials were able to come up with only ten buses to bring evacuees to Houston. By Thursday, although Bush assured the nation that assistance was on its way to New Orleans, local officials were clearly at their wits' end with the pace of the federal response. Mayor Nagin estimated that 50,000 survivors remained on rooftops or in shelters awaiting rescue and evacuation. He issued a plea for help, saying, "This is a desperate SOS. We are out of resources at the convention center."[58] Finally, on Friday—five days after the storm hit—more National Guard units and active-duty federal troops arrived to help regain control of the convention center and restore order to the city. Even as the Coast Guard continued to pluck survivors from rooftops and attics, a large contingent of buses arrived to evacuate residents. Fifty trucks carrying food and water and other supplies—the first to reach the storm victims at the convention center—rolled into the city. And the floodwaters began to recede. On Saturday President Bush ordered 7,000 active-duty soldiers from the 82nd Airborne and the 1st Cavalry divisions to the region, and they began arriving later that day.

The Aftermath

By Monday, September 5, a week after the hurricane struck, the rescue and relief operation appeared to be running relatively smoothly. The Corps had successfully patched the levee breaches at the 17th Street and London Avenue canals and were carefully pumping water out of the city into Lake Pontchartrain.[59] Residents of relatively dry neighborhoods were being allowed to return to their houses temporarily to see what they could salvage. Although conditions were improving, only around 10 percent of the city's pumping capacity was operational, and many neighborhoods remained under ten feet of water. On September 7, on orders from Mayor Nagin, New Orleans police officers, fire department officials, and military personnel began trying to compel the estimated 5,000 to 10,000 residents remaining in the city to leave, even those in undamaged homes, on the grounds that the risks posed by waterborne diseases and gas leaks were too great.

By the beginning of the third week after the storm, Coast Guard Vice Admiral Thad W. Allen had replaced FEMA Director Brown as overseer of the post-Katrina relief effort—a tacit admission of the federal government's culpability. Search-and-rescue missions continued. Flights were slated to resume in and out of the Louis Armstrong International Airport after a sixteen-day hiatus. And work was scheduled to begin on repairs to the extensively damaged I-10. Mail service was resuming in patches. Twenty-seven permanent pumps and forty-six temporary pumps were removing a total of 7 billion to 8 billion gallons of water daily;[60] as a result, the city was draining faster than expected and was likely to be dry within weeks, not months as originally forecast.

Nevertheless, the city faced numerous long-term challenges as it began to contemplate large-scale reconstruction. The water and sewer infrastructure had suffered massive damage: drinking water was leaking underground, probably because uprooted trees and fire hydrants had broken water mains, and pipes were likely full of toxic material. Untreated sewage was expected to seep from broken pipes for months; what sewage the city could collect it had to pump untreated into Lake Pontchartrain and the Gulf of Mexico. Restoring electricity was expected to be slow as well, as underground conduits and soaked transformers needed to be repaired, and then the wiring in each house and building had to be inspected. The transportation network was in disarray. Schools, police stations, and hospitals were moribund.[61]

Meanwhile, discussions were under way about how to more effectively help those rendered homeless by the storm. FEMA had begun to issue debit cards worth $2,000 to the 335,000 evacuees to enable them to pay for living expenses.[62] According to the Red Cross, some 36,000 Red Cross volunteers were providing food, shelter, and other emergency help to about 160,000 Katrina evacuees at 675 shelters in twenty-three states.[63] But more permanent solutions were needed, and to that end FEMA ordered 50,000 trailers and mobile homes to be placed in "trailer cities" throughout Louisiana. Proposals for massive trailer parks sparked resistance among officials from surrounding parishes, who feared the additional burden on their already strained finances.

While FEMA struggled to meet the pressing needs of displaced residents, officials at every level sought to assign blame for the debacle in New Orleans. The finger pointing began on Sunday, September 4, as soon as the initial panic abated. Homeland Security chief Chertoff told reporters that federal officials had not expected the damaging combination of powerful hurricane winds and levee breaches that flooded New Orleans—a claim that was belied by the warnings issued by the National Hurricane Center, the devastating results of the Hurricane Pam exercise, and the existence of a forty-page report, submitted via e-mail to the White House Situation Room at 1:47 a.m. on August 29, that made remarkably accurate predictions about Katrina's impacts.[64] Although FEMA officials were surveying the scene from helicopters within twenty-four hours of the storm, Chertoff had continued to insist that they were unaware of the scale of the devastation.[65] Rather than acknowledge their responsibility, federal officials disparaged Governor Blanco's leadership and faulted Mayor Nagin for failing to order a mandatory evacuation earlier, not delivering a more urgent and detailed request for assistance, and declining to commandeer buses to transport residents in the Superdome and convention center out of the city.

For his part, while acknowledging that some of the criticism was warranted, the mayor insisted that logistical hurdles made it difficult to use the available buses. He argued that there were not enough buses for the number of people remaining in the city, there were few places dry enough to stage the buses that were available, and flooded roads would have prevented the buses from leaving anyway. State and local officials insisted that FEMA was the real problem; it had not only failed to deliver urgently needed food and ice but had thwarted rescue and medical efforts by private citizens and officials from other agencies.

Even as New Orleans struggled to regain some semblance of normalcy, four weeks after Katrina hit, New Orleans faced a new menace: Hurricane Rita was brewing offshore and threatened to inflict more damage on the tattered city. After barreling across the gulf, Rita struck on Friday, September 23, bringing floodwaters back into New Orleans. Although its main impacts were felt in southwestern Louisiana, Rita's heavy rains and five-foot storm surge overwhelmed the patch on the Industrial Canal, reflooding the Lower Ninth Ward and St. Bernard Parish. Parts of Lakeview and Gentilly saw one to two feet of water, mainly because the pump stations that normally drained those areas were temporarily shut down to lighten the load on the Corps' makeshift dams on the 17th Street and London Avenue canals.

Fortunately, pump operators were able to drain reflooded areas relatively quickly, and by the following week commerce in New Orleans was picking up, as people began to return. Upon arriving, however, residents beheld a massive and daunting cleanup: the city had to dispose of an estimated 50 million cubic yards of debris. There were piles of rotting food and other foul-smelling garbage piled indiscriminately throughout flooded neighborhoods; smashed, waterlogged cars and stranded boats littered the streets; an estimated 300,000 refrigerators, freezers, stoves, and other "white goods" needed to be collected and recycled; and there were 5.5 million pounds of hazardous waste, from paint thinner to bleach, awaiting proper disposal.[66] A further concern was the thick cake of mud that coated almost the entire city. That mud had mixed with an unknown assortment of hazardous materials during the flood; it was potentially harmful as it dried and turned to dust and became airborne, or as people came into direct contact with it. The Louisiana Department of Environmental Quality insisted the residue was safe, but environmentalists remained concerned about elevated levels of lead and arsenic, pesticide residues, and hazardous chemicals generated by the incomplete combustion of petroleum products.

What Happened, and Why?

Efforts to sort out exactly why Katrina was such a debacle began within a week of the storm, and it quickly became a truism to say that Hurricane Katrina was a human-made, rather than a natural, disaster. New Orleans' levees had breached in more than a dozen locations, but discerning the causes at each site was complicated. Gradually, investigators obtained evidence that revealed how the city's flood-protection system had failed.[67] By all accounts, the Corps had made a host of design choices that reduced the ability of New Orleans' flood-control system to withstand a major hurricane. For example, the Corps decided to use flood walls, rather than wider earthen levees, to line the canals because it was reluctant to condemn property adjacent to the canals. "Usually, there are homes right up against the canal," explained Corps project manager Al Naomi.[68] "You have to relocate five miles of homes [to build a levee], or you can build a floodwall." Moreover, constructing a levee would have required building further into the canal itself, reducing the volume of water it could handle.

In many places the Corps also decided to use I-shaped walls instead of T-shaped walls, even though the latter, which have a horizontal section buried in the dirt, are generally stronger and more stable. But T walls are more expensive and, like levees, require additional land and a broad base of dense soil for support. In addition, the canal walls were built in ways that left them potentially unstable in a flood: some rose as high as eleven feet above the dirt berms in which they were anchored, even though a Corps engineering manual cautioned that such walls should "rarely exceed" seven feet because they can lose stability as waters rise.[69]

But the most serious flaws in the Corps' design and execution concerned the depth of the steel pilings and the soil into which they were driven. To save money, the sheet piling was driven only 17.5 feet deep at 17th Street and 16 feet deep at London Avenue.[70] Yet soil-boring data revealed a five- to twenty-foot-thick layer of spongy peat soil starting at fifteen to thirty feet beneath the surface.[71] Tests showed the peat soil to be unusually weak and to have a high water concentration, making it extremely vulnerable in a flood.[72]

Corps investigators concluded that even before Katrina's eye had crossed land, her storm surge had raised the water level in the Industrial Canal, forcing the flood walls outward and opening up a gap between the wall and its earthen base. As water coursed through the gap, the wall tipped over and water poured into eastern New Orleans and Gentilly to the west.[73] Despite the break, the water in the canal continued to rise, eventually spilling over both sides of the fourteen- to fifteen-foot levee (see Map 16.1).[74]

Then, as Katrina moved east of New Orleans, pushing a storm surge from the gulf, its winds shifted counterclockwise and drove the high water in Lake Pontchartrain south, reversing the flow in the drainage canals. Rising water put enormous pressure on both sides of the 17th Street Canal, pushing the walls outward. That movement opened a small space between the flood wall and its earthen levee. The gap quickly widened, as the pressure increased and the weak and unstable soils under the levee base became saturated. At around 10 a.m., the soil beneath the flood wall finally gave way, and eight fifty-foot concrete panels broke away with it.[75] An hour later, two breaches opened up in the London Avenue Canal, as soils gave way and flood walls collapsed.

In November 2005, Professor Raymond Seed, a civil engineer at the University of California, Berkeley and one of the heads of a team of experts financed by the National Science Foundation (NSF), told the Senate Homeland Security and Governmental Affairs Committee that the weakness and instability of the area's soils should have prompted the Corps to raise the safety factor it used in designing the Lake Pontchartrain and Vicinity levee system. In fact, the Corps used a safety factor of just 1.3—a standard that was appropriate for farmland but not for a densely populated urban area.[76] Robert Bea—also an engineering professor at the University of California, Berkeley, and cohead of the NSF team—added that malfeasance during construction of the flood-control system may have contributed to the levee failures, although the Corps' review, released in early June 2006, found "no evidence of contractor negligence or malfeasance."[77] The LSU team commissioned by the state similarly concluded that the conditions that caused the canal flood wall failures should have been obvious to the engineers that designed them. The Corps initially argued that Katrina had exceeded the forces the system was designed to withstand, but it admitted responsibility after federal meteorologists pointed out that the sustained winds over Lake Pontchartrain only reached 95 mph, well below those of a Category 3 hurricane.[78]

Although deeply culpable, the Corps did not bear sole responsibility for the weaknesses in the flood-control system; both the Louisiana congressional delegation and local officials—particularly the Orleans Levee Board and the Sewer and Water Board—had resisted measures to shore up the system and frequently supported diverting funds to projects that would yield more tangible economic benefits.[79] For instance, early on the Corps had proposed building gates to prevent water from the Gulf of Mexico from reaching Lake Pontchartrain and flooding the canals. That project was delayed by a lawsuit filed by the environmental group Save Our Wetlands, Inc., which contended the Corps had failed to study the project's ecological impacts. Most accounts blame environmentalists for blocking the floodgates, but many other entities—including state legislators, members of Congress, and *The Times-Picayune*—opposed the plan as well. Ultimately, the levee board supported the Corps' decision to abandon the floodgate approach and instead raise the levees along the lake and the Mississippi River and add flood walls on the canals. (The Corps' analysis subsequently confirmed that the levees would be as effective as gates and would be less expensive.)

The Corps had also recommended building butterfly gates at the end of each of the city's drainage canals. But in 1990 officials from the New Orleans Sewer and Water Board and the levee board vetoed that proposal, arguing that the gates would make it more difficult to pump water out of the city. They hired an engineer to devise an alternative approach that involved building higher walls along the canals—a plan they persuaded the Corps to adopt. In addition, the levee board convinced the Corps to employ a 100-year rather than a 200-year standard as the cost of the project escalated.[80]

Another egregious lapse that reflected badly on both local officials and the Corps was the lack of serious inspections and routine maintenance of the levee system. Shortly after the storm, *The Times-Picayune* reported that almost a year before Katrina hit, some residents near the 17th Street Canal levee had complained to the Sewer and Water Board that their yards were repeatedly filling with water.[81] Others had reported leaks or sand boils, both indications of water running under the surface of the levee. But either no one came to investigate or those who did declined to follow up. Similarly lackadaisical were the cursory annual inspections by officials from the Corps, the levee board, and the state Department of Transportation and Development. Those affairs usually lasted a mere five hours or less and consisted mostly of photo ops and fancy lunches.

Compounding the levee failures was the lack of preparedness of government at every level once the storm actually hit. At the local level, Colonel Terry Ebbert, director of New Orleans' Office of Homeland Security and Public Affairs, had decided to make the Superdome the city's only shelter, assuming that people would be taken to better-equipped shelters outside the metro area within forty-eight hours. As the water started to rise, however, it became clear that no provision had been made to transport the evacuees. Most of the city's RTA buses had been placed at a facility on Canal Street that officials (mistakenly) believed would stay dry. Some buses parked on the waterfront did escape flooding, but as the waters rose it became difficult to move them to designated staging areas. When it became apparent that FEMA had not wrangled any buses either, Governor Blanco's staff began scrambling to find some. But as news of violence and looting was broadcast, local officials began to resist lending New Orleans their school buses; in any case, many of the operators were afraid to drive in the rising waters. FEMA finally identified a supply of buses, but it took three days to put together a fleet that could begin moving people out; as a result, nearly a full week elapsed before the last people were evacuated from the convention center.

The complex relationships among local, state, and federal governments only made matters worse. According to journalist Eric Lipton and his colleagues, the crisis that began with the failures of the flood-control system deepened because of "a virtual standoff between hesitant federal officials and besieged authorities in Louisiana."[82] From the outset, negotiations among local, state, and federal officials were contentious and miscommunication was common. The main source of conflict was the question of who ought to have final authority over the relief effort.[83] Federal officials awaited direction from the city and state, while local officials, overwhelmed by the scale of the storm, were not only incapable of managing the crisis but unable to specify what they needed to deal with it.

For example, on August 29, when President Bush phoned Governor Blanco, she told him, "We need everything you've got"—a request that apparently was insufficiently precise to prompt him to order troops to the region. Two days later, when Blanco specifically asked the president for 40,000 soldiers to help quell the rising unrest in the city, Bush's advisers debated whether the federal government should assume control over the relief effort, ultimately concluding that Bush should try to seize control of the National Guard troops. The White House proceeded to send Blanco an urgent request, in the form of a memorandum of understanding, to allow Bush to take charge of the guard. The governor refused to sign, however, arguing that to do so would have prohibited the guard from carrying out law-and-order activities. According to journalist Robert Travis Scott, Blanco's advisers were concerned about the White House spin if the president took control.[84] For their part, the president's advisers worried about the political fallout of federalizing the relief operation. Ultimately, the White House decided to expedite the arrival of a large number of National Guard personnel, including many trained as military police, who would operate under the direction of the governor; only belatedly did the president order federal troops to the scene.

Beyond the tensions associated with federalism, organizational issues within FEMA hampered the relief effort. Local officials expected the federal government to provide rapid and large-scale aid, but FEMA was bogged down in legal and logistical questions; cumbersome rules, paperwork, and procedures stymied efforts by volunteers, the National Guard, and first-response teams to react. In large measure FEMA's organizational dysfunction was a product of its history. Created in 1979 by President Jimmy Carter, FEMA was a dumping ground for patronage appointments under Presidents Ronald Reagan and George H. W. Bush; as a result, it quickly earned a reputation for incompetence. President Bill Clinton broke with tradition: he appointed James Lee Witt, a highly regarded disaster management expert who had served as chief of the Arkansas Office of Emergency Services, to lead the agency, before elevating the FEMA head to a cabinet-level post in 1996. Witt proceeded to professionalize the agency and dramatically improve its performance. But President

George W. Bush returned to the practice of appointing political loyalists. As a consequence, when Katrina struck, five of the top eight FEMA officials had come to their posts with no experience in handling disasters.[85] Moreover, after the 2001 terrorist attacks, FEMA had been subsumed within the newly created Department of Homeland Security. No longer a cabinet-level agency, FEMA was diverted by terrorism threats and saw morale among its disaster-management experts plummet. Even the notoriously incompetent Brown had warned that FEMA was not up to the task of dealing with major disasters because of budget and personnel cuts.[86]

Rebuilding and Conflicting Ideas About Restoration

Savvy political actors are well aware that disasters are focusing events of the first order. They can open particularly wide policy windows in which advocates can attach their preferred solutions to newly salient problems.[87] So it was not surprising when, on August 30, the New Orleans *Times-Picayune* reported that Louisiana politicians were seizing on the damage wrought by Hurricane Katrina to press for long-requested federal assistance in shoring up the state's coastline.[88] In particular, they wanted emergency financing and fast-track permitting for a $34 billion hurricane-protection and coastal restoration plan that would cover all of southeast Louisiana.[89] Beyond the $14 billion coastal restoration long advocated by the state, the proposal included a system of levees tall enough to withstand the twenty-foot storm surges expected with a Category 5 hurricane; floodgates at the Rigolets and Chef Menteur passes into Lake Borgne, as well as on human-made waterways; gates at the mouth of the city's drainage canals; and relocation of pump stations to the lakefront combined with replacement of canals with underground culverts so lake surges could not penetrate the city. There was no need for the Corps to conduct its usual cost-benefit test, proponents argued; Katrina had already provided one.

Not everyone believed that a massive coastal restoration plan made sense. According to journalist Cornelia Dean, scientists disagreed about how much of a difference coastal marshes actually make in blunting hurricanes.[90] Geologists Robert Young and David Bush pointed out in an op-ed that neither more wetlands nor rebuilt barrier islands would have mitigated the damage from Katrina, most of which arose because of a storm surge from the east.[91] They also observed that the restoration plan did not address the root causes of wetlands loss: human-made alteration of the Mississippi River that has reduced the amount of sediment flowing into the marshes by as much as 80 percent, the saltwater allowed in by navigation canals that cut through the delta, and a lowering of ground levels throughout the region brought on by a combination of natural forces, urban drainage, and industrial activities. Young and Bush noted the irony of calling for higher levees, which exacerbate the loss of wetlands by preventing flooding that brings sediment. In any case, they added, with rising sea levels any recreated wetlands would soon be under water. Even the National Academy of Sciences, in its report on the restoration proposal, acknowledged that although the plan's components were scientifically sound, they would reduce annual wetland loss by only 20 percent.[92]

There was even more skepticism about the idea of a Category 5 hurricane-protection system. Some observers argued it was folly to try to armor the city, given what appeared to be its inevitable trajectory. They pointed to a 2005 article in the journal *Nature*, which reported that the city and its levees were sinking faster than previously believed—an inch a year in some places—and some parts of the levee system were three feet lower than intended.[93] Nevertheless, in November 2005, Congress agreed to spend $8 million for an "analysis and design" of an enhanced hurricane-protection system for Louisiana, as part of a $30.5 billion fiscal year 2006 spending bill for the Corps. Congress gave the Corps six months to submit a preliminary report on "a full range of flood control, coastal restoration, and hurricane-protection measures" capable of protecting southern Louisiana from a storm. A final report was due in twenty-four months. Most observers acknowledged, however, that neither the president nor Congress seemed to have the stomach for Category 5 protection; rather, they seemed inclined to favor Category 3 protection and incremental improvements over time.[94]

Most contentious of all, though, were debates over rebuilding the city itself. Shortly after the hurricane, a growing chorus of observers suggested that rebuilding New Orleans should involve more than simply raising levees, building new homes, and returning to the status quo. In an op-ed piece in *The New York Times*, geographer Craig Colten argued that "to rebuild New Orleans as it was on Aug. 29 would deal a cruel injustice to those who suffered the most in recent days. . . . Those who rebuild the city should try to work with nature rather than overwhelm it with structural solutions." To this end, he suggested, "The lowest-lying parts of the city where the waters stood deepest should be restored to the wetlands they were before 1700, absorbing Lake Pontchartrain's overflow and protecting the rest of the city."[95]Harvard geologist Dan Schrag told Cornelia Dean that "there has to be a discussion of what responsibility we have not to encourage people to rebuild their homes in the same way."[96] And *New York Times* business columnist Joseph Nocera argued that it was likely that the city would shrink, and that was a good thing. New Orleans' population, he said, was too large for the jobs its economy generated, which was one reason the city was so poor.[97]

Among the specific suggestions floated early on were massive landfilling, government seizure of property, and bulldozing of flood-prone neighborhoods.[98] Neighborhoods such as eastern New Orleans and Lakeview could be elevated to ten feet above the water line, making them less susceptible to flooding. All houses could be rebuilt to tougher building codes. The city as a whole could be modernized. Low-income housing, already blighted, could be razed. Underpinning such ideas was the notion that post-Katrina New Orleans was a clean slate, and there was a historic opportunity to undertake smarter planning.

Proposals to "rebuild smart" and "shrink the footprint" of New Orleans immediately encountered resistance, however, from historic preservationists, property rights activists, social justice advocates, and residents of neighborhoods that would be demolished. For example, geographer Michael E. Crutcher pointed out that a "smarter" New Orleans—that is, one that was denser and had more wetlands and a functional mass-transit network—would not belong to the people who lived there before.[99] Beverly Wright, a sociologist at Xavier University, reacted furiously to the suggestion that some neighborhoods should not be rebuilt, pointing out that there was no discussion of abandoning the Florida coast, which was hit every year by hurricanes.[100] Many displaced residents suspected that Mayor Nagin and his allies in the business community hoped to discourage New Orleans' poorest residents from returning. In response to the furor, the city council passed a defiant resolution saying, "All neighborhoods [should] be included in the timely and simultaneous rebuilding of all New Orleans neighborhoods."

In hopes of providing some guidance to the rebuilding, on September 30, 2005, Mayor Nagin appointed a seventeen-member Bring New Orleans Back Commission (BNOBC), led by Joseph Canizaro, a prominent conservative real estate developer with ties to the Bush White House, and charged it with developing a master plan for the redevelopment. At Canizaro's behest, the Washington, D.C.-based Urban Land Institute (ULI) agreed in mid-October to advise the commission pro bono.[101] In mid-November the thirty-seven-member ULI team traveled to New Orleans to tour the city and meet with residents; it also conducted town hall meetings in Houston, Baton Rouge, and other cities where evacuees were housed temporarily. In mid-November the panel issued its recommendations. It recommended that the city, having lost most of its tax base in the evacuation, turn its finances over to a municipal oversight board. It also advised creating a new redevelopment agency, the Crescent City Rebuilding Corporation, that could engage in land banking, buy homes and property, purchase and restructure mortgages, finance redevelopment projects, issue bonds, and help with neighborhood planning. They proposed restoring slivers of wetlands throughout the city, especially in low-lying areas, to enhance flood control. Most controversially, the ULI experts said the city should use its historic footprint, as well as lessons learned from Katrina, as guides in determining the most logical areas for redevelopment; the result would be to focus in the near term on rebuilding neighborhoods that suffered the least damage from post-Katrina flooding: the highest and most environmentally sound areas. For some of the lowest-lying areas—such as eastern New Orleans East, Gentilly, northern Lakeview, and parts of the Lower Ninth Ward—the city should consider mass buyouts and a transition to green space.[102]

Aware that the ULI report might cause a stir, in mid-December the BNOBC endorsed the idea of shrinking the city's footprint, but it modified its implementation to make it more palatable to homeowners who wanted to rebuild in low-lying areas.[103] At first, the commission floated the idea of allowing residents to rebuild anywhere and then, if a neighborhood was not developing adequately after three years, buying out the rebuilt homes and possibly condemning whole neighborhoods. The nonprofit watchdog group the Bureau of Governmental Research was strongly critical of this laissez-faire proposal and urged city leaders to come up with a realistic and smaller footprint on which to build New Orleans.[104] In response to such criticism, the BNOBC's final report—unveiled on January 11, 2006—suggested that Nagin put a moratorium on building permits in devastated areas and give residents four months to craft plans to revive them. A plan would have to be approved before residents could move back in. The commission also endorsed the idea of massive buyouts of residential property in neighborhoods that did not come up with an acceptable plan or attract sufficient development within a year; those areas would be transformed into open space. To help it succeed, each neighborhood would have access to teams of planners and other experts. (That said, it was never clear where residents would live while sorting out the fate of their neighborhoods, nor how they would rebuild without basic services.) The Louisiana Recovery Authority, a twenty-six-member commission established in mid-October by Governor Blanco to disburse $2.6 billion in federal rebuilding money, said the BNOBC plan struck the proper balance between residents' self-determination and tough choices.

But the BNOBC recommendations, and particularly a map that appeared on the front page of *The Times-Picayune* depicting neighborhoods slated for conversion as green dots, infuriated exiled residents, who pointed out that wealthy Lakeview—although badly flooded—did not receive a green dot. This plan will create a "whiter, richer, and less populated New Orleans that excludes the very kinds of people that give New Orleans its character and its culture," railed Martha Steward of the Jeremiah Group, a faith-based community organization.[105] Ultimately, the BNOBC planning exercise accomplished little beyond galvanizing opposition—but it did that with stunning effectiveness. Tulane geographer Richard Campanella describes the period that followed the BNOBC presentation as "one of the most remarkable episodes of civic engagement in recent American history."[106] Some residents had begun meeting within weeks of the storm. Starting in late January, however, scores of new organizations formed to take stock of their neighborhoods, while residents poured into meetings of existing organizations. "Despite their tenuous life circumstances," says Campanella, "New Orleanians by the thousands joined forces with their neighbors and volunteered to take stock of their communities; document local history, assets, resources, and problems; and plan solutions for the future."[107] Soon, umbrella organizations sprang up to coordinate the work of these ad hoc groups.

Political officials responded with alacrity to residents' ire. Mayor Nagin, who was facing reelection and was concerned that negativity would scare away business, all but disavowed the ULI/BNOBC report.[108] Two months earlier Nagin had made a firm commitment to rebuild the Lower Ninth Ward and New Orleans East after testimony before Congress suggested uncertainty about the future of those areas.[109] The city council also rejected the ULI approach, instead resolving that "[r]esources should be disbursed to all areas in a consistent and uniform fashion."[110] Former mayor Marc Morial weighed in as well, delivering a speech in early January that called for a return of all residents to the region.[111] In February, while the BNOBC was still working, the city council initiated its own neighborhood planning process, engaging Lambert Advisory, a Miami-based planning consultancy. The Lambert planning process focused exclusively on flooded neighborhoods.

That summer, in hopes of devising a plan that would pass muster with the Louisiana Recovery Authority, which was disbursing infrastructure funding, the city planning commission instigated yet another, more "democratic" process.

On August 1, backed by $3.5 million from the Rockefeller Foundation, the Greater New Orleans Foundation convened a series of public meetings in which groups representing more than seventy neighborhoods were asked to choose from among fifteen expert teams to help them craft rebuilding plans.[112] There were no comprehensive guidelines for the bottom-up process; instead, the idea was to weave the individual proposals into a citywide master plan. In late January 2007 the results of the citizen-driven planning process were released in the form of a 555-page Unified New Orleans Plan (UNOP).[113] This plan left all areas of the city open to redevelopment and proposed a host of new projects—libraries, schools, transportation, flood protection, and other amenities—valued at $14 billion. It left unclear, however, where the resources or capacity to carry out such an ambitious agenda would come from; moreover, its elements were not unified by a coherent vision of the city's future. Nevertheless, the city council approved the UNOP in June 2007, ultimately merging it with the Lambert plan.

In early December 2006, as citizens were finalizing elements of the UNOP, Mayor Nagin appointed Professor Edward J. Blakely, former chair of the Department of Urban and Regional Planning at the University of California, Berkeley, to coordinate recovery efforts for the city. In late March 2007 Blakely issued a third redevelopment plan that identified seventeen compact zones where the city would concentrate resources to stimulate reinvestment and renewal. Fourteen of the seventeen areas were in the more promising and less flooded western portion of the city. Priced at about $1.1 billion, Blakely's plan was notably more modest than its predecessors; its aim was to encourage commercial investments rather than to define particular areas as off limits.[114] Despite its relative modesty, even Blakely's plan faced daunting obstacles.

The proliferation of planning efforts notwithstanding, most of the redevelopment that subsequently occurred was piecemeal. Through the summer and fall of 2007 rebuilding continued in a haphazard way, resulting in precisely the "jack-o'-lantern" pattern of redevelopment that planners had hoped to avoid. Whole blocks in the Central Business District were quiet, and the downtown hospital complex remained shuttered. The poorest neighborhoods appeared abandoned. The state did not have the funding for Road Home, its federally financed homeowners' aid program, so by late 2007 only about one in five applicants had actually received money. In the spring of 2008, a full year after Blakely's plan was released, the city's designated redevelopment zones had changed hardly at all. Blakely, who had been given broad authority over a staff of 200 and jurisdiction over eight agencies, explained that federal money had been slow in arriving.[115] A year later, he resigned.

In addition to the challenge of obtaining funds, ongoing uncertainty about the security of the levee system complicated the rebuilding process. In late September 2005 the Corps created Task Force Guardian to oversee interim repairs to the levees and plan how to restore the system to its pre-Katrina level of protection, which was the limit of its authority. That project involved fixing 170 miles of damaged or destroyed levees, canals, and flood walls—more than half the system—by June 1, 2006, the start of the next hurricane season. By the two-year anniversary of Katrina the Corps had spent $7 billion on levee-system repairs out of a total of $15 billion it expected to spend by 2013.[116]

Uncertainty about the levee system in turn delayed FEMA's release of new floodplain maps designating which areas and structures the federal government would insure against floods.[117] Homeowners who wanted to rebuild before the final maps were issued could renovate without raising their floor levels—a costly process—if their homes met the "baseflood elevation" required in the 1984 maps or if they did not have "structural damage"—that is, if their home had not lost 50 percent or more of its pre-Katrina value. Those who followed these rules could not be dropped from the National Flood Insurance Program, and their flood insurance premiums could not rise more than 10 percent per year. In June 2007, after nearly two years of work, the Corps released a map showing block-by-block where flooding was likely to occur if a 100-year hurricane were to strike. The maps reflected improvements to the hurricane-protection system where they had been completed. Finally, in 2009, FEMA released its own maps detailing local flood risks. Still, because the levee work was incomplete, the new maps did not immediately affect flood insurance rates, nor did FEMA require any parish to implement building-elevation codes.[118]Although a great deal of progress has been made regarding improvements to New Orleans' flood control system since 2005, by no means is the work complete.

OUTCOMES

Hurricane Katrina resulted in 1,464 Louisiana deaths, with at least 135 people confirmed missing.[119] Total Katrina-related damage (throughout the Gulf Coast region) was estimated to exceed $81 billion, and overall economic losses associated with the storm were as high as $200 billion, making it the most expensive natural disaster in U.S. history.[120] By comparison, New York and Washington, D.C., experienced $87.9 billion in losses in the September 2001 attacks, which affected much smaller areas.[121] By 2010 the federal government had spent $142 billion on Gulf Coast recovery.[122]

In the five years after the storm hit, reconstruction of New Orleans was piecemeal and largely bottom-up. As journalist Campbell Robertson explains, many residents became "staunch advocates for their corners of the city, collecting local data, organizing committees, and even, in the case of the Vietnamese community, drawing up their own local master plan."[123] Despite the assistance of neighborhood organizations and out-of-state volunteers, numerous impediments faced individuals trying to rebuild, including astronomical insurance costs, grueling negotiations with insurance companies, dishonest contractors absconding with insurance money, and endless Road Home paperwork. For the city's poorest residents, such obstacles proved insurmountable, and many did not return. As a result, according to the Census Bureau, at 343,829 people, New Orleans was 29 percent smaller in April 2010 than it was a decade earlier; moreover, once more than two-thirds black, in spring 2010 the city was less than 60 percent black.[124]

On the other hand, some demographers estimated that by the summer of 2010, the New Orleans area had recovered 70 percent of its pre-Katrina jobs, and 79 percent of its pre-Katrina commercial activity.[125] One important legacy of the storm was a more vibrant civic life: according to the Brookings Institution and the Greater New Orleans Community Data Center, New Orleans residents had become much more likely than other Americans to attend a public meeting. The city's newly formed civic organizations—such as Citizens for 1 New Orleans, Women of the Storm, and numerous neighborhood-scale entities—could boast tangible achievements, including unified and professionalized regional levee boards; a watchdog inspector general, public contract reform, and a police monitor in New Orleans; and a single property assessor where previously there had been seven.[126]

The Brookings Institution and New Orleans have forged a biennial publication titled *The New Orleans Index at Five*. Its first publication in 2010 concluded that New Orleans was poised to become safer and more resilient—that is, better able to "absorb, minimize, bound back from, or avert future crises."[127] Among the key improvements after Katrina was the city's adoption of a master plan. In addition, the city's experiment in education, in which nearly two-thirds of the city's public school children attend charter schools, was producing encouraging results: as of 2010, nearly 60 percent of Orleans Parish children attended a public school that met state standards—twice as many as did before Katrina; on the Tenth Grade Graduate Exit Exam, the proportion of local students scoring at or above the basic level in English language rose from 37 percent in 2007 to 50 percent in 2010.[128] A network of community-health clinics had garnered widespread recognition: with $100 million in startup funds from the U.S. Department of Health and Human Services, the Louisiana Department of Health and Hospitals had created a network of twenty-five neighborhood health care providers that was running nearly ninety clinics throughout the city. As a result, less than one-third of respondents to a Brookings survey conducted in January 2010 lacked health care access due to cost—compared to 41 percent of adults across the United States.[129]

Furthermore, there were concerted efforts in some neighborhoods, notably the Lower Ninth Ward, to rebuild more sustainably. In 2010 the city council gave final approval to the city's new master plan, which featured sustainability as a prominent goal. But even before the plan's adoption a number of green initiatives were under way. The most high profile of these endeavors, the Make It Right Foundation headed by actor Brad Pitt, was building houses in the devastated Lower Ninth Ward that featured ample windows, solar panels, mold-resistant drywall, dual-flush toilets, and metal roofs that do not retain heat. The foundation aimed to build 150 homes and had completed thirty-nine as of August 2010. The nonprofit Global Green had become involved in building new schools that are energy efficient and equipped with solar panels. And Historic Green, a volunteer group, was helping residents install insulation and had built ten residential rain gardens.

On the other hand, municipal officials were still struggling to deal with 50,000 blighted and abandoned homes—about one-quarter of the city's housing stock—as well as 5,200 blighted commercial structures and 7,400 habitable but abandoned houses.[130] As journalist Bruce Nolan pointed out, "[I]n various neighborhoods, ugly scar tissue remains plainly visible in the form of vacant lots, empty houses, and the occasional rescue-team graffiti or dirty waterline."[131] Residents who lacked the means to rebuild faced rents that were almost 50 percent higher than before the storm, and the mixed-income developments that were supposed to replace demolished public-housing projects had not materialized; there was a list of 28,000 tenants waiting for subsidized housing.[132] Streets, water pipes, and sewer lines had also been restored in piecemeal fashion, leaving gaps throughout the system. Scores of city-owned properties—from community centers to playgrounds and libraries—were unoccupied.[133] The Charity and Veterans Affairs hospitals remained shuttered; in fact, only twelve of twenty-three hospitals in the metropolitan area were operating in 2010.[134]

Some other indicators were troubling as well. Chronic poverty remained a problem: average wages and median household income had risen but did so mostly because many of the city's poorest residents had not returned; even so, half of New Orleans residents were living within 200 percent of the federal poverty level.[135] Persistent crime and stagnant job growth had returned. The city remained heavily dependent on tourism, oil and gas, and shipping—three shrinking industries vulnerable to recession—and the city's workforce was ill-equipped for a more advanced economy. Elevated concentrations of arsenic and lead had been detected throughout the city, with the highest concentrations in soils from the poorer sections.[136] Although the administration of President Barack Obama took several steps to free up funding to facilitate improvements, federal assistance dwindled over time, and the city had to be creative in its efforts to become more resilient.[137]

Hurricanes Katrina and Rita also took a toll on the natural communities of Louisiana's Gulf Coast. Although ecologists expected some of the damage to be short term—coastal marshes evolved with hurricanes, after all—between them, the two storms shredded or sank about 217 square miles of marsh along the coastline, equivalent to 40 percent of the wetlands loss that had been expected to occur over fifty years; they wiped out the LaBranche wetlands, one of the state's most successful restoration projects; and they decimated the Chandeleur Islands, scraping 3.6 miles of sand from the chain.[138] In 2007 coastal scientists warned that the marshes protecting New Orleans would be gone by 2040 if action were not taken within ten years or less to restore them.[139] Recognizing the accelerated pace of wetlands loss, in 2012 the Louisiana legislature approved a fifty-year, $50 billion comprehensive master plan to shore up levees, restore barrier islands and marshes, and where necessary elevate homes and businesses to the level required by the National Flood Insurance Program. Developed by the Louisiana Coastal Protection and Restoration Authority, the plan included 109 projects to be implemented in two phases: the first from 2012 to 2031 and the second from 2032 to 2061.

To fund those projects, the plan relied primarily on three sources: about $170 million annually in federal excise taxes from oil development due to the Gulf of

Mexico Energy Security Act, available beginning in 2017; $75 million to $80 million annually from the federal Coastal Wetlands, Planning, Protection and Restoration Act, for which the state must provide a match of about $30 million per year; and the Coastal Protection and Restoration Trust Fund, which would furnish $30 million each year from royalties and severance taxes on mineral development. Bolstering the plan's finances in the near term were two pots of money: about $790 million dedicated by the Louisiana legislature out of federal money for Katrina rebuilding and $3 billion to $5 billion under the RESTORE Act, which divided fines levied against BP for the Deepwater Horizon oil spill under the Clean Water Act among the Gulf Coast states (see Chapter 11).[140] The Southeast Louisiana Flood Protection Authority created in the wake of Katrina had another idea for raising money: to the dismay of Governor Bobby Jindal and other state officials, in 2014 the board sued ninety-seven oil and gas companies charging them with causing more than one-third of the erosion of the Louisiana coastline and demanding compensation for that loss.[141]

Although the state struggled to fund coastal restoration, work on the structural aspects of the hurricane-protection plan was slated to be complete by the start of the 2011 hurricane season. By August 2010 the city's reinforced Hurricane and Storm Damage Risk Reduction System had been strengthened in a variety of ways: flood walls had been toughened with clay, soil had been reinforced with concrete, and I-shaped walls had been replaced with T-shaped ones. Critics noted, however, that the new hurricane-protection system was being built to withstand only a 100-year storm—a level many experts regarded as inadequate.[142] The Louisiana Coastal Protection and Restoration Study prescribed more extensive protection, including building much higher levees along existing levee alignments, constructing gates for the Chef Menteur and Rigolets passes into Lake Pontchartrain, and relocating or buying out homes in areas at the greatest risk of flooding. The Corps, which had backed away from using the Category 5 hurricane as the design standard, aimed to protect New Orleans from a "Katrina-like" event—a storm with a 1-in-400 chance of hitting in any given year. In light of the improvements made after Katrina, journalist Mark Schleifstein reported that by 2013 New Orleans boasted the best flood control system of any coastal community in the United States, though he noted that the system still had serious limitations.[143]

Federal and state governments have learned a great deal from their lack of preparedness from Hurricane Katrina. This served as an instrumental lesson for the October 2012 Hurricane Sandy, which affected many states in the Northeastern United States (e.g., New York, New Jersey, Connecticut). Although devastation occurred, effective response mechanisms were in place between FEMA and related agencies to act swiftly. However, this was not the case for Puerto Rico, a U.S. territory, where more than 3,000 people died in Hurricane Maria in 2017. Preparedness plans were not designed for hurricanes greater than a Category 1 in Puerto Rico. In 2018, the Florida panhandle encountered Category 4 Hurricane Michael. Florida government officials evacuated approximately 120,000 individuals, attempting to prevent the loss of lives.

CONCLUSIONS

New Orleans continues to be a place of massive social and economic inequity, which underlines the reality that it is the poor and disadvantaged who are particularly vulnerable to the impacts of environmental disaster; when disasters do strike, those who can least afford to relocate are often displaced, both internally and across national borders.[144] For environmentalists, the challenge is clear: attaining social justice is essential to achieving environmental sustainability and resilience, and any efforts to mitigate or adapt to the worst effects of climate change must be designed in ways that benefit the most vulnerable. One visible manifestation of the potential link between environmental sustainability and social justice is the trend toward green homebuilding that has emerged in New Orleans since the storm.[145] Nonprofit and social entrepreneurs regard post-Katrina New Orleans, and particularly its poorest neighborhoods, as incubators for more environmentally sustainable designs. As a result of their initiatives, according to Tulane University law professor Oliver Houck, New Orleans has become a greener, more user-friendly and financially stable city.

Ultimately, the foremost challenge is not predicting catastrophes, but preparing for them. In general, Americans underestimate the risks of flooding, do not anticipate recovering their investment in mitigation, or lack the upfront capital necessary to make improvements. As a result, they do a poor job of preparing for floods and other disasters.[146] As weather events have increased dramatically in frequency and severity, however, climate adaptation has gained a place on the agendas of many U.S. cities and counties. In recent years, many localities have made disaster preparedness a priority. Counties in Colorado, Florida, North Carolina, South Carolina, Texas, and Washington have constructed websites discussing their policies for management of natural disasters. Additionally, the National Association of Counties has developed an extensive report discussing the need for adequate disaster preparedness throughout the country.[147] This indicates that many localities are focusing on planning for natural disasters before they strike, increasing their likelihood of avoiding devastation of the magnitude seen in New Orleans following Hurricane Katrina.

Whereas the United States' approach tends to be reactive, the Netherlands has taken a far more precautionary and proactive approach to managing the natural hazards it faces. In 1953, after dikes and seawalls gave way during a violent storm that killed nearly 2,000 people and forced the evacuation of 70,000 others, the Dutch vowed to protect their country, more than half of which lies below sea level as a result of centuries of development. At a cost of some $8 billion over a quarter of a century, they erected a futuristic system of coastal defenses that could withstand a 10,000-year storm: they increased the height of their dikes to as much as forty feet above sea level, erected a type of shield that drastically reduces the amount of vulnerable coastline, and installed vast complexes of floodgates that close when the weather turns violent but remain open at other times so saltwater can flow into estuaries.[148] As journalist John McQuaid bitterly observed, "The Netherlands' flood defenses—a sculpted landscape of dunes, dikes, dams, barriers, sluices, and pumps designed to repel the twin threats of ocean storm surges and river flooding—are light years ahead of New Orleans' busted-up levee system."[149]

Although heavily engineered, the Dutch system is rooted in a philosophy of accommodating water and preserving natural flows where possible. "There's one important lesson we've learned as Dutch—we're fighting a heroic fight against nature, the sea and the rivers," said Ted Sluijter, a spokesperson for the Eastern Scheldt storm-surge barrier. "But if you fight nature, nature is going to strike back. Water needs space."[150] The Dutch also spend $500 million each year on inspection and maintenance to safeguard the system, which is known as the Delta Works. They are compulsive planners, constantly adjusting their approach to flood control in response to changing conditions. In hopes of learning from the masters, a series of workshops known as the Dutch Dialogues held at Tulane University in 2010 sought to brainstorm ways of rebuilding New Orleans that would put water at center stage.[151]

Snider argues that Hurricane Katrina has transformed Louisiana's approach to become a formidable leader in broader-scale coastal crisis preparedness. As coastal shorelines diminish due to climate change, Baton Rouge is the home to the Water Institute's new $60 million water campus. This serves as a flagship research center "where a sleek, glass-encased building that straddles the levee and floats out over the main channel of the Mississippi River" brings together researchers from across the globe to collectively offer solutions for coastal communities.[152]

Although it appears that hurricane preparedness has increased across the United States, many coastal areas remain vulnerable. It took Hurricane Katrina for Louisiana and the federal government to provide financial resources to attempt to prevent future crises. Major coastal cities such as Miami, Florida, and Galveston, Texas, await higher level protections. And, we often forget that Puerto Rico has been classified as a humanitarian crisis due to the lack of support from the U.S. federal government. For over a year, many families were without electricity and water. Many Americans do not even realize that Puerto Ricans are U.S. citizens.[153]

QUESTIONS TO CONSIDER

- How and to what extent have racial and income disparities thwarted recovery efforts in New Orleans?
- If you were an elected official, what advice would you provide to communities such as New Orleans on whether to rebuild or return the area to its natural habitat?
- Is the United States prepared to confront environmental injustices that are a direct result of racial and income disparities post-disaster?

NOTES

1. James K. Mitchell, "The Primacy of Partnership: Scoping a New National Disaster Recovery Policy," *Annals of the American Academy of Political and Social Science* 604 (2006): 228–255.
2. Many scientists believe that climate change will exacerbate the severity of hurricanes and other storms. The rise in the number of hurricanes in the early years of the twenty-first century was part of a natural cycle: from 1970 to 1994, there were very few hurricanes, as cooler waters in the North Atlantic strengthened the wind shear that tears storms apart; in 1995, however, hurricanes reverted to the active pattern of the 1950s and 1960s. In an article in the August 2005 issue of *Nature*, MIT meteorologist Kerry Emanuel argued that global warming may already have influenced storm patterns, but he acknowledged that the pattern in the Atlantic was mostly natural. See Kenneth Chang, "Storms Vary with Cycles, Experts Say," *The New York Times*, August 30, 2005, 1.
3. Raymond Burby, "Hurricane Katrina and the Paradoxes of Government Disaster Policy: Bringing About Wise Governmental Decisions for Hazardous Areas," *Annals of the American Academy of Political and Social Science* 604 (2006): 171–191; Rutherford Platt, *Disasters and Democracy: The Politics of Extreme Natural Events* (Washington, D.C.: Island Press, 1999). Geographer Gilbert White pointed out decades ago that reliance on structures such as levees for flood protection provides a false sense of security and actually lures development into harm's way; new development, in turn, demands more protection. See Craig Colten, *Perilous Place, Powerful Storms: Hurricane Protection in Coastal Louisiana* (Jackson: University of Mississippi Press, 2009).
4. Quoted in "Why We Don't Prepare for Disaster," *Time*, August 20, 2006. Available at http://www.time.com/time/magazine/article/0,9171,1229102,00.html.
5. The total cost of all natural disasters in the United States averaged $10 billion annually from 1975 to 1989, according to the National Academy of Sciences; between 1990 and 1998 the annual average cost was $17.2 billion. See John McQuaid and Mark Schleifstein, "Tempting Fate," *The Times-Picayune*, June 26, 2002, 1. In 1998 Roger Pielke Jr. and Christopher Landsea showed that hurricanes were getting worse primarily because there was more to destroy. See John McQuaid and Mark Schleifstein, *Path of Destruction: The Devastation of New Orleans and the Coming Age of Superstorms* (New York: Little Brown, 2006).
6. Dennis Mileti, *Disasters by Design: A Reassessment of Natural Hazards in the United States* (Washington, D.C.: Joseph Henry Press, 1999), 105–134.
7. Robert D. Bullard and Beverly C. Wright, "Introduction," in *Race, Place, and Environmental Justice After Hurricane Katrina*, ed. Robert D. Bullard and Beverly C. Wright (Boulder, Colo.: Westview Press, 2009), 1.
8. Susan L. Cutter and Christopher T. Emrich, "Moral Hazard, Social Catastrophe: The Changing Face of Vulnerability Along the Hurricane Coasts," *Annals of the American Academy of Political and Social Science* 604 (2006): 102–112.
9. Platt, *Disasters and Democracy*, 277.
10. "Why We Don't Prepare for Disaster."

11. Gary Rivlin, "White New Orleans Has Recovered from Katrina. Black New Orleans Has Not. August 29, 2016. Available at https://talkpoverty.org/2016/08/29/white-new-orleans-recovered-hurricane-katrina-black-new-orleans-not/.

12. Prior to Bienville's arrival, aboriginals had occupied the Mississippi Delta for some 800 years, but they did not build fixed settlements there.

13. A bayou is a slow-moving river that can reverse the direction of its flow as the tide goes in and out.

14. Richard Campanella, *Delta Urbanism: New Orleans* (Washington, D.C.: Planners Press, 2010). Levees are earthen berms that are typically built out of local soils.

15. Campanella, *Delta Urbanism.*

16. John Barry, *Rising Tide: The Great Mississippi Flood of* 1927 *and How It Changed America* (New York: Touchstone, 1997); Colten, *Perilous Place;* William R. Freudenburg, Robert Gramling, Shirley Laska, and Kai T. Erikson, *Catastrophe in the Making: The Engineering of Katrina and the Disasters of Tomorrow* (Washington, D.C.: Island Press, 2009). According to geographer Colten, the Corps accepted responsibility for flood control only reluctantly. Originally, the Corps was charged with maintaining the nation's navigation system. When Congress created the Mississippi River Commission, however, it broadened the Corps' mandate to include designing levees for flood control. See Colten, *Perilous Place.*

17. Quoted in John McPhee, *The Control of Nature* (New York: Noonday Press, 1989), 26.

18. Colten, *Perilous Place*.

19. The fortuitous introduction of A. Baldwin Wood's revolutionary screw pump in 1913 facilitated the drainage and development of vast tracts of marshland and swampland that became the city's interior.

20. By 1926 the New Orleans drainage system served 30,000 acres, with 560 miles of canals, drains, and pipes. See Campanella, *Delta Urbanism.*

21. In New Orleans, shipping interests exert particular influence because they play an outsized role in the local economy. Shipping interests control the Dock Board, which in turn shapes decisions about the city's development.

22. In Louisiana, counties are called parishes.

23. Matthew Brown, "Reasons to Go," *The Times-Picayune*, January 8, 2006, 1.

24. Freudenburg et al., *Catastrophe in the Making*.

25. Cornelia Dean, "Louisiana's Marshes Fight for Their Lives," *The New York Times*, November 15, 2005, F1.

26. Leslie Williams, "Higher Ground," *The Times-Picayune*, April 21, 2007, 1.

27. A hurricane is defined as a tropical cyclone in the North Atlantic with wind speeds above 75 miles per hour (mph). Hurricanes are categorized on the Saffir-Simpson Hurricane Wind Scale, which is based on barometric pressure, wind speeds, and other factors that indicate the likelihood of a storm surge. A storm surge is an abnormally high tide, on top of which the wind builds waves. A Category 3 hurricane has winds of

between 111 and 129 mph, a Category 4 hurricane has winds of between 130 and 156 mph, and a Category 5 storm has winds 157 mph and above.

28. Burby, "Hurricane Katrina." The Lake Pontchartrain and Vicinity Project sought to protect the portions of Orleans and Jefferson parishes that face the lake. Separate levee systems partially encircled urbanized areas and tied into the Mississippi River levees. A third levee system completely encircled New Orleans and St. Bernard parishes.

29. Michael Grunwald and Susan B. Glasser, "The Slow Drowning of New Orleans," *The Washington Post*, October 9, 2005, 1. The imaginary storm's central barometric pressure was 27.6 inches, its sustained winds were 100 mph, and it had an average storm surge of 11.5 feet—a blend of characteristics from Categories 2, 3, and 4 hurricanes.

30. A levee requires five to six feet of land at its base for every one foot of height. See Colten, *Perilous Place*. In many areas along the city's drainage canals, that much open land was not available, and the Corps would have had to condemn extensive tracts of private property to build levees. In those places, it built flood walls, which consist of concrete sections attached to steel-sheet pile drilled deep into the earth and fortified by a concrete and earthen base. The sections of a flood wall are joined with a flexible, waterproof substance that allows the concrete to expand and contract without cracking.

31. Journalists John McQuaid and Mark Schleifstein describe a decision by Frederic Chatry, chief engineer for the Corps' New Orleans District, upon learning that the existing levees had sunk below design heights. Raising hundreds of miles of levees would be extremely expensive, so Chatry decided to use the new elevation numbers only for projects not yet in the design phase. Completed projects, or those that were already under way, would be based on the old numbers—so some would be more than a foot shorter than their congressionally mandated height. See McQuaid and Schleifstein, *Path of Destruction*.

32. In 2005, after years of persistent lobbying, the Louisiana delegation managed to secure $570 million over four years for coastal restoration projects, more than before but still a fraction of the amount needed.

33. Campanella, *Delta Urbanism*.

34. Elizabeth Kolbert, "Storm Warnings," *The New Yorker*, September 19, 2006, 35.

35. Mark Fischetti, "Drowning New Orleans," *Scientific American*, October 2001, 78.

36. Andrew C. Revkin and Christopher Drew, "Intricate Flood Protection Long a Focus of Dispute," *The New York Times*, September 1, 2006, 16.

37. Grunwald and Glasser, "The Slow Drowning."

38. Mark Schleifstein, "Storm's Westward Path Puts N.O. on Edge," *The Times-Picayune*, August 27, 2005, 1.

39. Quoted in "In the Storm: A Look at How the Disaster Unfolded," *The Times-Picayune*, September 18, 2005, 21.

40. Bruce Nolan, "Katrina Takes Aim," *The Times-Picayune*, August 28, 2005, 1.

41. Quoted in "In the Storm."

42. There are three levels of evacuation orders: voluntary, recommended, and mandatory. According to historian Douglas Brinkley, only the third carries any real weight and places the responsibility for evacuation on state and local officials.

43. Nolan, "Katrina Takes Aim."

44. Joseph B. Treaster and Kate Zernike, "Hurricane Slams into Gulf Coast," *The New York Times*, August 30, 2005, 5.

45. Freudenburg et al., *Catastrophe in the Making*.

46. Critics point out that the RTA had approximately 360 buses available, each of which could hold up to sixty people. Only a fraction of those buses were deployed, however, and service was at first erratic and later nonexistent. They also note that Amtrak trains could have moved people out on Sunday without adding to the traffic clogging the highways, but no request was made; in fact, when Amtrak tried to offer seats on an unscheduled train being used to move equipment out on Sunday, it was unable to get through to the mayor. See Douglas Brinkley, *The Great Deluge: Hurricane Katrina, New Orleans, and the Mississippi Gulf Coast* (New York: William Morrow, 2006).

47. That many people did not evacuate should not have come as a surprise. Disaster management experts have long known that about 20 percent of residents will not evacuate. See William Waugh Jr., "Preface," *Annals of the American Academy of Political and Social Science* 604 (2006): 6–9. A survey conducted by LSU in 2003 found that 31 percent of New Orleans residents would stay in the city even in the face of a Category 4 hurricane. See Peter Applebome, Christopher Drew, Jere Longman, and Andrew C. Revkin, "A Delicate Balance Is Undone in a Flash," *The New York Times*, September 4, 2005, 1.

48. Cutter and Emrich, "Moral Hazard."

49. Amanda Ripley, "How Did This Happen?" *Time*, September 4, 2005, 54–59.

50. Initially, reports said that Katrina had sustained winds of greater than 140 mph, but in December the hurricane center revised its estimate down to 127 mph and labeled Katrina a strong Category 3 hurricane. See Mark Schleifstein, "Katrina Weaker Than Thought," *The Times-Picayune*, December 21, 2005; Brinkley, *The Great Deluge*.

51. Quoted in "In the Storm."

52. Rumors of levee breaches began circulating on Monday morning, but because communication systems were knocked out, they were difficult to confirm. According to *The Times-Picayune*, Mayor Nagin announced at 8 a.m. Monday that he had heard reports that levees had been breached. See "In the Storm."

53. The descriptions of conditions in the Superdome are horrific. According to a firsthand account by Lieutenant Colonel Bernard McLaughlin, "It is a hot, brutal day—the Dome is reeking of sweat, feces, urine, discarded diapers, soiled clothing, discarded food, and the garbage strewn about is almost frightening in its sheer volume. People are openly cursing and fighting, many are openly angry with Mayor Nagin. Our focus is maintaining order and keeping things in control until the buses arrive." Quoted in Brinkley, *The Great Deluge*, 420.

54. Joseph B. Treaster, "Life-or-Death Words of the Day in a Battered City: 'I Had to Get Out,'" *The New York Times*, August 31, 2005, 1.

55. Like reports of rape, reports of sniper fire circulated widely; they were particularly damaging because they made rescue personnel wary of entering an area where shots had been reported. According to historian Douglas Brinkley, the U.S. government purposely sought to downplay reports of violence in the storm's aftermath because they

were embarrassing. And, in fact, some reports were false or exaggerated. But credible eyewitnesses confirmed many others.

56. Dan Baum, "Deluged," *The New Yorker*, January 9, 2006, 50.

57. Although other reports confirmed that there were armed gangs roaming the streets, rampant looting, and violent incidents, subsequent investigations revealed the worst reports—particularly those about the Superdome and convention center—were overblown. During the week following the storm, there were four confirmed murders. After careful inspection, *The Times-Picayune* reported the toll was four dead in the convention center, one by violence, and six dead in the Superdome, none by violence. See "Hurricane-Force Rumors," *The Times-Picayune*, September 27, 2005, 1.

58. Quoted in "In the Storm."

59. Officials were pumping cautiously to avoid overtaxing the pumps: only 2 of the city's 148 drainage pumps were online at this point; they were supplemented by dozens of smaller pumps that had been brought into the city. There was some concern about the ecological impacts of the foul water—which was laced with raw sewage, bacteria, heavy metals, pesticides, and toxic chemicals—on Lake Pontchartrain. But most scientists were sanguine about the lake's long-term resilience. They suggested that bacterial contaminants would die off fairly quickly, organic material would degrade with natural processes, and metals would fall apart and be captured by the sediment. See Sewell Chan and Andrew C. Revkin, "Water Returned to Lake Pontchartrain Contains Toxic Material," *The New York Times*, September 7, 2005, 1.

60. Matthew Wald, "Engineers Say a Key Levee Won't Be Set for Months," *The New York Times*, September 14, 2005, 1.

61. On October 6 state officials declared that tap water was drinkable again across a broad swath of New Orleans' east bank. On October 12 the city was finally dry. And by October 18 raw sewage was being partially treated before being dumped.

62. To qualify for the grant, residents had to apply to FEMA, be displaced from their primary residence, and be living in a shelter, hotel, or motel, or with family or friends.

63. Ron Thibodeaux, "FEMA's New Shepherd Hopes to Unite Flock," *The Times-Picayune*, September 11, 2005, 5.

64. Bill Walsh, "Federal Report Predicted Cataclysm," *The Times-Picayune*, January 24, 2006, 1.

65. Despite protestations to the contrary by both Brown and Chertoff, it was clear that they had been informed of the severity of conditions in New Orleans. In October 2005 Marty Bahamonde, the lone FEMA official in New Orleans, told the Senate Homeland Security Committee that he had e-mailed FEMA leaders about the city's desperate need for medical help, oxygen canisters, food, and water. According to Bahamonde's testimony, on Monday, August 28, at 11 a.m., he received word that the 17th Street Canal had been breached, sending floodwaters into Lakeview and central New Orleans. He immediately phoned his superiors. At 7 p.m. he finally reached Brown, who promised to contact the White House.

66. Matthew Brown, "Health Risks in Wake of Storm Hard to Gauge," *The Times-Picayune*, November 11, 2005, 1. The cost of disposing of all this material threatened to impose a crippling financial burden on the city and state. Under federal disaster rules, state and local

governments are required to pay 25 percent of disaster-related costs. But because Katrina wiped out the region's tax base, President Bush agreed to pick up the entire cost of debris removal through November 26, 2005, and 90 percent of other costs associated with Katrina. He granted a thirty-day extension of the 100 percent reimbursement in late October.

67. There were four parallel investigations of the levee failures: one sponsored by the state of Louisiana, a second funded by the National Science Foundation, a third conducted by the American Society of Civil Engineers, and a fourth carried out by an Interagency Performance Evaluation Team (IPET) on behalf of the Corps. The Senate Homeland Security and Governmental Affairs Committee conducted a fifth, more wide-ranging investigation. In mid-December, at the behest of Defense Secretary Donald Rumsfeld, the National Academy of Sciences's National Research Council undertook a peer review of the Corps' analysis.

68. John McQuaid, "Floodwall Breaches Still a Mystery," *The Times-Picayune*, September 13, 2005, 1.

69. Christopher Drew and Andrew C. Revkin, "Design Flaws Seen in New Orleans Flood Walls," *The New York Times*, September 21, 2005, 4.

70. LSU researcher Ivor van Heerden concluded on the basis of ground sonar tests that the piling on the 17th Street Canal extended just ten feet below sea level, but the Corps disputed this. See Bob Marshall, "Short Sheeted," *The Times-Picayune*, November 10, 2005, 1.

71. The soil under New Orleans is soft, spongy, and unstable; it is composed of layers of peat soil, sand, silt, and soft clays laid down by the Mississippi River over hundreds of years, supplemented in places by garbage, trees, shells, and other materials. See John McQuaid, "Swamp Peat Was Poor Anchor, Engineer Says," *The Times-Picayune*, October 15, 2005, 1.

72. Investigators uncovered other factors that probably contributed to the flood walls' vulnerability as well. For example, in 1981 the Sewer and Water Board decided to increase the capacity of Pump Station No. 6 by dredging the 17th Street Canal. In doing so, it left the canal too deep for sheet pilings that were supposed to cut off seepage, reduced the distance water had to travel to reach the canal, may have removed some layers of clay that sealed the canal bottom, and reduced support for the wall on the New Orleans side. See Bob Marshall and Sheila Grissett, "Dredging Led to Deep Trouble, Experts Say," *The Times-Picayune*, December 9, 2005, 1.

73. Mark Schleifstein, "Corps Revises Cause of Levee Failure," *The Times-Picayune*, August 29, 2007, 1.

74. For the record, some of the investigating teams disagreed with this analysis. They suggested the main culprit was water flowing through the organic soils beneath the flood walls.

75. McQuaid and Schleifstein, *Path of Destruction.*

76. A safety factor describes the ability of a system to tolerate loads beyond what it was designed to hold or withstand. A safety factor of 2.0 is the norm for a dynamically loaded structure, such as a bridge. Adding even one or two tenths to a safety factor can dramatically increase a project's cost.

77. Bob Marshall, Sheila Grissett, and Mark Schleifstein, "Report: Flood Policy Flawed," *The Times-Picayune*, June 2, 2006, 1.

78. In any case, experts pointed out that the Saffir-Simpson scale was too crude to assess vulnerability in New Orleans, where flooding was the most serious concern. A more sophisticated analysis made clear that different parts of the city were differentially exposed to storm-surge waters. See John McQuaid, "Levee System Projections Flawed, Experts Say," *The Times-Picayune*, September 21, 2005, 1.

79. The authorization for the Lake Pontchartrain and Vicinity Project made the Corps responsible for hurricane-caused flooding. But the Sewer and Water Board, created in 1899, maintained the canals for drainage. And after the Mississippi River flood of 1927, the Orleans Levee District and its controlling board gained responsibility for flooding. This arrangement made for considerable jurisdictional confusion. For example, the Corps designs and supervises the construction of flood walls, but the levee district is responsible for ensuring the structural integrity of the levees that support those flood walls.

80. Grunwald and Glasser, "The Slow Drowning."

81. Bob Marshall, "Levee Leaks Reported to S&WB a Year Ago," *The Times-Picayune*, November 18, 2005, 1.

82. Eric Lipton, Christopher Drew, Scott Shane, and David Rohde, "Breakdowns Marked Path From Hurricane to Anarchy," *The New York Times*, September 11, 2005, 1.

83. The 1988 Robert T. Stafford Disaster Relief and Emergency Assistance Act formally made the federal government secondary to the states in responding to a disaster. But according to Richard Sylves, the newly adopted National Response Plan (NRP), which was activated when Homeland Security director Chertoff designated Hurricane Katrina "an incident of national significance," gave the Department of Homeland Security the power to mobilize and deploy federal resources, even in the absence of a request from the state. See Richard Sylves, "President Bush and Hurricane Katrina: A Presidential Leadership Study," *Annals of the American Academy of Political and Social Science* 604 (2006): 26–56.

84. Robert Travis Scott, "Politics Delayed Troops Dispatch to N.O.," *The Times- Picayune*, December 11, 2005, 1.

85. Sylves, "President Bush and Hurricane Katrina."

86. William L. Waugh, "The Political Costs of Failure in the Katrina and Rita Disasters," *Annals of the American Academy of Political and Social Science* 604 (2006): 11–25.

87. Thomas Birkland, *After Disaster: Agenda Setting, Public Policy, and Focusing Events* (Washington, D.C.: Georgetown University Press, 1997).

88. Bill Walsh, Bruce Alpert, and John McQuaid, "Feds' Disaster Planning Shifts Away from Preparedness," *The Times-Picayune*, August 30, 2005, 99.

89. Bob Marshall, "La. Wants to Speed Up Its Restoration Plans," *The Times-Picayune*, September 16, 2005, 4.

90. Cornelia Dean, "Some Question Protective Role of Marshes," *The New York Times*, November 15, 2005, F3.

91. Robert S. Young and David M. Bush, "Forced Marsh," *The New York Times*, September 27, 2005, 2.

92. Cornelia Dean, "Louisiana's Marshes Fight for Their Lives," *The New York Times*, November 15, 2005, F1.

93. Dan Baum, "The Lost Year," *The New Yorker*, August 21, 2006, 46.

94. Bill Walsh and Bruce Alpert, "Category 5 Protection Support Dries Up," *The Times-Picayune*, November 10, 2005, 1.

95. Craig Colten, "Restore the Marsh," *The New York Times*, September 10, 2005, 2. An 1878 map revealed that almost every place that was uninhabited in 1878 flooded after Katrina. See Gordon Russell, "An 1878 Map Reveals That Maybe Our Ancestors Were Right to Build on Higher Ground," *The Times-Picayune*, November 3, 2005, 1.

96. Quoted in Cornelia Dean, "Some Experts Say It's Time to Evacuate the Coast (for Good)," *The New York Times*, October 4, 2005, F2.

97. Joseph Nocera, "To Be Better, New Orleans, Think Smaller," *The New York Times*, September 24, 2005, C1.

98. Bill Walsh and Jim Barnett, "Some See Opportunity in Wake of Tragedy," *Times-Picayune*, September 4, 2005, 6.

99. Michael E. Crutcher, "Build Diversity," *The New York Times*, September 10, 2005, 3.

100. John Schwartz, Andrew C. Revkin, and Matthew L. Wald, "In Reviving New Orleans, a Challenge of Many Tiers," *The New York Times*, September 12, 2005, 5.

101. Founded in 1936, the ULI is a nonprofit with more than 25,000 members worldwide. Although it functions more like a university than a trade group, the ULI is known for its prodevelopment orientation.

102. Martha Carr, "Rebuilding Should Begin on High Ground, Group Says," *The Times-Picayune*, November 19, 2005, 1; Gary Rivlin, "Panel Advises New Orleans to Relinquish Purse Strings," *The New York Times*, November 19, 2005, 1. In developing its map, the ULI panel looked at land elevation, depth of flooding from Katrina and Rita, the number of days floodwaters inundated neighborhoods, inundation before and after flood walls broke and levees were breached, historic districts, structural damage, frequency of flooding, and vulnerability to future floods. Prior to the release of the ULI recommendations, the only person to publicly propose a method for determining the safest areas to rebuild had been Tulane University geographer Richard Campanella. See Martha Carr, "Experts Include Science in Rebuilding Equation," *The Times-Picayune*, November 25, 2005, 1.

103. Jeffrey Meitrodt and Frank Donze, "Plan Shrinks City Footprint," *The Times-Picayune*, December 14, 2005, 1.

104. Gordon Russell, "City's Rebuild Plan Draws Criticism," *The Times-Picayune*, December 23, 2005, 1.

105. Brian Friedman, "Group Says Plan to Rebuild Is Biased," *The Times-Picayune*, December 15, 2005, 99.

106. Richard Campanella, "Delta Urbanism and New Orleans: After," The Design Observer Group, April 1, 2010. Available at http://www.designobserver.com/places/entry.html?entry=12978.

107. Campanella, *Delta Urbanism: New Orleans*, 157–158.

108. Clifford J. Levy, "New Orleans Is Not Ready to Think Small, or Even Medium," *The New York Times*, December 11, 2005, D1.

109. Christine Hauser, "Mayor of New Orleans Vows to Rebuild 2 Devastated Areas," *The New York Times*, October 21, 2005, 2.

110. Adam Nossiter, "Fight Grows in New Orleans on Demolition and Rebuilding," *The New York Times*, January 6, 2006, 1.

111. Gwen Filosa, "Former Mayor Rejects Idea of a New Orleans Reduced in Size," *The Times-Picayune*, January 8, 2006, 1.

112. Nicolai Ouroussoff, "In New Orleans, Each Resident Is Master of Plan to Rebuild," *The New York Times*, August 8, 2006, E1.

113. Adam Nossiter, "All Areas Open in New Blueprint for New Orleans," *The New York Times*, January 31, 2007, 1.

114. Campanella, *Delta Urbanism: New Orleans*.

115. Adam Nossiter, "Big Plans Are Slow to Bear Fruit in New Orleans," *The New York Times*, April 1, 2008, 1.

116. Mark Schleifstein, "Army Corps Launching $4 Billion in Flood Projects," *The Times-Picayune*, January 11, 2009, 1. The Corps built gates at the mouth of the three major drainage canals that would eventually be replaced by pumps. At the Industrial Canal, the Corps replaced existing I-wall sheet piling with a stronger flood wall supported by concrete pilings. And at the 17th Street and London Avenue canals, engineers drove pilings that would serve as cofferdams while permanent repairs were being made. Eventually, the Corps planned to install more inverted T walls on the 17th Street and London Avenue canals; to do so, it would have to acquire at least 15 feet of property and as much as 150 feet along some stretches.

117. Even further complicating the rebuilding effort were insurance companies' efforts to avoid paying for damages from the storm. Most private homeowners' policies covered wind damage. But if the damage was the result of flooding, coverage was available only if the homeowner had separate flood insurance, purchased from the state or from a private insurer on behalf of the National Flood Insurance Program. Homeowners in "Special Flood Hazard Areas" are generally required to buy flood insurance. But many of the claims were coming from homeowners in low- to medium-risk zones. Moreover, homeowners with multiple insurance plans faced conflicts over who would pay, the private insurer or the government agencies that run disaster relief and flood insurance programs.

118. Chris Kirkham, Sheila Grissett, and Mark Schleifstein, "New Maps Detail Local Flood Risks," *The Times-Picayune*, February 6, 2009, 1.

119. Louisiana Department of Health and Hospitals, "Hurricane Katrina Deaths, Louisiana, 2005" August 28, 2008. Available at http://new.dhh.louisiana.gov/assets/docs/katrina/deceasedreports/KatrinaDeaths_082008.pdf.

120. Burby, "Hurricane Katrina."

121. Jan Moller, "Loss to State Could Top $1 Billion," *The Times-Picayune*, October 1, 2005, 11.

122. Bill Sasser, "Katrina Anniversary: How Well Has Recovery Money Been Spent?" *The Christian Science Monitor*, August 27, 2010, 1.

123. Campbell Robertson, "On Fifth Anniversary of Katrina, Signs of Healing in New Orleans," *The New York Times*, August 28, 2010, 1.

124. Campbell Robertson, "Smaller New Orleans After Katrina, Census Shows," *The New York Times*, February 3, 2011, 1.

125. Mark Schleifstein, "Even Greater Recovery Ahead, Demographer Says," *The Times-Picayune*, March 26, 2010, 2.

126. Bruce Nolan, "After Katrina," *The Times-Picayune*, August 15, 2010, 1.

127. Sasser, "Katrina Anniversary." The study purported to measure the city's resilience using twenty social and economic indicators.

128. Amy Liu and Nigel Holmes, "The State of New Orleans," *The New York Times*, August 29, 2010, 10; Stacy Teicher Khadaroo, "After Katrina, How Charter Schools Helped Recast New Orleans Education," *The Christian Science Monitor*, August 29, 2010.

129. Mark Guarino, "Four Ways New Orleans Is Better Than Before Katrina," *The Christian Science Monitor*, August 28, 2010.

130. Michelle Krupa, "Common Neighborhood Troubles Still Hold Up Recovery After Storm," *The Times-Picayune*, August 24, 2010, 1; Robertson, "On Fifth Anniversary of Katrina."

131. Nolan, "After Katrina."

132. Bill Sasser, "Five Ways New Orleans Is Still Struggling After Katrina," *The Christian Science Monitor*," August 28, 2010. In 2008 the U.S. Department of Housing and Urban Development authorized demolishing four of the city's public-housing complexes, eliminating more than 3,000 apartments. Only a handful of those properties' former residents subsequently moved into mixed-income neighborhoods.

133. Krupa, "Common Neighborhood Troubles."

134. In 2006 the city decided not to reopen Charity Hospital and instead build a $14 billion state-of-the-art medical complex on a seventy-acre site near downtown. The new complex is expected to open no earlier than 2014.

135. Bruce Nolan, "Storm Reshuffles Size, Wealth, Race in Region," *The Times-Picayune*, August 26, 2010, 1.

136. Both of these contaminants were present in high quantities before the storm, but research suggested that sediment deposition or flooded building materials had exacerbated the problem. See "Arsenic: New Research Reveals Hurricane Katrina's Impact on Ecological and Human Health," *Chemicals and Chemistry*, June 4, 2010.

137. Robertson, "On Fifth Anniversary of Katrina." In January 2010 the federal government released nearly half a billion dollars to rebuild the city's main public hospital and nearly $2 billion more for the schools. In addition, the Obama administration enabled FEMA to be more flexible in disbursing funds and appointed a cabinet official to reform the city's public-service sector.

138. Mark Schleifstein, "Hurricanes Katrina and Rita Turned 217 Square Miles of Coastal Land and Wetlands into Water," *The Times-Picayune*, October 11, 2006, 1.

139. Bob Marshall, "Last Chance: The Fight to Save a Disappearing Coast," *The Times-Picayune*, March 4, 2007, 1.

140. Bob Marshall, "Coastal Restoration Financing Is Uncertain, But Louisiana Has Ideas to Find $50 Billion," *The Lens*, April 2, 2014.

141. Jason Plautz, "As a State Wrangles, Its Coast Is Swept Out to Sea," *National Journal*, June 9, 2014.

142. John Schwartz, "Five Years After Katrina, 350 Miles of Protection," *The New York Times*, August 24, 2010, 1. A 2009 report by the National Academy of Engineering and the NRC confirmed that the Corps' $15 billion improvements were not guaranteed to withstand a Category 5 hurricane (a 1,000-year storm).

143. Mark Schleifstein, "Upgraded Metro New Orleans Levees Will Greatly Reduce Flooding, Even in 500-Year Storms," *The Times-Picayune*, August 16, 2013.

144. The International Organization for Migration projects there will be 200 million climate change migrants by 2050; others think the number will more likely be around 700 million. See "A New (Under) Class of Travellers," *The Economist*, June 27, 2009.

145. Mark Guarino, "After Katrina, New Orleans Housing Goes Green," *The Christian Science Monitor*, August 28, 2010.

146. Howard Kunreuther, "Disaster Mitigation and Insurance: Learning from Katrina," *Annals of the American Academy of Political and Social Science* 604 (2006): 208–226.

147. See, https://www.naco.org/resources/managing-disasters-county-level-focus-flooding-0.

148. William J. Broad, "High-Tech Flood Control, with Nature's Help," *The New York Times*, September 6, 2005, F1.

149. John McQuaid, "Dutch Defense, Dutch Masters," *The Times-Picayune*, November 13, 2005, 1.

150. Quoted in ibid.

151. Sheila Grissett, "Many Residents Say They'll Never Be the Same After Katrina," *The Times-Picayune*, August 29, 2010, 1.

152. Annie Snider, "Why America Hasn't Learned Any Lessons from Katrina." *Politico*, August 2017. Available at https://www.politico.com/magazine/story/2017/08/27/hurricane-harvey-katrina-lessons-louisiana-215543.

153. Brian Resnick and Eliza Barclay, "What Every American Needs to Know About the Puerto Rico Disaster." October 16, 2017. Available at https://www.vox.com/science-and-health/2017/9/26/16365994/hurricane-maria-2017-puerto-rico-san-juan-humanitarian-disaster-electricty-fuel-flights-facts.

CHAPTER SEVENTEEN

CONCLUSIONS
Politics, Values, and Environmental Policy Change

Having read the preceding cases, you do not expect that just because someone has identified an environmental problem and even figured out how to fix it, the government will necessarily address the problem or adopt the most "rational" solution. As the framework laid out in Chapter 1 made clear, policymaking is not a linear process in which government officials recognize a problem, deliberate based on all the available information, and select the optimal solution. Rather, policymaking consists of a series of engagements among advocates trying to get problems addressed in the venues they believe will be most hospitable to their concerns and in ways that are consistent with their values. Advocates of policy change struggle to create a context in which their definition of a problem dominates the debate, while awaiting a policy window in which to push their preferred solution. But defenders of the status quo have a crucial advantage: in American politics, it is much easier to prevent change than to enact it.

THE STRENGTH OF THE STATUS QUO

One of the features of the American environmental policymaking process that stands out most starkly in the preceding cases is the remarkable persistence of the status quo. The prevailing policy may be shielded by the president, well-situated members of Congress, the agencies that administer it, or the courts. The president can block change using a veto or veto threat or, more subtly, by simply failing to put forth a policy proposal and arguing instead for more research. Highly placed members of Congress can rebuff attempts to change policy as well: committee chairs can decline to send a bill to the floor; majority leaders in either chamber can refuse to schedule a vote on a bill. In the Senate a minority can filibuster legislation; in the House the majority party can use multiple referral, amendment-friendly rules, and other parliamentary maneuvers to derail a bill. Agencies can weaken a legislative mandate by delaying implementation or interpreting a statute in ways that are inconsistent with the apparent intent of Congress. And, of course, the courts may declare a law unconstitutional or, more commonly, find that an agency exceeded or acted in ways inconsistent with its statutory mandate.

The strength of the status quo manifests itself repeatedly in cases involving natural resources, where environmentalists have faced enormous resistance to reforms aimed at introducing their values into management practices. In natural resource policymaking, narrow, relatively homogeneous subsystems that are biased in favor of development historically have been extremely difficult for competing interests to penetrate. Although timber policy has become noticeably more environmentally protective in response to advocates' efforts, mining and grazing policies on public lands retain vestiges of century-old routines. As the case of snowmobiles in Yellowstone National Park makes clear, even the national parks struggle with a history of accommodation to commercial development. With egregious disasters like the Deepwater Horizon spill, we can fail to prompt major policy change when officials worry more about losing the revenues from resource extraction than they do about jeopardizing the local environment. The point is that the legacy of past policies shapes contemporary politics and policy.[1]

The strength of the status quo means not only that institutionalized interests have an advantage in policy debates, but also that policies tend to lag behind advances in the scientific understanding of the natural world. Many of the environmental laws passed in the 1970s reflect ecological notions of stability and balance that were anachronistic among practicing scientists even then, yet they continue to shape decision making. For example, the Endangered Species Act (ESA) focuses on conserving individual species and their habitats, even though ecologists and conservation biologists for decades have emphasized the importance of preserving ecosystem structures and functions and maintaining diversity at all levels of biological organization. Agencies have sought to update the ESA by changing how it is implemented, but the statute—last significantly modified in 1982—remains the main vehicle for biodiversity conservation in the United States.

Critics also charge that, because of their initial design, many environmental laws are inefficient or counterproductive. Economists have disparaged the Superfund program, for example, for diverting millions of dollars to what most public health professionals agree are relatively low-risk toxic dumpsites. Although neither the ESA nor Superfund is perfectly "rational" from many experts' point of view, both have stubbornly resisted reform—not just because some economic interests benefit from the laws as written but because environmentalists recognize that, once they are opened up for reconsideration, they are vulnerable to being weakened or rescinded altogether.

LEGISLATIVE POLICY CHANGE

Given the fragmentation of American politics—among both the levels and branches of government—and hence the numerous opportunities to stymie a challenge, dislodging the status quo requires a formidable effort. In legislative policymaking, two elements are almost always crucial to challengers' success: a large, diverse coalition and the backing of influential leaders. Therefore, advocates define environmental problems in ways they hope will both inspire leadership and facilitate coalition building. They are more likely to succeed in proffering the authoritative problem definition if they can tell a compelling causal story that is backed by credible science; their prospects are further enhanced if they have a ready-made solution that is certified by legitimate experts.

The Importance of Building Coalitions

One way that opponents of the status quo have tried to circumvent the arduous process of changing policy in Congress is by attaching riders to appropriations and omnibus budget bills. More commonly, however, proponents of fundamental policy change must assemble a broad supporting coalition. To build such a coalition, advocates have learned they must join forces with groups that may have different values but similar objectives. For example, environmentalists have allied with hazardous waste cleanup firms to resist changes in the Superfund program; gas, solar, and geothermal power companies, as well as insurance firms, have cooperated with environmentalists to press for cuts in greenhouse gas emissions; and labor unions and municipal governments have taken the side of environmentalists in opposing legislation to compensate property owners when regulation reduces the value of their land because they believe such laws will severely compromise government's ability to protect public (and workers') health and safety.[2]

Keeping such pragmatic coalitions together is no easy feat; in fact, it is difficult to maintain the cohesiveness of even a more homogeneous union. For environmentalists, it is hard to craft approaches that are palatable to mainstream interests without alienating the true believers that form the activist core. Moreover, when the vague ideas and symbols that bind members of an alliance together are translated into policies, rifts and factions often develop. For example, a schism appeared during climate change negotiations at The Hague in 2000, when some environmental groups supported a plan to allow the United States to get credit toward its carbon dioxide (CO_2) emissions limits for preserving forests, which absorb CO_2, while others opposed such a deal because they felt it allowed the United States to evade its responsibility for cleaning up its share of pollution. As the Cape Wind case makes vividly clear, proposals to site alternative energy plants can drive a wedge between environmental groups that ordinarily would see eye to eye.

Although it is easier to maintain cohesiveness among economic interests, those bonds can also be broken. In the acid rain case, for example, utilities' united front crumbled once some figured out how to clean up and so had an incentive to support limits on sulfur dioxide (SO_2) emissions. In the climate change case, business interests have diverged on the question of whether mandatory limits on CO_2 emissions are desirable. The dissolution of a coalition renders it vulnerable, creating opportunities for its opponents.

The Crucial Role of Leadership

Even if advocates succeed in building a supportive coalition for an environmental policy proposal, legislative majorities large enough to enact policy change do not simply arise spontaneously in response. Rather, one theme that emerges from the preceding cases is the importance of leaders in bringing about policy change. Just as well-placed individuals can block reform, talented leaders can greatly increase the likelihood of policy change by facilitating coalition building, brokering compromises that defuse apparently intractable conflicts, and shepherding solutions through the decision-making process. For example, after nearly a decade of stalemate over acid rain, Senator George Mitchell's skillful and dogged micromanagement of negotiations over the Clean Air Act Amendments of 1990 proved indispensable to the passage of that law. By conducting negotiations in private, away from the glare of the media and the demands of special interests, Mitchell was able to wring concessions from holdouts and forge an agreement among parties who for a decade had refused to bargain. Compare the outcome of that case with that of Hurricane Katrina, where public squabbling among different agencies and levels of government, unmediated by leadership, stymied rescue efforts, with disastrous consequences.

Leaders need not be "true believers." The stringency of the Clean Air and Clean Water acts of the early 1970s results from the efforts of President Richard Nixon and presidential hopeful, Senator Edmund Muskie, to trump one another and thereby garner the kudos of environmentalists. Twenty years later, Republican presidential candidate George H. W. Bush promised to be "the environmental president" in hopes of undermining support for Democrat Michael Dukakis, and one result was the Clean Air Act Amendments of 1990. Neither Nixon nor Bush was an avid environmentalist; Nixon, in particular, was contemptuous of environmentalists. Both were trying to garner political credit by capitalizing on the salience of environmental issues that advocates had worked so hard to create. Similarly, Governor Andrew Cuomo was responding to pressure from his core constituents, not personal conviction, when he banned fracking in the state of New York.

Elected officials are not the only people who can provide leadership; policy entrepreneurs outside government can also be effective. In the early 1970s, Ralph Nader's organization spurred Senator Muskie to propose more ambitious clean air and water laws than he originally intended. Three decades later, David Hunnicutt of Oregonians in Action launched a ballot initiative campaign that threatened to turn Oregon's land-use planning process on its head.

Using Information to Define Problems

Critical to advocates' success in building coalitions and inspiring leadership is their ability to devise and disseminate a compelling definition of the environmental problem. Although conventional tactics, such as lobbying legislators and donating money to political campaigns, remain essential means of influencing policy, scholars have recognized the importance of strategically defining problems in determining the outcome of policy contests. For environmentalists, those efforts entail framing scientific information in ways that emphasize the risk of inaction. When protection advocates succeed in translating the scientific explanation of a problem into a compelling story, as they did in the Love Canal, spotted owl, acid rain, Chesapeake Bay, fracking, and other cases, they are much more likely to attract support.

Among the attributes of problem definition that emerge from the cases is the political power of scientific explanations and predictions. According to a rational view of policymaking, "If predictive science can improve policy outcomes by guiding policy choices, then it can as well reduce the need for divisive debate and contentious decision making based on subjective values and interests."[3] As should be clear from the cases in this book, however, science cannot provide either a definitive or a value-free basis for policy. Scientific explanations are inherently uncertain because natural systems are not "closed" or static and therefore may respond to perturbations in fundamentally unpredictable ways.

In addition, scientists' ability to measure environmental phenomena is limited—by time, money, and an inability to control for complex interactions among variables in the real world. Scientists' explanations and predictions therefore reflect experts' assumptions and extrapolations, which in turn hinge on their values. Although they do not provide a certain or objective basis for policymaking, scientific explanations do furnish the elements of environmentalists' causal stories. Science-based stories that feature loathsome villains (such as large chemical companies), innocent victims (ideally children or fuzzy animals), and a simple connection between them are more likely to get sympathetic media coverage and therefore to prompt public concern. Examples abound: environmentalists have succeeded in generating public alarm about dolphins, turtles, owls, and caribou—all of which are charismatic species. They have less to show for their efforts to provoke public alarm about declining rangelands or climate change.

Another crucial attribute of stories that succeed in garnering public (and hence political) attention is an impending crisis. The cases of Love Canal and the New England fisheries illustrate the impact that dramatic events can have on the public, particularly if advocates manage to interpret those events in ways that reinforce their policy position. In the Love Canal case, advocates pounced on scientific studies relating chemicals in the air and soil to health effects to buttress their claims that their neighborhood was contaminated; journalists were their allies in this endeavor because of their propensity to focus on human interest stories and health risks. In New England the collapse of important groundfish stocks lent credibility to scientists' assessments and, relatedly, to environmentalists' legal claims that fishing needed to be curtailed.

Further buttressing environmentalists' efforts to define a problem is the support of credentialed scientists, who may have identified the problem in the first place. Periodic scientific bulletins serve as focusing events, keeping media—and therefore public and elite—attention trained on the issue. Scientific consensus reports, particularly if they are crafted by prestigious bodies such as the National Academy of Sciences, bolster the credibility of science-based stories. Individual scientists, acting as knowledge brokers, can be particularly persuasive proponents of policy change. The impact of activist scientists is especially obvious in the acid rain and climate change cases, but it is evident in many of the other cases as well.

Ultimately, for advocates of policy change to prevail, they must not only define a problem but also furnish a solution that is consistent with their definition of the problem, as well as be viable in terms of the budget and the political mood.[4] For instance, proponents of Cape Wind framed it as a solution to global warming; they also took advantage of the September 2001 terrorist attacks to portray installing offshore wind as a national security measure. Critics tried to undermine proponents' claims by depicting the project as a financial boondoggle.

While waiting for a policy window to open, policy entrepreneurs soften up their pet solutions in the expert community and among policymakers and the public. For example, academics had discussed the concept of emissions trading for more than two decades before it was finally incorporated into legislation to curb acid rain. During that period, economists expounded on the benefits of using market-based mechanisms to solve environmental problems in textbooks, classrooms, and scholarly and popular journals.

Even linking a ready-made solution to a compelling definition of the problem does not guarantee success, however, because opponents of policy change have devised a host of tactics to counter challenges to the status quo. Recognizing the importance of science to environmentalists' problem definition, opponents try to undermine the information and ideas on which protection advocates base their claims. The climate change case provides an especially vivid example of this tactic: the fossil fuel industry has funded contrarians, who reject the scientific consensus on global warming, as well as conservative think tanks and foundations, which disseminate these counterclaims.

In addition, opponents of regulation have tried to shift attention to the costs of policies proposed to address environmental concerns. Like scientific predictions of environmental outcomes, cost projections are highly uncertain and—as is made clear in the owl and acid rain cases—tend to overestimate the expense of complying with new regulations.[5] The reason, in part, is that such forecasts rarely take into account unpredictable events, technological innovations, or the market's response to regulation. Furthermore, like scientific projections, they are heavily shaped by modelers' assumptions. Nevertheless, cost projections can be compelling when interpreted as providing certainty about the future, particularly when—as in the owl case—they portend economic and social disaster.

Finally, opponents of protective regulations have become adept at crafting and disseminating causal stories of their own. Development advocates have defended the prerogatives of ranchers, fishermen, and farmers by portraying them as iconic Americans, heroic figures who rely on their rugged individualism and hard work to survive and would prosper but for the interference of overweening bureaucrats.

Tales that feature sympathetic individuals are particularly effective: the Portland case reveals the power of anecdotes in which average Americans suffer when government infringes on their private property rights.

ADMINISTRATIVE POLICY CHANGE

Although observers tend to focus on legislation, another avenue for major policy change in the American political system is administrative decision making. Agency officials have considerable discretion to interpret typically vague statutory mandates, and agencies' missions, standard operating procedures, and past interactions with client groups and congressional overseers shape day-to-day decision making. The Deepwater Horizon case provides one of the most vivid examples of how such forces can combine to cause agency officials to perceive their interests as closely aligned with those of the industry they are supposed to regulate, even when that industry's practices threaten human health and the environment.

As the cases in this volume make abundantly clear, simply changing the law is rarely enough to prompt a substantial shift in deeply entrenched administrative behavior. Upon confronting this reality, environmentalists have frequently turned to the courts. The "adversarial legalism" that results has, in turn, prompted efforts to try alternative approaches, such as collaborative problem solving and the use of nonregulatory tools, with mixed results.

The Courts

The courts have always been the forum for resolving intractable value conflicts; it is therefore not surprising that environmentalists and cornucopians find themselves so frequently making their cases before a judge. At the beginning of the modern environmental era, environmentalists filed lawsuits to prod the fledgling Environmental Protection Agency (EPA) to write regulations under the Clean Air and Clean Water Acts. Subsequently, they used the National Environmental Policy Act, the National Forest Management Act, the ESA, and other laws in the spotted owl case to redirect federal land management priorities. Efforts to control overfishing of cod, flounder, and haddock in New England were failing until a lawsuit forced the National Marine Fisheries Service (NMFS, pronounced "nymphs") to institute more protective rules.

Lawsuits like these, which involve judicial review of agency decision making, can have direct and indirect effects. They can prompt administrative policy change directly by narrowing the range of options an agency may consider and precluding the status quo altogether. Legal proceedings may have less immediate but equally potent impacts if they raise an issue's visibility and thereby facilitate reframing or result in the intervention of elected officials or their appointees. An underappreciated impact of lawsuits is the generation of new information, which in turn can form the basis for efforts to institute new policies. Another type of litigation—filing civil or criminal suits in hopes of proving liability—can prompt changes in the behavior of a polluter even in the absence of regulatory change. In the aftermath of the Deepwater Horizon spill, for example, advocates hope that steep fines under the Clean Water Act and the Oil Pollution Act of 1990 (the Natural Resource Damage Assessment) will cause oil and gas developers to operate more cautiously going forward.

Observing the success of environmentalists in court during the 1970s, those who oppose environmental regulations began resorting to litigation to achieve their ends. Snowmobilers persuaded a sympathetic federal judge to prevent a ban on using snowmobiles in Yellowstone National Park from taking effect. Natural gas producers (unsuccessfully) sued towns in New York for banning the practice of fracking within their borders. It has become common for both industry and environmental groups to sue the EPA over the same decision, with one litigant claiming the regulation is too strict and the other arguing that it is too lenient. By contrast with many other nations whose political cultures are less litigious, in the United States, jurisprudence—and the adversary relationships it fosters—has been a defining feature of environmental politics.

Ecosystem-Scale Collaboration and Nonregulatory Approaches

In an effort to avert the wave of lawsuits that seems to accompany every new environmental law or regulation, agencies have been experimenting with negotiated rulemaking, mediation, and other consensus-building procedures. Since the early 1990s a huge variety of place-based collaborative endeavors have emerged, particularly in the West, from citizen-led watershed protection programs to government-sponsored ecosystem-based management initiatives. Their proponents contend that such approaches yield outcomes that are both more environmentally protective and more durable than those produced by the adversarial approach. The work of the Malpai Borderlands Group to reconcile livestock grazing with environmentally sound management in the Southwest provides evidence to support this claim. Critics, however, point out that win-win solutions are not always possible and that compromise is often scientifically indefensible. They have suggested that collaborative efforts exclude stakeholders with "extreme" views and avoid tackling the most contentious issues in order to gain agreement. And they have questioned whether consensus is, in fact, an appropriate goal.

Also growing in popularity as a way to avoid controversy are nonregulatory tools, like providing information and "education" or providing positive inducements rather than establishing rules. In theory, such approaches promise to gain the voluntary cooperation of targets without provoking a backlash. But critics have also raised concerns about nonregulatory tools, particularly about their long-term accountability and the extent to which they shift the burden of environmental regulation from government to citizens. As the Chesapeake Bay case makes abundantly clear, because cooperative and nonregulatory approaches can be evaded without consequence, they may not yield the benefits that their proponents hoped they would.

ACKNOWLEDGING THE ROLE OF VALUES

A primary objective of this book is to identify patterns that help us understand the politics of environmental issues without obscuring the uniqueness of each case. We posited at the outset that two elements of American environmental politics remain relatively constant across the cases: advocates in environmental policy contests are almost always deeply divided over values. How these values translate into political choices depends heavily on the way a problem is defined and its solution characterized—in terms of science, economics, and risk. Direct discussions of competing values are rare. Although its adversaries caricature environmentalism as an effort by elites to impose their values on society, public opinion polls suggest that most Americans sympathize with the general aims of the environmental movement. Even its critics concede that environmentalism has become "the lens through which we look at our relationship to nature."[6]

That said, survey and behavioral evidence on American environmentalism is contradictory and suggests deep inconsistencies in people's attitudes and beliefs. Efforts to address issues such as climate change and sprawl expose the tensions latent in American environmentalism. Americans claim to value the natural environment, but most appear unwilling to relinquish their energy-consuming lifestyles—a reality that is underscored by the proliferation of SUVs and pickup trucks in U.S. cities and suburbs. Although individual behavior is clearly at the root of many of our environmental problems, political stories that cast ordinary citizens as the villains have fared poorly in the American political context. Similarly, stories that suggest direct human health effects have galvanized the public more consistently than those involving ecological decline.

In short, our commitment to environmental protection remains ambiguous, as does our willingness to sacrifice for the sake of sustaining or restoring natural systems. Our aversion to making trade-offs, in turn, enables opponents of environmental regulations to frame issues in ways that emphasize costs or inconvenience to individuals or that emphasize other core values—such as freedom, national security, or economic growth—and thereby stymie efforts to amass support for protective policies. For some, this dynamic may suggest that the prospects for genuine change in our approach to environmental protection are poor. But an alternative interpretation of events is possible: efforts in the West to find common ground among ranchers and local environmentalists suggest that, at least under some circumstances, concern about the future of a particular place may trump mutual distrust, and practical agreements may emerge even in the face of underlying value differences.

Most important, political science, like ecology, is not a predictive science: as is true of the natural world, the patterns we identify in human affairs are probabilistic and subject to change. New people and new ideas can affect the process in dramatic and unexpected ways. Recent empirical evidence on the sources of human well-being offers fertile ground for those who hope to persuade Americans to consider alternatives to the single-minded pursuit of economic growth.[7] Therefore, the lessons of past experience should not limit but rather should liberate both professional activists and concerned citizens who wish to challenge the status quo.

NOTES

1. Paul Pierson, *Politics in Time: History, Institutions, and Social Analysis* (Princeton, N.J.: Princeton University Press, 2004).
2. For a critical view of such coalitions, see Jonathan H. Adler, "Rent Seeking Behind the Green Curtain," *Regulation* 4 (1996): 26–34; Michael S. Greve and Fred L. Smith, eds., *Environmental Politics: Public Costs and Private Rewards* (New York: Praeger, 1992).
3. Daniel Sarewitz and Roger A. Pielke Jr., "Prediction in Science and Policy," in *Prediction: Science, Decision Making, and the Future of Nature,* ed. Daniel Sarewitz, Roger A. Pielke Jr., and Radford Byerly Jr. (Washington, D.C.: Island Press, 2000), 17.
4. John Kingdon, *Agendas, Alternatives, and Public Policies,* 2d ed. (New York: HarperCollins, 1995).
5. For example, in the late 1970s the chemical industry predicted that controlling benzene emissions would cost $350,000 per plant, but by substituting other chemicals for benzene, chemical manufacturers virtually eliminated pollution control costs. In 1993 carmakers estimated that the price of a new car would rise $650 to $1,200 as a result of regulations limiting the use of chlorofluorocarbons, yet in 1997 the actual cost was between $40 and $400 per car. Prior to the passage of the 1978 Surface Mining Control and Reclamation Act, estimates of compliance costs ranged from $6 to $12 per ton of coal, but in fact costs were in the range of $.50 to $1 per ton. See Eban Goodstein, "Polluted Data," *American Prospect,* November–December 1997. Available at http://www.prospect/print/V8/35/goodstein-e.html.
6. Charles T. Rubin, *The Green Crusade: Rethinking the Roots of Environmentalism* (Lanham, Md.: Rowman & Littlefield, 1998).
7. See, for example, Tim Jackson, *Prosperity without Growth: Economics for a Finite Planet* (London: Earthscan, 2009); Richard Layard, *Happiness: Lessons from a New Science* (New York: Penguin Press, 2005); Andrew C. Revkin, "A New Measure of Well-Being from a Happy Little Kingdom," *The New York Times,* October 4, 2005, C1.

INDEX